Thinking through the Environment

This Reader brings together material from ecological thought, environmental policy, environmental philosophy, social and political thought, historical sociology and cultural studies. The extracts tell the story of the way the natural environment has been understood in the modern world and how this has recently been questioned as contemporary societies are seen as characterized by uncertainty and complexity.

The literature guides the reader through the conventional grounds for thinking about rights and obligations in relation to future generations, non-human animals and the biotic communities, bringing each into question. This then leads into a critical examination of social and political theories and their capacity for drawing on ecological thought. Each of the seven sections of readings is introduced by the editor, who locates the set of readings within the specific themes and issues at the heart of each section.

This broad-reaching and thought-provoking set of readings stresses the diversity of response to environmental problems both within and between anthropocentric and ecocentric approaches, and will encourage the reader to examine how they are manifested in the areas of environmental ethics, policy analysis and social and political theory.

Mark J. Smith is Lecturer in Social Sciences at The Open University. His recent publications include *Ecologism: Towards Ecological Citizenship* (Open University Press, 1998), *Social Science in Question: Towards a Postdisciplinary Framework* (Sage, 1998), *Rethinking State Theory* (Routledge, 1999) and *Situating Hayek* (Routledge, 1999).

This Reader is part of a course, *Ecology, Justice and Citizenship* (D830), which is part of The Open University Masters in Social Sciences Programme.

The Open University Masters in Social Sciences

The Open University's unique, supported open learning Master's Programme in Social Sciences is designed to introduce the concepts, approaches, theories and techniques associated with a number of academic areas of study.

Structure of the MA

The MA enables students to select from a range of modules, to create a programme to suit their own professional or personal development across a range of subjects. On completion of the Postgraduate Foundation Module, students can choose from a range of social science modules to obtain an MA in Social Sciences, or may specialize in a particular subject area or 'study line'. At present there are three study lines leading to:

- an MA in Cultural and Media Studies
- an MA in Environment, Policy and Society
- an MSc in Psychological Research Methods.

Other planned study lines include an MSc in Psychology and an MA in Social Policy/MA in Social Policy and Criminology.
It is also possible to count study of other Master's modules towards the Masters in Social Sciences.

OU supported learning

The MA in Social Sciences programme provides great flexibility. Students study at their own pace, in their own time, anywhere in the European Union. They receive specially prepared course materials and benefit from tutorial support, thus offering them the chance to work with other students.

How to apply

If you would like to register for this programme, or simply find out more information, please write for the Masters in Social Sciences prospectus to the Course Reservations Centre, PO Box 724, The Open University, Walton Hall, Milton Keynes, MK7 6ZW, UK (Telephone 0(0 44) 1908 653232).

Thinking through the Environment

A Reader

Edited by
Mark J. Smith
at The Open University

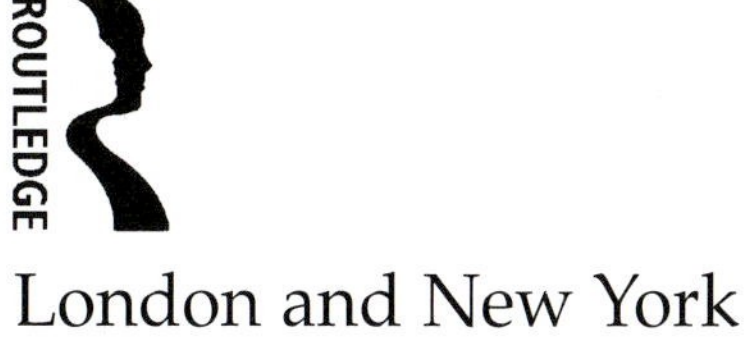

London and New York

First published 1999
by Routledge
11 New Fetter Lane, London EC4P 4EE

Simultaneously published in the USA and Canada
by Routledge
29 West 35th Street, New York, NY 10001

Routledge is an imprint of the Taylor & Francis Group

Typeset in Palatino by J&L Composition, Filey, North Yorkshire
Printed and bound in Great Britain by The Bath Press, Bath

British Library Cataloguing in Publication Data
A catalogue record for this book is available from the British Library

Library of Congress Cataloging in Publication Data
A catalogue record for this book has been applied for

ISBN 0–415–21171–9 (hbk)
ISBN 0–415–21172–7 (pbk)

Contents

Section 2 Rethinking obligations: future generations and intergenerational justice

Section 3 Animal welfare and conservation: expanding the circle?

Illustrations

Figures

Illustrations

Plates

Tables

Acknowledgements

This collection has been constructed to provide an introduction to debates on the environment. In particular, it provides a guide to the different ways that thinking through the environment has been conducted and it explores the close ties between ecology, ethics and social and political thought. It also provides illustrations of the impact of the various branches of ecological thought on environmental policy. Like all Open University texts, this book is the product of many collective discussions and a constant process of thinking through the fields of knowledge which feature in environmental discussions: from historical studies and environmental ethics to political thought and the sociology of the environment, to name a few.

Thanks are due to many friends and colleagues at The Open University for their comments and support during the editing and writing of this book. Particular thanks go to John Blunden, David Humphreys, Chris Nichols, Phil Sarre and Lynne Slocombe for their careful and considered reflections and many imaginative suggestions. I am especially grateful to the external assessor, Prof. David Pepper, for his supportive advice on this text and the Open University course of which it is a part. This collection would not have been possible without the effort and diligence of Penny Bennett and Gill Gowans, respectively the editor and co-publishing advisor at The Open University, as well as Sarah Lloyd, Sarah Carty and Goober Fox at Routledge, Sandra Jones, the copy-editor, and the secretarial support of Ros Shirley. Finally, the impetus behind the character of this collection sprang from my own situated experiences, so I dedicate this book to Jane, Harold and George, who made me take the future far more seriously.

Voyage into the unknown: ecological thought and human impacts

Mark J. Smith

We live in a time when environmental issues are constantly emerging as deeply contentious political issues. From oil spills on busy ocean waterways, slash and burn timber clearance in the remaining forested areas of the world and the disposal of toxic chemical and radioactive waste, to the maintenance of water quality, the control of ecologically disruptive gas emissions and the unknown consequences of human intervention in the gene pool, all of these activities and events now appear very close to our concerns and provide a constant diet for news programmes. The way we respond to these issues embodies our hopes and fears about the future. Our responses have not been confined to intergovernmental summits and the regulatory regimes devised by state authorities. We can see the results of emerging ecological awareness in all kinds of little ways, when we walk through supermarkets or when we deposit cans and bottles at recycling banks, or when we interpret the content of children's television. In children's cartoons, for instance, the leading heroic characters no longer save the world or civilization from some cunning and duplicitous villain, they 'save the planet' from some human-made ecological disaster.

When we place the human experience in terms of the lifespan of the planet, the dramatic ecological impact of humankind is a very recent phenomenon. For instance, assuming that our estimates of the earth's age are correct, if we imagine the lifespan of this planet in terms of the length of a single day, then the human species emerges at just two seconds to midnight. But look at the impact of the human species on the surface of the planet in this time: human beings have mined deep into the earth, they have stripped the surface of a massive acreage of forest habitat, many other species of animals and plants have become extinct. All of this, in the last 400 years at least, has been done in the name of human improvement and progress. All of these improving activities have a twofold impact – we take things from ecosystems but we also deposit wastes. Of course, all living things do this to survive, but human beings have managed to elevate this process to a point where ecosystems have been dramatically transformed in a relatively short period of time and, in the process, some have been strained to breaking point and beyond. In this vision of 'nature as a treasure chest', natural things become simply resources and they have little or no value independent of those attributed to them by the human beings who wish to own and use them.

Given the complexities and uncertainties involved in the ecological impacts of humankind, it is not surprising that there is considerable disagreement as to what the problems are, how they should be understood and the kinds of strategies that could be developed for resolving them. This collection of writings does not attempt to address the physical ecological impact of human activities, but instead provides a survey of the different ways in which natural things have been understood, so we will encounter a wide range of approaches associated with ecological thinking. It also explores key words such as ecocentrism and technocentrism, conservation and preservation, and environmentalism and ecologism. In addition, the readings in later sections include quite a few sources that defend anthropocentric ways of viewing ecosystems as well as those who criticize this as a manifestation of human chauvinism or speciesism. The selected readings will also offer you the opportunity to see how these concepts and categories are defined in different ways within different approaches. In short, this collection acknowledges the essentially contested character of ecological thinking but, more than that, offers the reader a chance to think through the relationship between standpoints in environmental ethics and ecological social and political theory as well as their implications for some of the areas of environmental policy-making. As you make your way through the seven sections of readings in *Thinking through the Environment*, you will find that there are very close connections between the way we think about the environment and the human practices and processes that affect it. The questions behind any discussion about human attitudes towards the environment are therefore twofold:

- How do we live with nature?
- How should we live with nature?

The point of a collection of readings like *Thinking through the Environment* is not to push one or another line of thinking or argument. The aim here is to plot a pathway through the discussions about the human relationship with the 'natural' world and establish how they are grounded in definite social and historical locations.

The ethical dimension

Once we accept that ecological thinking involves disagreement and argument between different positions, each with their own normative assumptions, we must be careful to establish what it means to recognize the ethical dimension. Taking an ethical standpoint on the environment means that you are making a moral judgement about the way human beings affect the environment; that is, to assess an action against some standard of what would be a 'good' decision. According to Peter Singer, an ethical question is not one concerned with self-interest but one about what makes a 'good life'; that we cannot justify an action simply in terms of how the person making that decision would gain but also in terms of the benefits such an action would generate for others. Singer makes use of the Shakespearean tragedy of Macbeth; that the murder of Duncan in terms of Macbeth's own 'vaulting ambition' to be 'king in his place' is not in itself ethically defensible. This, he argues, demonstrates that if such an act must be defended in ethical terms then it must generate some greater good (Singer, 1993: 10). Even the most individualist approaches within neo-liberal accounts of the social and political order (see Section 5), where self-interest is often prized as a virtue, make a case for self-interest in terms of its collective benefits for all members of the moral community. The interesting questions that follow from this are: 'How then is it possible to define a good decision and how is the moral community defined?'

Leaving aside, for the moment, the problem of defining the membership of the moral community, if we take the question of what is a good decision first, then we are led to address what kind of social arrangement can be seen as 'just'. In Sections 2–3, you will encounter a sequence of readings that raise questions about the possibility of defining justice and the implications of this for thinking about environmental issues and problems. Two different definitions predominate in the discussions. First, the view of justice in terms of the best possible outcome, usually *vis-à-vis* human welfare or collective happiness. The focus here is on the total consequences of a course of action and how these can be compared with the consequences of an alternative course of action. One of the problems with this is that it tends to limit our concern to human beings or those things that can be compared in some way to humankind, such as the capacity of non-human animals to experience suffering (discussed in Section 3). Moreover, by focusing on the total consequences of a course of action, the costs for some can be outweighed by the benefits for others. Second, justice can be associated with a set of rules that define the equal status of all members of the same moral community, such as the way we respect persons as 'subjects of life'. In this case, a just outcome would be one that does not violate these rules even if the best possible set of consequences is not achieved. In this view of justice the ends do not justify the means – all members of the moral community are ends in themselves.

Whilst this helps us to recognize the disputes on the concept of justice, it does not take us very far on environmental issues until we begin to consider the question of the moral community, its membership and, with that, its rules for inclusion and exclusion. The construction of the boundary of the moral community is closely connected to the way we define the relationships within the moral community. If we tie membership of the moral community to sentience, then this still excludes natural things that cannot experience a sense of well-being and suffering, like trees. Similarly, if membership depended on the presence of life, then plants would be entitled to consideration but mountains would be excluded. Historically, the boundary of the moral community has moved considerably and with that the recognition of the rights of the previously excluded. Throughout the modern period, for example, rights were denied to women on the grounds that they lacked the desired characteristics for full membership of the moral community, in this case that they lacked the capacity for rational thought. However, over time such objections were challenged so that the community entitled to rights progressively expanded to include more members of the human species, recently extending further to include non-human animals (see Figure 1). Full membership of the moral community has conventionally been understood as a reciprocal relationship between autonomous actors, so that whilst membership conveys rights on those included, these same members also have obligations to other members of the community.

We should also bear in mind that we can have obligations to people or things without them having some definite claim upon us as equal members of our moral community, that is, even if they do not possess rights. One of the difficulties ecological thinking poses for the application of conventional accounts of ethics to non-human animals or things is that this reciprocal relationship cannot always take place. For instance, the relationship between present and future generations is not reciprocal, nor is the relationship between plants and human beings. It may sound obvious but it is still important to stress that on these grounds, future generations of human beings, wild animals, trees, streams and other non-sentient natural things cannot be said to have obligations towards present generations of human beings. We can see how much of a challenge ecological thinking poses for this conception of a moral community when we look at Aldo Leopold's argument that we should enlarge 'the boundaries of the community to include soils, waters, plants, and animals, or collectively: the land' (Leopold, 1949: 204; see Reading 4.3b); which means that humans should acknowledge that they are part of a broader 'biotic community' and should act in ways that do not violate this community. Section 4 includes a variety of readings which develop arguments to extend the moral community to include trees, streams, mountains, forests and ecosystems.

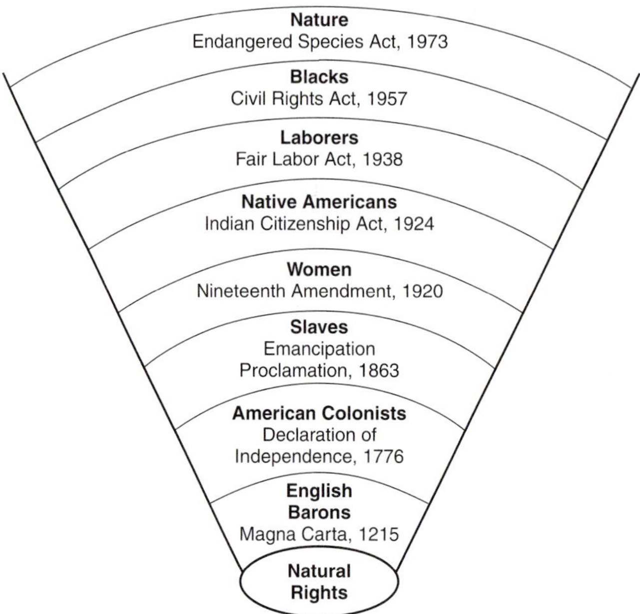

Figure 1 The expanding concept of rights
Source: R. T. Nash (1989: 7)

In Figure 2, you can see something of the way in which ethical consideration has already broadened and that ecological thinking raises important questions about how the moral community may be understood in the future.

The approaches considered in this reader explore the prospects for rethinking the grounds for ethical consideration and for reassessing the tendency to place human beings in a privileged ethical standpoint. So, as you work your way through the readings from Sections 2–4, you will see how debates about the environment can involve an expanding circle of ethical consideration. Once we accept that all members of the human species possess full membership of the moral community, it becomes harder to draw a definite line between human and non-human animals. Much depends on the criteria we develop to define membership, but in all those that have been devised to include all human beings, some non-human animals also have grounds for inclusion. The only defence against ethical expansion (besides blind and unthinking prejudice in favour of our own species) is to find some criterion which would limit membership within our own species. In so doing, by excluding children and people with disabilities, we are constructing a weaker ethical position for the most vulnerable in human societies. Not surprisingly, these ethical debates have been intense and will continue to be so because they involve very high stakes. The readings in Section 4 take the debate further by exploring not only the ways in which non-sentient things can be afforded greater protection within the terms of reference of existing ethical and legal rules, but also whether we should go much further and reconstruct the moral community in a radically new way. In Sections 5–7 of *Thinking through the Environment* we shall turn to the relationship between these debates on the make-up of the moral community

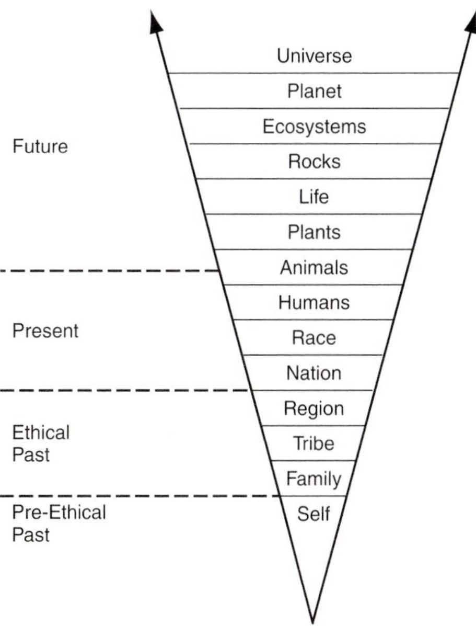

Figure 2 The evolution of ethics
Source: R. T. Nash (1989: 5)

and the construction of political, legal and social orders within which these ethical standpoints make sense. The question of how ethical approaches have concrete effects on the environment can only be approached once we recognize that they are located in definite social and historical locations, that is, in human societies where they are plausible and influential (Smith, 1998b). Ethical approaches are so important because they help us make sense of who we are and what we do.

The social and political dimension

Earlier, we saw how ethical debates do not take place in a vacuum; that they are situated in definite social and historical circumstances. It is also important to recognize that no society has ever been free from contestation about what should be the proper organization of the social and political order. We can see some of the implications of this for the environment when we look at the way these different approaches view the market economy. Ecological thinking has always been concerned with the effects of market-led economies on the environment, because of the way in which material growth has come to be seen as the primary goal of the political system. Indeed, for many societies, material growth is assumed to be tantamount to human progress. The side effects of the production, distribution and exchange of commodities of growing economies, both in terms of the extraction of raw materials and the waste produced prior to and after consumption, have often been ignored or relegated to the status of a little local difficulty. Some forms of ecological thinking ask such societies to think again about the core values and readdress the question of whether such growth is feasible and desirable. Other kinds of ecological thinking seek to harness market forces for the protection of the environment. Western societies experienced a longer period of industrialization than many other developing societies, which provided some scope for local ecosystems gradually to adapt to human activities – which is not to say that they remained unaltered or undamaged. However, the experience of the effects of rapid industrial growth in state socialist societies from the early and mid-twentieth century provides a clear warning of the possible consequences of recent growth trends throughout the world, given the ecological limits that exist for rapidly developing societies.

I shall shortly indicate the complications involved in associating ecological thinking with particular political ideologies. Liberal and conservative accounts of the environment (in the readings in Section 5) provide a range of approaches that wish to defend the role of both private property ownership and market forces in safeguarding the environment. One of the difficulties in pinning down the liberal views on the environment stems from the debate in liberal ethics between those concerned with outcomes and those concerned to establish a universal set of rules through which all actions can be assessed. Both kinds of argument can in fact be used to defend the instrumental uses of natural things, but once the moral community is enlarged beyond humankind, they can also provide a basis for establishing different ways of regarding nature. All the approaches considered in Sections 2–4, which draw upon the political language of rights, obligations and the moral community, draw from a liberal vocabulary that is very familiar in the Western world. More recently in the late twentieth century, neo-liberalism has attempted to reinvigorate the opportunities for finding market-led solutions to environmental problems; this is particularly voiced by Peter Saunders (1995; see Reading 5.5), who has combined a belief in the benefits of the 'growth machine' with a faith in the capacity of human ingenuity to find a technological fix to environmental problems. Against this we can see how conservative politics actually contains a much more sceptical account of the virtues of the market as potentially damaging to the social and moral fabric, which ensures that the environment is preserved for future generations. However, what unites all of these approaches is their underlying anthropocentrism (human-centredness), for in each case they argue that the environment should be conserved for human benefit.

So far, we have focused on the social and political theories that broadly endorse the social and political orders of Western societies. In Section 6, you will encounter a wide range of critical perspectives on Western societies, from different varieties of socialism to anarchism and feminism. In each case they diagnose the problems of these societies differently, whether this is because of the failures of markets (non-Marxist socialism), the capitalist exploitation of waged labour (Marxism), political oppression (anarchism), or the male-centredness of the social order (feminism). However, they all argue that it is only through the transformation of the social and political order that humankind can be emancipated from the various kinds of oppression and exploitation they highlight. These transformative accounts of the social and political order have enjoyed a close dialogue with ecological thinking throughout the last century, yet all have been found wanting by environmental activists.

Socialist accounts possess an underpinning materialism, whether this takes the form of an attachment to material growth as a way of ensuring a better environment or of improved quality of life for all members of society. The pressure of the mass labour movements in

Western societies has been significant in ensuring the establishment of regulatory regimes over some of the more visible kinds of industrial pollution (especially air pollution and water discharges, where a direct connection can be established between the pollutants and human illnesses). In actually existing Marxist societies in the twentieth century (as opposed to some kinds of Marxist theories), this has involved a quick march to a fully industrial society, regardless of the severe side effects on the local ecosystems, many of which have been damaged irreparably in a very short period of time. In the former Soviet Union, the effects on the health of the general populace have vastly exceeded those witnessed in capitalist societies, but in the absence of a democratic political system, the pollution continued unabated so that, even when we do not include the Chernobyl disaster, the effects of toxic chemical and radioactive wastes in the environment will affect both human beings and the wider biotic communities for centuries to come.

Of course, many Marxists are critical of the former state socialist political systems and have looked elsewhere for inspiration on how it is possible to create a society without both the exploitation they believe exists in capitalist economies or the oppressive regimes of the Soviet type. Anarchism provides a critique of the dangers of authority in all its forms and more radical ecological thinking has found many messages of use in both Marxist and anarchist accounts of the simpler life that their kind of social and political transformation would supposedly bring. Marx makes many aphoristic remarks about the simpler division of labour that would follow in a communist social order, while anarchists have provided a range of prototypes (although short-lived) that could be used as a starting-point for thinking about the development of small-scale, decentralized sustainable communities. But, even here, the size of an anarchist community and the spontaneous free will of its members in the absence of government is much less important than its approach towards natural things. Small can be ecologically destructive rather than beautiful, although the larger the scale of damaging practices, the more likely that the effects will be worse for the environment as well as longer lasting.

One of the most important developments in recent transformist accounts of the environment is the emergence of ecofeminism. The feminist account of social transformation can be just as concerned to maintain material growth as any of the other political ideologies considered in Sections 5 and 6. As a result, the emerging links between ecological thinking and feminist accounts of the social order have raised interesting questions for both. There are close connections between the emergence of instrumental accounts of nature in early scientific approaches with the institutional exclusion of women from culture, art, literature and the natural sciences. For ecofeminists the domination of women and nature associated with modern existence are closely related. The feminism critique of androcentric (male-centred) assumptions can be deployed against much of ecological thinking as well as the social and political theories raised so far. In addition, the ecofeminist critique of the use of 'masculine' technology to control female fertility, combined with their defence of the integrity of all living things, undermines many of the assumptions of the radical feminist approach developed since the 1960s. So, even here, the relationship between ecological thinking and existing social and political theories is fraught with difficulties and inconsistencies.

In the final selection of readings of the collection, Section 7, we return to consider how it is possible to think through the implications of redefining the moral community for the organization of the social and political order. Here, you will have the chance to work through some recent interventions on ecological change – the theories of ecological modernization and risk society – and consider how the relationship between society and nature is open to reassessment. These readings demonstrate how contemporary accounts of the environment are willing to acknowledge the existence of complexity, uncertainty and interconnectedness of humans, non-human animals, plants and non-sentient natural things. In addition, these closing readings ask us to rethink the way in which the moral community has been understood if we are to take the ethical consideration of the biotic community more seriously – by developing a conception of ecological citizenship which goes beyond the civil, political and social forms of citizenship characteristic of the ethical evolution of Western societies over the last 500 years (see Smith, 1998a).

References

Leopold, A. (1949) *A Sand County Almanac* (Oxford: Oxford University Press).
Nash, R. T. (1989) *The Rights of Nature: A History of Environmental Ethics* (Madison, Wisconsin: University of Wisconsin Press).
Saunders, P. (1995) *Capitalism: A Social Audit* (Milton Keynes: Open University Press).
Singer, P. (1993) *Practical Ethics*, 2nd edition (Cambridge: Cambridge University Press).
Smith, M. J. (1998a) *Ecologism: Towards Ecological Citizenship* (Milton Keynes: Open University Press).
Smith, M. J. (1998b) *Social Science in Question* (London: Sage).

SECTION 1

Situating the environment

Introduction

The opening reading by Adam Markham starts from the premise that human impacts on the environment come in a variety of forms. For Markham, the most insidious environmental problem is pollution. Yet like all dark clouds, it also has a silver lining as one of the most significant causes of change. The history of environmental policy-making has been one of responding to problems that have become unbearable. Markham also raises the crucial question of who is reponsible for pollution. Is it those people or organizations who make the things we use or grow the food we eat, or does the problem run much deeper than this? Throughout the readings in this section we shall address the belief that human beings have mastery over nature, that natural things have been seen as a resource for human welfare. Keith Thomas, in Reading 1.2, presents evidence that highlights the socially and historically situated character of attitudes towards the environment. Yet, despite the complex and contested character of the word 'nature', one thing stands out, namely the attempt to find ways of justifying the human use of natural things for human welfare. In particular, that all attempts to make the mastery of nature plausible and acceptable have been tied to the idea of 'human uniqueness'.

Such assumptions do not exist on their own, but within the broad fabric of human knowledge developed in a particular time and place. David Pepper, in Reading 1.3, highlights the importance of the uses of analogy and metaphor in the way the meaning of nature is communicated. In particular, Pepper demonstrates that how 'nature' has been defined should be understood in terms of its place in different knowledge systems, whether religious (through the metaphor of the chain of being) or scientific (with nature as a machine). Of special significance here is the emergence of technocentrism through which the empiricist view of the scientific method is used to control nature, as a means for achieving an improvement in human welfare. This is developed further in Reading 1.4, where Tim O'Riordan explores how technocentrism, as a potent blend of science and anthropocentrism, has remained a key characteristic of human knowledge of the environment. Many environmentalists retain their faith in the ability of humankind to understand and control events (as well as identify technological remedies for the problems caused by human interventions – the technofix). O'Riordan outlines the emergence of ecocentrism (in much recent ecological thought) and draws out the implications of challenging technocentrism as a solution to environmental problems and issues. This reading considers the scale of social organization, the materialism of human cultures and the role of ethical values. Ecological thought is diverse precisely because there are various ways in which these issues can be addressed.

In the subsequent readings we focus on some well-known studies of contemporary environmental impacts. Rachel Carson highlights how it is often difficult to anticipate the sorts of problems we are likely to face. In Readings 1.5a and 1.5b, drawn from her classic book *Silent Spring* (1962), Carson traces the emerging impact of a particular form of pollution, DDT. This illustrates how human impacts are often indirect and cumulative but that the effects can be just as dramatic as some of the more visible examples of pollution. Yet, there are few easy answers. How we respond to these problems is shaped by human experiences. This means that it is also important to bear in mind the way that cultural differences are represented through debates on the environment. In Reading 1.6, Frank Golley provides a historically informed exploration of how these debates have developed in North America. Here there is a greater concern with wilderness, and the differences between conservationist and preservationist strategies are highly visible. In the European context, the landscape has been comprehensively transformed through agricultural uses and urban development. In Reading 1.7, Richard Mabey provides an account of the British countryside where we can see how the discussion of human impacts is conducted through different values and concerns. This discussion has evolved through the experience of managing and transforming the countryside for human benefit. In these last two readings you will find your attention drawn not only to specific problems but also to the importance of the aesthetic appreciation of the natural landscape. Nevertheless, there are substantive differences between the largely untouched wilderness of the Rocky Mountains or the Canadian northern territories and the patchwork quilt of British farmland countryside.

1.1

Adam Markham

Why care? Pollution, nature and ethics

Clouds of stored summer rains
Thou shalt taste, before the stains
From the mountain soils they take,
And too unlucent for thee make.

(John Keats)

Should the polluter pay?

Caring about pollution in relation to the environment is a relatively modern phenomenon. Today's environmentalists are largely rooted in the Romantic tradition: they see beauty in nature, a wilderness worth preserving, or species, each with an inherent right to exist untouched by man or at least protected from his worst excesses. Pollution seems qualitatively different to other problems. Deforestation, dam-building, mining, urban expansion, however, are all physical threats to the unprotected natural world, and can be kept at bay by the creation of protected areas and the erection of fences. But pollution seeps under barbed wire and falls from the sky beyond the limits of the highest brick walls. It flows down rivers and it can be found in the deepest recesses of the oceans or in the snows of the uninhabited Antarctic wastes.

Epithets such as insidious, invisible, creeping and miasmal, give descriptions of pollution an almost anthropomorphic edge. In the twentieth century, pollution is seen as an embodiment of humankind's struggle against nature. Forests smitten by acid rain, coral reefs choked by sewage and wildlife decimated by pesticides. Western advocates of environmentalism have a Velcro-like attachment to the idea of pollution as a crime against nature. Concern about human health, while widespread in the populace, is only marginal to the concerns of many environmentalists. When Washington DC slapped a tap water ban on a million people in the district (for fear of *cryptosporidum* contamination) for five days in December 1993, there was hardly a peep out of the environmental groups. Yet Washington probably has the greatest density of anti-pollution activists of any city in the world. It's just that they were busy stopping global warming and marine pollution.

But pollution in the past has been a catalyst for major social change. Even before the word itself was in common usage, pollution was a source of complaint. Before there were anti-pollution campaigns there were complaints about smoke, smells and every sort of unsavoury menace. But it was not really before the nineteenth century and the age of social reformers that pollution issues were transformed into issues of the public good on a wide-scale basis. And even to achieve this, understanding of disease and public health had to have evolved.

The debate about pollution revolves around definitions, ethics and attitudes towards nature. What is pollution? Is it wrong to pollute? Who is responsible? Who or what is suffering and how? Where pollution is obvious and affects few individuals the answers to these questions can be reasonably simple. So that even in the fifteenth century, if a tannery's waste was leaking into a neighbour's water supply, a local tribunal could easily rule on the culprit and some compensation. Today, the complex nature of the pollutants, the web of environmental legislation and the need for proof of harm might tangle the same two litigants for weeks, months or even years in court. Ethics too are at issue. A business that pollutes in pursuit of profit for its owners or shareholders, may believe that what it is doing is right. Local farmers whose land and livestock suffer from emissions will hold a different set of beliefs.

Away from issues of local pollution, much argument centres on the use and abuse of the global commons. These are the natural resources that everyone uses and should benefit from – the oceans, the soil and the air. These resources are regarded as free goods by many. The so-called 'tragedy of the commons' is that while a few may benefit from the abuse of these resources, everyone must pay the price as no one individual or even nation bears overall responsibility. So as the oceans become more and more polluted, the industries that discharge their effluent, and the cities that pump sewage into the waters are acting as 'free-riders', sharing the costs of pollution with all those that share the seas. Problems like these have led to the development of international treaties and proposals such as the 'Polluter Pays Principle'. This latter concept suggests that those who cause pollution should be responsible for its clean-up, but is notoriously hard to implement.

Polluters often argue that it cannot be proven that the pollution in question originated with them, or that it caused the damage blamed upon it. If both these conditions can be satisfied, then arguments can continue for years about the financial implications of compensation. The mercury pollution in Minimata Bay, the Bhopal disaster and the wreck of the *Exxon Valdez* provide examples of where arguments over financial liability have dragged on for years. In Britain, the Central Electricity Generating Board and the Forestry Commission argued throughout the 1980s that acid rain from power stations was not the cause of forest die-back or lake acidification. The nuclear industry continually fights claims of radiation-related cancer; the pulp and paper industry employs scientists to argue that bleaching doesn't release dioxins to the environment; the pesticides industry spends millions to wage public relations war against those who say pesticides raise breast cancer rates and hormonally related birth defects. The business world has turned the burden of proof into an albatross around the necks of their opponents. The little guy, the co-user of the environment, hardly has a chance to win his arguments.

The primacy of economic growth, and the dominance of Western empirical science are partly to blame for this state of affairs. If business is forced onto the defensive over pollution, they have only to start threatening the loss of jobs and opposition starts to fall away. The 'polluter pays' often means that the consumer pays and people tend to care more about their expenditure budget than they do about pollution. Hence the failure of President Clinton's proposal for a tax on motor fuels in the US. Drivers did not want to pay even a few cents more on a gallon of petrol. In the world context this makes no sense as US per capita carbon emissions are higher than those of any other country, and their petrol prices far lower than anywhere else in the OECD. However, to an individual who has a five year loan on a new gas-guzzler stretching his income to the limit, the logic of fighting the tax is incontrovertible. And anyway, the oil companies have a handful of paid scientific consultants disputing, on TV and in the press, every new scientific finding on climate change, so Mr and Mrs America are not even convinced there is a problem. After all, what proof is there that climate change is going to happen?

Science, environment and Romanticism

The search for empirical proof that an environmental problem exists is rooted in the scientific developments of the sixteenth and seventeenth centuries. On the other hand, the modern environment movement depends for its support on the Western Romantic notion that nature is beautiful, spiritual and inherently worth preserving. Nature has a special value and contemplation or direct experience of it fulfils an inner human need. As Wordsworth said:[1]

> One impulse from the vernal wood
> Will tell you more of man
> Of moral evil and of good
> Than all the sages can.

For most of human history, nature has been something to be used or mastered, often feared and sometimes worshipped. Growing populations in Europe, along with scientific and philosophical advances leading up to the eighteenth century, brought about major changes in human relationships to the countryside. As the western world emerged from the Middle Ages, cities grew and trade strengthened among different cultures. Desire for new commodities like sugar, alcohol, pepper, glass and linen was springing up. There were more mouths to feed and meat consumption actually went down after the middle of the sixteenth century, but people wanted better bread and grain prices rose. Hunger was commonplace throughout Europe until the eighteenth century, and the majority of people were extremely poor and almost without possessions. According to Fernand Braudel 'the poor in the towns and countryside lived in a state of almost complete deprivation'.[2]

The slow climb from almost universal poverty which allowed the spread of household items such as chairs, tables, hearths and stoves, forks and even pottery, as well as a greater variety in diet, required the transformation of the landscape. In a world where food was scarce and belongings meagre, reflection on the beauty of nature was almost non-existent. Winter cold was a constant foe in Europe and the climate an arbiter of well-being in its influence on agriculture. Untouched nature, in the form of marshes, woodlands, mountains or wild rivers presented a series of obstacles to travellers, traders and farmers. The long and ancient ridgeway tracks of England, for instance, had been used for centuries to avoid woods and marshy valley bottoms. Fear of wild animals or bandits and the sudden calamities that could be brought by flood, drought or avalanche were widespread. As Grevel Lindop has succinctly stated, 'Nature is beautiful as long as we are safe within it.'[3]

Lindop identified three stages in the Romantic response to nature and pinned them down as being exemplified in the Lake District writings of Thomas Gray, Coleridge and Keats between 1769 and 1818. 'In this brief period', he writes, 'our culture's way of looking at the non-human world turned a corner.' Gray's view of Derwentwater as the 'sweetest scene I can yet discover in point of pastoral beauty' encapsulated the emerging idea that nature could be beautiful if viewed as a scene or picture. Thirty years later, Coleridge enjoyed a more moving experience looking at Bassenthwaite Water from his house. Says Lindop 'For Coleridge, the landscape (though still framed by the windows) has entered the mind: it is now an experience, a state of perception, a tranquillizing or intoxicating dream tasting of the creative imagination itself.' Finally Keats' assertion that the landscape's 'countenance or intellectual tone must surpass every imagination and defy every remembrance' speaks to Lindop of an implied intelligence in nature – 'The implication is there that we must learn from this; that it knows, in some sense, more than we do.'[4]

To the Romantics, a human being could but learn from and be spiritually reawakened by contact with nature. For example, extolling the splendour of dusk at Rydal Mount, Wordsworth wrote:[5]

> My soul though yet confined to earth
> Rejoices in a second birth.

Similar echoes can be heard in the voice of the so-called 'Peasant Poet', John Clare:[6]

> Nature thou truth of heaven if heaven be true
> Falsehood may tell her ever changeing lie
> But nature's truth looks green in every view
> And love in every landscape glads the eye.

The Romantic tide in thinking and the arts swept through Europe and America, and the ripples spread in many directions. There was not one stream of Romantic thought, but many. 'Nature is Imagination itself' said William Blake, and with others such as Rousseau and Alexander von Humboldt raised the appreciation of the wilder countryside to an almost religious level. However, as Keith Thomas has written, it was the English that excelled in the 'divinisation of nature'. They streamed into the newly accessible wildernesses of Snowdonia and Cumberland, and they formed the bulk of the new breed of alpinists that flocked to seek spiritual enlightenment while climbing and botanizing among the previously abhorred peaks of Switzerland, France, Italy and Spain.[7]

In America, the transcendental movement gained momentum. Centred upon the thoughts and writings of Ralph Waldo Emerson and Henry David Thoreau, they sought inspiration in the organization and force of nature. This veneration of nature was supported by the writings of Georges Buffon who used geology to suggest that fixed biblical time was false, and that the Earth might be 75,000 years old at least. It built, too, on the philosophical works of Immanuel Kant and Johann Gottfried Herder which for the first time offered unified cosmic histories that accepted evolution over unlimited time and a complex web of relationships between nature and society.[8]

In Thoreau's writings there is a profound sense of trust in nature as the means to balance man's empty materialistic urges. In *Walden* he writes, 'I went to the woods because I wished to live deliberately, to front only the essential facts of life, and see if I could not learn what it had to teach, and not, when I came to die, discover that I had not lived.'[9] Human solitude, he felt, could be turned to positive result in lifting the spirit in communion with nature – 'Yet I experienced sometimes that the most sweet and tender, the most encouraging society may be found in any natural object, even for the poor misanthrope and most melancholy man. There can be no very black melancholy to him who lives in the midst of nature and has his senses still.'[10] Emerson too, in *The American Scholar*, called for the recognition of the vital role of nature for man's being. 'The first in time and the first in importance of the influences on the mind is that of Nature.'[11]

The romantic revolution in cultural and spiritual thought was not taking place in a vacuum. A huge variety of outside influences was driving people towards a reassessment of their relationships with the natural world.

Notes

1 Quoted in Marshall, Peter, *Nature's Web: An Exploration of Ecological Thinking* (London: Simon and Schuster, 1992).
2 Braudel, Fernand, *The Structures of Everyday Life, Civilisation and Capitalism: 15–18th Century* (Volume 1) *The Structures of Everyday Life* (London: Fontana Press, 1985).
3 Lindop, Grevel, 'The Countenance of Nature: Towards an aesthetics of every thing, every place, every event', in *The Times Literary Supplement*, no. 4720, 17 September 1993).
4 Ibid.
5 Wordsworth, William, 'Ode Composed Upon an Evening of Extraordinary Splendor and Beauty', in Gill, Stephen (ed.), *The Oxford Authors: William Wordsworth* (Oxford: OUP, 1990).
6 Clare, John, in *The Oxford Authors: John Clare* (Oxford: OUP, 1984).
7 Thomas, Keith, *Man and the Natural World: Changing attitudes in England 1500–1800* (London: Penguin Books, 1984).
8 Marshall, Peter, *Nature's Web*, op. cit.
9 Thoreau, Henry David, *Walden and Civil Disobedience* (New York: Viking Penguin, 1986).
10 Ibid.
11 Quoted in Shabecoff, Philip, *A Fierce Green Fire: The American Environmental Movement* (New York: Hill and Wang, 1993).

Source: Adam Markham (1994) *A Brief History of Pollution*, London: Earthscan, pp. 24–30.

1.2

Keith Thomas

Human uniqueness

Inhibitions about the treatment of other species were dispelled by the reminder that there was a fundamental difference in kind between humanity and other forms of life. The justification for this belief went back beyond Christianity to the Greeks. According to Aristotle, the soul comprised three elements: the nutritive soul, which was shared by man with vegetables; the sensitive soul, which was shared by animals; and the intellectual or rational soul, which was peculiar to man.[1] This doctrine had been taken over by the medieval scholastics and fused with the Judaeo-Christian teaching that man was made in the image of God (Genesis i. 27). Instead of representing man as merely a superior animal, it elevated him to a wholly different status, halfway between the beasts and the angels. In the early modern period it was accompanied by a great deal of self-congratulation.

Man, it was said, was more beautiful, more perfectly formed than any of the other animals. He had 'more of divine majesty in his countenance' and 'a more exquisite symmetry of parts'.[2] Jeremiah Burroughes reminded his congregation that, when God saw his other works, he only said that they were 'good', whereas when he had made man he said '*very* good': 'Observe, it is never said "very good" till the last day, till man is made.'[3]

Even so, there was a marked lack of agreement as to just where man's unique superiority lay. The search for this elusive attribute has been one of the most enduring pursuits of Western philosophers, most of whom have tended to fix on one feature and emphasize it out of all proportion, sometimes to the point of absurdity. Thus man has been described as a political animal (Aristotle); a laughing animal (Thomas Willis); a tool-making animal (Benjamin Franklin); a religious animal (Edmund Burke); and a cooking animal (James Boswell, anticipating Lévi-Strauss). As the novelist Peacock's Mr Cranium observes, man has at one time or another been defined as a featherless biped, an animal which forms opinions and an animal which carries a stick.[4] What all such definitions have in common is that they assume a polarity between the categories 'man' and 'animal' and that they invariably regard the animal as the inferior. In practice, of course, the aim of such definitions has often been less to distinguish men from animals than to propound some ideal of human behaviour, as when Martin Luther in 1530 and Pope Leo XIII in 1891 each declared that the possession of private property was an essential difference between men and beasts.[5]

By Tudor times the amount of inherited law on the subject was already enormous. Since Plato a great deal had been made of man's erect posture: beasts looked down, but he looked up to Heaven.[6] Aristotle had developed the theme, adding that men laughed, that their hair went grey, and that they alone couldn't wiggle their ears.[7] In the early modern period differences in anatomy continued to impress. According to one early Stuart doctor:

> Man is of a far different structure in his guts from ravenous creatures as dogs, wolves, etc., who, minding only their belly, have their guts descending almost straight down from their ventricle or stomach to the fundament: whereas in this

> noble microcosm man, there are in these intestinal parts many anfractuous circumvolutions, windings and turnings, whereby, longer retention of his food being procured, he might so much the better attend upon sublime speculations, and profitable employments in Church and Commonwealth.[8]

In the late eighteenth century the aesthete Uvedale Price drew special attention to the nose. 'Man is, I believe, the only animal that has a marked projection in the middle of the face.'[9]

Three other human attributes were particularly stressed. The first was speech, a quality which John Ray described as 'so peculiar to man that no beast could ever attain to it'. It was through speech, said Ben Jonson, that man expressed his superiority to other creatures. Without it, agreed Bishop Wilkins, man would be 'a very mean creature'. Because beasts lacked language, explained the eighteenth-century economist James Anderson, their experience could not be transmitted to their posterity: man progressed, but every animal species had 'the same powers and propensities . . . that they had at the earliest period they were known'.[10]

The second distinguishing quality was reason. Man, as Bishop Cumberland put it, was 'an animal endowed with a mind'. Whether the difference was of kind or only of degree was a matter of debate. Some regarded animals as utterly irrational. Robert Lovell in 1661 divided the whole animal creation into two categories, 'rational' and 'irrational', putting only man in the former class. Gervase Markham reported the 'strongly held opinions' of 'many farriers' that horses had no brains at all; he himself had cut up the skulls of many dead horses and found nothing inside.[11] But most thought animals had elementary powers of understanding, albeit highly inferior ones. They had some practical intelligence, taught Aristotle, but they lacked the capacity for deliberation or speculative reason. From man's vast intellectual superiority, it was agreed, came his superior memory, his greater imagination, his curiosity, his sense of time, his sharper concept of the future, his use of numbers, his sense of beauty, his capacity for progress.[12] Above all, man could choose, whereas animals were prisoners of their instinct, guided only by appetite and incapable of free will.[13]

This distinctive human capacity for free agency and moral responsibility led on to the third, and, in the theologians' view, most decisive difference. This was not reason, which was, after all, shared to some extent by inferior creatures, but religion. Unlike animals, man had a conscience and a religious instinct.[14] He also had an immortal soul, whereas beasts perished and were incapable of an afterlife. This was no matter for regret: 'The life of a beast,' as a seventeenth-century preacher put it, was quite 'long enough for a beast-like life'. To suggest that animals might be immortal, said another in 1695, was an 'offensive absurdity'. Belief in the posthumous extinction of beasts was very important, he explained. It preserved the dignity of human nature, by showing an essential difference between the spirit of man and the souls of animals.[15]

In the seventeenth century the most remarkable attempt to magnify this difference was a doctrine originally formulated by a Spanish physician, Gomez Pereira, in 1554, but independently developed and made famous by René Descartes from the 1630s onwards. This was the view that animals were mere machines or automata, like clocks, capable of complex behaviour, but wholly incapable of speech, reasoning, or, on some interpretations, even sensation. For Descartes, the human body was also an automaton; after all, it performed many unconscious functions, like that of digestion. But the difference was that within the human machine there was a mind and therefore a separable soul, whereas brutes were automata without minds or souls. Only man combined both matter and intellect.[16]

This doctrine anticipated much later mechanistic psychology and contained the germs of the materialism of La Mettrie and other eighteenth-century thinkers. In due course, it would make it possible for scientists to argue that consciousness could be explained mechanically and that the whole of an individual's psychic life was the product of his physical organization. What Descartes said of animals would one day be said of man.[17] In the meantime, however, the Cartesian doctrine had the effect of further downgrading animals by comparison with human beings. Descartes himself seems to have modified his doctrine in later years and was unwilling to conclude that brutes were wholly incapable of sensation; for him the essential point was that they lacked the faculty of cogitation. He denied souls to animals because they exhibited no behaviour which could not be accounted for in terms of mere natural impulse.[18] But his supporters went further. Animals, they declared, did not feel pain; the cry of a beaten dog was no more evidence of the brute's suffering than was the sound of an organ proof that the instrument felt pain when struck.[19] Animal howls and writhings were merely external reflexes, unconnected with any inner sensation.

Today, this doctrine may seem to fly in the face of common sense. But it is not surprising that Cartesianism had its supporters at the time. An age accustomed to a host of mechanical marvels – clocks, watches, moving figures and automata of every kind – was well prepared to believe that animals were also machines, though made by God, not man.[20] Besides, Descartes was only sharpening a distinction already implicit in scholastic teaching. Aquinas, after all, had taught that the so-called prudence of animals was no more than divinely implanted instinct.[21] Moreover, Cartesianism seemed an excellent way of safeguarding religion. Its opponents, by contrast, could be made to seem theologically suspect, for when they conceded to beasts the powers of perception, memory and reflection, they were implicitly attributing to animals all the ingredients of an immortal soul, which was absurd; and if they denied that they had an immortal soul, even though they had such powers, they were by implication questioning whether man had an immortal soul either.[22] Cartesianism was a way of escaping both of these unequally unacceptable alternatives. It denied that animals had souls and it maintained that men were something more than mere machines. It was, thought Leibniz, an opinion into which its supporters had foolishly rushed 'because it seemed necessary either to ascribe immortal souls to beasts or to admit that the soul of man could be mortal'.[23]

But the most powerful argument for the Cartesian position was that it was the best possible rationalization for the way man actually treated animals. The alternative view had left room for human guilt by conceding that animals could and did suffer; and it aroused worries about the motives of a God who could allow beasts to undergo undeserved miseries on such a scale. Cartesianism, by contrast, absolved God from the charge of unjustly causing pain to innocent beasts by permitting humans to ill-treat them; it also justified the ascendancy of men, by freeing them, as Descartes put it, from 'any suspicion of crime, however often they may eat or kill animals'.[24] By denying the immortality of beasts, it removed any lingering doubts about the human right to exploit the brute creation. For, as the Cartesians observed, if animals really had an immortal element, the liberties men took with them would be impossible to justify; and to concede that animals had sensation was to make human behaviour seem intolerably cruel.[25] The suggestion that a beast could feel or possess an immaterial soul, commented John Locke, had so worried some men that they 'had rather thought fit to conclude all beasts perfect machines rather than allow their souls immortality'.[26] Descartes's explicit aim had been to make men 'lords and possessors of nature'.[27] It fitted in well with his intention that he should have portrayed other species as inert and lacking any spiritual dimension. In so doing he created an absolute break between man and the rest of nature, thus clearing the way very satisfactorily for the uninhibited exercise of human rule.

The Cartesian view of animal souls generated a vast learned literature, and it is no exaggeration to describe it as a central preoccupation of seventeenth- and eighteenth-century European intellectuals.[28] Yet, though Descartes's work was disseminated in England, the country threw up only half a dozen or so explicit defenders of the Cartesian position. They included the virtuoso Sir Kenelm Digby, who did not hesitate to declare that birds were machines, and that their motions when building their nests and feeding their young were no different from the striking of a clock or the ringing of an alarm.[29]. . .

Yet Descartes had only pushed the European emphasis on the gulf between man and beast to its logical conclusion. A transcendent God, outside his creation, symbolized the separation between spirit and nature. Man stood to animal as did heaven to earth, soul to body, culture to nature. There was a total qualitative difference between man and brute. In England the doctrine of human uniqueness was propounded from every pulpit. John Evelyn heard a sermon in 1659 on how man was 'a creature of different composure from the rest of animals; as both to soul and body; [and] how the one was to be the subject to the other'. In 1683 the Dean of Winchester conceded that animals had some human qualities, albeit 'in an inferior manner', but he denounced the idea that animals and men were therefore the same as a 'dangerous imagination'.[30] Throughout the eighteenth century the theme was reiterated. 'In the ascent from brutes to man,' declared Oliver Goldsmith, 'the line is strongly drawn, well marked, and unpassable.' 'How slender so ever it may sometimes appear,' wrote the naturalist William Bingley, 'the barrier which separates men from brutes is fixed and immutable.' The practical advantages of this distinction were clear, even if its theoretical rationale was elusive. 'Animals, whom we have made our slaves,' Charles Darwin would write, 'we do not like to consider our equal.'[31]

Notes

1 Aristotle, *De Anima*; C. S. Lewis, *The Discarded Image* (Cambridge, 1967), 152–65; Robert Burton, *The Anatomy of Melancholy* (EL, 1932), i. 154–5.
2 Aristotle, *Hist. An.*, 608[b]; Robinson, *Vindication of the Mosaick System*, 81; Hale, *Primitive Origination of Mankind*, 64.
3 Burroughes, *Gospel Reconciliation*, 6.
4 Aristotle, *Politics*, 1253[a]; *The Remaining Medical Works of Thomas Willis*, trans. S. P[ordage] (1681), 117; *Boswell's Life of Johnson*, ed. George Birkbeck Hill, revised by L. F. Powell (Oxford, 1934–64), iii. 245; v. 33n (cf. Claude Lévi-Strauss, *Le Cru et le Cuit* (Paris, 1964); *The Parliamentary History of England*, xvii (1813), 782; Thomas Love Peacock, *Headlong Hall* (1816), chap. 12.
5 Roy Pascal, *The Social Basis of the German Reformation* (1933), 161; *Church and State through the Centuries*, ed. Sidney Z. Ehler and John B. Morrall (1954), 326.
6 Plato, *Timaeus*, 90; Ovid, *Metamorphoses*, i. 84–6; Willet, *Hexapla in Genesin*, 107; Helkiah Crooke, ΜΙΚΡΟΚΟΣΜΟΓΡΑΦΙΑ. *A Description of the Body of Man* (2nd edn, 1631), 646; James Tyrrell, *A Brief Disquisition of the Law of Nature* (1692), 79; Herschel Baker, *The Image of Man* (New York, 1961 edn), 298 n18.
7 Aristotle, *De Part. An.*, 673[a]; *Hist. An.*, 518[a], 492[a]; H. C. Baldry, *The Unity of Mankind in Greek Thought* (Cambridge, 1965), 89–90.
8 Hart, *Diet of the Diseased*, 36 (following Pliny, *Nat. Hist.*, xi. 37).
9 Uvedale Price, *Essays on the Picturesque* (1810), iii. 223.
10 Ray, *Wisdom*, 191; Ben Jonson, *Timber* (Temple Classics, n.d.), 93; Pepys, *Diary*, viii. 554; James Anderson, *Recreations in Agriculture, Natural-History, Arts, and Miscellaneous Literature* (1799–1802), i. 9 (2nd pagination). Cf. Cicero, *De Inventione*, i. 4.
11 Richard Cumberland, *A Treatise of the Laws of Nature*, trans. John Maxwell (1727), 93; Robert Lovell, ΠΑΝΖΩΟΡΥΚΤΟΛΟΓΙΑ *Sive Panzoologico-mineralogia. Or a Compleat History of Animals and Minerals* (Oxford, 1661), intro.; Gervase Markham, *Markhams Maister-Peece* (1610), 57. Also Andrew Snape, *The Anatomy of an Horse* (1683), 105.
12 For representative opinions on this large subject see Aristotle, *Hist. An.*, 488[b]; Aquinas, *Summa Theologica*, i. 78. 4; Hobbes, *EW*, iii. 44, 48, 664; vii. 467; *id.*, *LW*, ii. 88–9; iii. 527; John Locke, *An Essay concerning Human Understanding*, ed. Peter H. Nidditch (Oxford, 1975), 159–60 (ii, chap. xi); Dunton, *Athenian Oracle*, i. 140; iii. 75; William Smellie, *The Philosophy of Natural History* (Edinburgh, 1790–99), ii. 457; [James Burnet, Lord Monboddo], *Antient Metaphysics* (1779–99), iii, appendix, chap. iii.
13 Aristotle, *Politics*, 1254[b]; Aquinas, *Summa Theologica*, ii(1). 6; ii(2). 95. 7; King, *Essay on the Origin of Evil*, 161–2.
14 *Proceedings in the Parliaments of Elizabeth I*, i, ed. T. E. Hartley (Leicester, 1981), 240; Richard Baxter, *Compassionate Counsel to all Young-Men* (1681), 69; John Howe, *The Living Temple* (1675), 22–3; George Berkeley, *Alciphron* (1732), 5th dialogue, 28; Vicesimus Knox, *Lucubrations* (1788), no. 135.
15 John Chishull, *Two Treatises* (1654), sig. A5; M[atthew] S[mith], *A Philosophical Discourse of the Nature of Rational and Irrational Souls* (1695), 21.
16 *Antoniana Margarita . . . per Gometium Pereiram* (Medina del Campo, 1554) (discussed by Pierre Bayle, *Dictionaire historique et critique* (2nd edn, Rotterdam, 1702), 'Pereira'); René Descartes, *Discours de la Méthode* (1637), v; *id.*, *Méditations métaphysiques* (1641), vi; *Œuvres de Descartes*, ed. Charles Adam and Paul Tannery (Paris, 1897–1913), iii. 85; iv. 574–5; v. 275–9. See in general Leonora D. Cohen, 'Descartes and Henry More on the Beast-Machine', *Ann. Sci.*, i (1936); Leonora Cohen Rosenfield, 'Un Chapitre de l'histoire de l'animal-machine (1645–1749)', *Revue de littérature comparée*, 17 (1937); *id.*, *From Beast-Machine to Man-Machine. Animal Soul in French Letters from Descartes to La Mettrie* (New York, 1941); Hester Hastings, *Man and Beast in French Thought of the Eighteenth Century* (1936), 19–63; Albert G. A. Balz, *Cartesian Studies* (New York, 1951), 106–57; Robert M. Young, 'Animal Soul', in *The Encyclopaedia of Philosophy*, ed. Paul Edwards (New York, 1967), i; J. S. Spink, *French Free-Thought from Gassendi to Voltaire* (1960), chap. xi; Thomas H. Huxley, 'On the Hypothesis that Animals are Automata, and its History', in *Method and Results* (1893).

17 See Aram Vartanian, *Diderot and Descartes* (Princeton, 1953); *id.*, *La Mettrie's L'Homme Machine* (Princeton, 1960); *id.*, 'Man-Machine from the Greeks to the Computer', *Dictionary of the History of Ideas*, ed. Philip P. Wiener (New York, 1973–4), iii.
18 *Œuvres de Descartes*, iv. 574–6; v. 276–8.
19 Anthony Le Grand, *An Entire Body of Philosophy, according to the Principles of the famous Renate Des Cartes*, trans. Richard Blome (1694), ii. 252.
20 As was noted by Descartes, *Discours*, v; *Œuvres*, iii. 121. Cf. Le Grand, op. cit., ii. 236; John Norris, *An Essay towards the Theory of the Ideal or Intelligible World* (1701–4), ii. 83–6.
21 Aquinas, *Summa Theologica*, i. 78. 4; ii(1). 17 (as pointed out by John Rodman, 'The Dolphin Papers', *The North American Rev.*, 259 (Spring 1974), 21).
22 Le Grand, *Entire Body of Philosophy*, i. 255–6; ii. 234–5.
23 Gottfried Wilhelm Leibniz, *Philosophical Papers and Letters*, trans. and ed. Leroy E. Loemker (2nd edn, Dordrecht, 1969), 588.
24 *Œuvres*, v. 279 (trans. in *Ann. Sci.*, i (1936), 53).
25 Norris, *An Essay*, ii. 74.
26 Bodl., MS. Locke f. 6, p. 26.
27 *Discours*, vi ('maîtres et possesseurs de la nature').
28 Cf. the bibliography by Ezra Abbot in William Rounseville Alger, *A Critical History of the Doctrine of a Future Life* (4th edn, New York, 1867), appendix.
29 Sir Kenelm Digby, *Two Treatises* (1645), i. 399–400 (though Bayle, *Dictionaire*, 'Rorarius', 2609, denied that Digby was of Descartes's opinion). For others see Henry Power, *Experimental Philosophy* (1664), sig. b2[v]; Tim. Nourse, *A Discourse upon the Nature and Faculties of Man* (1686), 77; Le Grand, *Entire Body of Philosophy*, part iii; Norris, *An Essay*, ii, chap. 2; [F.B.], *A Letter concerning the Soul and Knowledge of Brutes* (1721); Bernard Mandeville, *The Fable of the Bees*, ed. F. B. Kaye (Oxford, 1924), i. 181n. Cf. Wallace Shugg, 'The Cartesian Beast-Machine in English Literature (1663–1750)', *JHI*, 29 (1968).
30 Evelyn, *Diary*, iii. 234; Richard Meggott, *A Sermon preached at White-Hall* (1683), 11.
31 Goldsmith, iv. 203–4; Bingley, *Quadrupeds*, 2; Harold E. Gruber, *Darwin on Man* (1974), 447.

Source: Keith Thomas (1984) *Man and the Natural World 1500–1800*, London: Penguin, pp. 30–6.

1.3

David Pepper

The roots of technocentrism

Medieval and Renaissance cosmologies

The medieval (fifth to fifteenth century) view of the universe – what educated people thought about how it functioned and their position in it – was governed in its physical aspects by the ideas of Aristotle. These were integrated with evolving Judaeo-Christian ideas over a long period. The integration was a very close one, and was accomplished by the twelfth and thirteenth centuries. There was an almost perfect mapping of Aristotle's physical picture onto Christian theology – onto the Christian moral universe.

The cosmography was geocentric. The earth was at the centre of the universe, and, as all evidence suggested, it was solid, stationary, finite and spherical. The stars rotated around and were equidistant from Earth (see Figure 3). They were attached to the inner surface of a rotating sphere which looked like a dome from Earth, and marked the edge of the universe. Outside this sphere – beyond the universe's edge – was nothing, or *non ens* (non-being). This did not mean just empty space, it meant literally that nothing could exist. If one could travel to this boundary and stick one's hand beyond it, the hand would become non-existent. To put it another way, one could not ask questions about this region.

Within the sphere of the fixed stars the universe was divided into two zones, celestial and terrestrial, with the moon's orbit forming the boundary between them. The behaviour of the celestial objects was very predictable; they moved in circular orbits around the earth at constant speeds. But in the terrestrial region things moved randomly or in straight lines. Terrestrial things were born, died and decayed – they changed. This did not happen to celestial objects. Non-changingness suggested no need for change because of already-achieved perfection. Thus change meant imperfection. Circular motion suggested perfection – the perfect geometrical figure being a sphere – randomness suggested imperfection. So the celestial zone was one of perfection, the terrestrial zone was one of imperfection. This was an observational *and* a value difference. Observational (empirical) evidence and values were combined.

Between the moon and the sphere of stars were, in order, Mercury, Venus, the Sun, Mars, Jupiter and Saturn. Each of their orbits was part of a discrete sphere, so the arrangement was of spheres within spheres. When it came to explaining observed phenomena such as gravity, answers lay in a combination of this observed structure of the universe and the idea that there was *purpose and design* in it. This *teleological* view – that there was a distant goal towards which all events worked – meant that there was a *final cause* behind everything. Since the cosmology was a Christian one, the Christian God was the final cause. The universe was ruled by principles which helped to achieve God's purposes and he was the final cause of everything. These principles, or physical laws, were a function of God's design and were to be explained through understanding that design . . .

The Earth was made up of four elements, earth, air, fire and water, while heavenly objects – being in the region of perfection – must consist of a different, fifth, perfect element: the 'quintessence'. . . .

Being rational, however, God intended that the universe should return to its original state. No restoration was needed in the already perfect celestial region, but in the terrestrial zone all the objects were trying to fulfil God's desire. Thus if one held up a stone and then released it, it would fly towards the centre of the Earth, where it naturally belonged. It would go in a straight line – the shortest route. This explained gravity, and the well-known acceleration of objects as they descended was also explicable in terms of fulfilling God's purpose. Since the stone wanted to carry out God's desire, and since (as we all know) the nearer one gets to the desired object the stronger one's desire, the nearer the stone got to its natural place, the faster it went.

Thus behaviour of natural objects conformed with what they were mainly composed of and to their position in relation to where God wanted them to be. Subterranean air, water and fire travelled upwards with force (in volcanoes and springs and geysers) to get to their appointed positions, as did a candle flame point upwards. Water in the region which was appointed for air fell down. The explanations formed part of a remarkably logically consistent, coherent and complete system of theory. Several significant points may be noted about it.

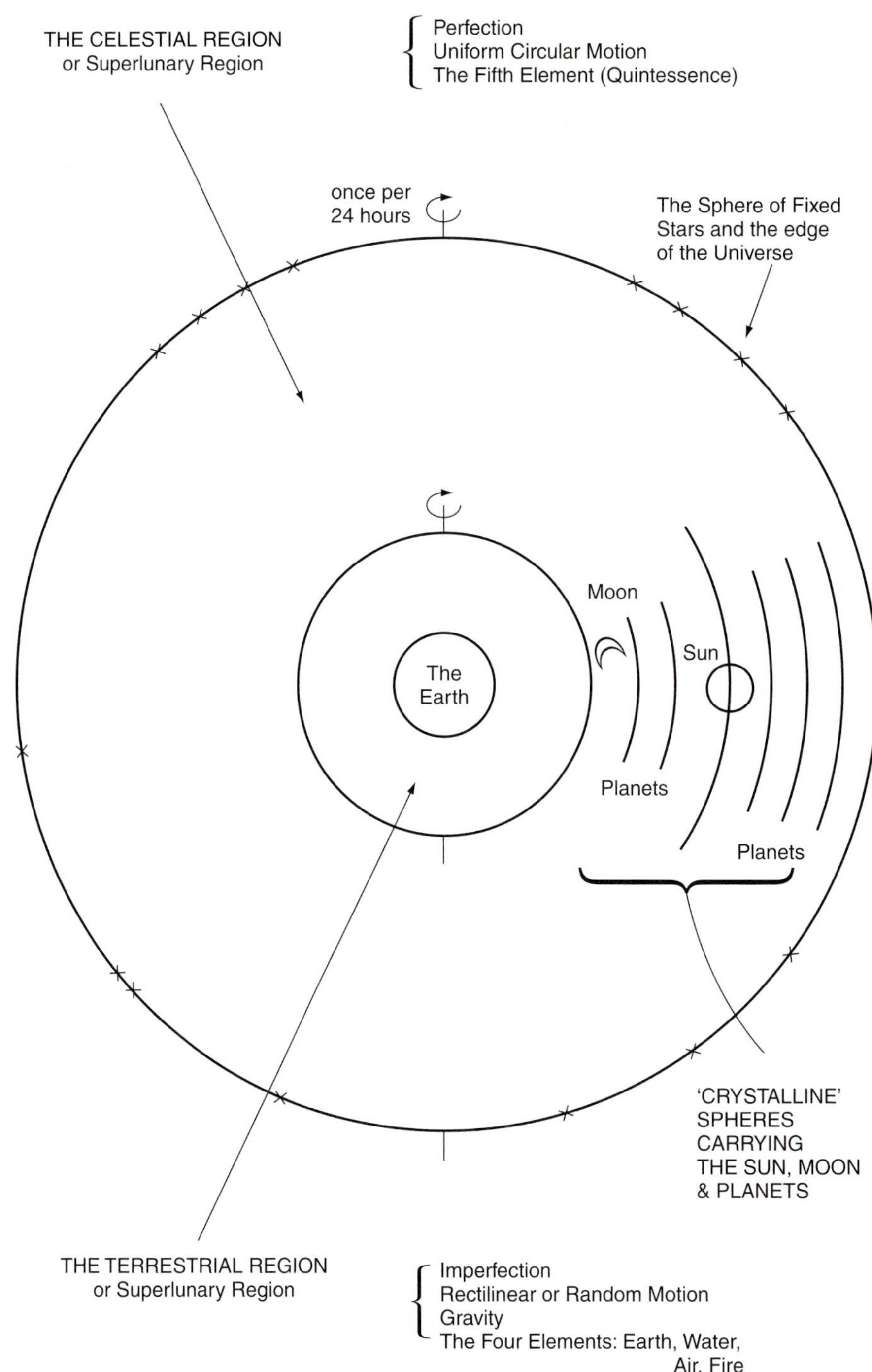
THE CELESTIAL REGION
or Superlunary Region
Perfection
Uniform Circular Motion
The Fifth Element (Quintessence)
once per
24 hours
The Sphere of Fixed
Stars and the edge
of the Universe
Moon
Sun
The
Earth
Planets
Planets
'CRYSTALLINE'
SPHERES
CARRYING
THE SUN, MOON
& PLANETS
THE TERRESTRIAL REGION
or Superlunary Region
Imperfection
Rectilinear or Random Motion
Gravity
The Four Elements: Earth, Water,
Air, Fire

Figure 3 The medieval cosmology

First, the explanations which scientists gave for physical phenomena followed from, and were compatible with, theology. The 'paradigm' for science was set by religion. This was inherent in the academic and power structure of the medieval universities. The Faculty of Theology was the senior one, dominating the others in the university hierarchy. . . . Second, despite the importance of God in the scheme, the medieval cosmology was *anthropocentric*. It was replete with human views imposed onto nature. Categories used to explain human experience were transferred to explain physical ones; hence the universe was said to be purposeful, in which objects desired to fit in with God's plan – purpose and desire being human attributes. Space had human values attributed to it. The celestial zone was most valued, but within the terrestrial zone the nearer one went to the Earth's centre the greater was the imperfection – with Hell at the centre. And on the Earth's surface there were areas more valued than others (sacred spaces, like Jerusalem, or the precincts of a cathedral). Scientific knowledge of these spaces, and of behaviour within them, could not be divorced from human values – the values of the society in which the science was done.

Third, and arising from this principle, the world could be seen very much in terms of analogy with human experience. Metaphors could be used to interpret it: it *was* a series of metaphors. Nature as a book was one of the principal medieval metaphors. It meant that since nature was a result of God's design and desires, one read two books, the Bible and nature, to discover God's purposes and act accordingly. So nature was not there just for people's sensuous enjoyment; it carried in it instructions and hidden meanings. The industry of the ant or the bee would be an example to us. The fly was a reminder of the shortness of life, and the glow-worm of the light of the Holy Spirit (Thomas 1983, 64). . . .

. . . In nature one read not only how to be saved but also, because God had created the Earth for humans, other knowledge of benefit to humans. So where a plant, stone or animal resembled in shape or colour or behaviour a human organ or disease, it could be used in healing. This was the doctrine of *signatures*, examples being that milky plants helped new mothers to produce milk, bony plants were good for the bones, summer plants helped to cure summer complaints (Mills 1982), spotted herbs cured spots, yellow ones healed jaundice and adder's tongue was good for snake bites (Thomas 1983).

If the book metaphor derived from human experience, so, too, did the *organic metaphor*. This was rife up to the eighteenth century, and it has reappeared today as a cornerstone of deep ecology. The medieval version saw the world as a huge animal with feelings, in which men and women, like intestinal parasites, lived:

> the Earth was perceived as a living body: the circulation of water through rivers and seas was comparable to the circulation of blood; the circulation of air through wind was the breath of the planet; volcanoes and geysers were seen as corresponding to the Earth's digestive system – eruptions were like belches and farts issuing out of a central stomach.
>
> (Gold 1984, 13)

Earth, which was female, also gave birth to stones and metals from her womb, having been fertilised by the (male) sky/heavens. Base metals then grew into silver and gold.

There was a spectrum of Renaissance organic philosophies. Common to them all were the notions that all parts of the living cosmos were connected into a unity, in mutual interdependence, and that they were all alive. This *vitalism* saw everything permeated by life – by the 'vital principle' – so that it was impossible to distinguish between living and non-living beings. The Earth was another living being among humans. Sometimes she nourished and nursed us, and for that she deserved respect and reverence. But like most females she also could be wild, passionate and frenzied. Men then argued that she needed to be tamed, says Merchant (1982), who draws parallels between the growth of attempts to control women through witch trials and the growth of attempts to control nature through classical science.

Merchant distinguishes three variants of the Renaissance organic view of nature, which also had implications for the view of human society (because nature was a macrocosm of the human microcosm). The first was that of a designed hierarchy – a Chain of Being – in nature (Figure 4), which also did and should exist in society. This clearly lent itself to conservatism, that is, maintenance of the existing social order. The second saw nature as an active unity of opposites in dialectical tension: a view which emphasised constant change and lent itself to revolutionary ideas (e.g. in Renaissance utopias or the utopian religious sects of the Civil War). The third considered nature as benevolent, peaceful and rustic; an 'arcadian' view prompting both romantic escapism and the notion of 'female' passivity – yielding itself up to manipulative ploughing, fertilisation and cultivation to provide a garden for the comfort and nurture of men exhausted by urban commerce.

The organic view stemmed from the pervasive cosmology which paralleled that of the physical world, whereby the medieval and Renaissance biological world was organised. That was the cosmology of the Great Chain of Being. In it the parallel was drawn between all elements in the universe, animate and inanimate, spiritual and material, and the links in a chain. They were joined together in a fixed hierarchy, and were interdependent. Everything in the Chain was alive, including stones, earth and 'slime'. For it followed that if everything had a sense of desire to fulfil God's purpose, then they had *living* attributes – desire and purpose are what living things show. They were given life by the overflowing of the Soul from the perfect God at the head of the hierarchy.

The idea of the Great Chain of Being originated with the Greeks, and was transmitted down to medieval writers, who adapted it to their cosmology. According to Lovejoy (1974), despite waning faith in metaphysics and increasing faith in Baconian empiricism the Chain of Being cosmology was widely diffused and accepted as late as the eighteenth century.

[. . .]

This Chain of Being metaphor and its related ideas tended to place people and nature into an intimate relationship, as may be true of medieval views in general, since each link of the Chain was vital for the whole chain's existence.

[. . .]

The Chain of Being gave people several good reasons to feel humble in relation to the rest of nature. First, removing any link from it would destroy the chain, hence all links were equally important for the Chain's completeness. Second, humankind, in medieval and Renaissance versions, was situated only in the middle of the chain, at the transition between being purely animal-like creatures of instinct and physical senses, and thinking beings with a soul who transcended the physical. Superior to humans were the intelligences of the angels. Third, humankind's distinctiveness from the creatures below was hardly chasm-like because of the continuity principle, which said that there was an almost imperceptible transition between each element of the chain.

[. . .]

The eternal – that which *is*, possessing alone (*seul*) the plenitude of its being.

↓

Powers
Dominations
Virtues
Choirs of Angels
Cherubim
Seraphim
Archangels
Angels
Man
  monkey
Quadrupeds
  flying squirrel
  bat
  ostrich
Birds
  amphibious birds
  aquatic birds
  flying fish
Fish
  eels and creeping fish
  water serpents
Reptiles
  slugs
Shellfish
  pond mussel
  lime-secreting worms
Insects
  worms
  polyp
  sensitive plants
  trees
  shrubs } plants
  herbs
  lichens
  moulds
  mushrooms and agaries (fungi)
  truffle
  stones composed of layers, fibres and filaments
  unorganised stones
Crystalline salts
  vitriols
Semimetals (non-malleable metals)
Malleable metals
  sulphur and bitumens
  compound earths (pure earths united with oils, salts, sulphurs etc.)
Pure earth
Water
Air
Ethereal Matter

(Side branches: tortoise, crocodile, sea lion, sea calf, hippopotamus, whales — ? — Fish; ? — crab, crayfish — Shellfish; ? — lizard, frog — Insects)

Figure 4 The Great Chain of Being, according to Charles Bonnett, naturalist (1720–93)
Source: Oldroyd (1980)

The scientific revolution and nature as a machine

The period of the 'scientific revolution' covered about 150 years, from the time of Copernicus, who published *On the Revolutions of the Heavenly Orbs* in 1543, to the end of the seventeenth century, with Isaac Newton's *Mathematical Principles of Natural Philosophy* (1687) and his *Opticks* (1703). During this time, the beginning of the 'modern' period, the principles of 'classical' science were established in contradistinction to the pre-modern cosmologies sketched above. These principles are sometimes known as the 'Newtonian paradigm', that is, the model derived from Newton's work, of what questions it was legitimate for science to ask of nature, how they should be asked, and what constituted valid answers.

Despite Einstein, quantum theory and all the other advances associated with twentieth-century physics, the Newtonian paradigm still constitutes conventional wisdom about science, and it is this conception which is therefore the basis of technocentrism. Indeed, the idea of society in technocentric, liberal political philosophy is that it is composed of discrete individuals interacting with other discrete individuals. The analogy with Newtonian physics' view of the physical world is clear, and not fortuitous: both pictures grew up together (Zohar and Marshall 1993).

In some ways the scientific revolution drew on preceding thought. Whitehead (1926, 15) considered that one great contribution which medievalism made to the scientific movement was 'the inexpugnable belief that every detailed occurrence can be correlated with its antecedents in a perfectly definite manner, exemplifying general principles'. Coupled with this belief in cause and effect – right back to the 'final cause' of all things, which was God's will – was the medieval insistence on God's rationality (compared with the Asian view of God as arbitrary and inscrutable). All this made the growing, naive, faith in the possibilities of rational science an unconscious derivative from faith in medieval theology, says Whitehead.

Despite this, the scientific revolution did fundamentally challenge the medieval cosmology, based as the latter was on divine purpose. Copernicus suggested a simple revision to the cosmography, swapping the positions of the earth and sun. But the implications of this were so enormous that it needed 150 years to construct the new cosmology which was required. For the new ideas were not only an intellectual challenge to the established science. They also ate away at the theology from which it stemmed – which in turn supported a particular social structure.

Indeed, were it not for social and economic challenges to that structure which were simultaneously occurring, the intellectual ideas represented by the Newtonian paradigm might not have triumphed in the eighteenth and nineteenth centuries as they did.

[. . .]

Johannes Kepler (1571–1630) sought universal principles in trying to explain planetary motion. He argued that the sun was the cause; both it and the planets functioned as magnets, and in rotating the sun pushed the planets around with it. So the sun was the driving force to a kind of machine:

> I am much occupied with the investigation of the physical causes. My aim in this is to show that the celestial machine is to be likened not to a divine organism but rather to a clockwork . . . insofar as nearly all the manifold movements are carried out by means of a single, quite simple magnetic force, as in the case of a

> clockwork all motions [are caused] by a simple weight. Moreover I show how this physical conception is to be presented through calculation and geometry.
> (Kepler, letter to Herwart von Hohenburg, 1605, cited by Holton (1956))

Thus Kepler was using a new metaphor for nature: the organism was replaced by a clock. This *mechanistic* conception of nature is a principal component of the classical scientific (technocentric) view. Kepler was religious, and he believed he could understand nature and, through it, God's intention, via mathematics and geometry:

> Why waste words? Geometry existed before the Creation, is co-eternal with the mind of God, *is God himself* (what exists in God that is not God himself?); geometry provided God with a model for the creation and was implanted into man, together with God's own likeness – and not merely conveyed to his mind through the eyes.
> (Kepler, *Harmonici Mundi*, 1619, cited by Koestler (1964))

This was, therefore, a conception of God as a Creator and nothing more. He was not ever-present and did not intervene in the workings of his creation. Rather, he was an engineer, using geometry to make a plan from which he had constructed a machine. He set the machine going and then left it.

This was an essentially *deterministic* view of nature. For the way a machine works is that its structure and past configuration determine its present behaviour. A clock is a series of cause-effect (deterministic) mechanisms (cogs and springs) linked together. And, most important, if you know enough about how the clock works, you can exactly predict its future behaviour. In this cosmology, the past now determines present and future, rather than, as in the medieval view of a final cause, a future goal (fulfilling God's will) determining present behaviour.

Kepler's view was developed by Galileo Galilei (1564–1642). He believed that the 'book of nature' was written in the language of mathematics and therefore had to be read and understood via mathematics. The physical problems of motion were to be treated as geometrical problems, and objects like falling stones as geometrical entities. This followed from the fact that God had structured the universe according to mathematical principles. So the way to understand it was through deduction from these principles in conjunction with observation and experimentation. The logical outcome of this mechanisation and mathematisation of nature was that when we look at nature we do not see what it really is (a machine). The true reality of nature is mathematical and regular. From this followed what has become a basic tenet of classical scientific philosophy: *what is truly real is mathematical and measurable, but what cannot be measured cannot have true existence.*

[. . .]

. . . Therefore the truth of scientific theory has less to do with how soundly or cleverly that theory is argued than with how *objective* is the knowledge from which it stems: *objective knowledge is 'true', and correct, while subjective knowledge is not.* It is in this distinction that classical science has its powerful appeal as being in some way independent. For it can give 'objective' knowledge of nature. Being objective rather than subjective, this knowledge must be free from the sectional interests of any particular person or group – it can therefore be *trusted. . .*

With this distinction between what is objectively 'out there' and what is subjectively in human perception arises also a fundamental distinction between society and nature. An enormous gulf, in fact, opens up between the two. The gulf was most starkly revealed by Descartes, the most radical of the thinkers of the scientific revolution and one of the creators of modern philosophy.

Developing on Kepler and Galileo, he argued that matter was nothing more than extension in space. It was geometry. It did not consist of sensible qualities like hardness, weight or colour, but was extension in breadth, length and depth. What was mathematical was real and what was real was mathematical. The implications of this were that the universe must be infinite (because the space of a geometer was infinite) and that since matter was extension in space the universe must be full: there could not be empty space.

The only way to have a full universe which also contained motion was to have matter divided into particles. For Descartes, matter was infinitely divisible, and the universe contained nothing except particles with size, shape, weight, motion and position. Explanations had to be in terms of matter – particles – in motion. No occult action, in terms of sympathy or desire, was possible.

Descartes extended the concepts of nature as a machine and the unreality of that which was not measurable. He viewed animals, and the human body too, as machines. They were automata, and their workings could be fully known by reducing them to matters of physics and chemistry, which in turn would be understood in terms of mathematics. This is a *reductionist* view, whereby, through analysis (breaking down of things into component parts), everything can be eventually reduced to the same basic quantities and qualities, all of which are measurable. . . .

Modern ecocentrics criticise classical science for its mechanistic and reductionist tendencies. These tendencies made it possible for scientists to argue that consciousness could be explained mechanically and that a person's entire psychic life 'was the product of his physical organisation. What Descartes said of animals would one day be said of man' (Thomas 1983, 33).

Reductionism also poses the question: 'If everything consists of the same basic stuff, with the same basic form (atoms), in what way, if any, are humans distinguishable from the rest of nature?'

[. . .]

. . . Descartes reasoned that the very *act* of doubting was a process of thinking, that the one thing that could not be doubted was that *he* was thinking and that this thinking was the act which established the fact of human existence. Humans were therefore definable as no more nor less than thinking beings. What separated them from the rest of nature, and from their own bodies, was this thought process. Whereas the body could be analysed into its component parts, the mind could not. Descartes thereby had introduced a most fundamental dualism in modern thought: that between mind and matter. Whereas matter was composed of primary, objectively knowable, qualities, the mind was subjective and it attributed secondary qualities to nature.

Thus this *Cartesian dualism* involved mind and matter, subject and object, and it had a profound implication for the society–nature relationship because nature became composed of *objects metaphysically separated from humans*. These objects had primary qualities and no others. They were reducible to atoms, whose unthinking, machine-like behaviour was universally the same and explicable in terms of mathematical laws. Humans, by contrast, were defined as rational thinking beings – subjects who observed objects, including nature, and could impart secondary qualities to them.

Ecocentrics (Capra 1982, Skolimowski 1992) now consider that this dualism paved the way for a society–nature separation in which the former was conceived of as *superior* to the latter. 'Cartesian' thinking is regarded as a prime culprit in creating a scientific world view where humans are separate from and above nature, a mere machine. Descartes is even cited as a villain of the piece by animal rights protagonists (e.g. Singer 1983, 218–20).

[. . .]

The Baconian creed and its high priests

[Francis] Bacon [(1561–1626)] was the first figure in the scientific revolution to draw out the full implications for the society–nature relationship of the emerging principles of the new science. As we have noted, pre-modern cosmology asserted the unity of society and nature; Cartesian dualism effectively separated the two. But it was Bacon who asserted the creed that *scientific knowledge equals power over nature.*

[. . .]

Bacon's scientific method can be visualised as pyramidal – founded on a broad base of *empirical* (based on observation and experiment) knowledge, and building on to it laws of nature of increasing generality; the whole pyramid crowned with the universal law that will explain everything. In practice, classical science has followed Newton, by mixing and combining both deductive and inductive approaches. But the importance of Bacon's induction lay in its implications that 'truth is the daughter of time' and that science is progressive.

For collecting empirical knowledge took a long time, hence inductive science was based on the steady accumulation of data from experimentation and from direct ('field') observations of nature. New knowledge was built upon old knowledge. Thus total scientific knowledge increased with time; it could not be encompassed by any one individual or age. It involved communal activity, where the community of scholars worked towards a future goal – establishing truth (although the ultimate truth could never be known, for this was the exclusive province of God).

Such activity could not be accomplished within the existing scientific institutions, since the medieval tradition was too strong in them whereby argument rather than objectively assessed facts constituted the hallmark of valid knowledge. Hence Bacon advocated a new start for science, in separate research institutions that would be state supported.

Bacon further argued that science was central to human endeavour, because of its goals and the motives associated with doing science. First and foremost, the true understanding of nature that only inductive method could achieve was the means of glorifying God, who had created nature.

But the case for state support required more justification, and Bacon embodied this in his definition of the purpose of scientific activity. For its second aim was to relieve 'man's estate'. It was a *philanthropic* activity where the scientist should assume the moral duty of improving society's material lot. This could come about by understanding how the machine of nature worked. This was the first stage in using the laws of nature for society's benefit. So science's purpose was to 'command nature in action'.

References

Capra, F. (1982) *The Turning Point*, London: Wildwood House.
Gold, M. (1984) 'A history of nature', in Massey, D. and Allen, J. (eds), *Geography Matters!*, Cambridge: Cambridge University Press, 12–33.
Holton, G. (1956) 'Johannes Kepler's universe: its physics and metaphysics', *American Journal of Physics*, 24, 340–51.
Koestler, A. (1964) *The Sleepwalkers*, Harmondsworth: Penguin.
Lovejoy, A. (1974) *The Great Chain of Being*, Cambridge, Mass.: Harvard University Press.
Merchant, C. (1982) *The Death of Nature: women, ecology and the scientific revolution*, London: Wildwood House.
Mills, W. (1982) 'Metaphorical vision: changes in Western attitudes to the environment', *Annals of the Association of American Geographers*, 72(2), 237–53.
Oldroyd, D. R. (1980) *Darwinian Impacts: an introduction to the Darwinian revolution*, Milton Keynes: Open University Press.
Singer, P. (1983) *Animal Liberation: towards an end to man's inhumanity to animals*, Wellingborough: Thorsons.
Skolimowski, H. (1992) *Living Philosophy*, London: Arkana.
Thomas, K. (1983) *Man and the Natural World: changing attitudes in England, 1500 to 1800*, London: Allen Lane.
Whitehead, A. N. (1926) *Science and the Modern World*, Cambridge: Cambridge University Press. London: Free Association Books edition 1985 edited by Robert Young.
Zohar, D. and Marshall, I. (1993) *The Quantum Society: mind, physics and a new social vision*, London: Bloomsbury.

Source: David Pepper (1996) *Modern Environmentalism*, London: Routledge, pp. 126–45.

1.4

Tim O'Riordan

Ecocentrism and technocentrism

The evolution of modern environmentalism

The contradictions that beset modern environmentalism reflect the divergent evolution of two ideological themes which arose at the birth of the conservation movement (McConnell, 1971), although their intellectual antecedents lie deeper in history. One line of thought can be identified as the *ecocentric mode*, described by McConnell (1965, p. 190) as 'resting upon the supposition of a natural order in which all things moved according to natural law, in which the most delicate and perfect balance was maintained up to the point at which man entered with all his ignorance and presumption'. Thereafter the 'web of life' was broken by a degenerative succession of 'disturbed harmonies' leading ultimately to the destruction of man himself. The other viewpoint is the *technocentric mode* characterised by Hays (1959,

p. 2) as the application of rational and 'value-free' scientific and managerial techniques by a professional elite, who regarded the natural environment as 'neutral stuff' from which man could profitably shape his destiny.

The two perspectives differ not just in their attitudes to nature but also in their morality that tempers action. Ecocentrism preaches the virtues of reverence, humility, responsibility, and care; it argues for low impact technology (but is not antitechnological); it decries bigness and impersonality in all forms (but especially in the city); and demands a code of behaviour that seeks permanence and stability based upon ecological principles of diversity and homeostasis. Until recently ecocentrism was more of a moral or spiritual crusade, its proponents generally preferring to shun the political arena in favour of the world of rhetoric and contemplation. The technocentric ideology, by way of contrast, is almost arrogant in its assumption that man is supremely able to understand and control events to suit his purposes. This assurance extends even to the application of theories and models to manipulate and predict changes in value systems and behaviour, while the exercise of science to 'manage' nature has been assumed for some time.

Ecocentrism is concerned with *ends* and the proper kind of means, whereas technocentrism focuses more on *means per se*, particularly the utilisation of managerial principles, since its optimism about the continued improvement of the human condition allows it to be rather less troubled about the evaluative significance of its achievements. The technocentrist admires the comforting power of technology and is usually found among an urban-dwelling elite who thrive on the sophisticated communications of the electronic global village and the jet age invisible college. Technocentrists tend to be politically influential for they usually move in the same circles as the politically and economically powerful, who are soothed by the confidence of technocentric ideology and impressed by its presumption of knowledge.

But we should avoid the temptation to divide the world neatly into an ecocentric camp of environmentalists and a technocentric camp of manipulative professionals and administrators. In real life the boundaries are much more blurred. There is every reason to believe that each one of us favours certain elements of both modes, depending upon the institutional setting, the issue at hand, and our changing socioeconomic status. The engineer, the planner, the administrator, and the technician are not insensitive to the beauties of nature, nor are they unaware of the consequences of 'tampering' with ecosystems. They differ from the ecocentric environmentalist largely in the degree of responsibility their job demands of them and in the extent to which a 'natural morality' guides their actions. Sometimes the very nature of the roles they play may channel them into decision paths from which they have little escape. For example, Sax (1970, pp. 34–35) describes how the noted conservationist Stanley Cain (when an Assistant Secretary in the US Department of the Interior) signed an order permitting the destruction of a small area of wildfowl habitat, a decision he later admitted was 'based first on political considerations and second on the feeling that the values [that were lost] were not great in the area [i.e. the wildfowl nesting area] to be filled'. Sax's (1970, p. 56) commentary on this act illustrates the dilemma:

> The greatest problems are often the outcome of the smallest-scale decisions precisely because the ultimate, aggregate impacts of these decisions are so difficult to see and the pressures so difficult to cope with from the perspective of the insider. It is much easier to tell a developer that he cannot dam up the Grand

> Canyon than to tell each real estate investor, one by one over time, that he cannot fill an acre or two of marshy 'waste' land.

The ecocentrist is not immune from this duality. Henry Thoreau, probably the best known ascetic to shun the 'comforts' of civilised life for a more 'natural' existence in the New England woodland, depended upon Bostonian society to publish and proselytise his views and eventually returned to that society which had given him the education and status he so very much enjoyed. The British naturalist Fraser-Darling (1971, pp. 79–80, 97) is quick to defend his Scottish Highlands home even though he admits that he can only continue to live there and be effective because of the technologically sophisticated and energy-intensive life-support systems which he criticises.

The duality of the technocentric and ecocentric modes is probably evident to most of us and need not be further elaborated. Suffice it to conclude that coexistence does not necessarily produce compromise, though in thoughtful people it should lead to better understanding. The reader is urged to peruse the delightfully readable account by McPhee (1972) of the conflict between the two ideologies in which he describes the confrontation between David Brower, former executive director of the Sierra Club, and various developers. The fascinating feature of McPhee's account is the apparent irreconcilability of the two viewpoints. The developers accept population pressure and economic growth as inevitable forces that can and must be accommodated by proper multiple-purpose management and the application of sensitive planning controls. They claim that the result is a better habitat for man – more rewarding, no less beautiful, yet more accessible to more people. To the environmentalist, however, growth is not a deterministic force but a process that can and must be controlled to save man from his damaging excesses. To leave the natural landscape alone is an act of faith, worship, and moral justice, for the beauty that is lost can never be replaced. . . .

The ecocentric mode

The ecocentric root of modern environmentalism is nourished by the philosophies of the romantic transcendentalists of mid-nineteenth century America. Burch (1971, pp. 67–108) contends that their philosophies provided a powerfully credible account of the relationship between nature and society, an interpretation that straddled the fierce individualism (freedom without equality) of the frontier and the bland homogeneity (equality without freedom) of the agrarian landscape and the suburb. The frontiersman was less concerned with obligation and more interested in self-centred freedom, while the rural or suburban 'yeoman' was willing to lose some of his individualism in order to conform to the comforting and law-abiding values of his neighbours. The transcendentalists sought to blend these two images through the symbol of nature. Nature, they claimed, enjoyed its own morality which, when understood, could lead the sympathetic and responsive human being to a new spiritual awareness of his own potential, his obligations to others, and his responsibilities to the life-supporting processes of his natural surroundings.

The power of transcendentalism lay not so much in its naturalism as in its intense social morality about democracy, truth, beauty, and a respect for nature. Their rudimentary understanding of the ecosystem led the transcendentalists to believe that democracy could only be attained by imitating what they understood as the lesson of nature – the pursuit of self-actualisation and creative diversity within mutually sustaining communities. 'I can conceive of no flourishing and heroic elements of Democracy maintaining itself at all', wrote Whitman (1955, p. 692), 'without the Nature element forming the main part – to be its health element and beauty element – to really underlie the whole politics, sanity, religion and art of the New World.' The nature metaphor carried with it reformist connotations, for the symbiotic relationships among organisms symbolised attempts by minorities to gain political recognition.

The transcendentalists regarded the city as the antithesis of Nature, for it bred the values of individualism, competition, snobbery, and social rigidity that they despised. No kind of freedom would ever be found under such conditions, only an increasing autocratic control and a deadening conformity to rules and restrictions.

The transcendentalist philosophies sired two subsequent lines of thought which have influenced modern environmental policies. One is the case for a 'bioethic' advocated by 'nature moralists' such as Muir (see Wolfe, 1945) and Leopold (1949) . . . The other is the theme of self-reliance within the context of a small, recognizably interdependent community, a theme which has stimulated the imagination of many a utopianist ever since.

Bioethics

Supporters of the bioethic line of reasoning seek to protect the integrity of natural ecosystems, not simply for the pleasure of man but as a *biotic right*. Nature, they contend, contains its own 'purpose' which should be respected as a matter of ethical principle. Natural architecture has a grandeur which both humbles and enobles [sic] man, and stimulates him to emulate it. Wild nature is not to be regarded merely as a convenient respite from the stresses of 'civilised living', but as an integral companion to man. 'Thousands of tired, nerve-shaken, overcivilized people', observed Muir (1971, p. 32), 'are beginning to find out that going to the mountains is going home; that wildness is a necessity; and that mountain parks and reservations are useful not only as fountains of timber and irrigating rivers, but as fountains of life.'

There are a number of political, legal, and moral ramifications of the philosophy of bioethics. Morally, it forces man to be more conscious of his rights and responsibilities toward nature. As Fleischman (1969) correctly observes, nature is not 'in balance', nor is it 'wounded', by man's misdeeds. Man differs from all other members of the global ecosystem in two significant respects, a code of morality and the power of conscious reason. 'There is no biological justification for conservation', he concludes (Fleischman, 1969, p. 26), 'Nature will not miss whooping cranes or condors or redwoods any more than it misses the millions of other vanished species. Conservation is based on human value systems. Its validation lies in the human situation and the human heart.'

The moral implications of the bioethic principle are beginning to be recognised in environmental policy making. Although the early conservation period was noted for its battles to designate national parks and wilderness areas, the political motivations at the time were based less on bioethical notions than on commercial considerations of tourism and the promotion of national prestige (Runte, 1976). Before being dedicated, the first national parks in North America were carefully scrutinised for their forestry and mineral wealth and found wanting. But in modern times bioethical rhetoric is employed to protect wilderness areas, national parks, and wildlife habitats from the pressures of exploitation.

[. . .]

. . . A bioethical approach to the management of wilderness areas is also recommended by two Forest Service social scientists, Hendee and Stankey (1973). They feel that natural ecological processes should be left undisturbed even to the point of allowing destruction by wildfire. The old management philosophy of 'rearranging' natural processes to protect culturally preferred artificial habitats, they feel, should be abandoned.

An extension of this bioethical development in the area of economics is the introduction of a new theory of discounting, using specially weighted interest rates to reflect the uniqueness and irreplaceability of natural areas (Krutilla, 1973; Smith, 1974). . . . The cost-benefit calculus is thus loaded to favour the long-term protection of special ecosystems whose values should increase as technical options widen. In fact, this is a good example of the overlap between the rationalist scientific and the ecocentric ideologies in environmental management, for Krutilla and his colleagues are manipulating one of the more controversial technocentric tools to justify biocentric principles.

The legal ramifications of the bioethic argument could also be quite profound. Some environmental lawyers (e.g., Stone, 1972, see Reading 4.7) have suggested a common law of biotic rights, a doctrine first proposed by US Supreme Court Justice Douglas in his dissenting opinion over the Mineral King case (see Sax, 1973). Referring to Aldo Leopold's well known plea for a land ethic, Justice Douglas commented that 'contemporary public concern for protecting nature's ecological equilibrium should lead to a conferral of standing upon environmental objects to sue for their own preservation' (Environmental Reporter Cases, 1972, p. 2044). Currently this doctrine is being tested through attempts by American and Canadian environmentalists to alter the definition of legal standing. . . .

The political implications of the bioethic argument are significant in that the normal rules of compromise which usually guide political bargaining may not be followed. Wandesforde-Smith (1971, p. 481) describes the problem:

> Treating the environmentalists like any other interest group does not seem to work. . . . The environmentalists appear intransigent, extremely difficult to bargain with, and unwilling to accept a compromise. To them it is an all-or-nothing proposition because wilderness values are irreplaceable and priceless; not the kind of values that can be traded-off under the rubric of multiple use or according to the principles of professional forestry.

Combined with new legal tools, this approach could open up a pattern of political conciliation and decisionmaking quite different from that which has been the case heretofore.

The bioethic motif is of importance to modern environmentalism in that it stresses the essential humility of man in the face of natural forces. Nature produces 'resistances' which man ignores at his cost or peril, but which he can accept and understand to his inestimable benefit. Thus, bioethics incorporate the notion of *limits*, or nonnegotiable barriers to certain uses of natural areas. There is of course a long-standing controversy over just where and how demanding these limits are, an issue which permeates the growth–nongrowth debate. The bioethical viewpoint is that 'limits' establish their own kind of morality upon man, a challenge to his ingenuity and 'humanness' which constantly demands recognition and response.

[. . .]

The self-reliant community

To many utopian writers the method of linking self-actualisation to a sense of collective responsibility lay in the establishment of small, self-sustaining communities where nature still was very much in evidence. . . .

[They] were profoundly disturbed by the dehumanising and desocialising effects of rapid industrialisation and urbanisation, especially the impersonal and alienating atmosphere of the megalopolis, the intellectually deadening aspects of occupational specialisation, and the increasingly wasteful diversion of scarce resources (energy, skilled manpower, time and organisational talent) into the maintenance and administration of excessively large industrial, social, and governmental organisations. The result, they concluded, was socially counterproductive; the 'conventional wisdom' of economies of scale – proximity, jointly supplied services, variety of jobs, housing and entertainment opportunities, diminishing marginal costs – simply does not apply beyond a certain size. In fact size creates its own disadvantages – low productivity because of job dissatisfaction, delinquency due to alienation and frustration, family disruption caused by boredom and lack of communication.

[. . .]

A fine example of this 'communal' philosophy is provided by Schumacher (1973, pp. 50–58) in his delightful essay entitled 'Buddhist Economics'. For the Buddhist work gives man the opportunity to utilise and develop his faculties while accepting limits to his egocentricity by joining with others in a communal task. 'The Buddhist sees the essence of civilization not in the manipulation of wants, but in the purification of the human character' observes Schumacher (1973, p. 52). 'Character, at the same time, is formed primarily by a man's work. And work, properly conducted in conditions of human dignity and freedom, blesses those who do it and equally their products.' Given these circumstances the worker regards his labour as a matter of pride and utility, not as dreary occupational time that must be suffered in order to have the necessary income to buy happiness elsewhere. And to the employer, the worker becomes an indispensable part of the creative process, rather than a human tool to watch dials and press buttons.

To the ecocentric mind this kind of morality can never be found in the city, the result, according to Roszak (1973, p. 384) of 'freakish historical departure' which has produced 'a passion for megatechnics and artificiality'. Salvation can only lie in the re-emergence of the 'tribe' and the 'personalised collectivity': damnation is the way of the city, the bureaucracy, and the multinational corporation. With the transformation of scale comes the transformation of ethos, and the transmutation of the technocentric to the ecocentric.

The self-reliant community theme has traditionally been apolitical. Santmire (1973, p. 67) claims that this in part accounts for its lack of effectiveness in stopping the drift towards centralisation since, with its preoccupation with new forms of social morality, the ecocentric community had 'little psychic energy left for sustained intellectual and moral involvement in the practical political arena, whether that be with a view to upholding, transforming, or overthrowing the inherited order'.

But Santmire goes much further in his critique of the modern ecocentric fad for returning to nature, which he terms the 'cult of the simple rustic life'. He regards it as a political 'cop out', a selfish response on the part of the wealthy who can afford to buy their own private natural amenity, and alienated modern youth, the offspring of the wealthy, who seek to reject totally the materialistic values and the psychological stresses of the 'rat race' by retreating to nature-orientated communes. Both responses are socially reprehensible according to Santmire (1973, p. 78), who remarks that

> the cult of the simple rustic life, like the nineteenth century religion of nature, brings with it an implicit – sometimes explicit – social irresponsibility. It would be too much to say that the contemporary cult has been consciously developed in order to divert public attention from the pressing urban problems of the day . . . [but] there can be little doubt that [it] does reinforce a prior commitment to the *status quo*, especially in the ranks of the small town, suburban and affluent urban citizenry.

Hence the nature image can function as an *escape* from an uncertain future and a *refuge* from a decadent, unjust society, purposes it has served since the days of ancient Israel. Insofar as the rustic movement remains peripheral to the forces of social change, it will never provide a credible alternative to the powerful forces of centralisation. So despite its faddish popularity the tiny, isolated commune is not the answer to our present problems even though many individuals may choose to adopt this form of lifestyle.

[. . .]

. . . [E]cocentrism has influenced modern environmentalism in a number of ways. First, it provides a *natural morality* – a set of rules for man's behaviour based upon the limits and obligations imposed by natural ecosystems. Taken to its extreme, this natural morality displaces the humanistic morality that is intrinsically derived through man's cultural institutions. . . . But even in more moderate form, ecocentrism provides checks to the headlong pursuit of 'progress' which, by and large, is the objective of the technocentric mode. Second, it talks of *limits* (blurred perhaps but recognisable) of energy flows or productive capacity, and of the costs of organisation and system maintenance. These limits impose restrictions upon man's activities and hence influence the compass (if not the direction) of 'progress'. Third, it talks in ecosystem metaphors of *permanence and stability*, diversity, creativity, homeostasis, and the protection of options. These have important policy ramifications for the preservation of unique ecological and/or cultural habitats and 'lifestyles'. Fourth, it raises questions about *ends and means*, particularly the nature of democracy, participation, communication among groups holding conflicting yet legitimate convictions, the distribution of political power and economic wealth, and the importance of personal responsibility. These questions have profound political overtones, few of which are fully understood even by the politicians themselves. Finally, ecocentrism preaches the virtues of *self-reliance and self-sufficiency*. Here again the political ramifications of this trend are not properly understood. Anarchist environmentalists aim to develop decentralised institutional arrangements to facilitate flexible and adaptable responses to changing circumstances by individuals and communities alike, and to avoid the vulnerability of dependence upon trade, food supplies, large corporations, of key occupational groups.

The technocentric mode

Man's conscious actions are anthropocentric by definition. Whether he seeks to establish a system of biotic rights or to transform a forest into a residential suburb, the act is conceived by man in the context of his social and political culture. The distinction between technocentrism and ecocentrism relates to the *values* that are brought to bear on those acts and the estimation of the likely consequences. The technocentric mode is identified by *rationality*, the 'objective' appraisal of means to achieve given goals, by *managerial efficiency*, the application of organisational and productive techniques that produce the most for the least effort, and by a sense of *optimism and faith* in the ability of man to understand and control physical, biological, and social processes for the benefit of present and future generations. Progress, efficiency, rationality, and control – these form the ideology of technocentrism that downplays the sense of wonder, reverence, and moral obligation that are the hallmarks of the ecocentric mode. Many commentators believe that these ideological roots can be traced to the biblical exhortation to 'be fruitful and multiply and replenish the earth and subdue it: and have dominion over the fish of the sea, and over the fowl of the air and over every living thing that moveth upon the earth' (*Genesis*, chapter 1, v.28). This is a controversial issue . . . Others (e.g. Hays, 1959; Weisberg, 1971; Leiss, 1972) contend that technocentrism is a function of a particular kind of economic and political existence, and therefore has emerged most forcefully with the concentration of economic and political power that took place, ironically, during the first American conservation movement at the turn of this century.

Technocentrism and conservation

The early conservationists preached a managerial ethic and created particular resource development institutions that many today regard as serious impediments to harmonious environmental management. The first American conservation movement was promoted by a group of professional resource managers who enjoyed the attention of their political masters but who were neither accountable nor responsible to 'lay' public opinion. These men were proud of their scientific knowledge and their technical mastery. So conservation became a utilitarian notion, the orderly exploitation of resources for the greatest good to the greatest number over the longest time, the prevention of waste and the control of the earth for the good of man. Because rational management required order and control, governments quickly established regulatory mechanisms to stabilise prices and apportion supplies, since it was believed that the improvement of social welfare could best be achieved through the promotion of private interests regulated by public 'scrutiny'. In their passion to achieve rational means, the early conservationists lost sight of the political and ecological consequences of their actions. Weisberg (1971, pp. 23–24) comments that

> the legislation issued in the Progressive Era was not motivated by a questioning of the distribution or ownership of wealth or resources, but by a question of method: the problem of finding the most reasonable technique to promote efficient growth on the part of those who already controlled land and resource patterns. The conservation movement in fact was built around the difficulties of management, rather than ecological diversity and stability.

References

Burch, W. R., Jr., 1971, *Daydreams and Nightmares: A Sociological Essay on the American Environment* (Harper and Row, New York), 175 pp.

Environmental Reporter Cases, 1972, *An Annual Compilation of Environmental Cases* (Bureau of National Affairs, Washington).

Fleischman, P., 1969, 'Conservation, the biological fallacy', *Landscape*, 18, 23–27.

Fraser-Darling, F., 1971, *Wilderness and Plenty* (Ballantine Books, New York), 112 pp.
Hays, S. P., 1959, *Conservation and the Gospel of Efficiency* (Harvard University Press, Cambridge, Mass.), 277 pp.
Hendee, J. C., and Stankey, G. H., 1973, 'Biocentricity in wilderness management', *BioScience*, 23, 535–538.
Krutilla, J. V. (ed.), 1973, *Natural Environments: Theoretical and Applied Analyses* (Johns Hopkins University Press, Baltimore), 360 pp.
Leiss, W., 1972, *The Domination of Nature* (George Braziller, New York), 231 pp.
Leopold, A., 1949, *A Sand County Almanac* (Oxford University Press, New York).
McConnell, G., 1965, 'The conservation movement: past and present', in *Readings in Resource Management and Conservation*, eds I. Burton, R. W. Kates (University of Chicago Press, Chicago), pp. 189–201.
McConnell, G., 1971, 'The environmental movement: ambiguities and meanings', *Natural Resources Journal*, 11, 427–436.
McPhee, J., 1972, *Encounter with the Archdruid* (Sierra Club-Ballantine, New York), 215 pp.
Muir, J., 1971, 'In wilderness is the preservation of the world', in *Americans and Environment*, ed. J. Opie (D. C. Heath, Lexington, Mass.), pp. 32–40.
Roszak, T., 1973, *Where the Wasteland Ends: Politics and Transcendence in Post-industrial Society*, 2nd edition (Doubleday, Garden City, NY), 451 pp.
Runte, A., 1976, 'Wealth, wilderness and wonderland', in *The American Environment: Perceptions and Policies*, eds J. W. Watson, T. O'Riordan (John Wiley, Chichester), pp. 47–62.
Santmire, H. P., 1973, 'Historical dimensions of the American crisis', in *Western Man and Environmental Ethics*, ed. I. G. Barbour (Addison-Wesley, Reading, Mass.), pp. 66–92.
Sax, J., 1970, *Defending the Environment: A Strategy for Citizen Action* (Knopf, New York), 252 pp.
Sax, J., 1973, 'Standing to sue: a critical review of the Mineral King decision', *Natural Resources Journal*, 13, 76–88.
Schumacher, E. F., 1973, *Small Is Beautiful: Economics as if People Really Mattered* (Harper Torchbooks, New York), 290 pp.
Smith, K., 1974, *Technical Change, Relative Prices and Environmental Resources Evaluation* (Johns Hopkins University Press, Baltimore), 116 pp.
Stone, C., 1972, 'Should trees have standing – toward legal rights for natural objects', *Southern Californian Law Review*, 45, 450–488.
Wandesforde-Smith, G., 1971, 'The bureaucratic response to environmental politics', *Natural Resources Journal*, 11, 479–488.
Weisberg, B., 1971, *Beyond Repair: The Ecology of Capitalism* (Beacon Press, Boston), 184 pp.
Whitman, W., 1955, *Leaves of Grass and Prose Works*, ed. M. van Doren (Viking Books, New York).
Wolfe, L. M., 1945, *Son of Wilderness: The Life of John Mair (Knopf, New York).*

Source: Tim O'Riordan (1981), *Environmentalism*, London: Pion Books, pp. 1–12.

1.5a

Rachel Carson

A fable for tomorrow

There was once a town in the heart of America where all life seemed to live in harmony with its surroundings. The town lay in the midst of a checkerboard of prosperous farms, with fields of grain and hillsides of orchards where, in spring, white clouds of bloom drifted above the green fields. In autumn, oak and maple and birch set up a blaze of colour that flamed and flickered across a backdrop of pines. Then foxes barked in the hills and deer silently crossed the fields, half hidden in the mists of the autumn mornings.

Along the roads, laurel, viburnum and alder, great ferns and wildflowers delighted the traveller's eye through much of the year. Even in winter the roadsides were places of beauty, where countless birds came to feed on the berries and on the seed heads of the dried weeds rising above the snow. The countryside was, in fact, famous for the abundance and variety of its bird life, and when the flood of migrants was pouring through in spring and autumn people travelled from great distances to observe them. Others came to fish the streams, which flowed clear and cold out of the hills and contained shady pools where trout lay. So it had been from the days many years ago when the first settlers raised their houses, sank their wells, and built their barns.

Then a strange blight crept over the area and everything began to change. Some evil spell had settled on the community: mysterious maladies swept the flocks of chickens; the cattle and sheep sickened and died. Everywhere was a shadow of death. The farmers spoke of much illness among their families. In the town the doctors had become more and more puzzled by new kinds of sickness appearing among their patients. There had been several sudden and unexplained deaths, not only among adults but even among children, who would be stricken suddenly while at play and die within a few hours.

There was a strange stillness. The birds, for example – where had they gone? Many people spoke of them, puzzled and disturbed. The feeding stations in the backyards were deserted. The few birds seen anywhere were moribund; they trembled violently and could not fly. It was a spring without voices. On the mornings that had once throbbed with the dawn chorus of robins, catbirds, doves, jays, wrens, and scores of other bird voices there was now no sound; only silence lay over the fields and woods and marsh.

On the farms the hens brooded, but no chicks hatched. The farmers complained that they were unable to raise any pigs – the litters were small and the young survived only a few days. The apple trees were coming into bloom but no bees droned among the blossoms, so there was no pollination and there would be no fruit.

The roadsides, once so attractive, were now lined with browned and withered vegetation as though swept by fire. These, too, were silent, deserted by all living things. Even the streams were now lifeless. Anglers no longer visited them, for all the fish had died.

In the gutters under the eaves and between the shingles of the roofs, a white granular powder still showed a few patches; some weeks before it had fallen like snow upon the roofs and the lawns, the fields and streams.

No witchcraft, no enemy action had silenced the rebirth of new life in this stricken world. The people had done it themselves.

This town does not actually exist, but it might easily have a thousand counterparts in America or elsewhere in the world. I know of no community that has experienced all the misfortunes I describe. Yet every one of these disasters has actually happened somewhere, and many real communities have already suffered a substantial number of them. A grim spectre has crept upon us almost unnoticed, and this imagined tragedy may easily become a stark reality we all shall know.

What has already silenced the voices of spring in countless towns in America?

Source: Rachel Carson (1991) *Silent Spring*, London: Penguin, pp. 21–2.

1.5b

Rachel Carson

And no birds sing

Over increasingly large areas of the United States, spring now comes unheralded by the return of the birds, and the early mornings are strangely silent where once they were filled with the beauty of bird song. This sudden silencing of the song of birds, this obliteration of the colour and beauty and interest they lend to our world have come about swiftly, insidiously, and unnoticed by those whose communities are as yet unaffected.

From the town of Hinsdale, Illinois, a housewife wrote in despair to one of the world's leading ornithologists, Robert Cushman Murphy, Curator Emeritus of Birds at the American Museum of Natural History.

> Here in our village the elm trees have been sprayed for several years [she wrote in 1958]. When we moved here six years ago, there was a wealth of bird life; I put up a feeder and had a steady stream of cardinals, chickadees, downies and nuthatches all winter, and the cardinals and chickadees brought their young ones in the summer.
>
> After several years of DDT spray, the town is almost devoid of robins and starlings; chickadees have not been on my shelf for two years, and this year the cardinals are gone too; the nesting population in the neighbourhood seems to consist of one dove pair and perhaps one catbird family.
>
> It is hard to explain to the children that the birds have been killed off, when they have learned in school that a Federal law protects the birds from killing or capture. 'Will they ever come back?' they ask, and I do not have the answer. The elms are still dying, and so are the birds. *Is* anything being done? *Can* anything be done? Can *I* do anything?

A year after the federal government had launched a massive spraying programme against the fire ant, an Alabama woman wrote:

> Our place has been a veritable bird sanctuary for over half a century. Last July we all remarked, 'There are more birds than ever.' Then, suddenly, in the second week of August, they all disappeared. I was accustomed to rising early to care for my favourite mare that had a young filly. There was not a sound of the song of a bird. It was eerie, terrifying. What was man doing to our perfect and beautiful world? Finally, five months later a blue jay appeared and a wren.

The autumn months to which she referred brought other sombre reports from the deep South, where in Mississippi, Louisiana, and Alabama the *Field Notes*, published quarterly by the National Audubon Society and the United States Fish and Wildlife Service, noted the striking phenomenon of 'blank spots weirdly empty of virtually *all* bird life' [*Auduban Field Notes*]. The *Field Notes* are a compilation of the reports of seasoned observers who have spent many years afield in their particular areas and have unparalleled knowledge of the normal bird life of the region. One such observer reported that in driving about southern Mississippi that autumn she saw 'no land birds at all for long distances'. Another in Baton Rouge reported that the contents of her feeders had lain untouched 'for weeks on end', while fruiting shrubs in her yard, that ordinarily would be stripped clean by that time, still were laden with berries. Still another reported that his picture window, 'which often used to frame a scene splashed with the red of forty or fifty cardinals and crowded with other species, seldom permitted a view of as many as a bird or two at a time'. Professor Maurice Brooks of the University of West Virginia, an authority on the birds of the Appalachian region, reported that the West Virginia bird population had undergone 'an incredible reduction'.

One story might serve as the tragic symbol of the fate of the birds – a fate that has already overtaken some species, and that threatens all. It is the story of the robin, the bird known to everyone. To millions of Americans, the season's first robin means that the grip of winter is broken. Its coming is an event reported in newspapers and told eagerly at the breakfast table. And as the number of migrants grows and the first mists of green appear in the woodlands, thousands of people listen for the first dawn chorus of the robins throbbing in the early morning light. But now all is changed, and not even the return of the birds may be taken for granted.

The survival of the robin, and indeed of many other species as well, seems fatefully linked with the American elm, a tree that is part of the history of thousands of towns from the Atlantic to the Rockies, gracing their streets and their village squares and college campuses with majestic archways of green. Now the elms are stricken with a disease that afflicts them throughout their range, a disease so serious that many experts believe all efforts to save the elms will in the end be futile. It would be tragic to lose the elms, but it would be doubly tragic if, in vain efforts to save them, we plunge vast segments of our bird populations into the night of extinction. Yet this is precisely what is threatened.

The so-called Dutch elm disease entered the United States from Europe about 1930 in elm burl logs imported for the veneer industry [Swingle *et al.*, 1949]. It is a fungus disease; the organism invades the water-conducting vessels of the tree, spreads by spores carried in the flow of sap, and by its poisonous secretions as well as by mechanical clogging causes the branches to wilt and the tree to die. The disease is spread from diseased to healthy trees by elm bark beetles. The galleries which the insects have tunnelled out under the bark of dead trees become contaminated with spores of the invading fungus, and the spores adhere to the insect body and are carried wherever the beetle flies. Efforts to control the fungus disease of the elms have been directed largely towards control of the carrier insect. In community after community, especially throughout the strongholds of the American elm, the Midwest and New England, intensive spraying has become a routine procedure.

What this spraying could mean to bird life, and especially to the robin, was first made clear by the work of two ornithologists at Michigan State University, Professor George Wallace and one of his graduate students, John Mehner [Mehner and Wallace, 1959]. When Mr Mehner began work for the doctorate in 1954, he chose a research project that had to do with robin populations. This was quite by chance, for at that time no one suspected that the robins were in danger. But even as he undertook the work, events occurred that were to change its character and indeed to deprive him of his material.

Spraying for Dutch elm disease began in a small way on the university campus in 1954. The following year the city of East Lansing (where the university is located) joined in, spraying on the campus was expanded, and, with local programmes for gypsy moth and mosquito control also under way, the rain of chemicals increased to a downpour.

During 1954, the year of the first light spraying, all seemed well. The following spring the migrating robins began to return to the campus as usual. Like the bluebells in Tomlinson's haunting essay 'The Lost Wood', they were 'expecting no evil' as they reoccupied their familiar territories. But soon it became evident that something was wrong. Dead and dying robins began to appear on the campus. Few birds were seen in their normal foraging activities or assembling in their usual roosts. Few nests were built; few young appeared. The pattern was repeated with monotonous regularity in succeeding springs. The sprayed area had become a lethal trap in which each wave of migrating robins would be eliminated in about a week. Then new arrivals would come in, only to add to the numbers of doomed birds seen on the campus in the agonized tremors that precede death.

'The campus is serving as a graveyard for most of the robins that attempt to take up residence in the spring,' said Dr Wallace. But why? At first he suspected some disease of the nervous system, but soon it became evident that

> in spite of the assurances of the insecticide people that their sprays were 'harmless to birds' the robins were really dying of insecticidal poisoning; they exhibited the well-known symptoms of loss of balance, followed by tremors, convulsions, and death.
>
> [Wallace, 1959]

Several facts suggested that the robins were being poisoned, not so much by direct contact with the insecticides as indirectly, by eating earthworms. Campus earthworms had been fed inadvertently to crayfish in a research project and all the crayfish had promptly died. A snake kept in a laboratory cage had gone into violent tremors after being fed such worms. And earthworms are the principal food of robins in the spring.

A key piece in the jigsaw puzzle of the doomed robins was soon to be supplied by Dr Roy Barker of the Illinois Natural History Survey at Urbana [Barker, 1958]. Dr Barker's work, published in 1958, traced the intricate cycle of events by which the robins' fate is linked to the elm trees by way of the earthworms. The trees are sprayed in the spring (usually at the rate of 2 to 6 pounds of DDT per 50-foot tree, which may be the equivalent of as much as 23 *pounds per acre* where elms are numerous) and often again in July, at about half this concentration. Powerful sprayers direct a stream of poison to all parts of the tallest trees, killing directly not only the target organism, the bark beetle, but other insects, including pollinating species and predatory spiders and beetles. The poison forms a tenacious film over the leaves and bark. Rains do not wash it away. In the autumn the leaves fall to the ground, accumulate in sodden layers, and begin the slow process of becoming one with the soil. In this they are aided by the toil of the earthworms, who feed in the leaf litter, for elm leaves are among their favourite foods. In feeding on the leaves the worms always swallow the insecticide, accumulating and concentrating it in their bodies. Dr Barker found deposits of DDT throughout the digestive tracts of the worms, their blood vessels, nerves, and body wall. Undoubtedly some of the earthworms themselves succumb, but others survive to become 'biological magnifiers' of the poison. In the spring the robins return to provide another link in the cycle. As few as eleven large earthworms can transfer a lethal dose of DDT to a robin. And eleven worms form a small part of a day's rations to a bird that eats ten to twelve earthworms in as many minutes.

Not all robins receive a lethal dose, but another consequence may lead to the extinction of their kind as surely as fatal poisoning. The shadow of sterility lies over all the bird studies and indeed lengthens to include all living things within its potential range. There are now only two or three dozen robins to be found each spring on the entire 185-acre campus of Michigan State University, compared with a conservatively estimated 370 adults in this area before spraying. In 1954 every robin nest under observation by Mehner produced young. Towards the end of June, 1957, when at least 370 young birds (the normal replacement of the adult population) would have been foraging over the campus in the years before spraying began, Mehner could find *only one young robin* [Hickey and Hunt, 1960a]. A year later Dr Wallace was to report:

> At no time during the spring or summer [of 1958] did I see a fledgling robin anywhere on the main campus, and so far I have failed to find anyone else who has seen one there.
>
> [Wallace, 1959]

Part of this failure to produce young is due, of course, to the fact that one or more of a pair of robins dies before the nesting cycle is completed. But Wallace has significant records which point to something more sinister – the actual destruction of the birds' capacity to reproduce. He has, for example,

> records of robins and other birds building nests but laying no eggs, and others laying eggs and incubating them but not hatching them. We have one record of a robin that sat on its eggs faithfully for twenty-one days and they did not hatch. The normal incubation period is thirteen days. . . . Our analyses are showing high concentrations of DDT in the testes and ovaries of breeding birds [he told a congressional committee in 1960]. Ten males had amounts ranging from 30 to 109 parts per million in the testes, and two females had 151 and 211 parts per million respectively in the egg follicles in their ovaries.
>
> [Wallace, 1961]

Soon studies in other areas began to develop findings equally dismal. Professor Joseph Hickey and his students at the University of Wisconsin, after careful comparative studies of sprayed and unsprayed areas, reported the robin mortality to be at least 86 to 88 per cent [Hickey and Hunt, 1960b; Hickey, 1961]. The Cranbrook Institute of Science at Bloomfield Hills, Michigan, in an effort to assess the extent of bird loss caused by the spraying of the elms, asked in 1956 that all birds thought to be victims of DDT poisoning be turned in to the institute for examination. The request had a response beyond all expectations. Within a few weeks the deep-freeze facilities of the institute were taxed to capacity, so that other specimens had to be refused. By 1959 a thousand poisoned birds from this single community had been turned in or reported. Although the robin was the chief victim (one woman calling the institute reported twelve robins lying dead on her lawn as she spoke), sixty-three different species were included among the specimens examined at the institute.

The robins, then, are only one part of the chain of devastation linked to the spraying of the elms, even as the elm programme is only one of the multitudinous spray programmes that cover our land with poisons. Heavy mortality has occurred among about ninety species of birds, including those most familiar to suburbanites and amateur naturalists. The populations of nesting birds in general have declined as much as 90 per cent in some of the sprayed towns. As we shall see, all the various types of birds are affected – ground feeders, tree-top feeders, bark feeders, predators.

It is only reasonable to suppose that all birds and mammals heavily dependent on earthworms or other soil organisms for food are threatened by the robins' fate. Some forty-five species of birds include earthworms in their diet. Among them is the woodcock, a species that winters in southern areas recently heavily sprayed with heptachlor. Two significant discoveries have now been made about the woodcock. Production of young birds on the New Brunswick breeding grounds is definitely reduced, and adult birds that have been analysed contain large residues of DDT and heptachlor.

Already there are disturbing records of heavy mortality among more than twenty other species of ground-feeding birds whose food – worms, ants, grubs, or other soil organisms – has been poisoned. These include three of the thrushes whose songs are among the most exquisite of bird voices, the olive-backed, the wood, and the hermit. And the sparrows that flit through the shrubby understory of the woodlands and forage with rustling sounds amid the fallen leaves – the song sparrow and the white-throat – these, too, have been found among the victims of the elm sprays.

Mammals, also, may easily be involved in the cycle, directly or indirectly. Earthworms are important among the various foods of the raccoon, and are eaten in the spring and autumn by opossums. Such subterranean tunnellers as shrews and moles capture them in some numbers, and then perhaps pass on the poison to predators such as screech owls and barn owls. Several dying screech owls were picked up in Wisconsin following heavy rains in spring, perhaps poisoned by feeding on earthworms. Hawks and owls have been found in convulsions – great horned owls, screech owls, red-shouldered hawks, sparrowhawks, marsh hawks. These may be cases of secondary poisoning, caused by eating birds or mice that have accumulated insecticides in their livers or other organs [Walton, 1928; Dexter, 1951; Wright, 1960].

Nor is it only the creatures that forage on the ground or those who prey on them that are endangered by the foliar spraying of the elms. All of the tree-top feeders, the birds that glean their insect food from the leaves, have disappeared from heavily sprayed areas, among them those woodland sprites the kinglets, both ruby-crowned and golden-crowned, the tiny gnatcatchers, and many of the warblers, whose migrating hordes flow through the trees in spring in a multi-coloured tide of life. In 1956, a late spring delayed spraying so that it coincided with the arrival of an exceptionally heavy wave of warbler migration. Nearly all species of warblers present in the area were represented in the heavy kill that followed. In Whitefish Bay, Wisconsin, at least a thousand myrtle warblers could be seen in migration during former years; in 1958, after the spraying of the elms, observers could find only two. So, with additions from other communities, the list grows, and the warblers killed by the spray include those that most charm and fascinate all who are aware of them: the black-and-white, the yellow, the magnolia, and the Cape May; the oven-bird, whose call throbs in the May-time woods; the Blackburnian, whose wings are touched with flame; the chestnut-sided, the Canadian, and the black-throated green. These tree-top feeders are affected either directly by eating poisoned insects or indirectly by a shortage of food.

The loss of food has also struck hard at the swallows that cruise the skies, straining out the aerial insects as herring strain the plankton of the sea. A Wisconsin naturalist reported:

> Swallows have been hard hit. Everyone complains of how few they have compared to four or five years ago. Our sky overhead was full of them only four

> years ago. Now we seldom see any. . . . This could be both lack of insects because of spray, or poisoned insects.

Of other birds this same observer wrote:

> Another striking loss is the phoebe. Flycatchers are scarce everywhere but the early hardy common phoebe is no more. I've seen one this spring and only one last spring. Other birders in Wisconsin make the same complaint. I have had five or six pair of cardinals in the past, none now. Wrens, robins, catbirds and screech owls have nested each year in our garden. There are none now. Summer mornings are without bird song. Only pest birds, pigeons, starlings and English sparrows remain. It is tragic and I can't bear it.
>
> [Coordination of Pesticides Programs, 1960]

The dormant sprays applied to the elms in the autumn, sending the poison into every little crevice in the bark, are probably responsible for the severe reduction observed in the number of chickadees, nuthatches, titmice, woodpeckers, and brown creepers. During the winter of 1957–8, Dr Wallace saw no chickadees or nuthatches at his home feeding station for the first time in many years. Three nuthatches he found later provided a sorry little step-by-step lesson in cause and effect: one was feeding on an elm, another was found dying of typical DDT symptoms, the third was dead. The dying nuthatch was later found to have 226 parts per million of DDT in its tissues [Wallace, 1959].

[. . .]

Various scientific studies have established the critical role of birds in insect control in various situations. Thus, woodpeckers are the primary control of the Engelmann spruce beetle, reducing its populations from 45 to 98 per cent, and are important in the control of the codling moth in apple orchards. Chickadees and other winter-resident birds can protect orchards against the cankerworm [MacLellan, 1961; Knight, 1958].

But what happens in nature is not allowed to happen in the modern, chemical-drenched world, where spraying destroys not only the insects but their principal enemy, the birds. When later there is a resurgence of the insect population, as almost always happens, the birds are not there to keep their numbers in check. As the Curator of Birds at the Milwaukee Public Museum, Owen J. Gromme, wrote to the Milwaukee *Journal*:

> The greatest enemy of insect life is other predatory insects, birds, and some small mammals, but DDT kills indiscriminately, including nature's own safeguards or policemen. . . . In the name of progress are we to become victims of our own diabolical means of insect control to provide temporary comfort, only to lose out to destroying insects later on? By what means will we control new pests, which will attack remaining tree species after the elms are gone, when nature's safeguards (the birds) have been wiped out by poison?

References

Audubon Field Notes. 'Fall Migration – Aug. 16 to Nov. 30, 1958', Vol. 13 (1959), No. 1, pp. 1–68.

Barker, Roy J., 'Notes on Some Ecological Effects of DDT Sprayed on Elms', *Jour. Wildlife Management*, Vol. 22 (1958), No. 3, pp. 269–74.

'Coordination of Pesticides Programs', *Hearings*, H.R. 11502, 86th Congress, Com. on Merchant Marine and Fisheries, May 1960, pp. 10, 12.

Dexter, R. W., 'Earthworms in the Winter Diet of the Opossum and the Raccoon', *Jour. Mammal.*, Vol. 32 (1951), p. 464.
Hickey, Joseph J., 'Some Effects of Insecticides on Terrestrial Birdlife', *Report* of Subcom. on Relation of Chemicals to Forestry and Wildlife, State of Wisconsin, January 1961, pp. 2–43.
Hickey, Joseph J. and Hunt, L. Barrie, 'Songbird Mortality Following Annual Programs to Control Dutch Elm Disease', *Atlantic Naturalist*, Vol. 15 (1960a), No. 2, pp. 87–92.
Hickey, Joseph J. and Hunt, L. Barrie, 'Initial Songbird Mortality Following a Dutch Elm Disease Control Program', *Jour. Wildlife Management*, Vol. 24 (1960b), No. 3, pp. 259–65.
Knight, F. B., 'The Effects of Woodpeckers on Populations of the Engelmann Spruce Beetle', *Jour. Econ. Entomol.*, Vol. 51 (1958), pp. 603–7.
MacLellan, C. R., 'Woodpecker Control of the Codling Moth in Nova Scotia Orchards', *Atlantic Naturalist*, Vol. 16 (1961), No. 1, pp. 17–25.
Mehner, John F. and Wallace, George J., 'Robin Populations and Insecticides', *Atlantic Naturalist*, Vol. 14 (1959), No. 1, pp. 4–10.
Swingle, R. U., *et al.*, 'Dutch Elm Disease', *Yearbook of Agric.*, U.S. Dept of Agric., 1949, pp. 451–2.
Wallace, George J., 'Insecticides and Birds', *Audubon Mag.*, January–February 1959.
Wallace, George J., 'Another Year of Robin Losses on a University Campus', *Audubon Mag.*, March–April 1960.
Wallace, George J., *et al.*, *Bird Mortality in the Dutch Elm Disease Program in Michigan*. Cranbrook Inst. of Science Bulletin 41 (1961).
Walton, W. R., *Earthworms as Pests and Otherwise*, U.S. Dept of Agric. Farmers' Bulletin No. 1569 (1928).
Wright, Bruce S., 'Woodcock Reproduction in DDT-Sprayed Areas of New Brunswick', *Jour. Wildlife Management*, Vol. 24 (1960), No. 4, pp. 419–20.

Source: Rachel Carson (1991) *Silent Spring*, London: Penguin, pp. 100–9.

1.6

Frank B. Golley

Environmental attitudes in North America

Introduction

North America has been an arena for a variety of environmental dilemmas which, while not unique, represent different problems and responses to those in Europe, Africa or Asia. North America was essentially unknown to Europeans until the end of the fifteenth century and settlement occurred relatively slowly. The Spanish explored the southern and western sections and occupied the southwest and California. The most successful settlements were by largely disaffected groups from the British Isles, with additional contingents from Germany and France and from Africa. North America has continued to receive immigrants from all parts of the world.

The initial dilemma caused by this process concerned the interaction of Europeans with the existing inhabitants of North America. After all, the continent was fully inhabited by a rich assortment of native people who practised a variety of technologies. New settlers were faced with occupying other people's land, with developing ways to interact with the existing inhabitants and learning how to survive under new environmental conditions. This effort of adaptation was very important in the first years of settlement, but became less important as the self-confidence of the European settlers increased and in due course became replaced by a contempt, hatred and purposeful aggression against native Americans. We will consider a case example of this phenomenon as our first environmental dilemma.

Settlement not only involved interaction with the inhabitants and owners of the land but use of the natural resources to establish European patterns of life in the New World. The continent was rich in natural resources which could be exploited by new technologies. The natural resource which was of special importance at the beginning of settlement was the forest (Lillard, 1947). Forests provided the wood for Britain's navy, especially when access to Baltic wood was prevented, as during the first Dutch War of 1652. The King's foresters marked great pine trees for ship masts and bent live oaks for timbers of naval vessels; special ships were constructed to take these resources to England. The forests provided much of the material and energy for the emerging American industrial revolution, and the lake states' forests and western forests provided the lumber for expansion of settlement. Overcutting, exploitation without a reforestation plan, destruction of streams, and great wild fires were all associated with the lumber industry. Public concern at the turn of the nineteenth century led to the creation of the US Forest Service by President Theodore Roosevelt in 1905, with Gifford Pinchot as chief forester. Pinchot was a utilitarian conservationist who believed that wise use of forests, not preservation, was the best policy (Pinkett, 1970). Conservationists, such as John Muir, disagreed with Pinchot's policies (Fox, 1981). The conflict over forest management continues today and represents our second case study of an environmental dilemma.

European explorers and colonists encountered a continent that was immense, seemingly inexhaustible, and dangerous. It required hard work, technical skill and luck to succeed in the wilderness. Settlers replicated successfully European lifestyles and social organization, especially in New England, in Virginia tidewater areas and in areas settled by Germans from the Rhine Palatinate. However, the American landscape, after the revolution, gave birth to a new individual, the frontiersman or pioneer. The frontiersman became the mythical hero who opened new lands to settlement through his courage, skill and derring do.

[. . .]

At the end of the nineteenth century Frederick Jackson Turner, a major figure in American historical scholarship, declared that the frontier was no more (Turner, 1893), although pressure to continue the advance persisted and became international in the Spanish American War of 1898 and various invasions of Mexico and Central America. However, the frontier as a testing ground for character had not truly disappeared. National parks, forests and later wilderness areas were established where young people, especially, could encounter the challenges of nature. Wilderness was a concept created largely by Americans to refer to those natural areas which provide experience of the conditions that were thought to have moulded the American character (Nash, 1982).

The environmental dilemma comes from the conflict between the wilderness advocates who want to preserve nature as it was when the frontier existed and the users who represent other frontier types, such as the logger, cowboy and miner, and see the wilderness as an economic resource. . . . While wilderness is quintessentially an American concept, it illustrates a dilemma that is universal.

[. . .]

Encounter with native Americans

At the time Europeans first visited North America as explorers and settlers, the native American Indian populations were a vigorous and culturally dynamic people distributed across all environments of the continent; their cultures were finely adapted to the resources of the environments they inhabited, and where possible these people created cities, fortresses, apartment houses, complex governments and other characteristics of so-called advanced civilizations. Unfortunately for them, American Indians were not resistant to many European diseases and their numbers were decimated and their social systems were destroyed by death even before face-to-face encounters with Europeans occurred (Crosby, 1986). Diseases swept before the Europeans, weakening native people and making them unable to resist the invasion of their homelands.

Nevertheless, the initial encounters between Indians and Europeans were usually peaceful, and the Indians were eager to trade and helped the Europeans adapt to new environments (Cronan, 1983). However, the two cultures could not co-exist. The newcomers were obsessed with land as property and hence with the boundaries of property; Indians perceived land as a communal environment where boundaries fluctuated with season and habitat. While Indians quickly adopted European technology, they seldom adjusted to the idea that the environment could be divided up by individuals and treated in any way an individual desired. Even today, the reservations are considered the common property of the tribe, where individuals have rights to use resources, but only in the context of the whole tribe.

The European was puzzled by the Indian. The Spanish debated whether Indians were animals or men. Various Europeans, such as Samuel Penn, the Quaker, attempted to understand the Indians and treat them honourably, in the way that Europeans conceived honour. However, the vanguard of Europeans who penetrated Indian territory to trap beaver, graze cattle and hunt were frequently of Celtic origin, highly adapted to frontier life, violent and ruthless in defence of their self-defined prerogatives, and skilled in the extensive use of natural resources (McWhiney, 1988). The frontiersman became a mythical hero and after the American revolution, when the government came under the influence of frontiersmen, the relation with the Indians was translated from one of legal co-existence into one of annihilation.

An example of this is the fate of the Cherokees who lived in the southern Appalachian mountains and Piedmont region of Georgia, North and South Carolina and Tennessee. These people, through contact with missionaries and with government assistance, had become sophisticated in a European sense, with a written language, a newspaper, living in wooden houses and farming lands, some even with African slaves (Vipperman, 1978, 1989). Gold was discovered in Cherokee land in 1829 and the US Government signed a fraudulent treaty with a renegade, unofficial group of Cherokees which agreed to move the Cherokees to Oklahoma, paying them five million dollars for their land and improvements in the southern Appalachians, and giving them land west of the Mississippi River. The leaders of the Cherokee nation protested against this false treaty but the US Government sent troops to enforce it. . . .

The ethical dilemma is this: how can we accept human diversity in our environment, when the other individuals view the environment in a fundamentally different way than we do? How does one find a compromise with people who do not divide the world into property and who do not convert nature into a resource that can be exploited to extinction on a theory of economic convertibility? To do so requires a self-confidence and self-knowledge that is frequently missing in contemporary Euro-American culture. I see no simple answer to this question. But it requires an answer, especially as the world grows more crowded and we literally rub shoulders with people who do not share our environmental and social perceptions.

Encountering the great forest

The forest encountered by European settlers in America was incredible. The words of John Bartram, a botanical explorer of Georgia in 1773, illustrate what one encountered in many places and can still see in a few National Parks (van Doren, 1928). In the following quotation Bartram is referring to a landscape near Augusta, Georgia:

> Leaving the pleasant town of Wrightsborough, we continued eight or nine miles through a fertile plain and high forest, to the north branch of Little River, being the largest of the two, crossing which, we entered an extensive fertile plain, bordering on the river, and shaded by trees of vast growth, which at once spoke its fertility. Continuing some time through these shady groves, the scene opens, and discloses to view the most magnificent forest I had ever seen. We rose gradually a sloping bank of twenty or thirty feet elevation, and immediately entered this sublime forest. The ground is perfectly a level green plain, thinly planted by nature with the most stately forest trees, such as the gigantic black oak (*Q. tinctoria*), *Liriodendron, Juglans nigra, Platanus, Juglans exaltata, Fagus sylvatica, Ulmus sylvatica, Liquidambar styraciflua*, whose mighty trunks, seemingly of an equal height appeared like superb columns. To keep within the bounds of truth and reality, in describing the magnitude and grandeur of these trees, would, I fear, fail of credibility; yet, I think I can assert, that many of the black oaks measured eight, nine, ten, and eleven feet diameter five feet above the ground, as we measured several that were above thirty feet girth, and from hence they ascend perfectly straight, with a gradual taper, forty or fifty feet to the limbs; but below five or six feet, these trunks would measure a third more in circumference, on account of the projecting jambs, or supports, which are more or less, according to the number of horizontal roots that they arise from: the tulip tree, liquidambar, and beech, were equally stately.

However, these great forests were frightening to Europeans. In their letters and diaries they frequently remarked how wonderful it was to remove enough trees to see the stars or, even better, to see a neighbour and know that you are in a humanized landscape (Lillard, 1947). The forests were so vast and so large that few thought that they could ever be cut down. A peculiar type of male labourer, the logger, evolved and the mythical figure of the giant logger, Paul Bunyon, came to represent the heroic battle of man against the forest (Lillard, 1947). Stories of Paul Bunyon and his giant blue axe are still popular subjects of folk tales and songs in the Lake Region and the Pacific Northwest.

However, the forests were felled and, in some regions, so many trees were removed that there was not enough wood left to cook food, so that people had to move to find fuel wood. In other areas the wastes left on the land caught fire. In Wisconsin and Michigan these fires became extensive conflagrations which burned forests, and caused large loss of human life. The Peshtigo, Wisconsin fire of 1871 was one of the worst, burning a million and a quarter acres and killing over 1100 people (Lillard, 1947). Eventually, the public recognized the environmental damage caused by logging and supported the organization of a US Forest Service, which became responsible for the forest and range land still under government jurisdiction. Even so, the Forest Service was faced with many claims on the use of forest land, and it could not adopt policy that was unacceptable locally.

The environmental dilemma in this instance involves management with multiple conflicting objectives. The opposing user groups (loggers, hunters, recreationists, conservationists) have few basic ideas or needs in common. Frequently they do not even live in the same region. The resources available in the groups to manipulate public support differ widely, and the Forest Service often has had to take a position contrary to the goals and needs of specific user groups. Where there are consistent patterns of the Service supporting the goals of one group over long periods of time, such as logging interests, other interests become frustrated and can result in violence. For example, Earth First has recently emerged as a force for protection of the environment in the western USA (Scarce, 1990). Earth First members drive metal spikes in trees to prevent them from being sawed in mills, sabotage logging equipment, and remove survey stakes for roads and development; they justify this economic violence as an ethical activity. While the number of Earth First members is probably very small, there is widespread support and approval of their actions among the public who perceive loggers as people who have had special help from the taxpayer through access to trees planted and managed by public funds and who want to destroy old-growth forest for private gain.

The environmental dilemma illustrated by this case study is the problem of satisfying multiple interests in a social and political environment where there has been no history or mechanism of compromise. In countries such as Sweden, where most citizens share environmental attitudes and offical commissions involve all the parties of interest, compromise positions can be developed and frustration is reduced among interest groups. However, in the USA the solutions are obtained through an adversarial relationship that pits one side against the other, often in a court of law. This approach to environmental disagreement fits the penchant of American society to cast interaction in competitive, economic terms. The presence of abundant resources on the frontier has permitted the society to avoid the most serious costs of the approach. Frequently, the costs of competition are paid by the environment or passed on to future generations. With abundant resources the losers could move to another place and try again. However, as resources become limited through population growth, over-use and changed demands, pressures on them increase and there is now growing frustration as larger and larger segments of the society are denied opportunities. A new approach to resource allocation is needed in American society.

In this situation, the land ethic of Aldo Leopold has been a steady compass for many environmentalists. Leopold declared that actions that preserved the integrity, stability and beauty of natural communities were ethical and good (Leopold, 1949). Actions that did otherwise were unethical and bad. While Leopold's emphasis on maintaining natural order is influenced by deterministic ecological concepts, such as the succession and climax theories of Frederic Clements, which were widespread at the time he was active, his ethos provides useful guidance in the resolution of this environmental dilemma.

Wilderness and national parks

My final case study involves a particularly American obsession: the concept of wilderness. Wilderness represents an extensive area where human development is absent and where natural forces are allowed to operate uncontrolled. Wilderness represents in an entirely mythical form the American continent before European settlement. It ignores the activity of the American Indian, the fact that wilderness areas are tightly coupled to larger landscape regions, and the fact that the environment is continually changing. However, it does provide an opportunity to experience nature in a way that is otherwise impossible and represents a uniquely unmanipulated landscape where humans can escape humans for a moment.

As the American landscape was brought under private ownership or governmental management, the public recognized the need to set aside certain areas of great natural beauty which had no obvious economic value. Yellowstone National Park was among the first of these reserves, and the political debate over its formation was intense (Chase, 1986). Nevertheless, public pleasure in the preservation of Yellowstone, and the growing number of other parks amply justified a policy of park expansion. However, here too the National Park supervisors had problems of management for multiple objectives, and shortly after World War II, a commission headed by A. Starker Leopold of California, the son of Aldo Leopold, examined this problem and recommended that the parks should be managed for those patterns of wildlife, vegetation, and ecological condition that existed when the parks were first visited by Europeans (Chase, 1986).

The consequences of this decision were enormous. Vegetation and wildlife managers attempted to return the ecological condition to that described in the diaries or journals of explorers, interpreted by ecologists and historians. Natural processes were allowed to operate only so far as they recreated earlier conditions. Human use of parks and wilderness were managed to fit these objectives. However, many of the park environments were exceedingly fragile, many tended to fluctuate widely, and the goals conflicted more and more with the ecologists' understanding of a dynamic, non-static, non-equilibrium, natural system. Park management policy began to change, allowing natural processes to operate uncontrolled. However, this more modern form of management resulted in conflict with public and private demands. Allowing mountain goats to die of disease in Glacier National Park, and wildfire to burn uncontrolled in Yellowstone National Park, resulted in management crises. The response was to bureaucratize management. In Yellowstone, rangers are required to file daily fire reports, and the supervisors must guarantee that they have the resources to contain a fire. The process of park management, while always political, has become so politicized as to destroy the enthusiasm of the employees and drive able people to other professions. Politicians respond to local public pressure based on partial and often misinformed opinion, and their compromise benefits no one but their own re-election.

The environmental dilemma created in this case study involves the development of management objectives that represent our best knowledge of how nature functions in a public arena which has multiple, conflicting private interests, imperfect knowledge of the system, and is tied to the political process. Ironically, a policy of managing for a mythical past condition fits public interest and an outdated deterministic ecology better than does a more modern policy, which would allow natural processes to operate within broad limits. The ethical problem of setting aside land as wilderness, which only the physically fit can enjoy, in a form which ignores natural and social reality, is difficult to resolve. How far should a park be developed to provide universal access and urbanized entertainment? How can public education teach the history of the country and natural science so that citizens can express their private views within realistic boundaries? How can local needs and national demands be harmonized?

References

Chase, A. (1986) *Playing God in Yellowstone: the Destruction of America's First National Park*, Atlantic Monthly Press, Boston.

Cronan, W. (1983) *Changes in the Land; Indians, Colonists and the Ecology of New England*, Hill and Wang, New York.

Crosby, A.W. (1986) *Ecological Imperialism, The Biological Expansion of Europe 900–1900*, Cambridge University Press, Cambridge.

Fox, S. (1981) *John Muir and his Legacy, the American Conservation Movement*, Little, Brown, Boston.

Leopold, A. (1949) *A Sand County Almanac* and *Sketches Here and There*, Oxford University Press, Oxford.

Lillard, R.G. (1947) *The Great Forest*, Alfred A. Knopf, New York, pp. 65, 106, 210.

McWhiney, G. (1988) *Cracker Culture, Celtic Ways in the Old South*, University of Alabama Press, Tuscaloosa.

Nash, R. (1982) *Wilderness and the American Mind*, 3rd edn, Yale University Press, New Haven.

Pinkett, H.T. (1970) *Gifford Pinchot, Private and Public Forester*, University of Illinois Press, Urbana.

Scarce, R. (1990) *Eco-Warriors, Understanding the Radical Environmental Movement*, The Noble Press, Chicago.

Turner, F.J. (1893) 'The significance of the frontier in American history'. Printed in *The Frontier in American History*, 1920, New York, pp. 1–38.

Van Doren, M. (ed.) (1928) *Travels of William Bartram*, Dover, New York.

Vipperman, C.J. (1978) ' "Forcibly We Must", the Georgia case for Cherokee removal, 1802–32', *J. Cherokee Studies*, Spring, 103–9.

Vipperman, C.J. (1989)'The bungled treaty of New Echota: the failure of Cherokee removal, 1836–38'. *Georgia Hist. Q.*, 73, 540–58.

Source: Frank B. Golley (1992) 'Environmental attitudes in North America', in R.J. Berry (ed.) *Environmental Dilemmas: Ethics and Decisions*, London: Chapman & Hall, pp. 209–32.

1.7

Richard Mabey

Perspectives on the British countryside

I often pulled my hat over my eyes to watch the rising of the lark, or to see the hawk hang in the summer sky and the kite take its circles round the wood. I often lingered a minute on the woodland stile to hear the woodpigeons clapping their wings among the dark oaks. I hunted curious flowers in rapture and muttered thoughts in their praise. I loved the pasture with its rushes and thistles and

> sheep-tracks. I adored the wild, marshy fen with its solitary heronshaw sweeping along in its melancholy sky. I wandered the heath in raptures among the rabbit burrows and golden-blossomed furze. I dropt down on a thymy mole-hill or mossy eminence to survey the summer landscape. . . . I marked the various colours in flat, spreading fields, checkered into closes of different-tinctured grain like the colours of a map; the copper-tinted clover in blossom; the sun-tanned green of the ripening hay; the lighter charlock and the sunset imitation of the scarlet headaches; the blue corn-bottles crowding their splendid colours in large sheets over the land and troubling the cornfields with destroying beauty; the different greens of the woodland trees, the dark oak, the paler ash, the mellow lime, the white poplars, peeping above the rest like leafy steeples, the grey willow shining in the sun, as if the morning mist still lingered on its cool green. . . . I observed all this with the same raptures as I have done since. But I knew nothing of poetry. It was felt and not uttered.
>
> (John Clare, from *The Autobiography*)[1]

It is ironic that, of all the descriptions and celebrations that have come down to us from the English countryside's supposed golden past, it is the works of John Clare that we are turning to more and more for some kind of solace in its troubled present. Clare was a self-taught farm labourer, not a scientist nor campaigning politician. He was writing nearly 150 years ago – certainly a time of great agricultural upheaval, but hardly bothered by the kind of land-use conflicts we face today. Yet his writings touch with uncanny accuracy our own affections and worries. There is the detail of the landscape: the heron in a fenland sky, the gorse in blossom on the heath. There is the image of a pattern of agriculture still of a piece with the natural world. And there is that underlying sense of poignancy and loss, not just of a particular countryside, but of a spontaneous, unfettered experience of it that seems – almost by the act of recollection – to belong irretrievably in the past.

Clare lived through the statutory enclosure of his native village. He witnessed the end of Helpston's open-field system, the fencing and ploughing of the common fens and pastures of Emmonsales Heath, and the break-up of a pattern of life that had survived relatively intact since the establishment of the manorial system 800 years before. Out of that traumatic experience came what is perhaps the most powerful of all his poems, the elegy for his lost homeland which he called 'Remembrances'. It still reads like a battle hymn. One by one the casualties are listed: 'Langley bush . . . old eastwell's boiling spring . . . pleasant Swordy well'. And bitterly, sharply, the culprit – 'the axe of spoiler and self interest' – is identified:

> And crossberry way and old round oaks narrow lane
> With its hollow trees like pulpits I shall never see again
> Inclosure like a buonaparte let not a thing remain
> It levelled every bush and tree and levelled every hill.
> And hung the moles for traitors – though the brook is running still
> It runs a naked stream and chill

It is in this poem, perhaps more than any other, that we can trace the source of our sympathy with Clare. It is not just that what is under attack seems to be precisely the same now as it was then. I think that what we recognize more is the sense of affront, of an invasion of personal territory by forces beyond our control. Clare was describing a transformation in the fabric of rural England in terms of the destruction of specific features by identified agents. Yet they are not just particularized, they are *personalized*. Those were *his* trees, *his* brooks, *his* moles – not by virtue of ownership, but of *familiarity*. And it is this that joins us across a century and a half, for with losses that close to the heart it makes not a jot of difference if the cause is an enclosing landlord or a new motorway.

[. . .]

As I grew up I learnt all the received justifications for this: world starvation, a national economic recession, an expanding population. Yet all I saw then was the rich landowners getting richer, the poorer farmers having to sell up, and a stabilizing population whose national obsession seemed to be finding ways of eating less. I understand the real force behind some of these pressures now (and the private interests concealed in others). Yet I find it no easier to accept the way they so often casually ride over the interests not just of different individuals but of whole communities. And though we are now beginning to realize the wider importance of the natural world – even the small fragments remaining on these islands – I do not feel that this makes the *personal* case for nature conservation any less important. There are, after all, many millions of people who have such a case.

The greatest shock in the present transformation is that it has come about not so much from an invasion by urban sprawl or industrial development, but from insidious and often unobserved changes in the internal workings of the countryside itself. We were not prepared for this. The attacks upon the countryside that accelerated in the expansionist years after the Second World War had made the conflict seem a very clear and traditional one. It was the green fields of St George's England versus the dragons of Mammon and industry: open-cast mining, oil refineries, nuclear power stations, new airports. Seeing the fields apparently changing colour of their own accord produced a considerable sense of confusion not only about the morality of conservation but about the particular objectives towards which it was directed. It was one thing to question the need for new motorways when oil looked like running out by the end of the century; but quite another to press for what seemed like obstructions to the production of more food for a hungry world or to ways of easing the labour of men who worked the land. If we wanted to keep our wildlife, would we have to begin regarding it as another article of trade between countryside and consumer – perhaps even begin *paying* for it, as we did for bread and beer and newsprint? When it was put that baldly it seemed an alien and offensive suggestion, an infringement of a natural right that would be on a par with having to pay for air or sunshine. It also seemed, in some less definable way, to debase one of the intrinsic values of nature, which is – and it is not a truism to say so – its *naturalness*.

Nature has always had this double meaning for us, being a matter of style as much as of content, a way of being as much as a collection of particular living things. For much of our history the two had run parallel, and the 'natural things' – the bluebells and the butterflies – had appeared 'naturally', without effort or planning. That was part of their appeal. But the radical changes brought about by the new agriculture made us sharply aware that what we had regarded as a natural landscape was a much more complex product of growth and husbandry. The turf of the southern chalk downs, which W. H. Hudson christened 'the living garment', turned out to be the product of intensive sheep-grazing. A good deal of it had once been under the plough and before that the site of Celtic forts and townships. The wild sweeps of moorland in the Scottish Highlands had been created by a massive programme of forest clearance. The Norfolk Broads were the flooded remains of medieval open-cast peat-mines. Heaths and reed-beds would cease to exist if they were not deliberately cut or burnt. Even the new protectiveness itself seems to compromise the natural world with a slight hint of preciousness. Could we really use the word 'natural' about orchids which depended on artificial pollination for their survival, or butterflies that had to be bred in tented enclosures?

These new realizations raised many fundamental questions. If part of what we valued about nature was its wildness and spontaneity, was deliberate nature conservation a contradiction in terms? Was the familiar picture of British wildlife we had inherited, and which was captured so perfectly by Clare, already an anachronism, on a par with the perennial nostalgia for some pastoral Golden Age? Had the time come for us, as products of late-twentieth-century industrialization, to 'grow out' of the natural world? We could always save its bits and pieces in zoos and botanic gardens if we wished. If we preferred to be thoroughly modern we could even make them out of plastic, as was seriously suggested in a report produced by a senior official in the Department of the Environment:

> The city of the future must be clean and green, safe and sound. . . . I suggest that a landscape architect visits Kew Gardens and selects typical specimens of endangered species, e.g. elm and beech with prevalent urban species, e.g. the London plane. Casts could then be taken of the selected specimens – trunks and boughs would be moulded glass-reinforced plastic. Foliage would be polythene. The tree would be bonfire and vandal proof. Foliage would be self-extinguishing.[2]

This line of argument leads us, quite clearly, into a dead-end. Seeing the structures that traditionally supported our wildlife slipping away – apparently for the very best of reasons – we have been casting about for some new, *contemporary* role for our wildlife, perhaps even for a new wildlife. We have looked at the possibilities of an urbanized nature, a fascinating world of opportunist organisms in its own right, but one which contains very few of the plants and animals we are most familiar with. We are thinking about creating entirely new 'natural' landscapes in the corners left over by intensive agriculture. We have invented the notion of the 'living museum', which, whilst it would not literally contain plastic trees, nevertheless suggests a depressing idea that our remaining wildlife might become a collection of living fossils. We have become aware that the survival of the network of wildlife on the planet may be inextricably connected with our own survival, and we argue this very persuasively, though we know it is an intellectual defence, a kind of 'nature in the head'. And none of these alternatives comes anywhere near to making up for what we are losing. The dispiriting procession of transient, unfamiliar, distant experiences they offer makes us realize that what we want from conservation is not a museum of nature, a remote collection of undifferentiated wild 'things', but the community of distinct, familiar forms that is part of our cultural history.

And this brings us back full circle to Clare. Clare was not concerned about the maintenance of a viable breeding population of moles, or the extinction of *Quercus robur*, but about the brutal appropriation of living things that were not just part of his private experience but of the community in which he lived. And, unlike us, he was close enough to what was happening not to be fooled by promises that it was all for his own good. He took for granted what we are only just beginning to appreciate, that conservation is concerned ultimately with *relationships*, between man and nature, and man and man.

[. . .]

. . . One hears complaints from time to time about the elaborate measures taken in this country to prop up minute populations of plants and animals – lady's slipper orchids and marsh harriers, for instance – which seem to be on the very edge of their range here and incapable of surviving without the wildlife equivalent of intensive care. Ian Prestt, the Director of the Royal Society for the Protection of Birds (RSPB), which has been responsible for the preservation (and re-establishment) of many exceptionally rare British breeding birds, has a firm answer to these criticisms. In an aptly military metaphor, he talks about the importance of 'holding the line'. The 'natural range' of a creature is, precisely, wherever it naturally occurs. If it is having difficulty in surviving somewhere in that range, it is as likely as not because of some kind of human pressure. (Lady's slipper orchid, once comparatively widespread in limestone woods in northern England, has been reduced to a single site almost solely by the activities of collectors.) The wider a species is spread – which usually implies not just physical dispersal, but a certain amount of genetic variety as well – the better equipped it is to escape overall extinction. If you do not bother to 'hold the line' at any point, you diminish that species' overall defences against unpredictable environmental changes. You have also, along the way, set a very bad example to those areas that may be even more important population centres.

Nothing illustrates this better than the changing fortunes of our peregrine falcons. Before the Second World War Britain had something of the order of 700 pairs, a significant proportion of the probable European population at that time, and a great joy to those who were privileged to live near an eyrie – and in the 1950s they were still nesting on the cliffs of Dover.[3] But not everyone agreed about either their numbers or their delightfulness. In 1959 pigeon fanciers began a concerted publicity campaign against the peregrine, labelling it 'the bandit of the skies' and blaming it for the indiscriminate slaughter of exhausted racing birds. Correspondents in *Racing Pigeon* went as far as suggesting that there might be as many as 100,000 peregrines in Britain, and that the government should take immediate action to control them for the vermin they were. In spite of the slight note of hysteria in this propaganda, the Home Office responded and commissioned a census of British peregrine populations. The results are now part of conservation history. Far from multiplying, the birds were found to be in a state of catastrophic decline. In 1962 the population was down to about half of its pre-war level, and in the whole country only sixty-eight pairs were known to have reared young successfully. The next year Derek Ratcliffe published his classic paper linking this massive slump (which was also being observed in other predatory birds) with the accumulation in the birds of toxic agricultural pesticides, passed down the food chain from dressed grain to seed-eater to bird-of-prey.[4] The results of this survey had far-reaching effects upon the control of agricultural chemicals in this country and played an important role in the ecological awakening that happened in the sixties.

Over the next decade the sales of the most insidious and persistent chemicals were gradually reduced. Slowly the peregrine population began to climb back, and more birds began to bring off young. A wardening system was established for the more vulnerable eyries (one of the ironic results of the bird's scarcity being its increased attractiveness to falconers, who raided newly-reoccupied eyries for the young).

[. . .]

From whatever point of view it originates – political, scientific, aesthetic, ethical – it is hard to see how conservation can even begin unless there is a kind of Hippocratic commitment to the maintenance of species, and a belief that absolute extinction, like the loss of life in human medicine, is the ultimate sign of failure. Yet this *is* only a beginning (and not even a consistent one; you will not find many voices supporting the presence of dry-rot or head-lice on the planet) and a conservation policy that confined itself to rescuing plants and animals from the verge of extinction would be tantamount to reducing a whole health service to an emergency operating theatre. A species that is rare or endangered is no longer, by definition, contributing much either to the workings of the ecosystem or to our well-being. That is not to suggest that it is 'useless'. It has its own significance as an irreplaceable component of the earth's gene pool, as a potential crop plant perhaps, or the source of unexplored medicinal chemicals. . . . Rare species may also be of great scientific interest, particularly if they are located in isolated pockets that may represent the last remnants of more widespread populations. And none of us is immune to the excitement and sense of privilege at meeting what another generation of naturalists called 'curiosities'.

But it is the *common* species that keep the living world ticking over and provide most of our everyday experiences of wildlife, and I would argue that maintaining the abundance of these is as important a conservation priority as maintaining the existence of rarities. It would be invidious to draw up hierarchies of importance for a system whose complexity we have scarcely even begun to understand; yet it is clear, for instance, that the myriads of small animals that maintain the fertility of the soil and prevent the seaward fringes of the land from physically fraying away have a conservation importance of a different order from a scarce bush cricket; and that, in spring, a small and rather plain waterweed on the mud by a Surrey pond does rather less to rejoice the spirit than an abundance of bluebells and buttercups. I am not suggesting that conservation priorities should be based on popularity ratings (though I don't see why these shouldn't be taken into account just as much as the equally subjective 'nuisance' ratings are). But it is worth bearing in mind that, whilst our views about the ecological or economic importance of many endangered species are based on a mixture of faith and speculation, we know for certain about one of the roles of bluebells: people *like* them, and that, even by IUCN's[5] exacting philosophical criteria, makes them a 'useful' species. This is not an argument for downgrading the attention given to rare or little known species, but for paying more heed to the common.

Yet this forces us back to those contradictions and confusions we explored at the beginning of this chapter. The deliberate conservation of common species like the bluebell sounds superfluous, even extravagant. They are 'natural'. They will take care of themselves. Our historically conditioned view of nature (abetted by communications media that have a predilection for the plight of glamorous species in far-off lands, and often, it must be said, by conservation agencies themselves faced with difficult choices about where to allot their limited resources) has pushed conservation – both in its public image and in its practice – into a potentially dangerous preoccupation with the exotic and the rare.

Notes

1 John Clare, *The Autobiography* (quoted in *Rainbows, Fleas and Flowers*, ed. Geoffrey Grigson, John Baker, 1971).
2 *Sunday Times*, 11 June 1978.
3 John Parslow, *Breeding Birds of Britain and Ireland*, T. and A. D. Poyser, 1973.
4 Derek Ratcliffe, 'The Status of the Peregrine in Great Britain', *Bird Study*, 10, 1963.

5 IUCN originally stood for International Union for the Conservation of Nature and Natural Resources. It is now called IUCN-World Conservation Union. The classification of the species has since changed. To find out more you can contact The World Conservation Monitoring Centre in Cambridge (http://www.wcmc.org.uk/).

Source: Richard Mabey (1980) *The Common Ground*, London: J. M. Dent, pp. 1–9.

SECTION 2

Rethinking obligations: future generations and intergenerational justice

Introduction

Many of the frequent justifications used to defend the environment – the conservation of minerals, the preservation of the world's forests and against the stockpiling of nuclear waste – draw upon the idea of intergenerational justice in one way or another. In this section you will encounter readings that think through whether it is possible, and if so what it means, to have obligations to future generations. In addition, these readings also provide opportunities to consider different approaches to this issue vis-à-vis specific areas of environmental concern like water quality or deforestation. This debate, while conducted in largely anthropocentric terms, also involves high stakes. Our decision on whether to fulfil obligations to the future can have a dramatic impact on our own quality of material life.

The opening extract by Michael Golding (Reading 2.1) focuses on the relationship between present and future generations. In particular, he presents a case for discriminating in favour of those future generations that are close to us (especially our immediate successors) with whom, it is presumed, we share our social ideals. Golding argues that our descendants can place a claim upon us if they share the same conception of a 'good life' which underpins our moral community. More distant generations are seen as less likely to share our social ideals – thus our obligations have a cut-off point. This is directly challenged by Daniel Callahan's case (Reading 2.2) for taking the claims of distant future generations more seriously. Callahan argues that future generations should be seen as members of our moral community and he presents an alternative set of ethical standards to secure the existence and rights of the not-yet-born.

Whereas Golding and Callahan explore what such obligations could involve if we endorsed them, the next two readings are concerned with the possibility of thinking through justice between generations. Mark J. Smith (Reading 2.3) explores how different assumptions about intergenerational justice are tied to quite specific views about what constitutes a moral community and, by implication, what would make a particular social and political order acceptable. In addition, Smith introduces a key debate between the philosophers John Rawls and Robert Nozick about whether we should take future generations into account when making decisions in the present. Brian Barry (Reading 2.4) develops this further by exploring how the existing ethical standpoints and conceptions of justice have proved to be inappropriate or ill-equipped to incorporate the interests of potential people. Barry provides one of the most lucid discussions of the relations of inequality between present and future generations – inequalities in terms of both power and knowledge. Whilst, as the present generation, we have enormous power to affect our descendants, we also have limited knowledge of the effects of our actions on their lives. Bob Goodin's exploration of 'discounting' (Reading 2.5) provides an important illustration of what it means in practice if we do not acknowledge the interests of future generations on an equivalent basis to those who are living – that through the technique of discounting the value of costs and benefits which follow from our present decisions in the future, we are in danger of discriminating against our descendants. In response to these issues, in Reading 2.6, Gregory Kavka and Virginia Warren examine the implications of trying to find ways of representing the interests of future generations in existing political institutions.

In the last three readings you will find a more explicit focus on contemporary environmental issues and problems – water quality, deforestation and nuclear waste. Each of these readings explores the relationship between present generations and future impacts. In some instances there is a concern with the future impact of existing practices or how present problems could be changed in the lives of our immediate successors or beyond. In Reading 2.7, John Blunden focuses on what is likely to be the most important environmental issue of the twenty-first century, viz. water quality. In Reading 2.8, David Humphreys demonstrates how thinking about the needs and interests of the not-yet-born cannot be separated from concerns about justice in forest policy today. Finally, in Reading 2.9, Andrew Blowers explores one of the problems we are storing up for distant future generations, the safety of nuclear waste disposal – a time-bomb with a very long fuse. Each of these practical readings has implications for what we mean by intergenerational justice and what this means for present decisions.

2.1

Michael P. Golding

Obligations to future generations

. . . [T]he notion of obligations to future generations . . . finds increasing use in discussions of social policies and programs, particularly as concerns population distribution and control and environment control. Thus, it may be claimed, the solution of problems in these areas is not merely a matter of enhancing our own good, improving our own conditions of life, but is also a matter of discharging an obligation to future generations.

Before I turn to the question of the basis of such obligations – the necessity of the plural is actually doubtful – there are three general points to be considered: (1) Who are the individuals in whose regard it is maintained that we have such obligations, to whom do we owe such obligations? (2) What, essentially, do obligations to future generations oblige us to do, what are they aimed at? and (3), To what class of obligation do such obligations belong, what kind of obligation are they? . . .

. . . Obligations to future generations are distinct from the obligations we have to our presently living fellows, who are therefore excluded from the purview of the former, although it might well be the case that *what* we owe to future generations is identical with (or overlaps) what we owe to the present generation. However, I think we may go further than this and also exclude our most immediate descendants, our children, grandchildren and great-grandchildren, perhaps. What is distinctive about the notion of obligations to future generations is, I think, that it refers to generations with which the possessors of the obligations cannot expect in a literal sense to share a common life. . . .

But if their inner boundary be drawn in this way, what can we say about their outer limits? Is there a cut-off point for the individuals in whose regard we have such obligations?

Here, it seems, there are two alternatives. First, we can flatly say that there are no outer limits to their purview: all future generations come within their province. A second and more modest answer would be that we do not have such obligations towards any assignable future generation. In either case the referent is a broad and unspecified community of the future, and I think it can be shown that we run into difficulties unless certain qualifications are taken into account.

Our second point concerns the question of what it is that obligations to future generations oblige us to do. The short answer is that they oblige us to do many things. But an intervening step is required here, for obligations to future generations are distinct from general duties to perform acts which are in themselves intrinsically right, although such obligations give rise to duties to perform specific acts. Obligations to future generations are essentially an obligation to produce – or to attempt to produce – a desirable state of affairs *for* the community of the future, to promote conditions of good living for future generations. . . . If we think we have an obligation to transmit our cultural heritage to future generations it is because we think that our cultural heritage promotes, or perhaps even embodies, good living. In so doing we would hardly wish to falsify the records of our civilization, for future generations must also have, as a condition of good living, the opportunity to learn from the mistakes of the past. . . .

To come closer to contemporary discussion, consider, for example, population control, which is often grounded upon an obligation to future generations. It is not maintained that population control is intrinsically right – although the rhetoric frequently seems to approach such a claim – but rather that it will contribute towards a better life for future generations, and perhaps immediate posterity as well. (If population control were intrinsically anything, I would incline to thinking it intrinsically wrong.) On the other hand, consider the elimination of water and air pollution. Here it might be maintained that we have a definite duty to cease polluting the environment on the grounds that such pollution is intrinsically bad[1] or that it violates a Divine command. Given the current mood of neo-paganism, even secularists speak of the despoilment of the environment as a sacrilege of sorts. When the building of a new dam upsets the ecological balance and puts the wildlife under a threat, we react negatively and feel that something bad has resulted. And this is not because we necessarily believe that our own interests or those of future generations have been undermined. Both views, but especially the latter (Divine command), represent men as holding sovereignty over nature only as trustees to whom not everything is permitted. Nevertheless, these ways of grounding the duty to care for the environment are distinguishable from a grounding of the duty upon an obligation to future generations, although one who acknowledges such an obligation will also properly regard himself as a trustee to whom not everything is permitted. Caring for the environment is presumably among the many things that the obligation to future generations obliges us to do because we thereby presumably promote conditions of good living for the community of the future.

The obligation . . . is not an immediate catalogue of specific duties. It is in this respect rather like the responsibility that a parent has to see to the welfare of his child. Discharging one's parental responsibility requires concern, seeking, and active effort to promote the good *of* the child, which is the central obligation of the parent and out of which grow the specific parental obligations and duties. The use of the term "responsibility" to characterize the parent's obligation connotes, in part, the element of discretion and flexibility which is requisite to the discharging of the obligation in a variety of antecedently unforseeable situations. Determination of the specific duty is often quite problematic even – and sometimes especially – for the conscientious parent who is anxious to do what is good for his child. And, anticipating my later discussion, this also holds for obligations to future generations. There are, of course, differences, too. Parental responsibility is enriched and reinforced by love, which can hardly obtain between us and future generations.[2] (Still, the very fact that the responsibility to promote the child's good is an obligation means that it is expected to operate even in the absence of love.) Secondly, the parental obligation is always towards assignable individuals, which is not the case with obligations to future generations. There is, however, an additional feature of likeness between the two obligations which I shall mention shortly.

The third point about obligations to future generations – to what class of obligation do they belong? – is that they are *owed*, albeit owed to an unspecified, and perhaps unspecifiable, community of the future. Obligations to future generations, therefore, are distinct from a general duty, when presented with alternatives for action, to choose the act which produces the greatest good. Such a duty is not owed to anyone, and the beneficiaries of my fulfilling a duty to promote the greatest good are not necessarily individuals to whom I stand in the moral relation of having an obligation that is owed. But when I owe it to someone to promote his good, he is never, to this extent, merely an incidental beneficiary of my effort to fulfill the obligation. He has a presumptive *right* to it and can assert a claim against me for it. Obligations to future generations are of this kind. There is something which is due to the community of the future from us. The moral relation between us and future generations is one in which they have a claim against us to promote their good. Future generations are, thus, possessors of presumptive rights.

This conclusion is surely odd. How can future generations – the not-yet-born – *now* have claims against us? This question serves to turn us finally to consider the basis of our obligations to future generations. I think it useful to begin by discussing and removing one source of the oddity.

It should first be noticed that there is no oddity in investing present effort in order to promote a future state of affairs or in having an owed obligation to do so. The oddity arises only on a theory of obligations and claims (and, hence, of rights) that virtually identifies them with acts of willing, with the exercise of sovereignty of one over another, with the pressing of demands – in a word, with *making* claims. But, clearly, future generations are not now engaged in acts of willing, are not now exercising sovereignty over us, and are not now pressing their demands. Future generations are not now making claims against us, nor will it be *possible* for them to do so. (Our immediate posterity are in this last respect in a different case.) . . .

. . . [T]here is a distinction to be drawn between *having* claims and *making* claims. The mere fact that someone claims something from me is not sufficient to establish it as his right, or that he has a claim relative to me. On the other hand, someone may have a claim relative to me whether or not he makes the claim, demands, or is even able to make a claim. (This is not to deny that claiming plays a role in the theory of rights.) Two points require attention here. First, some claims are frivolous. What is demanded cannot really be claimed as a matter of right. The crucial factor in determining this is the *social ideal*, which we may provisionally define as a conception of the good life for man. It serves as the yardstick by which demands, current and potential, are measured. Secondly, whether someone's claim confers an entitlement upon him to receive what is claimed *from me* depends upon my moral relation to him, on whether he is a member of my *moral community*. It is these factors, rather than any actual demanding, which establish whether someone has a claim relative to me.

[. . .]

Who are the members of my moral community? (Who is my neighbor?) The fact is that I am a member of more than one moral community, for I belong to a variety of groups whose members owe obligations to one another. And many of the particular obligations that are owed vary from group to group. As a result my obligations are often in conflict and I experience a fragmentation of energy and responsibility in attempting to meet my obligations. What I ought to desire for the members of one of these groups is frequently in opposition to what I ought to desire for the members of another of these groups. Moral communities are constituted, or generated, in a number of ways, one of which is especially relevant to our problem. Yet these ways are not mutually exclusive, and they can be mutually reinforcing. This is a large topic and I cannot go into its details here. It is sufficient for our purpose to take brief notice of two possible ways of generating a moral community so as to set in relief the particular kind of moral community that is requisite for obligations to future generations.

A moral community may be constituted by an explicit contract between its members. In this case the particular obligations which the members have towards each other are fixed by the terms of their bargain. Secondly, a moral community may be generated out of a social arrangement in which each member derives benefits from the efforts of other members. As a result a member acquires an obligation to share the burden of sustaining the social arrangement. Both of these are communities in which entrance and participation are fundamentally a matter of self-interest, and only rarely will there be an obligation of the sort that was discussed earlier, that is, a responsibility to secure the good of the members. In general the obligations will be of more specialized kinds. It is also apparent that obligations acquired in these ways can easily come into conflict with other obligations that one may have. Clearly, a moral community comprised of present and future generations cannot arise from either of these sources. We cannot enter into an explicit contract with the community of the future. And although future generations might derive benefits from us, these benefits cannot be reciprocated. (It is possible that the [biologically] dead do derive *some* benefits from the living, but I do not think that this possibility is crucial. Incidentally, just as the living could have obligations to the distant unborn, the living also have obligations to the dead. If obligation to the past is a superstition, then so is obligation to the future.)[3] Our immediate posterity, who will share a common life with us, are in a better position in this respect; so that obligations towards our children, born and unborn, conceivably *could* be generated from participation in a mutually beneficial social arrangement. This, however, would be misleading.

It seems, then, that communities in which entrance and participation are fundamentally matters of self-interest, do not fit our specifications.

[. . .]

So far, in the above account of the generation of my moral community, the question of membership has been discussed solely in reference to those towards whom I initially have the sentiments that are identified with fellow-feeling. But we can go beyond this. Again we take our cue from the history of the development of rights. For just as the content of a system of rights that are possessed by the members of a moral community is enlarged over time by the pressing of claims, demanding, so also is the moral community enlarged by the pressing of claims by individuals who have been hitherto excluded. The claiming is not only a claim for something, but may also be an assertion: "Here I am, I count too." The struggle for rights has also been a counter-struggle. The widening of moral communities has been accompanied by attempts at exclusion. It is important for us to take note of one feature of this situation.

The structure of the situation is highlighted when a stranger puts forward his demand. The question immediately arises, shall his claim be recognized as a matter of right?[4] Initially I have no affection for him. But is this crucial in determining whether he ought to count as a member of my moral community? The determination depends, rather, on what he is like and what are the conditions of his life. One's obligations to a stranger are never immediately clear. If a visitor from Mars or Venus were to appear, I would not know what to desire for him. I would not know whether my conception of the good life is relevant to him and to his conditions of life. The good that I acknowledge might not be good for him. Humans, of course, are in a better case than Martians or Venusians. Still, since the stranger appears as strange, different, what I maintain in my attempt to exclude him is that my conception of the good is not relevant to him, that "his kind" do not count. He, on the other hand, is in effect saying to me: Given your social ideal, you must acknowledge my claim, for it *is* relevant to me given what I am; your good is my good, also.[5] If I should finally come to concede this, the full force of my obligation to him will be manifest to me quite independently of any fellow-feeling that might or might not be aroused. The *involuntary* character of the obligation will be clear to me, as it probably never is in the case of individuals who command one's sympathy. And once I admit him as a member of my moral community, I will also acknowledge my responsibility to secure this good for him even in the absence of any future claiming on his part.

With this we have completed the account of the constitution of the type of moral community that is required for obligations to future generations. I shall not recapitulate its elements. The step that incorporates future generations into our moral community is small and obvious. Future generations are members of our moral community because, and insofar as, our social ideal is relevant to them, given what they are and their conditions of life. I believe that this account applies also to obligations towards our immediate posterity. . . . Underlying this account is the important fact that such obligations fall into the area of the moral life which is independent of considerations of explicit contract and personal advantage. Moral duty and virtue also fall into this area. But I should like to emphasize again that I do not wish to be understood as putting this account forward as an analysis of moral virtue and duty in general.

As we turn at long last specifically to our obligations to future generations, it is worth noticing that the term "contract" has been used to cover the kind of moral community that I have been discussing. It occurs in a famous passage in Burke's *Reflections on the Revolution in France*:

> Society is indeed a contract. Subordinate contracts for objects of mere occasional interest may be dissolved at pleasure – but the state ought not to be considered as nothing better than a partnership agreement in a trade of pepper and coffee, calico or tobacco, or some other such low concern, to be taken up for a little temporary interest, and to be dissolved by the fancy of the parties. It is to be looked upon with other reverence; because it is not a partnership in things subservient only to the gross animal existence of a temporary and perishable nature.

> It is a partnership in all science; a partnership in all art; a partnership in every virtue, and in all perfection. As the ends of such a partnership cannot be obtained in many generations, it becomes a partnership not only between those who are living, but between those who are living, those who are dead and those who are to be born.
>
> Each contract of each particular state is but a clause in the great primaeval contract of eternal society, linking the lower with the higher natures, connecting the visible and invisible world, according to a fixed compact sanctioned by the inviolable oath which holds all physical and all moral natures, each in their appointed place.[6]

The contract Burke has in mind is hardly an explicit contract, for it is "between those who are living, those who are dead and those who are to be born." He implicitly affirms, I think, obligations to future generations. In speaking of the "ends of such a partnership." Burke intends a conception of the good life for man – a social ideal. And, if I do not misinterpret him, I think it also plain that Burke assumes that it is relatively the same conception of the good life whose realization is the object of the efforts of the living, the dead, and the unborn. They all revere the same social ideal. Moreover, he seems to assume that the conditions of life of the three groups are more or less the same. And, finally, he seems to assume that the same general characterization is true of these groups ("all physical and moral natures, each in their appointed place").

Now I think that Burke is correct in making assumptions of these sorts if we are to have obligations to future generations. However, it is precisely with such assumptions that the notion of obligation to future generations begins to run into difficulties. My discussion, until this point, has proceeded on the view that we *have* obligations to future generations. But do we? I am not sure that the question can be answered in the affirmative with any certainty. I shall conclude this note with a very brief discussion of some of the difficulties. They may be summed up in the question: Is our conception – "conceptions" might be a more accurate word – of the good life for man relevant[7] to future generations?

It will be recalled that I began by stressing the importance of fixing the purview of obligations to future generations. They comprise the community of the future, a community with which we cannot expect to share a common life. It appears to me that the more *remote* the members of this community are, the more problematic our obligations to them become. That they are members of our moral community is highly doubtful, for we probably do not know what to desire for them.

[. . .]

. . . One might go so far as to say that if we have an obligation to distant future generations it is an obligation not to plan for them. Not only do we not know their conditions of life, we also do not know whether they will maintain the same (or a similar) conception of the good life for man as we do. Can we even be fairly sure that the same general characterization is true both of them and us?

The . . . more distant the generation we focus upon, the less likely it is that we have an obligation to promote its good. We would be both ethically and practically well-advised to set our sights on more immediate generations and, perhaps, solely upon our immediate posterity. After all, even if we do have obligations to future generations, our obligations to immediate posterity are undoubtedly much clearer. The nearer the generations are to us, the more likely it is that our conception of the good life is relevant to them. There is certainly enough work for us to do in discharging our responsibility to promote a good life for them. But it would be unwise, both from an ethical and a practical perspective, to seek to promote the good of the very distant.

And it could also be *wrong*, if it be granted – as I think it must – that our obligations towards (and hence the rights relative to us of) near future generations and especially our immediate posterity are clearer than those of more distant generations. By "more distant" I do not necessarily mean "very distant." We shall have to be highly scrupulous in regard to anything we do for any future generation that also could adversely affect the rights of an intervening generation. Anything else would be "gambling in futures". We should, therefore, be hesitant to act on the dire predictions of certain extreme "crisis ecologists" and on the proposals of those who would have us plan for mere survival. In the main, we would be ethically well advised to confine ourselves to removing the obstacles that stand in the way of immediate posterity's realizing the social ideal. This involves not only the active task of cleaning up the environment and making our cities more habitable, but also implies restraints upon us. Obviously, the specific obligations that we have cannot be determined in the abstract. This article is not the place for an evaluation of concrete proposals that have been made. I would only add that population limitation schemes seem rather dubious to me. I find it inherently paradoxical that we should have an obligation to future generations (near and distant) to determine in effect the very membership of those generations.[8] . . .

. . . It appears that whether we have obligations to future generations in part depends on what we do for the present.

Notes

1 See the remarks of Russell E. Train (Chairman of the Council on Environmental Quality), quoted in *National Geographic*, 138 (1970), 780: "If we're to be responsible we must accept the fact that we owe a massive debt to our environment. It won't be settled in a matter of months, and it won't be forgiven us".

2 Cf. the discussion of *Fernstenliebe* (Love of the Remotest) in Nicolai Hartmann, *Ethics*, trans. by Coit, II (New York: The Macmillan Co., 1932), 317ff.

3 Paraphrasing C. S. Lewis, *The Abolition of Man* (New York: The Macmillan Co., paperback ed. 1969), p. 56: "If my duty to my parents is a superstition, then so is my duty to posterity."

4 When Sarah died, Abraham "approached the children of Heth, saying: I am a stranger and a sojourner with you; give me a possession of a burying-place with you, that I may bury my dead out of my sight" (Genesis 23:3, 4). A classical commentary remarks that Abraham is saying: If I am a stranger, I will purchase it, but if I am a sojourner it is mine as a matter of right.

5 Cf. T. H. Green, *Lectures on the Principles of Political Obligation* (New York and London: Longmans, 1959; Ann Arbor: University of Michigan Press, 1967), Sec. 140. I here acknowledge my debt to Green, in which acknowledgment I was remiss in my article on Human Rights.

6 *Reflections on the Revolution in France* (London: Dent. 1910), pp. 93–94.

7 The author at last begs pardon for having to use such an abused word.

8 On this and other arguments relating to the problem, see Martin P. Golding and Naomi H. Golding, "Ethical and Value Issues in Population Limitation and Distribution in The United States", *Vanderbilt Law Review*, 24 (1971), 495–523.

Source: Michael P. Golding (1972) 'Obligations to future generations', *The Monist*, 56, pp. 85–99.

2.2

Daniel Callahan

What obligations do we have to future generations?

The problem which this paper poses immediately raises some questions. How can we say anything at all about our obligations to future generations, those generations yet-to-be-born? Is it not the case, precisely because they do not yet exist, that it becomes meaningless to even conceive of obligations toward them? Since the future does not exist, and is in that respect nothingness, how can we possibly have obligations toward it? Moreover, since it is possible that a contemplation of our obligations to future generations may result in posing moral dilemmas for those of us already alive – what if we have to give up something in their behalf? – is it not just a way of asking for trouble to raise the question at all?

Let me attempt an answer to the first question. If future generations do not exist for us, we can, nonetheless, be certain that we *will have existed for them* – as part of their heritage, some marks of which are bound to still be around even thousands of years in the future. This is only to say that just as we now exist, they will also exist – the nothingness of future generations is a pregnant nothingness, needing only time to come to birth. As for a contemplation of our obligation toward future generations being understood as an evasion of obligations to present lives, there are some senses in which this could well be true. In particular, when it is said that we should immediately introduce coercive population control for the sake of the unborn, at the price of a loss of freedom and the likelihood of injustice for those already born, one can only wonder about the priorities of human life thus expressed. This is a point to which I shall return.

The most obvious reason for raising the question about our obligations to future generations stems from a simple perception: that what we do now will have consequences, good or ill, for those who come after us. Just as the actions, choices and thinking of our ancestors, close and distant, influence the way we live our lives, what we do will influence the lives of those for whom we will be ancestors. That we do not know how or by what particular chain of events this influence will exert or express itself a hundred, or a thousand or ten thousand years from now is beside the point for the moment. What matters at the outset is to recognize that there will be some influence. Just as we can trace the roots of our culture back at least 3,000 years, future generations will be able to trace theirs. To proceed as if there will be no relationship between the now and the then would be, at the very least, silly; there is no reason to presume a complete break in the chain of generations and the continuing transmission of everything from genes, to ideas, to cultures. More critically, to act as if future generations will not exist and will not be the heirs of what we do could be to act in a most irresponsible way. Modern technology, weaponry and pollution have given mankind an unprecedented power to influence the lives of those who will live in the future. It may be optimistic to think that this power will be used for good, or even that it can be; but it is only realistic to think that it can be used for evil, that we already have the power to exert a very baneful influence on the lives of those who will follow us.

Let me briefly sketch some of the more unpleasant possibilities. If we choose to take seriously the demographers, we have to recognize that present rates of population growth will, unless checked, produce a vastly more crowded world for our children and grandchildren than we now live in. The present doubling time for world population is 37 years. We may be as hopeful as we like about agricultural miracles, better systems of population distribution, and more ingenious technologies of production, consumption and reclamation – in short, we may argue that the world can cope with 7 to 10 to 20 billion people. But even if we are all that hopeful we should admit that life a few generations hence will be different; perhaps no worse, but still different. And it will be our actions today which will in many critical ways make that difference. For it is we who by our rate of reproducing are preparing the demographic base from which future generations will proceed. We can proclaim ourselves innocent and non-responsible for the present state of affairs. After all, in reproducing we are doing nothing different from what our ancestors did; and after all as well, if we are going to hold any generation responsible for the population problems being stored up for future generations, why not point the finger at our parents' generation, or at our grandparents', and so on, well back into the past? Just as we have had to take our chances with the population size our ancestors gave us, why should not future generations have to take their chances with the size we bequeath to them?

I do not think it quite so easy to wash our hands of the matter. There is one difference between our generation and those which went before. We *know* there is a certain rate of population growth at present, and we can see its implications; that cannot be said for earlier generations, Malthus notwithstanding. Moreover, we know why it is growing so rapidly: better nutrition, reduced infant mortality rates, longer life-spans.

[. . .]

A few more examples. If the ecologists are correct, what we are now doing to our natural resources and our environment may well be irreparably harmful. It is not just that we may be ruining things for ourselves; we may be ruining things for all of those who follow us. The animals we poison into extinction will not exist in the future; that is what extinction means. The lands we ruin will not bear fruit for our heirs. The lakes we pollute will not be available for our children, or for theirs. To take another type of example, the cities we plan now will be lived in by future generations; the technologies we devise will condition the ways and meaning of life of those who proceed from us.

One can go on in this vein, but what are the essential issues? I would like to deal with three of them here. First, the problem of the nature of the obligation we owe to future generations. Second, the relationship between our obligations to the present generation and our obligations to future generations. Third, the problem of developing some appropriate norms for our present behavior where that behavior has implications for the lives of the unborn.

1. *What is the nature of the obligation we owe to future generations?* Professor Martin P. Golding, the only person I know of who has recently written on this subject, has argued cogently that what is distinctive about

> the notion of the obligation to future generations is . . . that it refers to generations with which the possessors . . . cannot expect in a literal sense to share a common life.[1]

In this respect, he is making a distinction between our immediate descendants, with whom our own generation will overlap, and our far-distant descendants. The question he poses is how we can have obligations toward those with whom we will not share a common life – who will not, as he puts it, apparently be part of our own moral community, as those living now are. If future generations are to have a claim on us, if they are able in a sense to make demands upon us, need we take these claims seriously, as a matter of rights? For Golding, the solution to this question depends on whether the claim can be seen as part of our present "social idea."

[. . .]

Up to a point, I find Golding's argument compelling. In essence, he is saying that if we are to establish a bond of claims, rights and obligations with future generations, it has first to be shown that, somehow, they are one with us – that, though remote in time, they are part of our own moral community. And he develops a way of showing how in principle this could be the case.

I would want to supplement Golding's argument in two ways. First, it seems critical that, when we talk of "our moral community," the phrase be understood to encompass the entire human community. Otherwise the ground is set for hazardous exclusions, of a kind which has ever plagued the human community: the judgment by one human group that another is not worthy of respect or protection. What was the institution of slavery other than the result of a judgment that African blacks were not part of "our moral community" – "our" in that case meaning the white Western community, which presumed the right to make such determinations? . . . To be sure, to state that we have moral obligations to "the community of all human beings" introduces its own problems. One of them turns on the practical impossibility of effectively discharging obligations to "all human beings." I grant the problems there and will assist the skeptics on that point by adding in their behalf that the difficulties mount immeasurably if future humans are added to the list of those toward whom we have obligations.

A second way I would supplement Golding's argument would be by laying a greater stress on the relationship between obligations to the past and to the future. The immediate problem here turns on the way in which, if at all, we are to establish an obligation – any obligation – toward other human beings, born or to-be-born. Allow me to evade that larger question, and take up a related one more directly pertinent to the question being considered in this article – our obligation to future generations. Golding quotes a passage from Burke's *Reflections on the Revolution in France* which characterizes "society" as "a partnership not only between those who are living, but between those who are living, those who are dead and those who are to be born."[2] But neither that passage nor the citation as a whole shows why anyone of us is obliged to become participants in the partnership, to take up the burdens and obligations which it entails. None of us, to put the matter baldly, was given any choice about whether we would be born or not, or be thrust into that partnership which is society. Why then ought we to feel any obligation to the partnership – we did not create it nor were we asked if we wanted to be part of it?

My own response, which can only be sketched here, is that it cannot be otherwise, either biologically or morally. That we exist at all puts us in debt to those who conceived us – our parents – and in debt to that society in which we were born, without which we might have been conceived but could not have survived (for our parents were not sufficient unto themselves). We could not be asking or discussing any of these questions if we did not exist; and we would not exist had not someone and some society taken some responsibility for our welfare. The condition for raising the question of whether we owe the past anything at all is that the past (concretized in the form of parents and society) took upon itself an obligation toward us – to bring us into existence and to sustain us. If we value our own life at all, then we must value and feel some obligation toward those who made that life possible; we did not arrive in this world on our own, nor did we come to maturity on our own. . . .

From our obligation to the past stems our obligation to the future. On the one hand, we owe to those coming after us at least what we ourselves were given by those who came before us: the possibility of life and survival. On the other hand, we also owe to the future an amelioration of those conditions which, in our own life (and by our own lights), lessened our possibilities for living a full human life.

[. . .]

Where I would only want to supplement Golding's way of developing the basis of our obligation to future generations, I want to differ on the problem of how problematic our obligations to remote future generations are. Golding asks whether *our* conception of the good life for man is *relevant* to future generations. As long as we can show the pertinence of our social ideal to distant generations, and thus show that they are members of our moral community, all is well – obligations can be established. But, he says,

> it appears to me that the more *remote* the members of this community are, the more problematic our obligations to them become. That they are members of our moral community is highly doubtful, for we probably do not know what to desire for them.[3]

Golding is arguing that, on the one hand, it is possible to show the theoretical basis for obligations to future generations; but is saying, on the other, that conditions necessary to activate these obligations do not exist. Because remote future generations are so distant in time, we do not know what to desire for them . . .

. . . Golding, at the outset of his paper, talks about our obligations to future generations.

> Obligations to future generations are essentially an obligation to produce – or attempt to produce – a desirable state of affairs *for* the community of the future, to promote conditions of good living for future generations.[4]

I am struck by the fact that he casts this obligation solely in terms of positive obligations, things we must do to *enhance* their life. I should think it no less obvious that we would also be obliged to refrain from doing things which might be harmful to future generations. Quite apart from trying to produce good we should also avoid causing harm.[5] And the two are not necessarily identical. While our ignorance of the desires of future generations may make it practically impossible to know what to work for positively on their behalf – and thus relieve us of some of our obligations – we cannot claim total ignorance when it comes to knowing what might be very harmful to them. We know enough about radiation hazards, for instance, to be sure that a widespread testing of nuclear devices would have harmful genetic consequences for future generations. Unless we suppose they might actually desire those consequences – which would be capricious on our part – we could hardly excuse our nuclear weapons testing on the grounds of our ignorance of what would be "relevant" to the life of those generations. Thus it would seem that, in some circumstances, our knowledge of harmful long-range consequences would be sufficient to restrain us from certain kinds of acts.

The pertinence of this consideration comes to the fore also when we consider ecological destruction and excessive population growth. While it may be impossible for us to know the desires of future generations, we know it is possible – though remote – that the present generation could destroy the environment in some irrevocable way. And it is at least conceivably possible that the present generation could so heavily populate the earth that it would be impossible for that earth to sustain the large numbers of people which could result from a continuation of high population growth rates. If we do nothing about these hazards, then we would have at least a moral certainty that our actions would be storing up evils for future generations – breakdown of culture, overcrowding, critical shortages – evils of a kind that they would not have brought upon themselves and that they would not be able to cope with. Once again, assuming we could surmise the existence of these hazards from our present behavior, we would have a situation which would seem to demand that we refrain from certain kinds of acts for the sake of future generations. . . .

Golding is quite right to say that, because of our ignorance, we should not plan for distant generations. But this does not relieve us of our obligation to make certain (a) that there will be future generations – which is a way of reaffirming the value we attribute to our own life; and (b) that the possibility of those generations planning for themselves is not irrevocably destroyed by our failure now to refrain from those acts which could have evil consequences for them; we have no right to preempt their choices.

[. . .]

From a moral point of view, the problem is not whether we can peer into the future and determine what future human beings will need and desire. It is possible to conceive – as distinguished from imagine – that future generations will be so different that our social ideal will no longer pertain. Golding may, in that sense, be right. I am only asserting that since we cannot *know* what their social ideal will be, we should act on the assumption that it will not be all that dissimilar from our own; we have no special reason to think otherwise. Hence, the course of responsible behavior in this generation would be to take what we do know, and can reasonably project, and act accordingly.

[. . .]

2. *What is the relationship between our obligations to future generations and our obligations to the present generation?* If it can be agreed that we do have real obligations to distant generations, we are left with the problem of determining how we are to balance those obligations against our obligations to the present generation. In existing generations, moral dilemmas arise because of real or apparent conflicts among or between obligations. We do not know how to serve fully two or more goods at the same time, doing equal justice to all of them. It is conceivable that the same kind of dilemmas can arise across the generations, where the discharging of our obligations to the living could require acting in ways which would have harmful consequences for future generations. While I would aggressively not include myself among them, there are some, we should recall, who feel that if the price for maintaining our present Western way of life is a world-wide nuclear war, then that price should be paid, regardless of the environmental and genetic consequences for future generations. They really believe that it is better to be dead than red, and not only us, but, if necessary, all those who come after us. To choose death rather than slavery for oneself is one thing; to choose it for one's children quite another – the former is more easily justified than the latter. This seems to me a social ideal gone berserk, but for some it is the ideal.

Arguments of this type seem, then, to presume that our main obligation is to the living or those who will immediately come after us.

At the other extreme are those who would apparently give our obligations to future generations a greater weight than those we owe to the living. Thus some biologists and ecologists speak as if our obligation to preserve the human species, or our obligation to preserve forests and wildlife, takes precedence over all other obligations. . . .

I have presented some extreme positions, but it is possible to discover others which present what I think are genuine as distinguished from apocalyptic ethical dilemmas. Let us return to the contention that our present rate of consumption of natural resources combined with our present, and constantly expanding, rate of pollution of the environment may permanently despoil the earth. This will not happen at once, and for that matter our children and grandchildren may be able to survive; but sooner or later, a limit will be reached, a point of no return. We will not personally suffer, but others will, especially if what we are doing now will, if not stopped, make it impossible for future generations to survive, much less to live a decent life. One major argument against genetic engineering is that our present attempts at intervention may introduce irrevocable changes, which future generations will not be able to undo. Thus we would, in effect, have imposed ourselves upon them, leaving them no option but to accept our legacy.

These concerns are to be taken seriously. If it is the case – and that of course needs to be established – that our present behavior can prove harmful to future generations, then, according to my earlier argument, we have an obligation to refrain *as far as possible* from that behavior. The key phrase is "as far as possible." For it may well be the case that, in order to do justice to the present generation, some acts may be required which will jeopardize either the life or the quality of life of future generations. Let us imagine an improbable though ethically suggestive chain of events. Let us imagine that the nations of the world gave up the arms race and decided to turn all their efforts to the alleviation of hunger and disease, the development of the underdeveloped nations, and the establishment of a minimally decent standard of living for all. And let us assume they wanted to do this at once. However, if they wanted to do it at once – to save those now living – they would by and large have to do so with the tools of existing technologies. They would not have time to invent fertilizers which could guarantee a non-pollution of the earth, or time to invent factories which do not emit noxious fumes, or time to devise substitutes for nonrenewable resources, or time to find ways of raising crops which would not require the destruction of some wildlife and some primitive forest areas. All those steps could require years. At that point, of course, the ecologists would become violently alarmed, for they could easily foresee what the long-term price for the short-term salvation of individual lives could be. . . .

I think there is a partial way of resolving this dilemma. One thing that can be said for the present generation over . . . future generations is that they have *existing* rights: the right to life, liberty and the pursuit of happiness (to choose one familiar formulation). They are here and the future generations are not.

[. . .]

3. What ethical norms would be appropriate for that behavior of our present generation which has implications for the lives of future generations? Let me try to suggest some summary rules concerning our obligations toward future generations:

(a) Do nothing which could jeopardize the very existence of future generations.

(b) Do nothing which could jeopardize the possibility of future generations exercising those fundamental rights necessary for a life of human dignity.
(c) If it seems necessary, in the interests of the existing rights of the living, to behave in ways which could jeopardize the equivalent rights of those yet to be born, do so in that way which would as far as possible minimize the jeopardy.
(d) When trying to determine whether present behavior will in fact jeopardize future life, calculate in as responsible and sensitive a manner as one would in trying to determine whether an act with uncertain consequences would be harmful to one's own children. If you would not conjure up the possibility of magical solutions occurring to save your own children at the last moment from the harmful consequences of your gambling with their future, do not do so even with future generations.

. . . Of necessity, the parent's interpretation of the child's needs will be conditioned by the parent's understanding of life, and this will be time- and culture-bound. While the parent may be aware of this, there is little he can do about it; he has no place to begin except where he is, hoping that the choices he must make in the child's behalf will not, should they prove wrong, be irredeemable.

If we extend the same model to let it serve as a prototype of our obligations toward future generations, then two conclusions appear inescapable: first, that we must use our own and present understanding of human life as the basis for any projections into the future – no other is available; and, second, that there are no grounds for introducing, in judging that behavior of ours with implications for future generations (close or distant), any norms sharply at variance with those we would employ in judging our obligations to those presently alive.

Notes

1 Martin P. Golding, "Obligations to Future Generations," in the "Philosophy and Public Policy" issue of *The Monist* (January 1972), p. 86.
2 E. Burke, *Reflections on the Revolution in France* (London: Dent, 1910), pp. 93–94: cited in Golding.
3 Golding, op. cit., p. 97.
4 Ibid., p. 86.
5 I am not implying that the avoidance of doing harm constitutes the full range of ethical responsibilities. I am only saying that in the case of our obligations to future generations, where our duty to enhance their welfare is problematic, we can and should have recourse to the limited ethical goal of avoiding harm.

Source: Daniel Callahan (1971) 'What obligations do we have to future generations?,' *American Ecclesiastical Review*, 164, pp. 265–80.

2.3

Mark J. Smith

Intergenerational justice

The idea of justice provides social life with its underlying normative order, for it acts as the point of reference for defining appropriate behaviour within that community. The concept of justice rests upon the idea of regard for and fairness towards other members of the same moral community. Justice is an essentially contested concept and, as such, is ineradicably evaluative (Lukes, 1974). The concept of justice serves as an important reference point for establishing the core values of a moral community or, for that matter, a particular social theory. There is a clear difference between the arguments in favour of social justice and advocating individual justice. So, rather than seeing justice as something fixed, we can examine the various ways in which it is defined in order to highlight the limits and possibilities of each approach. When we examine justice we are, in effect, examining the horizon of our moral considerations. This can be illustrated by focusing upon the way in which John Rawls develops the argument for a 'just savings principle'. This serves as a means of resolving some of the difficulties in dealing with the impact of human activities on the environment and of understanding the legacy we leave for our descendants.

The concept of justice provides a way of working out the principles upon which a social contract between all members of a moral community can be established. In *A Theory of Justice* (1971), Rawls focuses our attention on the question of the sort of society in which we would like to live. He constructs a scenario in which all members of a moral community are placed in a situation before any social arrangements have been fixed. This position he describes as the 'original position', behind the 'veil of ignorance'. Initially we will assume that the moral community is composed of present generations. In this original position, Rawls asks us to consider 'what sort of society would we consider to be just?', before we have any idea about what position or attributes we would hold in this society. In this hypothetical mind game, we are put in a position where we have to describe in advance how egalitarian our own society should be even though we have no idea whether we would be placed at the top or the bottom of the society we choose to construct. In such a situation, rational actors would opt for a social contract which is broadly egalitarian just in case they are placed at the bottom of that society. However, this could pose some difficulties if we created a society in which we all received the same rewards despite the variation in the form and intensity of the effort we put in. In a strictly equal society, there are few incentives for self-improvement. In order to address this without contradicting the first decision, he introduces the 'difference principle'. This permits individual members of the moral community to pursue self-interested actions, but only in so far as these actions also benefit the least well-off. A 'just social contract', according to Rawls, would be one in which the benefits of material progress were spread throughout society and in which the lowest in the social hierarchy were not left out. Rawls is, in effect, identifying a redistributive social order as a just one, rather than the inegalitarian social orders which tend to characterize Western industrial societies.

In considering whether our actions are just in relation to future generations, it only remains to extend the moral community from actually existing generations to include future generations as well. In this way we should consider what sort of relationship present generations should have with future generations, if we did not know which generation we would be in. The argument developed by Rawls stops short of this, for he only goes so far as recognizing the ties of sentiment between present and future generations. Rawls devises the 'just savings principle' to address the ways in which present actions could harm the interests of future generations. The idea of 'just savings' refers to the setting aside of natural assets in the same way as responsible decisions by a generation of a family should ensure the security of the next generation. This means that each generation should not start off in a situation worse than the previous generation. In this way, present generations should leave a share of finite natural resources and an environment which is largely unspoilt for future generations to enjoy. However, it is difficult to decide exactly what share of finite resources can be exploited for present needs and what should be left for future generations. Much depends upon how many generations we should wish to consider in this calculation. The more future generations we include, the more the share of finite natural resources which we can presently exploit shrinks. In this account of justice, we should refrain from actions which are likely to have adverse effects on others.

The view of justice developed by Rawls has been most notably challenged by the anarchist individualism of Robert Nozick. In *Anarchy, State and Utopia* (1974), Nozick argues that a just outcome is one which follows from a series of individual voluntary contracts regardless of whether the outcome is broadly equal or not. This approach assumes that the entitlements enjoyed by individuals are inviolable. Nozick draws upon the natural rights of life, limb and property identified by John Locke. The acquisition and possession of private property is identified as an important component of a free individualistic social order. To demonstrate the extent of each individual's obligations to others, Nozick uses the analogy that every individual is an island. The impact which each individual has upon others is conveyed through the metaphor of sea currents between islands. The activities in each case may create a range of impacts and the side-effects may be carried from one place to another. A key characteristic of the condition of anarchy is the assumption that each individual is free from authority. However, it is possible to establish authority, if it is based entirely on the consent of all those involved, and for all members to be able to renounce it, if their consent is no longer forthcoming. In addition, it should not infringe natural liberties such as private property. Consequently, the impact of environmental pollution places no obligation on polluters to clean up the effects of their activities, unless it infringes such natural liberties. In such situations, where damage to private property has occurred from pollution, the owner of the affected property is entitled to seek some form of compensation. In this account, present generations have no obligations towards future generations. For Nozick, no principle of justice can be legitimately established which would require that a proportion of existing resources should be set aside for our successors to use. Since potential people are not yet alive and cannot be said to be owners of property, they possess no entitlement which we can respect and claims for compensation are not valid.

So far we have contrasted two approaches towards intergenerational justice with very different assumptions about what constitutes a social contract. In the case of Rawls, we are justified in recognizing obligations to future generations, especially if we include future generations in the 'original position'. Rawls applies the contractarian assumptions of Kant, that if something is to be applied at all, it must apply to all members of the moral community. However, in Nozick's account, obligations to future generations are not binding because future human generations fall beyond the restricted criteria for membership of the moral community. It remains possible to justify obligatory commitments to a limited number of generations on a variety of grounds. If we retain the individual as the focus of analysis and regard obligations as limited to our own immediate descendants, it can be argued that each successive generation represents a 50 per cent reduction in our own genetic legacy and a corresponding reduction in obligations. Similarly, arguments for limited obligations based on the protection of family inheritance are usually left to a single blood line rather than all future offspring. The principle of primogeniture establishes that the eldest male has certain entitlements with regard to the inheritance of the assets of a family and, with that, a corresponding series of obligations to manage those assets for the benefit of future generations.

References

Lukes, S. (1974) *Power: A Radical Approach* (Basingstoke: Macmillan).
Nozick, Robert (1974) *Anarchy, State and Utopia* (New York: Basic Books).
Rawls, John (1971) *A Theory of Justice* (London: Oxford University Press).

Source: Mark J. Smith (1998) *Ecologism: Towards Ecological Citizenship*, Milton Keynes: Open University Press, pp. 22–5.

2.4

Brian Barry

Justice between generations: power and knowledge

> Suppose that, as a result of using up all the world's resources, human life did come to an end. So what? What is so desirable about an indefinite continuation of the human species, religious convictions apart?[1]

My object in this paper is to ask what if anything those alive at any given time owe their descendants, whether in the form of positive efforts (e.g. investment in capital goods) or in the form of forbearance from possible actions (e.g. those causing irreversible damage to the natural environment). We scan the 'classics' in vain for guidance on this question, and, I think, for understandable reasons. Among human beings, unlike (say) mayflies, generations do not succeed one another in the sense that one is off the scene before the next comes into existence. 'Generations' are an abstraction from a continuous process of population replacement. Prudent provision for the welfare of all those currently alive therefore entails some considerable regard for the future. The way we get into problems that cannot be handled in this way is that there may be 'sleepers' (actions taken at one time that have much more significant effects in the long run than in the short run) or actions that are on balance beneficial in the short run and harmful in the long run (or vice versa).

More precisely the problem arises (as a problem requiring decision) not when actions actually have long-run effects that are different in scale or direction from their short-run effects but when they are *believed* to do so. The increased salience of the problem for us comes about not just because we are more likely to have the opportunity, thanks to technology, of doing things with long-run consequences not mediated by similar short-run consequences but also because there is more chance of our knowing about it. A useful new technology that we have no reason to believe has adverse long-term effects does not present any problem of decision-making for us, even if in fact, unknown to us, it has the most deleterious long-run consequences. Conversely, new knowledge may suggest that things we have been doing for some time have harmful long-term effects. Even if people have been doing something with adverse long-term effects for hundreds or thousands of years, so that we are currently experiencing the ill effects in the form of, say, higher disease rates or lower crop yields than we should otherwise be enjoying, it may still require some break-through in scientific understanding to show that the current situation has been brought about by certain human practices.

In recent years, we have all been made aware by the 'ecological' movement how delicately balanced are the processes that support life on the earth's surface and how easily some disequilibrium may ramify through a variety of processes with cumulative effects: The stage is set for some potentially very awkward decisions by this increased awareness that apparently insignificant impacts on the environment may, by the time they have fully worked themselves through, have serious consequences for human life. We may, any day, be confronted with convincing evidence for believing that something on which we rely heavily – the internal-combustion engine, say – is doing irreversible damage to the ecosystem, even if the effects of our current actions will not build up to a catastrophic deterioration for many years (perhaps centuries) to come.

If we ask what makes our relations with our successors in hundreds of years' time so different from our relations with our contemporaries as to challenge the ordinary moral notions that we use in everyday affairs, there are two candidates that come to mind, one concerned with power and one with knowledge. I shall consider these in turn.

A truistic but fundamental difference between our relations with our successors and our relations with our contemporaries, then, is the absolute difference in power. The present inhabitants of Britain may believe that, although they have some discretion in the amount of aid they give to the people of Bangladesh, they have little to hope or fear from the present inhabitants of Bangladesh in return. But they cannot be sure that later geopolitical events may not change this in their own lifetime. We can be quite certain, however, that people alive in several centuries' time will not be able to do anything that will make us better off or worse off now, although we can to some degree make them better off or worse off.

Admittedly, our successors have absolute control of something in the future that we may care about now: our reputations. It is up to them to decide what they think of us – or indeed whether they think about us at all. And presumably what, or whether, they think of us is going to be in some way affected by the way that we act towards them. I must confess, however, to doubting that this does much to level up the asymmetry of power between us and our successors, for two reasons. First, although they control a resource which may matter to us, we have no way of negotiating an agreement with them to the effect that they will treat our reputations in a certain way if we behave now in a certain way. We therefore have to guess how they will react to the way we behave, and in the nature of the case such guesses are bound to be inexact. Second, and more important, although individuals are undoubtedly moved by thoughts of posthumous fame for their artistic achievements or political records, it does not seem plausible to suppose that the same motivation would lead a mass electorate to support, say, measures of energy conservation. Altogether, therefore, I do not think that the fact of later generations determining our reputations deserves to be given much weight as an offset to the otherwise completely unilateral power that we have over our successors.

How important is this asymmetry of power between us and our successors – the fact that we can help or hurt them but they can't help or hurt us? It is tempting to say at once that this cannot possibly in itself make any moral difference. Yet it is perhaps surprising to realize that a variety of commonly held views about the basis of morality seem to entail that the absence of reciprocal power relations eliminates the possibility of our having moral obligations (or at any rate obligations of justice) to our successors.

There is a tradition of thought running from Hobbes and Hume to Hart and Warnock according to which the point of morality is that it offers (in Hobbes's terms) 'convenient articles of peace': human beings are sufficiently equal in their capacity to hurt one another, and in their dependence on one another's co-operation to live well, that it is mutually advantageous to all of them to support an institution designed to give people artificial motives for respecting the interests of others.

It seems plain that such a view cannot generate the conclusion that we have moral obligations to those who will not be alive until long after we are dead. Thus, G. J. Warnock . . . offers two reasons for saying that moral principles should have universal application rather than being confined to particular groups.

> First, everyone presumably will be a non-member of some group, and cannot in general have any absolute guarantee that he will encounter no members of groups that are not his own; thus if principles are group-bound, he remains, so to speak, at risk. . . . Second . . . if conduct is to be seen as regulated only *within* groups, we still have the possibility of unrestricted hostility and conflict *between* groups . . .
>
> (Warnock 1967: 150)

Obviously, neither of these reasons carries weight in relation to our successors, since we do precisely have an absolute guarantee that we shall never encounter them and cannot conceivably suffer from their hostility to us. It should be added in fairness to Warnock that he himself suggests that morality requires us to take account of the interests of future generations and also of animals. But my point is that I do not see how this squares with his premises.

It is, indeed, possible to get some distance by invoking the fact with which I began this paper, that the notion of 'successive generations' is an artificial one since there is a continuous process of replacement in human populations. Once we have universalized our moral principles to apply to everyone alive now, there are because of this continuity severe practical problems in drawing a neat cut-off point in the future. In the absence of a definite cut-off point, it may seem natural to say that our moral principles hold without temporal limit.

But could what is in effect no stronger force than inertia be sufficient to lead us to make big sacrifices for remote generations if these seemed to be called for by atemporal morality? Surely if morality is at basis no more than mutual self-defence, we would (whether or not we made it explicit) agree to ignore the interests of those coming hundreds of years after us.

There is an alternative line of argument about the basis of moral obligations, also involving reciprocity, from which the denial of obligations to future generations follows directly. This view is seldom put forward systematically though it crops up often enough in conversation. This is the idea that by living in a society one gets caught up in a network of interdependencies and from these somehow arise obligations. . . . Obviously, this more parochial view, which makes obligations depend on actual rather than potential reciprocal relationships, rules out any obligations to subsequent generations, since there is no reciprocity with them.

. . . [I]t is very close to basing obligations on sentiments. This further move is made in one of the very few papers addressed explicitly to the present topic [Golding 1972][2] . . . and permits some consideration to be given to future generations – but in a way that I personally find more morally offensive than a blunt disregard of all future interests. According to Golding, obligations rest on a sense of 'moral community'. Whether or not we have any obligations to future generations depends on whether we expect them to live in ways that would lead us to regard them as part of our 'moral community'. If we think they will develop in ways we disapprove of, we have no obligations to them. This view is obviously a diachronic version of the common American view that famine need only be relieved in countries with the right attitude to capitalism.

A third view which appears to leave little room for obligations to future generations is the kind of Lockean philosophy recently revived by Robert Nozick in *Anarchy, State and Utopia* (1974). . . . Indeed, it is scarcely accidental that the uniquely short-sighted destruction of trees, animals and soil in the U.S.A. should have been perpetrated by believers in a natural right to property. According to Nozick, any attempt to use the state to redistribute resources among contemporaries in order to bring about some 'end state' is illegitimate, so presumably by the same token any deliberate collective action aimed at distributing resources over time would fall under the same ban. Provided an individual has come by a good justly, he may justly dispose of it any way he likes – by giving it away or bequeathing it, trading it for something else, consuming it, or destroying it. No question of justice arises in all this so long as he does not injure the rights to property and security from physical harm of anyone else. Since we have a right to dispose of our property as we wish, subsequent generations could not charge us with injustice if we were to consume whatever we could in our own lifetimes and direct that what was left should be destroyed at our deaths. (Having one's property destroyed at death has been popular at various times and places and could presumably become so again.) It would clearly be, on Nozick's view, unjust for the survivors to fail to carry out such directions.

Once again, we can see that the problem is the lack of bargaining power in the hands of later generations. Those without bargaining power may appeal to the generous sentiments of others but they cannot make legitimate moral demands, as Nozick's examples of the men on their desert islands vividly illustrates. He asks us to imagine a number of men washed up on desert islands with the possibility of sending goods to each other and transmitting messages by radio transmitter but no means of travelling themselves. Sternly resisting the temptation to comment further on the outlook of a man for whom the paradigm of human relations is a number of adult males on desert islands, let us ask what moral obligations they have to each other. Nozick's answer is simple: none. Even if one has the good fortune to have landed on an island flowing with milk and honey while his neighbour is gradually starving on a barren waste, there is no obligation on one to supply the other's needs. Where could such an obligation possibly come from? To get a parallel with the relations between generations all we have to do is imagine that the islands are situated along an ocean current. Goods can be dispatched in one direction only, down the current. Even if those further down the line could call for help (as later generations in fact cannot) they could make no moral claims on those higher up.

I have so far concentrated on one potentially relevant fact about our relations with our successors: the asymmetry of power. The second one, which is invariably mentioned in this context, is the fact that we have less and less knowledge about the future the more remote the time ahead we are thinking about. . . .

. . . Of course, it may be held that we have *no* knowledge of the way in which our present actions will affect the interests of those who come after us in more than *k* years' time – either because we don't know what effects our actions will have on the state of the universe then or because we can have no idea what their interests will be. In that case, it obviously follows that our accepting an obligation to concern ourselves with their interests does not entail our behaving any differently from the way we would behave if we did not accept such an obligation. We can decide what to do without having to bother about any effects it may have beyond *k* years' time. But the obligation still remains latent in the sense that, if at some future date we do come to believe that we have relevant information about the effects of our actions on people living in more than *k* years' time, we should take account of it in determining our actions. The obligation would have been activated.

Ignorance of the future may be invoked to deny that obligations to remote descendants have any practical implications so that we can ignore them with a good conscience in deciding what to do. Thus John Passmore (1974) . . . canvasses among other possibilities the rigorous atemporal utilitarianism put forward by Sidgwick [1962] according to which 'the time at which a man exists cannot affect the value of his happiness from a universal point of view' . . . . But he says that, because of the existence of uncertainty, even Sidgwick's approach would lead us to the conclusion that 'our obligations are to *immediate* posterity, we ought to try to improve the world so that we shall be able to hand it over to our immediate successors in a better condition, and that is all' (p. 91, italics in original).

I think this all too convenient conclusion ought to be treated with great mistrust. Of course, we don't know what the precise tastes of our remote descendants will be, but they are unlikely to include a desire for skin cancer, soil erosion, or the inundation of all low-lying areas as a result of the melting of the ice-caps. And, other things being equal, the interests of future generations cannot be harmed by our leaving them more choices rather than fewer.

Even more dubious, it seems to me, is the habit (especially common among economists for some reason) of drawing blank cheques on the future to cover our own deficiencies. The shortages, pollution, over-population, etc., that we leave behind will be no problem for our successors because, it is said, they will invent ways of dealing with them. This Micawberish attitude of expecting something to turn up would be rightly considered imprudent in an individual and I do not see how it is any less so when extended to our successors.

[. . .]

. . . Presumably the whole idea of talking about obligations is to put to us a motive for doing things that we would (at least sometimes) not be inclined to do otherwise. But the demands of universal utilitarianism – that I should always act in such a way as to maximize the sum of happiness over the future course of human (or maybe sentient) history – are so extreme that I cannot bring myself to believe that there is any such obligation.

At the same time, I find it impossible to believe that it can be right to disregard totally the interests of even remotely future generations, to the extent that we have some idea of the way in which our current actions will affect those interests. If I am correct in saying that it is an implication of the three theories of morality that I briefly considered earlier that there are no obligations to distant future generations, they too have to be rejected.

But if we dump mutual self-protection, entitlement, and community . . . what are we left with? Unless we are prepared to fall back on an appeal to intuitions (and it may come to that), the only general approach remaining is as far as I can see some sort of ideal contractarian construction: what is required by justice is that we should be prepared to do what we would demand of others if we didn't know the details of our or their situation.

The name of Rawls naturally, and rightly, springs to mind here. But it should be recognized that the ideal contractarian formula is open-ended and does not have to be identified with Rawls's use of it. Nevertheless, Rawls's *A Theory of Justice* (1972) . . . is the obvious place to start and I shall therefore now set out and criticize Rawls's contribution to the problem of justice between generations. The first point to notice is that Rawls discusses the problem only in the context of the 'just savings rate' and this imposes two limitations on the generality of his conclusions. The obvious one is that if we concentrate on the question how much we are obliged to make our successors better off, we miss the whole question whether there may not be an obligation to avoid harm that is stronger than any obligation to make better off. This is after all a common view about relations among contemporaries.

The second, and ultimately perhaps more serious, limitation is that investment has a characteristic that enables discussion of it to dodge the most awkward difficulties. The only way in which we can leave people in *n* years (where *n* is a large number) more productive capital than they would otherwise have had is to create the additional capital now and hope that the intervening generations will pass it on, or more precisely to create it now and hope our immediate descendants and their successors will each pass on a larger total to *their* successors than they would have done had we left them less ourselves. If they do, then members of remote future generations will indeed be better off than they would otherwise have been thanks to our efforts. But there is no way in which we can be confident that our efforts will have any net effect because everything depends on the behaviour of the intervening generations, whom we have no way of binding.

Although it does not strictly follow from all this, it is easy to reach the conclusion if we concentrate on the 'just savings rate' that the problem of relations between generations can be reduced to the question of the relations between one generation and its immediate successors. There is no way of making remoter generations better off by making savings now that does not involve making nearer generations better off and, conversely, if we make our immediate successors better off by making savings now we at any rate make it possible for them to make *their* successors better off than they would otherwise have been.

Obviously, it might still be held that if we take account of remoter generations this should lay on us a greater obligation to build up capital now than would arise if we knew that our immediate successors would be the last generation ever. But since we have no way of ensuring that our immediate successors will not go on a binge and run down the capital we leave them, this must surely weaken the case for our having to make extra efforts to save merely so as to make it *possible* for our immediate successors to pass on more than they would otherwise have done.

When, by contrast, we look at the bads rather than the goods that we have the opportunity of passing on to our successors, we can see that the same convenient assumption . . . is not generally applicable. True, resource depletion has something of the same characteristic. The only way in which we can leave more to our remote successors is to leave more to our immediate successors; and if we make extra efforts to conserve resources so as to give our immediate successors more scope to pass on resources in their turn, we take the risk that they will simply blow the lot anyway.

But this is not necessarily the case with other bads that we might pass on. There could in principle be some ecological sleeper-effect that we set off now with no ill effects for some hundreds of years and then catastrophic effects. And there are in any case real examples (such as the use of fluorocarbon sprays) of things that we do now that may well have continuous and irreversible ill effects during the rest of the period during which there is life on the planet and that can either not be counteracted at all or only counteracted at great cost or inconvenience.

Of course, our successors may make things even worse for remotely future generations, adding further ecological damage to that done by us. And if we refrain from causing some kind of ecological damage, there can be no guarantee that our successors will not cause it themselves. But this does not suffice to destroy the distinction between investment, which has the property that our successors can choose whether to pass on the benefits we leave them, and ecological damage, which has its own adverse effects on remote future generations whether or not our successors add to it. It is, I think, because of the reduction of the problem that follows from taking investment as the paradigm of relations between generations that Rawls is satisfied with a solution that would otherwise be manifestly inadequate.

He postulates throughout *A Theory of Justice* (without, I think, ever adequately explaining why) that the people in the 'original position' (whose choices from behind the 'veil of ignorance' are to constitute principles of justice) know that they are all contemporaries, although they do not know to which generation they belong. The obvious problem that this raises is a sort of *n*-generation prisoner's dilemma. Generation *k*, who happen to be behind the veil of ignorance, may be willing to save on condition that their predecessors have saved. But there is no way in which they can take a conditional decision of this kind because there is no way of reaching a binding agreement (or indeed any agreement) with their predecessors. All *they* can do is to decide themselves whether to save or not. As Rawls says, setting out the problem: 'Either previous generations have saved or they have not; there is nothing the parties [in the original position] can do to affect it' (p. 292). How can we escape this difficulty?

It might appear that the obvious way out of the difficulty is to drop the postulate that the people in the 'original position' are contemporaries, and this is I believe the path that Rawls should have taken to be true to his own theory. But he does not take it. Instead, the tack that he takes is, he says, to 'make a motivational assumption'. The 'goodwill' of the parties in the original position 'stretches over at least two generations'. We may, though we need not, 'think of the parties as heads of families, and therefore as having a desire to further the welfare of their nearest descendants'. He concludes as follows:

> What is essential is that each person in the original position should care about the well-being of some of those in the next generation, it being presumed that

> their concern is for different individuals in each case. Moreover for anyone in the next generation, there is someone who cares for him in the present generation. Thus the interests of all are looked after and, given the 'veil of ignorance', the whole strand is tied together. (All quotes from pp. 128–9.)

One slightly technical objection that must be made to this is that the conditions stated by Rawls as necessary for the interests of all to be looked after are unnecessarily strong. Given the veil of ignorance, it is not necessary for each party to *know* that there is someone in the next generation he cares about. He will have a motive to support principles giving weight to the welfare of the next generation provided he knows that he will probably care about somebody in the next generation. Similarly, there is no need for everybody in the next generation to have someone who cares for him in this one so long as the uncared-for cannot be identified as a category and thus made the object of discriminatory principle-choosing from behind the veil of ignorance. And as far as I can see, they are pretty safe from that risk.

This, however, is just a skirmish. There are two powerful objections to the use Rawls makes of his 'motivational assumption'. The first, which I have already foreshadowed, is that even if it does everything Rawls wants it to do, that is still not enough. The really nasty problems (to some extent actual but even more potential) involve obligations to remote descendants rather than immediate descendants and on these Rawls has nothing to say. It has been suggested that we might boost the extension of concern into the future derivable from sentiment (which is what Rawls's derivation amounts to) by pointing to the fact that if we care about our grandchildren and they care about their grandchildren, we should care about our grandchildren's grandchildren, and so on *ad infinitum*. But those who base themselves on sentiment must follow where it leads, and if primary concern is as short-winded as Rawls suggests, it is scarcely plausible that secondary concern will alter the picture much. Certainly, by a few centuries' time it would be asymptotically approaching zero.

The second objection, which seems to me decisive, is that the derivation of obligations to future generations from the 'motivational assumption' is a pretty thin performance. The only justification offered for the 'motivational assumption' is that it enables Rawls to derive obligations to future generations. But surely this is a little too easy, like a conjurer putting a rabbit in a hat, taking it out again and expecting a round of applause. What it comes to is that we impute to the people in the original position a desire for the welfare of their descendants, on the basis of this we 'deduce' that they will choose principles requiring some action in pursuit of that welfare, and on the basis of the general theory that what would be chosen in the original position constitutes principles of justice, we say that the principle governing savings that they would choose is the 'just savings principle'. But if it is acceptable to introduce desires for the welfare of immediate descendants into the original position simply in order to get them out again as obligations, what grounds can there be for refusing to put into the original position a desire for the welfare of at least some contemporaries?

For the whole idea – and the intellectual fascination – of 'justice as fairness' is that it takes self-interested men, and, by the alchemy of the 'original position', forces them to choose principles of universal scope. In relation to subsequent generations, the postulate of self-interest is relaxed to allow concern for successors, but this naturally limited sympathy is not forced by the logic of the 'original position' to be extended any further than it extends naturally. Our limited sympathies towards our successors are fed into the sausage-machine of 'justice as fairness' and returned to us duly certified as obligations. We come seeking moral guidance and simply get our existing prejudices underwritten – hardly what one would expect from a rationalist philosopher.

The alternative route out of Rawls's difficulties is to pursue the logic of his own analysis more rigorously. This entails scrapping the part of the construction specifying that all the people in the 'original position' are contemporaries and know that they are. We should now have to imagine that there is a meeting to decide on intergenerational relationships at which all generations are represented. Clearly, the 'veil of ignorance' would be required to conceal from them which generation each of them belonged to. Otherwise, an earlier generation would always have the whip-hand over a later one in the negotiations.

[. . .]

Certainly, if I try to analyse the source of my own strong conviction that we should be wrong to take risks with the continuation of human life, I find that it does not lie in any sense of injury to the interests of people who will not get born but rather in a sense of its cosmic impertinence – that we should be grossly abusing our position by taking it upon ourselves to put a term on human life and its possibilities. I must confess to feeling great intellectual discomfort in moving outside a framework in which ethical principles are related to human interests, but if I am right then these are the terms in which we have to start thinking. In contrast to Passmore, I conclude that those who say we need a 'new ethic' are in fact right. It . . . should surely as a minimum include the notion that those alive at any time are custodians rather than owners of the planet, and ought to pass it on in at least no worse shape than they found it in.

Notes

1 Wilfred Beckerman, 'The Myth of "Finite" Resources', *Business and Society Review* 12 (Winter 1974–5), 22.

2 Also reprinted in this volume, Reading 2.1.

References

Golding, Michael P. (1972) 'Obligations to future generations', *Monist*, vol. 56, pp 85–99.
Nozick, Robert (1974) *Anarchy, State and Utopia* (New York: Basic Books).
Passmore, John (1974) *Man's Responsibility for Nature* (London: Duckworth).
Rawls, John (1972) *A Theory of Justice* (London: Oxford University Press).
Sidgwick, H. (1962) *Methods of Ethics* (London: Macmillan).
Warnock, G. J. (1967) *The Object of Morality* (London: Methuen).

Source: Brian Barry (1978) 'Justice between generations', in P. Hacker and J. Raz (eds) *Law, Morality and Society*, Oxford: Clarendon Press, pp. 268–84.

2.5

Robert E. Goodin

Futurity: an analysis of discounting

Another crucial component of the environmentalist ethic is concern for the further future. 'How can we induce people and institutions to think in terms of the long-range future, and not just in terms of their short-range selfish interest?' asks the umbrella group for American greens, who reckon that a 'future focus' ought to figure conspicuously among 'key green values'.[1] The 1983 Manifesto of the German Greens echoes these themes:

> The pillage of nature brings about long-term damage, part of which can never be restored. This is accepted in the interest of short-term profit. The very basis of people's lives is endangered by nuclear installations, by air, water and soil pollution, by storage of dangerous waste products and by the squandering of raw materials. . . . We stand for an economic system geared to the vital needs of human beings and future generations, to the preservation of nature and a careful management of natural resources.[2]

This is no mere matter of worrying about one's own future or that of one's own immediate family. There are good grounds for that, too, environmentalists would argue. Pollution is imprudent. It amounts to fouling our own nests, and we or those whom we care about will (or may well) ultimately suffer in consequence.

Such concerns stretch only a little way into the future, however. They are consistent with activities that create environmental time bombs of colossal proportions, so long as they have moderately long fuses.[3] Full-blooded environmentalists would worry about those, too, however long their fuses. Greens are concerned with entire future generations, not just with their own progeny, and with distant generations, not just their own children and their children's children. Their concern, as they standardly say, is with the long-term future of life on earth.[4]

In this as in so many other matters, the environmentalist's principal opponent in policy debates is the economist. On the question of the proper treatment of the further future, the environmental ethic is counterpoised most directly to the standard economic practice of discounting the future.[5] Economists standardly advise us to weigh future pay-offs (costs or benefits) less heavily – indeed, disproportionately less heavily – the further in the future they come. Technically expressed, they discount future income streams according to a discount function that is exponential (that is, geometrical) in form.[6]

The consequence of that practice is obvious. Costs and benefits that are relatively near to hand weigh relatively heavily. The £1 million that will accrue next year has a present value of £952,400, assuming a (relatively modest) 5 per cent discount rate. But the further in the future a cost or benefit is, the disproportionately less heavily will it weigh with us now. Thus, for example, the present value of the same £1 million twenty years away is reckoned to be merely £376,000, and the same sum a century away appears in current accounts with a paltry present value of only £761.

(The cynical may say that sounds about right. Inured as we all are to the fact of inflation, we rightly regard large sums in the distant future as 'funny money'. But all these calculations – and all that follow later in this section – are expressed in terms of 'constant prices'. The effects of inflation have already been factored out. The suggestion is that, even after inflation has been taken into account, we should regard £1 million next century as equivalent to no more than £761 today.)

Discounting in that way is commended, virtually as a hallmark of rationality itself, in just about every economics text and development manual in print.[7] But is it so obviously rational to ignore almost completely what the consequences of our present actions will have beyond a few generations? That is not a long time, even in terms of human history much less in terms of geological time. On economistic calculations of present values, though, consequences that come in a century or two – however large or however certain they may be – are treated as being simply of no proper concern to us.

There are many ways to resist such conclusions, of course. Perhaps the most obvious strategy – and certainly the one most often employed by writers on ethics in general, and on environmental ethics in particular – is to shift the terms of discourse. Essentially, the aim is to concede the low ground of pragmatics to economists, while claiming for oneself the high moral ground.

Perhaps, this line goes, it is peculiarly efficient – peculiarly rational, even, in some narrowly economistic sense of that term – to discount the future as economists would recommend. Here as elsewhere, though, a trade-off must be made between equity and efficiency, between the maximization of total social utility and its just distribution. Hence, philosophers deploying this device would say, it may be economically efficient for us to short-change the future; but it is nonetheless wrong for us to do so. It would amount to unjust treatment of future generations.

Of course, those taking this tack must then provide some theory of intergenerational justice strong enough to sustain such a claim. Many avenues have been explored but most, it is probably fair to say, are variations on one basic theme. Justice, it is standardly said, ought to be blind to facts about people that are truly arbitrary from a moral point of view. Among those arbitrary facts is the precise timing of one's birth. Indeed, the century of one's birth is as irrelevant from a moral point of view as the precise microsecond. Justice therefore requires that we weigh equally in our present decisions the interests of all generations, whenever they come in human history.[8]

If that sort of argument goes through, the upshot is that high morality requires that we [do] not discount – or, as economists would put it, that we apply a 'zero discount rate' – whereas the low morality of economic rationality requires some positive rate of time discounting. How well the argument goes through, in the first place, depends on the relative weight of arguments for both moralities (high and low) on these questions.

Even if the philosophers' arguments go through, though, the form of their rejoinder has conceded some important ground to the economists. There is a trade-off to be made, they will have implicitly agreed. And that is in itself a costly concession. Maybe the moralistic arguments against discounting will prevail in nine cases out of ten, or ninety out of a hundred. But in casting their rejoinder as the other side of an important value trade-off, philosophers are conceding that there is always going to be something to be said for the economists' case. That being so, it will inevitably sometimes prevail – if only occasionally, and if only at the margins.[9]

Environmentalists would therefore be well advised to look for some better way of resisting the economists' devil-may-care conclusions about the further future. Ideally, they should try to meet the economists on their own ground, and to undermine their case for discounting from within. On inspection, it turns out that there is indeed some considerable scope for exposing obvious flaws in the justifications that economists themselves offer for discounting.

Those justifications are, of course, many and varied.[10] Some hardly count as justifications at all. Perhaps the most standard among them appeals to nothing more than people's blind prejudice in favour of the present over the future, in a way that many economists themselves cannot quite countenance.[11]

Other economistic arguments for discounting the further future turn on just plain sloppy thinking. They confuse the argument in favour of discounting for time *per se* – which is what they are seeking – with arguments in favour of discounting future pay-offs on account of things that merely correlate (and then only imperfectly) with the passage of time. Thus, it may well be true that the further we are looking into the future, the more uncertain we will be what the pay-offs actually will be and whether we will ourselves be around to experience them; or the later the pay-offs come the more resources we are likely to have to cushion us against, or at least to compensate us for, their evil effects. But all those things – increased uncertainty, technology, wealth – are only contingently connected with the passage of time. They therefore provide no reason for supposing that it will necessarily be proper to discount later pay-offs.

Furthermore, and in a way more importantly, none of these considerations – not even the first-mentioned brute psychological fact – provides any reason at all for the rapidly progressive form that the economist's geometric discount function ordinarily takes. There is no reason to suppose that uncertainty or technological progress or wealth or even people's psychological attachment to the future alters at some fixed rate of *r* per cent per year, compounding continually on into the further future. Yet, of course, if the rate varies then the case for compounding collapses and (depending on the details of the formula we use instead) the further future may well loom larger in our present calculations than the standard economic formula suggests.

Environmentalists can get a long way towards undermining disregard for the further future merely by attacking the bad arguments that economists themselves give for discounting. Economists do, however, have one good argument to justify discounting in that powerfully exponential form. And it is this argument that environmentalists, and friends of the further future in general, must be principally concerned to address.

This one good argument for exponential discounting of future costs and benefits treats discounting as a form of compound interest in reverse. A smaller sum now ought, on this argument, be seen as equivalent to – ought be deemed the 'present value' of – a larger sum later, simply because if you put that smaller sum into the bank earning compound interest now it will actually have grown into that larger sum by the later date. Thus, £761 now ought be seen as equivalent to – the present value of – £1 million next century, simply because £761 invested now at 5 per cent per year, compounded, will amount to £1 million in a hundred years' time.[12] So money, or anything that can be bought and sold for money, ought be discounted at a rate that is equal to the long-term interest rate.[13]

That way of summarizing the conclusion, however, only serves to highlight what is most wrong with this argument. First, it requires us to guess what the interest rate really will be over the (possibly very) long term. Real interest rates have, in fact, been highly variable, so there is no good way of knowing what discount rate to use.[14] Second and more important is the fact that that argument for discounting applies, first and foremost, to money and things that can be bought and sold in exchange for money.

The argument for discounting the future value of things that can be bought and sold goes like this. We should be indifferent between being given £37,600 today or the guarantee of a £100,000 house in twenty years' time, because we can put that smaller sum on compound interest and (assuming a constant 5 per cent interest rate over the entire period[15]) have in hand at the end of twenty years £100,000 with which to buy that exact house. By the same token, we should be indifferent between being given a £37,600 house today or the guarantee of a £100,000 house in twenty years' time, because (ignoring transaction costs and making the same heroic assumptions about interest rates) we could sell the cheaper house today, put the money on compound interest in the bank, and buy the more expensive house later.

But obviously that argument only works for things, like houses, that really can be bought and sold. Consider, by way of contrast, the case of something that cannot: human lives. That is not to deny that lives have 'monetary equivalents'. Of course they do, in all sorts of ways. We might look at the sum for which people insure their lives, or at the compensation payments that courts order when someone has been killed. Suppose such measures suggest that the going rate for the life of someone like Mr Smith is £100,000.

Even if we can, in such ways, come up with a monetary equivalent of non-tradable commodities, it would be fallacious to use those sums in any scheme of discounting justified in the way sketched above.[16] Adapting the house example, devout discounters might reason as follows. Mr Smith should be indifferent between getting £37,600 now or medical treatment to remove a latent tumour that will cause his certain death in twenty years' time; that would be equivalent to conferring on him a benefit worth £100,000 (what his life is worth) in twenty years, and £37,600 invested at 5 per cent compound interest will grow to £100,000 by the end of twenty years.

But what is not true in the case of the life, in a way that it is in the case of the house, is that invested proceeds of the smaller sum could later be used to buy the greater good. Mr Smith may have an extra £100,000 in the bank on the day he dies, thanks to the workings of compound interest. But he cannot use it to buy off the Grim Reaper, in a way that the person in the earlier example could use it to buy exactly the same house.

What makes discounting rationally mandatory, in the case of things like houses that genuinely can be bought and sold, is that money put on compound interest can buy exactly the same things later. Then it really would be irrational – it would be to distinguish between things that are indistinguishable – categorically to refuse the offer of a smaller sum now which will yield exactly the same goods later.

In the case of things that cannot be bought and sold, though, that is not what will happen. Money invested on compound interest will not buy us exactly the same thing later. It will buy us something that is 'equivalent' or 'as good', perhaps. But what it will buy us will undeniably be something that is different.

It follows from that fact alone that it is not necessarily irrational to reject discounting for non-tradable commodities. It is not irrational, at least, in the sense that it does not amount to drawing distinctions between things that are literally indistinguishable. Clearly, in the case of things that cannot be bought and sold, we will be getting things that are different – albeit 'equivalent' – if we discount than if we do not.

That is to say that discounting is not necessarily rational, in such cases. That is not to say, though, that discounting is necessarily not rational. Everything depends, in these cases, on just how close is the 'equivalence' between what we would be getting and what we would be losing if we discounted the present value of future losses of non-tradable commodities.[17]

At this point, then, the case against economistic discounting of the further future links up with previous discussions of irreplaceability. The losses that, on the argument just sketched, we may rightly refuse to discount are losses of things for which there are no good substitutes or equivalents. If something is replaceable, or if there are good substitutes for it, then there is no reason in principle not to discount the prospect of its future loss. It is only the irreplaceable whose future is potentially immune to the solvent of compound interest calculations.[18]

A theory of irreplaceability is therefore what is needed to resist the economists' strongest argument for discounting the future. . . . [T]he green theory of value gives us grounds for supposing that at least some things produced by natural forces are irreplaceable, precisely because they have a history of having been produced by those natural forces. The things might be replicated artifically. But history cannot be so replicated.

. . . Suppose . . . our concern is with natural types rather than with mere tokens. We do not much care about the deaths of individual animals. We do care powerfully about the loss of whole species.

Then we might be prepared to tolerate a certain limited discounting of future streams of resources. In this discounting, we would be limited to trading like for like – whales for whales, baboons for baboons. So long as what we care about is types rather than tokens – the natural order, rather than the particular animals – we should regard them as good substitutes for one another. And so long as those populations have the characteristics of an interest-bearing resource, growing with time if not destroyed now (as biological populations obviously do), then the compound-interest-in-reverse case for discounting applies. We should care less about saving a single baboon in the future than in the present simply because, if the one in the present lives and mates, then it will produce many more in the future.

This sort of argument, notice, applies only to particular sorts of environmental assets. It applies only to those akin to interest-bearing resources. It discounts each resource according to a different rate, depending on the growth rate of that resource itself. And since no resource (no biological population, even) continues growing at exponential rates indefinitely into the future, neither can our discounting of such resources. The structure – the formula – as well as the rate at which we discount these sorts of resources must match the pattern of growth in those resources themselves. Hence, discounting justified in these resource-specific ways cannot display the marked unconcern for the further future that geometric discounting implies.

If our concern is for the general shape of the natural order, a second consequence for resource futures follows. Suppose our basic aim is, in so far as possible, to 'leave no footprints'. Suppose that we are therefore trying to preserve distinct natural types into the indefinite future, not worrying too much about the fate of particular tokens of those types (individual animals, as distinct from whole species). Then we may cream off only as much of the resource flow as is consistent with leaving enough to reproduce at least as much in the future.[19] In biological terms, we should take only as much as is consistent with leaving a breeding population. In economic terms, we should be striving to 'maximize sustainable yield'.[20]

Notes

1 Green Committees of Correspondence 1986, item 10.
2 Die Grünen (1983, p. 9).
3 As Barry (1977, p. 277), for example, rightly objects against Rawls's (1971, sec. 44) derivation of intergenerational justice.
4 Thus, for example, Ophuls (1977, p. v) dedicates his *Ecology and the Politics of Scarcity* to 'the posterity that has never done anything for me'. See similarly Routley and Routley (1978) and Page (1977, chs 7–10).
5 There have always been some economists opposing those practices, of course (see e.g. Page 1977). With the 'second environmental crisis' and growing concerns about global warming or destruction of the ozone layer, more and more are coming around to such longer-term perspectives (D'Arge, Schultze and Brookshire 1982; Nordhaus 1982).
6 Technically, the standard formula is $PV = X/(1 + r)^t$ where PV is present value, X is the sum that will accrue in t years and r is the discount rate.
7 For just two of the most influential examples, see the OECD manual on *Project Appraisal and Planning for Developing Countries* (Little and Mirrlees 1974, secs 1.7 and 4.1) and the textbook for the required course on policy analysis at Harvard's Kennedy School of Government (Stokey and Zeckhauser 1978, ch. 10).
8 Such arguments are of course inspired by, though typically go well beyond, Rawls (1971, sec. 44). See e.g. Barry (1977; 1978; 1983; Sikora and Barry (1978); Feinberg (1974); Page (1977, ch. 9); Partridge (1981); and MacLean and Brown (1983). For a rather different derivation, see Goodin (1985, pp. 169–79).
9 If they have further consented to the economistic way of putting their point against discounting – as an argument for a 'zero discount rate' – then there will be yet more room for compromise between the contending principles, each of which has something to be said for it. If moral principle requires a zero discount rate and economic efficiency a 10 per cent discount rate, and economic efficiency is say one-tenth as important as moral principle in the case at hand, then the obvious compromise is to apply a discount rate one-tenth of the way between what moralists and economists recommend – that is, a 1 per cent discount rate.
10 These arguments are elaborated in Goodin (1982a). See also Parfit (1983; 1984, pp. 480–7).
11 This is what economists call 'pure time discounting'. Some economists would themselves reject discounting grounded in no more than this: Pigou (1932, pt 1, ch. 2, sec. 3) lambasts pure time discounting as a mere failure of the 'telescopic faculty', and Ramsey (1928, p. 543) says plainly that it 'is ethically indefensible and arises merely from the weakness of the imagination'. Others bemoan that as 'an authoritarian rejection of individual preferences', arguing that if individuals discount the future for no good reason in this way, then so too should democratic governments committed to reflecting faithfully their preferences (Marglin 1962, p. 197; 1963, p. 97). What that argument overlooks is that the people themselves, who would now have us discount future pay-offs, would wish come the time the pay-off occurs that we had not done so; it is not a case of substituting planners' judgement for citizens' own, but rather a case of planners choosing which of citizens' conflicting judgements to track in their plans (Sen 1957; 1961, p. 482; Goodin 1982a, p. 55; 1982b, ch. 3). As Sen (1961, p. 482) astutely puts it, 'If the difference is only due to the distance in time, then the position is symmetrical. A future object looks less important now, and similarly, a present object will look less important in the future. . . . [T]here is no necessary reason why to-day's discount of to-morrow should be used, and not to-morrow's discount of today.'
12 Marglin (1967, pp. 59–61); Little and Mirrlees (1974, p. 11); Stokey and Zeckhauser (1978, ch. 10).
13 Environmental costs and benefits are standardly treated in this way. Similarly, the standard resource-economic advice is to exhaust exhaustible resources up to the point at which their increase in scarcity value equals the market interest rate (Hotelling 1931; Marglin 1962; Page 1977).
14 In a way, it might not matter much. The nature of exponential decay functions of this sort is such that, just about whatever rate you use, costs and benefits in the distant future will virtually disappear from present calculations. But that is true only so long as you can be reasonably certain that on average interest rates will at least be positive over the relevant period. (Were they

negative, then the present value of £1 tomorrow would be more – not less – than the value of £1 received today.) If the darker fears of the eco-doomsayers prove well founded, even that may well not be true. There are some circumstances under which we should contemplate a negative interest rate (Stokey and Zeckhauser 1978, p. 175).

15 Or – what is more realistic in some ways, but less so in others – a constellation of varying interest rates over the period with the same practical consequences.

16 It is no mere 'straw man' that is here being attacked. Notice, for example, health economists writing, 'The reason for discounting future life years is precisely that they are being valued relative to dollars, and since a dollar in the future is discounted relative to a present dollar, so must a life year in the future be discounted relative to a present dollar' (Weinstein and Stason 1977, p. 720).

17 In the case of £100,000 for Mr Smith's life, it is essentially a question of whether he would be willing to commit suicide for that sum.

18 That point is sometimes phrased, somewhat misleadingly, in terms of 'sustainability' – i.e. the sustainability of the particular resource flows which we regard as irreplaceable and non-substitutable. That is, I think, the sense in which Achterberg (1991), for example, sees a broadly overlapping consensus organized around ensuring the sustainability of the Dutch environment for future generations.

19 'May', notice – not 'should'. This argument permits discounting, of this distinctly limited sort. It does not require it – or, indeed, provide any positive justification at all for it.

20 Williams (1978).

References

Achterberg, Wouter (1991) 'Can liberal democracy survive the environmental crisis?' Paper presented to the Workshop on Green Political Theory, ECPR Joint Sessions, Colchester UK, March 1991.

Barry, Brian (1977) 'Justice between generations'. In P. M. S. Hacker and J. Raz (eds), *Law, Morality and Society*, Oxford: Clarendon Press, 268–84. Reprinted in Barry (1989).

—— (1978) 'Circumstances of justice and future generations'. In R. I. Sikora, and Barry, Brian (eds), *Obligations to Future Generations*, Philadelphia, Pa.: Temple University Press, 204–48.

—— (1983) 'Intergenerational justice in energy policy'. In Douglas MacLean and Peter G. Brown (eds), *Energy and the Future*, Totowa, NJ: Rowman and Littlefield, 15–30. Reprinted in Barry (1989).

—— (1989) *Democracy, Power and Justice: Essays in Political Theory*. Oxford: Clarendon Press.

D'Arge, Ralph C., Schultze, William D. and Brookshire, David S. (1982) 'Carbon dioxide and intergenerational choice'. *American Economic Review (Papers and Proceedings)* 72: 251–6.

Die Grünen [German Greens] (1983) *Programme of the German Green Party*, trans. Hans Fernbach. London: Heretic Books.

Feinberg, Joel (1974) 'The rights of animals and unborn generations'. In William T. Blackstone (ed.), *Philosophy and Environmental Crisis*, Athens: University of Georgia Press, 43–68.

Goodin, Robert E. (1982a) 'Discounting discounting'. *Journal of Public Policy* 2: 53–72.

—— (1982b) *Political Theory and Public Policy*. Chicago: University of Chicago Press.

—— (1985) *Protecting the Vulnerable*. Chicago: University of Chicago Press.

Green Committees of Correspondence (Green CoC) (US) (1986) *Ten Key Values*. Kansas City, Mo.: CoC.

Hotelling, Harold (1931) 'The economics of exhaustible resources'. *Journal of Political Economy* 39: 137–75.

Little, I. M. D. and Mirrlees, J. A. (1974) *Project Appraisal and Planning for Developing Countries*. London: Heinemann.

MacLean, Douglas and Brown, Peter G. (eds) (1983) *Energy and the Future*. Totowa, N.J.: Rowman and Littlefield.

Marglin, Stephen A. (1962) 'Economic factors affecting system design'. In Arthur Maass *et al.*(eds), *Design of Water Resource Systems*, Cambridge, Mass.: Harvard University Press, 159–225.

—— (1963) 'The social rate of discount and the optimal rate of investment'. *Quarterly Journal of Economics* 77: 95–111.
—— (1967) *Public Investment Criteria*. London: Allen & Unwin.
Nordhaus, William (1982) 'How fast should we graze the global commons?' *American Economic Review (Papers and Proceedings)* 72: 242–6.
Ophuls, William (1977) *Ecology and the Politics of Scarcity: Prologue to a Political Theory of the Steady State*. San Francisco, Calif.: W. H. Freeman.
Page, Talbot (1977) *Conservation and Economic Efficiency*. Baltimore, Md: Johns Hopkins University Press (for Resources for the Future).
Parfit, Derek (1983) 'Energy policy and the further future: the social discount rate'. In Douglas MacLean and Peter G. Brown (eds), *Energy and the Future*, Totowa, NJ: Rowman and Littlefield, 31–8. Reprinted in Parfit 1984.
—— (1984) *Reasons and Persons*. Oxford: Clarendon Press.
Partridge, Ernest (ed.) (1981) *Responsibilities to Future Generations*. Buffalo, NY: Prometheus.
Pigou, A. C. (1932) *The Economics of Welfare*, 4th edn. London: Macmillan.
Ramsey, F. P. (1928) 'A mathematical theory of savings'. *Economic Journal* 38: 543–59.
Rawls, John (1971) *A Theory of Justice*. Cambridge, Mass.: Harvard University Press.
Routley, Richard and Routley, Val (1978) 'Nuclear energy and obligations to the future'. *Inquiry* 21: 133–79.
Sen, Amartya (1957) 'A note on Tinbergen on the optimum rate of saving'. *Economic Journal* 67: 745–48.
—— (1961) 'On optimising the rate of savings'. *Economic Journal* 71: 479–96.
Sikora, R. I. and Barry, Brian (eds) (1978) *Obligations to Future Generations*. Philadelphia, Pa.: Temple University Press.
Stokey, Edith and Zeckhauser, Richard (1978) *A Primer for Policy Analysis*. New York: Norton.
Weinstein, M. C. and Stason, W. B. (1977) 'Foundations of cost-effectiveness for health and medical practices'. *New England Journal of Medicine* 296: 716–21.
Williams, Mary B. (1978) 'Discounting versus maximum sustainable yield'. In R. I. Sikora and Brian Barry (eds), *Obligations to Future Generations*, Philadelphia, Pa.: Temple University Press, 169–79.

Source: Robert E. Goodin (1992) *Green Political Theory*, Cambridge: Polity Press, pp. 65–73.

2.6

Gregory S. Kavka and Virginia Warren

Can future generations be represented?

We shall assume that it makes sense to speak of those currently living having obligations to future generations, and that we do in fact have certain obligations of this kind, for example, to preserve a planetary environment that can support human life. Objections to this, based on the claim that we cannot have obligations to those who do not exist, or to those whose lives will not overlap our own, are sometimes raised. But these objections have been adequately rebutted elsewhere.[1]

We wish here to consider objections to the idea that a democratic nation's obligations to future generations can be, or should be, fulfilled (in part) by according future generations representation within the political process.[2] Setting aside more practical objections . . . the main theoretical objections to political representation for future generations fall roughly into two categories. Firstly, there are objections based on the existence of other classes of persons, besides future people, whose interests are gravely affected by decisions made in the political process, but who are not represented in that process. . . . Secondly, there are a number of objections of the form, 'It does not make sense to talk of representation for future people. The nature of representation is such that future generations *cannot* truly be represented.' In this section, we shall deal with objections of this sort, and shall attempt to show that none is convincing.

The general issue concerns the *prerequisites* (i.e., necessary conditions) of one party representing another. In particular, two questions arise. What features must the members of a class of entities possess, in order for it to be logically possible for someone to represent them? And what must the relationship be between one party and another, if the latter is to be a genuine representative of the former? Different answers to these questions yield different versions of the claim that the notion of representation for future generations is incoherent.

Firstly, it might be suggested that present existence is itself a prerequisite of being represented. If this is so, it necessarily follows that future generations cannot be represented. When put in the rhetorical form 'How could you possibly represent what does not exist?', this suggestion may sound plausible. But it is clearly refuted by some of the legal practices of modern nations, in which lawyers, trustees of estates, etc., often represent deceased persons, or the interests of particular future persons (e.g., unborn descendants for whom trusts are established).

Though readily disposed of, the above objection is at the root of two others. Each involves the claim that there are certain prerequisites of being represented that future people fail to satisfy because they do not exist yet. If one thinks of paradigm cases of representation – such as a lawyer representing his client or a congressman representing his district's voters – it may seem that choosing one's representative, and instructing or advising him, are prerequisites of being represented. But again, if we take into account the range of representation within modern legal systems, we see that this is not so. Children, and mentally incompetent adults, are often legally represented by parents or guardians they have never chosen; while indigent criminal defendants are frequently represented by public defenders appointed by the court. As to instructions and advice, young children and certain incompetent adults are unable to supply these to their legal representatives. And deceased persons cannot advise or instruct their legal representatives with regard to contingencies not foreseen by them while they were alive.

The positive doctrine implied by these legal examples is one suggested by Joel Feinberg:[3] that the main, and perhaps sole, prerequisite of being representable is *having interests*, being capable of faring well or faring ill. (As the representability of deceased persons indicates, we must interpret Feinberg's notion of 'having interests' in the atemporal sense of having interests at some time – past, present, or future.) Acceptance of this plausible 'interest principle', as Feinberg calls it, does not, however, immediately disarm all conceptual objections to political representation for future generations. One counter-move is to claim that representatives not selected and/or instructed by a group may, at best, represent that group's interests, but not the group itself; or, at least, they cannot be the group's political representatives for, in order to be a political representative of a group, one must be selected for that role by the group. In our view, though, this counter-move is merely a nonsubstantial verbal manoeuver. We are interested in defending the substantive proposal that, within democratic governments, there be specially designated representatives charged with protecting and promoting the interests of future generations. Whether they are called 'political' representatives, and whether they are regarded as representing future people or, instead, their interests (if this distinction even makes sense) is of no consequence.

A more serious objection to the interest principle concerns the tasks which a representative must carry out. It may be expressed as follows. If, as the interest principle implies, the essence of representing a group is to protect, promote or speak for its interests, then future generations cannot be represented. For how is a 'political representative' to know future people's interests, if future people are not around to tell him what they want, and he does not know what they, their lives, and their living conditions will be like? Further, even if he did know what future people's interests will be, he would not know how to act at present to promote these interests, given how uncertain the long-term effects of government policies and programmes are.

It may be true that if one party has no reasonable beliefs at all about what another party's interests are, or has no idea at all about how to promote what he believes to be the other party's interests, it would be impossible for the former to represent the latter, except by acting on the latter's instructions. But surely we do know a significant amount about the interests of future people, for at least a large number of generations. These people will have the same (or very similar) biological needs for food, clean air and water, shelter, etc. that we have, and it will be in their interests if these things, and the means of procuring them – including arable land, reliable supplies of energy, and various minerals – are available.[4] And while there are grave uncertainties involved in predicting the long-term effects of present social actions and policies, we can make reasonable and nonrandom projections. We can predict, for example, that while future technological developments *might* alleviate our main environmental problems, it is likely that pursuing policies which degrade our biological environment – for example, those which emphasize nuclear fission as a main power source, or which allow industrial pollution of our air and water – will have a marked negative impact on the quality of life for some (or all) future generations. The ability to make such probabilistic, nonrandom judgments about the effects of policies on the interests of a given individual or group is all that is necessary for representation. In particular, certainty, or something approaching it, is not required. If it were, paradigm cases of representation, such as elected congressmen representing their constituents (or lawyers representing their clients, or even people representing themselves), would not qualify as genuine representation! For typically, congressmen do not know with certainty, or near certainty, what the interests of their constituents are, or how various proposed policies would affect the fulfillment or frustration of those interests.[5]

To summarize, only two plausible logical prerequisites of representation have emerged from our brief discussion. Firstly, the represented party must (atemporally) have interests. Secondly, the representative must either be instructed by the represented party, *or* must know enough about the likely interests of the represented party and the means of promoting those interests, to be able to make better than random judgments about how alternative policies are likely to affect the interests of the represented party. But surely both of these prerequisites can be satisfied in the case of present people representing future generations within a democratic political system. For future people will have interests,[6] and we can have well-grounded and reasonable beliefs about some things that can now be done to protect those interests. If this is so, there are no sound reasons to suppose that future generations cannot, in principle, be represented in a democratic political system.

[. . .]

Is representation for future generations practical?

[. . .]

Despite the curious ring to the title 'Representative of Future Generations', when considering political representation it is natural to think first of the legislature. We argued, [earlier] that it is theoretically possible to represent the interests of future people, since being represented requires neither that one currently exist, nor that one choose and instruct one's representative.

The more troublesome issue is to determine how many such representatives there should be. While proportional representation is the rule for existing citizens, there are two reasons that, in the case of future citizens, this rule must be broken. Firstly, no one knows how many future citizens will actually exist, or even for how many generations the nation will continue. (In fact, the number of future people might differ according to whether or not we did directly represent their interests.) Secondly, if we assign representatives proportioned to an estimate of the number of future people there are likely to be, representatives of future generations would vastly outnumber those of present citizens. We think it doubtful that present people are morally obligated to submerge their interests to this extent; and, in any case, they certainly would not do so. So, on grounds both of practicality and of morality, a decidedly less-than-proportional representation for future generations would be required.

A modest and workable form of less-than-proportional representation would be to have a handful of representatives for future people, or even a single one in each house of congress. Their function would be to further the interests of future people by participating in debate, working out compromises, proposing legislation, serving on committees, voting, etc. Their impact could be appreciable, despite their small numbers.

There are several possible methods of choosing such representatives. They could be appointed by the president, the only elected official (besides the vice president) who represents the country as a whole. Or, a national election could be held, in which existing citizens chose from among candidates nominated by various political parties.

It might be objected that however such representatives were selected, they might be opposed to giving any (or much) weight to the interests of future generations. Their opposition might either be kept secret during the campaign, or it might even be the cornerstone of their campaign! We admit that these things could happen. But we doubt that this objection is critical, since our whole political system is alive with similar possibilities. The president could select Supreme Court Justices who did not believe that judges should follow the law, or a C.I.A. director who wished to disband that agency. In fact, citizens could elect to office a slate of anarchists who were bent on dismantling the government. The most that this objection shows is that unless a sufficient amount of public approval exists, arrangements for legislative representation of future generations will not accomplish their purpose.

A second alternative is to have the interests of future people represented within the executive branch, either by a cabinet member, or by a director (comparable to the F.B.I. director) whose tenure was not necessarily tied to a specific administration. This alternative might prove to be more expensive than having legislative representatives, since a whole department or agency would be created. But, then, a department or agency could also conduct research, help to educate the public, have consultants to supply information to any government officials needing it, and issue reports on the predicted effects of government action on the welfare of future generations (analogous to current environmental impact reports). Having a separate body would be preferable, we think, to having 'future generations officers' (an analogy to affirmative action officers) spread throughout various agencies. For it is psychologically difficult to be a single outsider, a devil's advocate, in a bureaucratic organization; one is likely to lose sight of one's separate goal, or to be ignored and ineffective.

It surely will be objected to either of these alternatives that adding more government is simply too expensive. . . . Still, implementing either alternative would cost something. But, there are economic, as well as moral, reasons for doing so. In the long run, it would probably cost more money to ignore future persons' needs than to take care of them (e.g., cleaning up pollution is usually more expensive than preventing it). There are also other long-term costs of ignoring future generations which are not, or not merely, financial. War can result from short-sighted policies about food production, population, energy sources, etc.; and war is paid for not only in money, but also in death and suffering and, less tangibly, in the stifling of progress in the sciences and arts.

One final objection to our thesis is that representatives of future generations are not needed, because the welfare of future people is already being taken care of by some other group – by private enterprise, by existing government officials and agencies, and by special-interest groups. Firstly, we think it not at all realistic to expect action from private enterprise. For businesses operate for profit, and it is generally not profitable for a business to care about events occurring after it no longer exists.[7] The situation is even more dire, however, since businesses frequently do not act with much foresight even to protect their own interests in the not-so-distant future. For example, American automobile companies for many years fought strenuously against government regulations designed to increase gasoline mileage. Secondly, elected officials tend also to be relatively short-sighted, in order to gain re-election. Thirdly, existing government agencies, such as the Nuclear Regulatory Agency and the Environmental Protection Agency, do handle some of the work we are claiming needs to be done. But much more ought to be done. Also, a central department could develop a comprehensive policy, so that the efforts of existing agencies are coordinated and do not undercut each other. In addition, we think that the debate over the environment, etc., is at present often distorted. We frequently hear that we must choose between the interests of a few intrepid backpackers who would use the wilderness, and the interests of the many who would not. That is, the interests of future generations are often downplayed. If, however, there were official representatives to place those interests in the forefront of the debate, the attention of politicians and the general public could be captured. Fourthly, interest groups, such as the Sierra Club and Zero Population Growth, are not now, and not soon likely to be, capable of adequately defending the interests of future people, nor are these interests their sole concern.[8]

Having outlined two workable alternatives – adding a few congressmen, or a special department or agency, to represent future generations – we would like to comment on the public acceptability of these alternatives. The following is a false dilemma: that direct representation for future generations is either impossible (because people are not concerned about the well-being of future generations) or else superfluous (because people are concerned, and so they already indirectly represent the interests of future generations adequately). A middle ground may eventually emerge, wherein a sufficient number of people are sufficiently concerned to effectively institute direct representation, although many others still do not identify with, or think about, future generations.

To raise public awareness about the effects of our political decisions on future generations, to the level required for changing the political system as we have suggested, would take the combined efforts of concerned parents, educators, scientists, artists, etc., over a period of time. While the task is formidable, we should remember that, a mere decade ago, the public was hardly aware of environmental issues (concern about which largely overlaps with concern for future generations). We do think that direct representation of the interests of future people is needed, and would be workable in either of the forms we have proposed. But, we would be pleased indeed if, a generation from now, people cared so much about future generations, that indirect representation would suffice.

Notes

1 For example see, Richard and Val Routley, 'Nuclear Energy and Obligations to the Future', *Inquiry* 21 (1978): secs. 2–3. Cf. Gregory S. Kavka, 'The Futurity Problem', in *Obligations to Future Generations*, ed. Richard Sikora and Brian Barry (Philadelphia: Temple University Press, 1978).

2 For the purposes of discussing representation within a national political system, we shift focus from our *generation's* obligations to posterity, to a *nation's* obligations to its posterity (i.e., the obligations of a nation's present citizens to its future citizens).

3 Joel Feinberg, 'The Rights of Animals and Unborn Generations', in *Rights, Justice, and the Bounds of Liberty* (Princeton: Princeton University Press, 1980), pp. 159–84.

4 The possibility of basic human needs being changed by genetic engineering, or dependence on earthly resources being reduced by economic exploitation of space, cannot be ruled out in the very long run. But there will be many intervening generations dependent on earthly resources to supply their material needs, in any case.

5 Congressmen have the advantage of being able to monitor some of the effects of policies on their constituents and modify the policies accordingly. Representatives of future generations may gain further *evidence* about the likely effects of policies on future generations, but will not be able to directly observe the effects. This is one reason that representing future people would be more difficult than representing present people.

6 Assuming that they will exist. Of course, we need not be *certain* that they will exist, to have reason to act to promote their interests, anymore than a bachelor has to be certain of living to retirement in order to have reason to contribute to his pension fund. On the question of whether we have moral reasons for seeing to it that future people exist, see Kavka, 'The Futurity Problem', sec. 4, and Jonathan Bennett, 'On Maximizing Happiness', secs. 5–10, in *Future Generations*, ed. Sikora and Barry.

7 There is at least one exception. If the public intensely desired the protection of the interests of future generations, then having (and advertising) such a policy might increase sales.

8 Even if such groups represented future people only, this would not weaken the case for representation for future people within the government. After all, the existence of farmers' lobbies, for example, does not substitute for farmers having legislative representation.

Source: Gregory S. Kavka and Virginia Warren (1983) 'Political representation for future generations', in R. Elliot and A. Gare (eds) *Environmental Philosophy*, Milton Keynes: Open University Press.

2.7

John Blunden

Fresh water: a natural resource issue for the twenty-first century?

'Water, like energy in the 1970s, will probably become the most critical natural resource issue of the next century.' This somewhat apocalyptic comment was made in a keynote paper delivered to an international conference on fresh water and the environment held in Dublin in 1992 which attempted to survey the future of the world fresh water environment for the United Nations Environment Programme (Koudstaal *et al.*, 1992, p. 278). Any attempt to examine its validity needs to address a series of interrelated issues (see Figure 5), all of which have to be considered against a background of the needs of a growing world population.

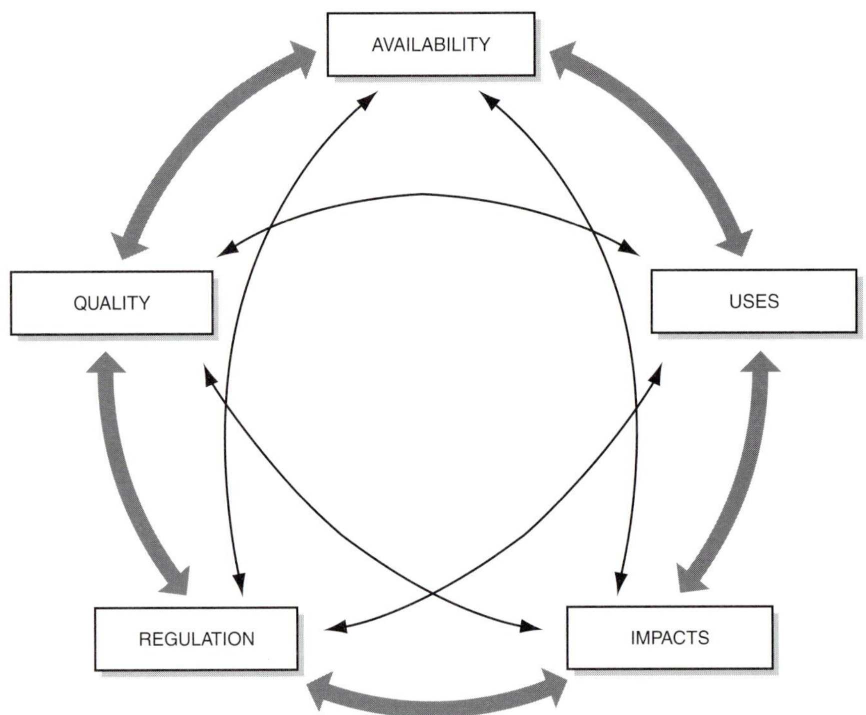

Figure 5 Fresh water: five issues to consider
Source: John Blunden, Open University, Milton Keynes

Availability

Perhaps the key issue is that of availability, in so far as the main source of fresh water, rainfall, does not always occur where it is needed. The largest body of fresh water, containing one-fifth of the world's available supplies in an area roughly the size of Belgium, is Lake Baikal. But this is in Siberia, remote from centres of population and 'off the map'. Moreover, while Iceland gets enough rain every year to fill a small reservoir for each member of its 250,000 inhabitants, Kuwait, with seven times the number of people, scarcely receives any at all. Its only redress is to employ a plant which desalinates sea-water, even though operations of this type can cost up to ten times the cost of fresh water from conventional sources. Nevertheless, the Kuwait operation is now but one of the 7,500 desalination plants in operation around the world, although in total, they only supply one-thousandth of its consumption of fresh water.

Present uses

Self-evidently, the desire for water in sufficient quantities, at least for the purposes of agriculture, is greatest where its availability cannot be assured by adequate rainfall at the right season. Here crop production has to be supported by irrigation which, in fact, makes use of around 75 per cent of the world's fresh water resources. The reason for this is not merely the prevalence of irrigated agriculture, but the notorious inefficiency of most surface irrigation systems with around half the water lost by seepage or evaporation even before it reaches the fields. Thus, whilst the production of one kilogram of wheat when fed by normal rainfall requires about 500 litres of water to reach maturity, one kilogram of irrigated rice uses over 1,800 litres.

The demands made by industry for water in the developed countries are also considerable. Where the concentration of industry is particularly apparent, regional water shortages, even in countries which are generally amply supplied, have led to major catchment developments, some of which involve inter-basin water transfers. In the north-east of England, for example, the head waters of the River Tyne, which encompass Keilder, the largest reservoir in Europe, have been partly diverted by tunnel southwards into the River Tees to help supply the major industrial developments in Cleveland. The magnitude of such schemes is perhaps less surprising when the needs of particular manufacturing processes are appreciated. For example, the manufacture of one tonne of cement requires 3,600 litres of water, one tonne of steel 8,000 to 12,000 litres, one tonne of paper over 27,000 litres, and a car 38,000 litres. Even 'a tonne of bread produced industrially consumes 2,100 to 4,200 litres'. However, 'in the UK 1,000 litres of beer has needed up to 4,200 litres of water for its production. In the USA the figure for beer appears to be larger at 15,000 litres of water. American beer is either cleaner or weaker than its British equivalent we may suppose' (Simmons, 1989, p. 274). Notwithstanding this minor puzzle, the evidence certainly suggests that, amongst the industrial countries, the USA has the greatest per capita consumption of water. Thus, although Phoenix, Arizona, has the same amount of rainfall as Lodwar in the north of Kenya, it uses 20 times as much water per head. Even its domestic consumption of water for washing, cooking and the flushing away of wastes is four times greater than that of the UK at over 600 litres a day. Only in the luxury hotels of the EU is such a figure reached in Europe.

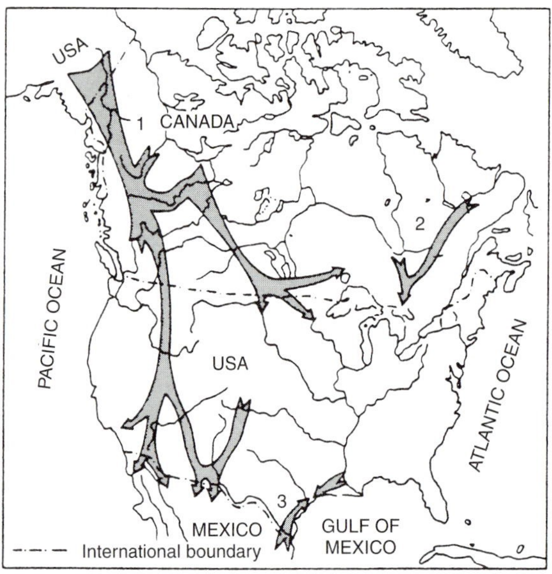

Figure 6 An engineering *'folie de grandeur'*? Three major North American projects for long-distance inter-basin water transfer from regions of abundance to those of shortage. (1) North American Water and Power Alliance (NAWAPA); (2) Grand Canal; (3) Texas River Basins.

Note: None of these schemes has been realized, partly because such transfers can have dramatic environmental repercussions. The North American Water and Power Alliance suggested a plan for taking what is called 'unused' water from the northern rivers of Canada and feeding it as far as the Great Lakes, New York, Los Angeles and Mexico. Nearly all of the western rivers of North America would be reservoirs with, as its main artery, a flooded Rocky Mountain trench in British Columbia and Montana. Engineering problems apart and the question of intrusion into the most scenic of recreation areas of the North American west, there would not be a wild river left and much of their associated habitats would also be destroyed.

Source: Goudie (1986: 156) based on Shiklomanov (1985, figs 12.6, 12.9 and 12.11).

However, in the 1990s even the USA has recognized that further growth in levels of fresh water consumption is not sustainable and that the use made of this flow resource, at least in certain regions, such as the South-West, must be brought under control through the imposition of rigorous management regimes (United States National Research Council, 1992). The time is well past when the profligate use of river water could be supplemented by pumping supplies from deep-lying aquifers such as the Ogallala, which stretches under eight Great Plains states from South Dakota to Texas. Once containing as much water as Lake Superior, it is said that one out of every six ears of grain grown in the USA depended on it. But this stock source of fossil water, like others, is now seriously depleted and at present levels of use the Ogallala can last no more than a further 20 years. The replenishment of an aquifer like this can take many centuries. Soon, the USA will be totally dependent on flow water resources, which will need careful management in the interests of meeting both basic human needs and those of the environment.

In some parts of southern California this is beginning to happen. Economical irrigation methods which involve the use of sprinklers or drip methods are now being put in place, increasing efficiency by up to 40 per cent; the largest domestic supplier in the region has at last begun a programme of demand-side management by more than doubling prices since the beginning of the 1990s; and there are city authorities that now demand that the housing stock's water using appliances incorporate water-saving technologies. However, in many parts of the South-West water consumers are living on borrowed time.

Quality and regulation

Unfortunately, some problems of water distribution and use are further aggravated by pollution since, for every litre of dirty waste water that is discharged into a lake or river, many more become, to some extent, contaminated. Indeed, the World Health Organization in a report suggests that as many as 10 per cent of the world's rivers that have been monitored are polluted (World Health Organization, 1992–93). In the USA, a developed country with a mature industrial manufacturing base, all of its water-using factories discharge considerable quantities of liquid waste. Most of this effluent passes into rivers or lakes where, in the course of the breakdown of its organic content through natural biological processes, it consumes the oxygen present in the receiving water body. This is sometimes to the detriment of aquatic life or sometimes its complete destruction. However, much of the waste is in the form of chemicals which, when discharged to other water bodies, can create a hostile environment of considerable local toxicity to both plants and animals unless dilution is very rapid.

These high levels of effluent discharge prevail in spite of the 1972 Federal Water Pollution Control Amendments Act, which took over responsibility for water quality from the individual states and provided a comprehensive set of powers to attack both industrial and municipal polluters. Although 1985 was a target date for the achievement of clean effluent discharges, this has not been realized. Amending legislation in 1977, weakening the powers of the original Act; a series of legal challenges brought by industrial interest groups; and cuts in the budgets and staffing of the Environmental Protection Agency, during the Reagan presidency, have meant that the present standards fall well short of those originally envisaged.

Other developed countries have enacted legislation to control the quality of effluent discharged to water bodies – the European Union, for example, operates a system of uniform emission limits which it is attempting to tighten. However, only rarely do industries either totally recycle or purify their process water. As they have successfully claimed in the USA, the cost is too great. In the meantime, the European Commission remains frustrated with progress over pollution control. Indeed, it has expressed 'disappointment that the state of the aquatic environment in the European Community has not improved to the extent expected', and argues that future economic growth in the Union will be constrained by the contamination of fresh water, not only from industrial, but also from agricultural sources (Ecotec, 1992).

As for Europe at large, one of the worst records for the discharge of liquid industrial effluent is to be found in Poland, where about 75 per cent of its rivers are now too contaminated even for further industrial use. Where Russia is concerned, it was revealed in a report published in April 1993 that three-quarters of its surface water was unfit to drink and a third of its underground water supplies contaminated. As the report in question put it: 'water is a prime victim of decades of the abuse of Russia's environment' (Boulton, 1993). Equally discouraging is the situation in the developing world where industrialization has occurred. More than two-thirds of China's rivers are seriously polluted, while 40 of Malaysia's rivers are reported to be biologically dead.

Certainly, both in the developed and developing countries, the contamination of water supplies by agricultural chemicals used as fertilizers – such as nitrogen, phosphate and potassium – is becoming increasingly serious. The problem is that these are, more often than not, delivered in such quantities that the plants that make up the field crop can rarely use all of them. This situation particularly prevails in those regions of the developed countries where large-scale grain production is practised and in the developing world in those areas where the 'green revolution', based as it is on the use of high-yielding strains of rice, has brought about substantial increases in production.

Applications of artificial fertilizer to field crops grown by high output farmers in Europe frequently result in 40 to 50 per cent of the nitrates entering the run-off waters. The same is true of phosphorus, but to a much lesser extent since most of it found in fresh water arises from the discharge of sewage and detergents. The addition of these nutrients can produce a water body that rapidly suffers eutrophication. This is characterized by the presence of algal blooms whose decay robs the water of oxygen. Evidence of the growing environmental impact of nitrates during the 1970s and 1980s has been found around the world. Concentrations in the River Tomo in Japan rose during these two decades by 14 per cent, in the Rhine by 29 per cent, and in the River Wear in England by nearly 50 per cent. Nitrogenous agricultural fertilizers also appear to be the main source of nitrates present in the ground waters used to supply regional domestic consumers. At concentrations above 10 to 12 milligrams a litre, they can cause health problems, especially to bottle-fed babies. This raises a more important general point concerning the kind of effects that water pollutants are having on the present generation of people, especially those in the developing countries.

Present impacts and future options

United Nations (UN) agencies have reported the presence of organochlorine pesticides in rivers in agricultural areas of developing countries at levels considerably higher than in European rivers. In Colombia, Malaysia and Tanzania, levels have been such that their ingestion by fish and their concentration within the fish presents a health hazard to those who consume these as a major part of their diet. However, in the developing countries water is often additionally affected by organisms, some of them derived from sewage, which give rise to a range of diseases. These include typhoid, food poisoning and hepatitis. Water diseases present a particular problem for women. Because of their domestic role, which includes their function as water collectors and the part they play in producing the staple foods for their families, women as a group are the most vulnerable to water-related illness. This was recognized by the United Nations Committee for the Implementation of the International Drinking Water Supply and Sanitation Decade, which not only paid particular attention to the vulnerability of women in this respect, but also incorporated the UN International Research and Training Institute for the Advancement of Women (INSTRAW) in the membership of the Committee. According to Pietila and Vickers (1990), a task force on women, water supply and sanitation has since been established to provide guidelines on future policies in this field with regard to their full participation in attempts to resolve these problems.

Not surprisingly, the problems of water contamination remain high on the agenda in spite of UN initiatives. It is estimated that contamination of this kind, which still affects 25 per cent of the world's population, is a major factor in more than five million deaths a year. Thus the provision of clean drinking water is a major task for many developing countries – in Tanzania alone this involves 20 million rural inhabitants – and any attempts to achieve such an end can mean competition with the provision of water for irrigated agriculture. Certainly the magnitude of the contaminated water problem is brought home by the realization that if all of this water could be purified, it would be equal to half the current demands of the industrial nations.

Meanwhile, demand for fresh water is rising, linked as it is with population growth in developing countries and the spread of agriculture. UNEP has been reported as stating that the world's use of fresh water has 'increased nearly fourfold in the last 50 years to 4,130 cubic km a year' (Maddox, 1993). In using three-quarters of this total, the area of irrigated land has increased more than a third in the last two decades. The growth in Asian demand for water is the fastest. In 1993 it used just over half the world's available fresh water, but by the year 2000 UNEP expects this to rise to nearly two-thirds, mostly to support irrigation.

However, the quantitative significance of water used for irrigation should not, perhaps, overshadow the need for the demand-side management of water, which can be effectively employed even in the short term both in relation to manufacturing industry and in the domestic context (see Table 1). Such measures may have considerable regional significance, especially those which enjoy a high degree of economic development. Similarly, the problems of water pollution need to be addressed. As the Director of the UN Natural Resources and Energy Division made clear at the United Nations Conference on Environment and Development in June 1992, it is also necessary that 'far reaching measures . . . be taken to preserve the quality of water bodies . . . involving new approaches, such as integrated pollution control, inside an international framework for effective collaboration' (Pastizzi-Ferenic, 1992, pp. 6–7).

Without such progress in fresh water management, and given the likely increases in population by 2020, it is expected that another 40 nations will have joined the 26 nations, mainly in Africa and the Middle East, that already have inadequate fresh water resources. The number of people thus affected would be expected to grow tenfold from the present 300 million to 3 billion – one-third of the projected population of the world.

Summary

- If the world's needs for water are to be met adequately in the next century, both water resources and water demand will need to be managed.
- At present there is no single water crisis operating at a global level, only a series of looming crises scattered amongst different regions, some in the developed world, but predominantly and most significantly in the developing world.
- Attempts to come to terms with water shortages have often resulted in over-pumping and mining of ground water resources.

Table 1 Responding to the looming water crisis

Increasing levels of difficulty in meeting regional needs	What can be done in the short term?	What can be done in the medium to longer term?	What can be done in the very long term?
Use of deep-lying aquifers leading to resource depletion	Nothing – total depletion. Water can only be replenished over hundreds of years		
Rivers and lake waters which are threatened by			
(1) excessive water extraction	Legislation to keep water at levels which maintain aquatic biota		
(2) agricultural pollution from fertilizers and pesticides	Removal of subsidies favouring industrial farming systems	Use of costly 'stripping' techniques to remove nitrogen from ground water. Use of computers and satellite guidance to control fertilizer and pesticide applications to match precisely plant needs	
(3) industrial pollution from waste discharges	Tightening of treatment standards with rigorous enforcement procedures	Development of new technologies with lower pollution propensity	
(4) inadequate sewage treatment	Improved investment in standard technology at sewage works	Domestic recycling	
Problems of water deficient regions aggravated by:			
(1) irrigation requirements	Replacement of open channels with sprinkler and drip techniques designed to deliver precise quantities to each plant	Transfer to dry-farming methods	Development of: • low-cost (high-)volume desalination plants
(2) high levels of industrial activity	All water treated and recycled		• inter-basin water transfers
(3) increasing household demands	Water metering Domestic recycling Design and installation of household equipment with low water consumption levels	Separation of household supplies into potable and non-potable sources	• establishment of national water grids

- Pollution from industries, cities and agriculture is prevalent and remains outside attempts to make the necessary link between economic development, properly managed water resources, and social and environmental well-being.
- Probably the single most significant danger for many regions will be that of resource management failure since this could push water towards a scarcity value and an emotional intensity resembling that of oil in the period after 1974.

References

Boulton, L. (1993) 'An unhealthy drink for a nation', *Financial Times*, 7 April.

Ecotec (1992) Research and Technological Development for the Supply and Use of Freshwater Resources, ref. no. EUR 14723, Luxembourg, EC Office of Publications.

Goudie, A. (1986) *The Human Impact on the Natural Environment*, 2nd edition, Oxford: Basil Blackwell.

Koudstaal, R., Rijsberman, F. and Savenije, H. (1992) 'Water and sustainable development', abridged from a report for the International Conference on Water and the Environment – Development Issues in the 21st Century, Dublin, January 1992, *National Resources Forum*, Vol. 16, No. 4, pp. 277–90.

Maddox, B. (1993) 'The world's tap seizes up', *Financial Times*, 17 March.

Pastizzi-Ferenic, D. (1992) 'Natural resources and environmentally sustainable development', *National Resources Forum*, Vol. 16, No. 1, pp. 3–10.

Pietila, H. and Vickers, J. (1990) *Making Women Matter: the Role of the United Nations*, London: Zed Books.

Shiklomanov, I. A. (1985) 'Large-Scale water transfers' in J. C. Rodda (ed.) *Facets of Hydrology II*, Chichester: John Wiley & Sons.

Simmons, I. G. (1989) *Changing the Face of the Earth – Culture, Environment, History*, Oxford: Blackwell.

US National Research Council (1992) *Water Transfers in the West: Efficiency, Equity and Environment*, Washington, DC: National Academy Press.

World Health Organization (1992–93) *Bi-annual Report of the Director General to the World Health Assembly and the United Nations*, Geneva: WHO.

Source: Newly commissioned.

2.8

David Humphreys

Forest policy: justice within and between generations

Forests can be found in almost every geographical region of the world. Broadly speaking there are three forested zones. Boreal forests are found in climatic zones with snowy winters and short summers; these principally coniferous forests grow in the far north of the northern hemisphere in the region of the Arctic Circle. Temperate forests are to be found in the climatic zones between the polar and the tropical; most of the world's temperate forests are held by Russia, the United States and Canada. With the exception of a small area of Queensland, Australia, all the world's tropical forests are to be found in the tropical latitudes of Latin America, Africa and Asia. Within these three climatic zones a diverse range of different ecosystems can be found.

Why do we value forests?

Forests provide a diverse range of ecological, social and economic goods and services. Forests supply products such as timber, rattan, fruits, nuts, berries and rubber, and ecological services such as soil conservation (deforested land is more prone to flooding and erosion) and watershed management (streams and rivers frequently dry up after deforestation has occurred). Most of the world's biological diversity is to be found in tropical forests, including many plants of medicinal value to humans. In addition to biological diversity, forest conservation helps to maintain cultural diversity, with forests serving as customary ancestral habitat for many indigenous and tribal peoples, especially in the tropics. Forests also play a role in climatic regulation. Forests are an important sink of carbon dioxide, one of the principal greenhouse gases, which is taken up from the atmosphere by photosynthesis when trees grow. However, when deforestation occurs due to forest burning, or when trees are felled and left to rot, carbon dioxide is released into the atmosphere, thus contributing to global warming. In principle, if forests are sustainably managed over generations, they can provide a small, but steady, return of these goods and services in perpetuity.

What are the forces behind deforestation?

Deforestation may be said to occur for two broad reasons. First, many actors with a stake in forest use do not use the forest sustainably. This is particularly the case with timber companies, most of which extract timber at a rate faster than the forest can naturally regenerate. Indeed some timber companies practise clear felling of forests, that is they fell all trees in a particular area, making no attempt to practise selective logging of mature trees. Furthermore, in order to extract timber from the forests, timber companies often build roads, which act as a migratory channel for the rural poor to enter the forest in search of land for agriculture.

The second reason why deforestation occurs is that not only do forests provide goods desired by people, but so too does deforested land. For example, deforestation releases land for agriculture, both for the landless poor from outside the forest in search of land for subsistence agriculture, as well as for commercial agriculturalists. In parts of Latin America, land is deforested to graze cattle for the northern beef market, the so-called 'hamburger connection'. The forest fires that have ravaged southern Borneo and the Brazilian Amazon in recent years have occurred partly owing to the search for land for agriculture by the rural poor, although it must be stressed that far more forest has been burnt by companies clearing land for commercial agriculture, to cover up timber poaching and even for insurance fraud (World Wide Fund for Nature 1997). In Ecuador and Nigeria, deforestation has occurred as a result of oil exploration by transnational corporations. It is not just legally sanctioned trade which causes deforestation; in parts of Southeast Asia and South America forests have been cleared to grow poppies and coca plants for the illegal drugs trade (International Narcotics Control Board 1992). There are therefore numerous potentialities for conflict between actors over the use to which forests, or alternatively deforested land, should be put. It is this conflict that lies at the heart of forest use as a political issue. Deforestation occurs because actors with a stake in unsustainable forest use or in forest destruction have more economic and political power than actors with a stake in sustainable forest use or in forest conservation.

Frameworks for forest policy

Forest policy may be defined as policy that affects an area of forest. This definition encompasses a wide range of actors, with some policy-makers seeking to conserve an area of forest, while others exploit forests in an unsustainable manner, destroying or degrading a forested area. The actors involved in forest policy will inevitably vary according to the country involved, the nature of its government, the dependence of its economy on forest exploitation, and the type of forest involved. A further variable will be the level at which policy is made; the actors involved in policy-making will depend on whether policy is made at the international, regional, national or local level.

The international tropical timber trade

At the international level, most policy statements and principles are negotiated by governments. One such example is the International Tropical Timber Organization (ITTO), a United Nations forum of the world's leading tropical timber producers and consumers. Nominally the forum is an intergovernmental one, although two other groups of actors are involved. First, individuals from conservation non-governmental organizations (NGOs) may attend as observers and in some cases have been appointed to national delegations to advise on conservation issues. Second, many timber industrialists attend the ITTO as observers. Timber industrialists have also gained entry to national delegations, indeed in far greater numbers than the conservation NGOs. The close relationship between governments and timber trade organizations, especially in Southeast Asia, is especially visible at the ITTO (Humphreys 1996).

For example, the delegations to the ITTO of two leading tropical timber-producing countries, namely the Philippines and Indonesia, have been led not by government ministers or civil servants, but by leading timber industrialists. Timber industrialists have also funded many ITTO forest development projects and have been elected to chairs of ITTO committees. The influence of the timber trade at the ITTO is thus far greater than the influence of the conservation NGOs. Nominally therefore the ITTO is an intergovernmental forum, although the structure of interests represented is far wider than purely governments, with the organization bestowing a legitimacy upon timber trade interests (Lee, Humphreys and Pugh 1997). Consequently many conservation NGOs have, since the early 1990s, ceased attending the ITTO in protest at the organization's perceived bias in favour of the trade at the expense of conservation.

In order to understand how effective forest conservation can be achieved, it is necessary to consider the question of justice. Three dimensions to justice can be discerned in forest policy-making: justice between countries; justice within countries; and justice between generations. These will now be considered in turn.

Justice between countries

Here it is instructive briefly to consider two rounds of intergovernmental negotiations that sought to establish a consensus for a forest convention. These negotiations took place in the preparatory discussions prior to the United Nations Conference on Environment and Development (1990–92) and under the auspices of the Intergovernmental Panel on Forests (1995–97). These negotiations were essentially polarized along North–South lines, with the North represented by the European Union, Canada, the United States and Japan, and the South represented by the Group of 77 Developing Countries (G77). The different views of North and South with respect to rights and duties were central to the negotiations. The North has sought to frame forest conservation as an issue where the South has a moral duty to conserve the world's tropical forests for the good of all humanity. The South has responded that it has the right, enshrined in international law, to use its forests in line with national development policies.

One of the arguments made by the South is that the North, which has chopped down most of its forests and is consuming most timber felled in the South, bears an additional responsibility for global forest destruction. It is therefore, argues the South, not only unjust, but an assumption of supranational rights for the North to attempt to stipulate that the South should conserve its forests. In response to the view of some Northern countries that forests should be seen as a 'global common' in which all humanity has a stake, Southern countries have asserted that the North has a moral duty to provide compensation to the South for the opportunity cost forgone if the South is to desist from exploiting its forest resources. For example, if the South were to agree to conserve its forests, it would lose the foreign exchange that would accrue from forest development. Claims for compensation have taken the form of demands by the South for external debt relief, technology transfers from North to South, and increased aid flows. The principal reason why, after two rounds of negotiations in the 1990s, negotiations have not been launched for a forest convention is that the North has not been prepared to meet these forest-related demands of the South.

Justice within countries

North–South negotiations involve principally governments, and the underlying assumption of such negotiations is that a government may speak for all its citizens. However, and as we have seen in the case of the ITTO, governments may align themselves with some interest groups at the expense of others. In many cases indigenous forest peoples have suffered loss of land at the hands of powerful political and economic interests from outside the forests. In Brazil and Sarawak, to name just two examples, forest burning and logging have robbed indigenous peoples of their tribal homelands. While this policy may benefit the national economy and may help to promote national development, the cost in terms of cultural destruction and the loss of the habitat of humans and other species has been huge.

Indigenous forest peoples are gradually becoming better organized in response to the threats to their way of life posed by outside interests. In 1992 indigenous peoples from all the world's main tropical forested regions formulated a joint declaration in Penang which asserted '[r]espect for our autonomous forms of self-government, as differentiated political systems at the community, regional and other levels' (World Rainforest Movement 1992). In response to claims that forests should be seen as a 'global common' as some Northern governments have asserted, or that they are a 'national resource' as the G77 claim, indigenous forest peoples and other forest communities assert that forests should be seen as a 'local common'. In the words of the 1992 Penang declaration, 'we declare that we are the original peoples, the rightful owners and the cultures that defend the tropical forests of the world' (World Rainforest Movement 1992). Most indigenous peoples wish to be left alone to practise their traditional lifestyles. Even the possibility of their receiving financial compensation for the destruction of their forests is an unattractive proposition for most indigenous peoples, as money is seen to be as destructive of traditional lifestyles every bit as much as deforestation.

Justice between generations

As with other environmental issues, long-term beneficiaries are rarely present when decisions are made, and future generations are never present. As noted above, forests fill a diverse range of social and economic functions. If there is to be justice between generations, all forest values must be passed on to future generations more or less intact. However, the policy of some actors has been to conserve some forest values while destroying others. Let us consider the question of plantations. Some timber industrialists may clear an area of forest and replace it with a plantation of (in economic terms) high-value timber species. Plantations may successfully pass on to future generations certain forest values; future generations may receive the same carbon dioxide sink capacity and stock of timber as their predecessors inherited. However, the fact that plantations are usually composed of just one timber species (so called monocultures) means that they are not ecologically representative of natural forest cover: they will be unable to support the same level of biodiversity as natural forests; they will be unable to provide the same returns of non-timber products such as fruits and nuts; and they will not serve the same cultural and spiritual functions as natural forests. Hence while plantations may be a satisfactory policy response with respect to some aspects of intergenerational equity, they are incapable of satisfying a broad notion of intergenerational equity that passes on to the next generation all values found in a given area of forest.

In effect therefore a plantation policy best serves the interests of timber industrialists at the expense of future generations. The immediate successors of today's plantation owners will *ceteris paribus* inherit a financially healthy company. They thus effectively benefit from the policy of their predecessors by which the ecological capital of future generations was converted into the private capital of a business concern. However the immediate successors of the present generation (and this includes those with a stake in the plantation company in question) will inherit an area of degraded forest land that is devoid of those other forest values that have been lost by the replacement of natural forest by plantations.

Conclusion

Those with the greatest impacts upon the forests invariably tend to be powerful outside interests, such as industrial companies backed by economic and political power who displace the traditional forms of land control of local communities and indigenous peoples. Migrating landless farmers may also enter the forest in search of land for subsistence agriculture if they have lost, or been evicted from, land elsewhere. Deforestation is therefore the result of incursions into the forest by both the powerful who displace and the powerless who are displaced. As such, a recognition of the rights and concerns of disempowered local communities is not only an environmental imperative, it is also one of social justice within the present generation. Only when the question of injustice within generations has been addressed, so that those with a stake in forest conservation or in sustainable forest use have a greater influence in the policy-making process than actors with a stake in short-sighted and ecologically unsustainable economic exploitation, will the present generation be able to bequeath to its successors those forest values it inherited, thus satisfying the imperative of intergenerational justice.

References

Humphreys, D. (1996) *Forest Politics: The Evolution of International Cooperation* (London: Earthscan).

International Narcotics Control Board (1992) *Report of the International Narcotics Control Board for 1992* (New York: United Nations).

Lee, K., Humphreys, D. and Pugh, M. (1997) 'Privatisation in the United Nations System: Patterns of Influence in Three Intergovernmental Organisations', *Global Society: Journal of Interdisciplinary International Relations*, vol. 11, no. 3, pp. 339–57.

World Rainforest Movement (1992) 'Charter of the Indigenous Tribal Peoples of the Tropical Forests', Penang, Malaysia, 15 February.

World Wide Fund for Nature (1997) *The Year the World Caught Fire* (Godalming: World Wide Fund for Nature).

Source: Newly commissioned.

2.9

Andrew Blowers

Radioactive waste: an inescapable legacy

A source of fear

Radioactivity engenders more fear than perhaps any other hazard of the modern world. The dangers from routine, deliberate or accidental releases are well known. They include short-term radiation sickness, the long-term chronic effects of exposure resulting in cancers, reproductive failure, birth or genetic defects, and the wholesale contamination and destruction resulting from reactor melt-down or the devastation wrought by a nuclear war.

The fear of radioactivity does not discriminate between statistical levels of risk or probability of danger; it is perceived as a risk whatever its source. Radioactivity is recognized as an involuntary but irreversible risk, moreover one that is socially-induced and, therefore, to a degree avoidable. The risk is pervasive and persistent.

The nuclear industry had its origins in warfare, the awesome destructive power of the bomb demonstrated in the appalling destruction of Nagasaki and Hiroshima. Later nuclear energy was promoted as a clean, safe and seemingly inexhaustible source of electricity ('too cheap to meter'). Confidence was eroded as costs mounted and accidents occurred culminating in the disaster at Chernobyl in 1986. By the 1980s, attention had turned to the rear-end of the nuclear cycle, the problem of safely managing the accumulation of wastes that are the inevitable consequence of the nuclear programme. Radioactive wastes present a unique and intractable problem for contemporary society – how to protect the environment and life for a period extending over an immeasurable period of time.

Nuclear waste: the scale of the problem

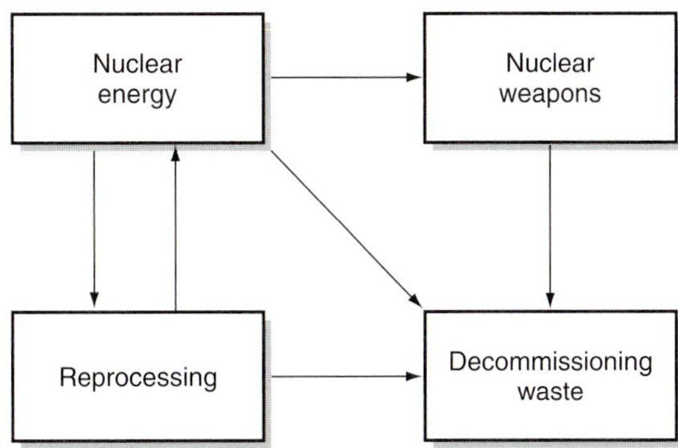

Figure 7 Producing nuclear waste

1 *Nuclear weapons.* Seven nations – the USA, Russia, the UK, France, India, Pakistan and China – have developed the capacity for making nuclear warheads. In addition it is widely known that some other states, notably Israel, possess this capability and several other states have achieved the technical basis for making nuclear weapons. Moreover, the risk of clandestine trade in nuclear materials for bomb making is considerable, particularly as a result of the break-up of the Soviet Union. With the ending of the Cold War there is the problem of dealing with the nuclear stockpile and the clean-up of test sites and plants formerly engaged in weapons manufacture.

2 *Nuclear energy.* There are over thirty nations which have commercial nuclear power plants (see Table 2). The largest concentration of nuclear reactors is in Europe (including the former Soviet Union) with over 200 reactors and North America with 130 (USA 109). Outside these areas the most important producers are found in Japan (51 reactors), Korea (11) and India (10). The leading producers of nuclear electricity are the USA, France and Japan followed by Germany, Russia, Canada, the Ukraine and the UK. There is very little nuclear production in Africa, Australia or Latin America. In most countries the

Table 2 Nuclear power status around the world

	In operation		*Under construction*
	No. of units	*Total net MWe*	*No. of units*
Argentina	2	935	1
Armenia	1	376	
Belgium	7	5,527	
Brazil	1	626	1
Bulgaria	6	3,538	
Canada	21	14,907	
China	3	2,167	
Czech Republic	4	1,648	2
Finland	4	2,310	
France	56	58,493	4
Germany	20	22,017	
Hungary	4	1,729	
India	10	1,695	4
Iran			2
Japan	51	39,917	3
Kazakhstan	1	70	
Korea, Rep. of	11	9,120	5
Lithuania	2	2,370	
Mexico	2	1,308	
Netherlands	2	504	
Pakistan	1	125	1
Romania			2
Russian Federation	29	19,843	4
South Africa	2	1,842	
Slovak Republic	4	1,632	4
Slovenia	1	632	
Spain	9	7,124	
Sweden	12	10,002	
Switzerland	5	3,050	
United Kingdom	35	12,908	
Ukraine	16	13,629	5
United States	109	98,784	1
World total*	437	343,712	39

* The total includes Taiwan, China where six reactors totalling 4884 MWe are in operation.

Notes to table: Data are subject to revision. During 1995 two reactors were shut down (including Bruce-2 in Canada, which could restart in the future); seven reactors were connected to the grid and the construction of three reactors was temporarily suspended (in Romania).

Source: *IAEA* Bulletin, 1/1996, p. 53

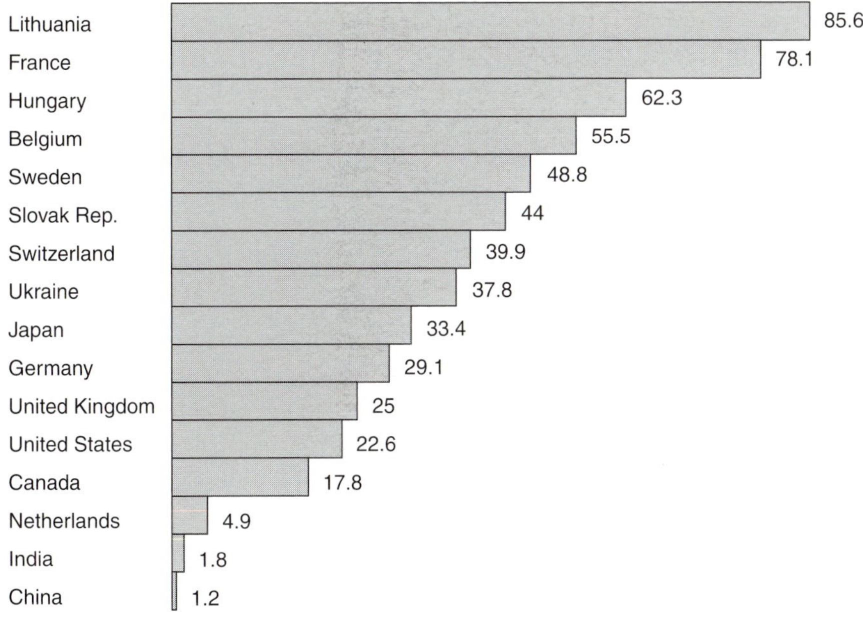

Figure 8 Nuclear share of electricity generation in selected countries
Source: Adapted from *IAEA Bulletin*, 1/1996, p. 53

development of nuclear energy has ceased though there is still growth in South East Asia with the prospect of future development in countries such as Indonesia. All commissioned reactors create high level waste (HLW) in the form of spent fuel, intermediate level wastes (ILW) including fuel cladding, control rods, filters, sludges and resins from reactor vessels and large volumes of low level wastes (LLW) arising from contamination of materials during nuclear processes.

3 *Reprocessing*. Some countries are engaged in reprocessing of spent fuel to recover uranium and plutonium. Most military reprocessing operations have ceased, notably in the United States and Russia but reprocessing for civil purposes continues, notably in France and the UK giving rise to a substantial trade in nuclear materials. Reprocessing results in large and complex volumes of nuclear wastes.

4 *Decommissioning*. There remains the problem of what to do with the waste from all the nuclear operations in various sites across the world. The legacy is inescapable. It is a characteristic of radionuclides that they decay over time but the period of decay (measured in half-lives) varies widely. While much of the voluminous low level and short-lived intermediate waste with short half-lives will decay to safe levels within a foreseeable future, there will remain highly dangerous radionuclides with half-lives extending in some cases to hundreds of thousands and even millions of years. Plutonium, one of the most dangerous substances known, has a half-life of 24,000 years and will remain dangerous for an unimaginable time-span. It is calculated that the total world inventory of plutonium is about 1,239 tonnes (Royal Society, 1998) of which around 100 tonnes is in the UK and the global total is currently growing at 50 tonnes per year.

The search for a technical solution

LLW and some ILW is managed in most countries in shallow or surface repositories. But for the longer-lived ILW and HLW there is no permanent solution. HLW is either kept as spent fuel at power stations or cooled and transformed into solid form through vitrification. In the UK the bulk of HLW and ILW wastes are kept in stores at Sellafield (with smaller volumes at Dounreay). The early Magnox reactors in the UK produce much larger volumes of spent fuel than later designs which results in the UK having a waste burden second only to the USA amounting to nearly 17 per cent of the global total (see Figure 9).

There is a broad consensus that the ultimate solution for these longer-lived wastes is deep geological disposal. As long ago as 1976 the Flowers Report (RCEP) pronounced that any expansion of the nuclear programme should be dependent on demonstrating that safe containment of long-lived highly radioactive wastes could be achieved. This inaugurated a frantic search for possible disposal sites in the UK. One by one the options, both on land and at sea, were foreclosed as opposition barred the way, leaving the industry's bastion at Sellafield as the site chosen for a potential repository for IL/LLW (Blowers *et al.*, 1991). While Sellafield looked the best political option, it had to be proved to be technically acceptable in terms of long-term safety. But, after an extensive programme of investigation, a proposal for an underground laboratory at the site was turned down in 1997 by John Gummer, Secretary of State, who found the proposal 'seriously premature' and was 'concerned about the scientific uncertainties and technical deficiencies in the proposals'. Thus, a twenty-year search for sites had proved abortive and the UK has to start all over again.

Similar frustration in the search for permanent repositories has occurred elsewhere. In

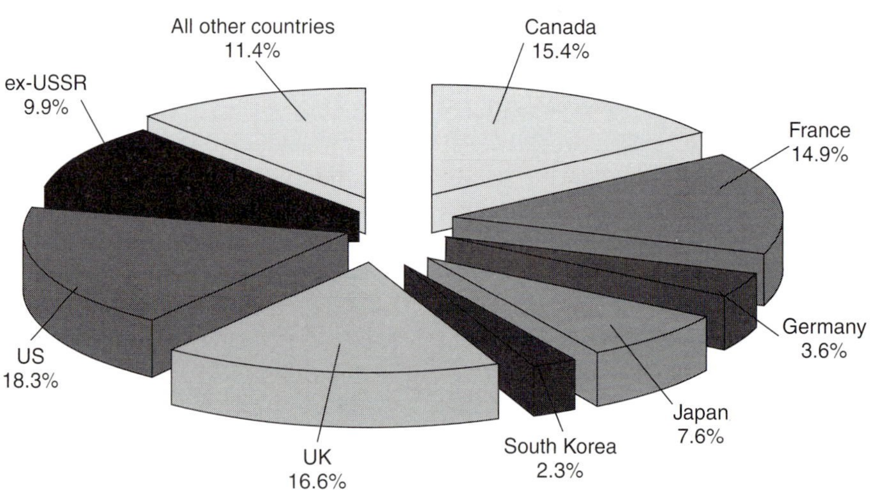

Figure 9 Nuclear waste worldwide. Lifetime spent fuel accumulations of reactors operating or under construction in 1995, based on International Atomic Energy Agency estimates. The total amount is estimated to reach 447,000 metric tons of heavy metal
Source: *Physics Today*, June 1997, p. 57

the United States, the search for a HLW disposal site has been narrowed down to Yucca Mountain in the Nevada Desert, where both political resistance and technical uncertainty have yet to be overcome. In Germany there have been fierce protests against shipments to the interim storage site at Gorleben and against the proposed disposal site in a nearby salt dome. France and Spain have embarked on a process of seeking a site acceptable to the local population without any certainty of success. In Sweden there is a central storage site for spent fuel and various options for deep disposal have run into protest by potential host communities. Proposals for storage or disposal sites have also been opposed in Japan, Switzerland and other countries.

So far the only repositories in operation are for short-lived ILW and LLW (for example, under the Baltic Sea coast in Sweden, above ground in France and Spain, sub-surface in the United States and UK), while in New Mexico in the US there is a deep disposal site intended initially for retrievable defence wastes. Despite the scale and urgency of the problem, everywhere the industry has been confronted with technical problems and political resistance in its search for host communities.

Nuclear oases – bearing the burden

The search for sites has consistently focused on 'nuclear oases', communities where resistance is likely to be weakest. They are peripheral communities in the sense of being economically monocultural, politically powerless, culturally backward, environmentally degraded and relatively remote (Blowers and Leroy, 1994). Some, like Sellafield, are already dependent on the nuclear industry; others, Gorleben for example, are emerging centres for nuclear waste; others are greenfield locations where former industries are in decline (e.g. the phosphate mining centre of Carlsbad, New Mexico) or areas with a history of environmental degradation, such as the Nevada Test site where Yucca Mountain is located. Even in these disadvantaged areas – perceived to be the most vulnerable to persuasion – protest has been vigorous from local groups fearful of the dangers to present and future generations and from environmental groups committed to closing down the nuclear industry.

In an effort to achieve more equitable solutions, some countries (for example, the USA, France, Canada and Switzerland) have sought volunteer host communities by offering various forms of compensation such as tax breaks, investment, economic regeneration, infrastructure development and participation in decision-making. But, such incentives can only be considered equitable in an unequal situation. Compensation can only be considered fair if all options are considered and incentives applied not to attract volunteers, but as a necessary compensation for the disruption and blight experienced by the community that is finally selected.

Provision for meeting the costs of decommissioning and restoration of sites is met by discounting. In principle, discounting is a calculation of what needs to be invested now to achieve a specific sum at a specified future date. The investment will depend on the discount rate (percentage of present value of the costs) and time period chosen. The problem is that discounting over long time periods, such as 100 years, tends to undervalue the future costs – in other words, it transfers costs from the present to the future without any guarantee of the presumed benefits of the nuclear industry.

Sustainable development and future generations

The criterion for sustainable development requires that '[r]adioactive waste shall be managed in such a way that will not impose undue burdens on future generations' (IAEA, 1995, Principle 5). This is used to justify finding a permanent solution to managing the wastes as soon as practicable. The timing is a matter of debate. Early disposal that avoids long-term storage and surveillance is favoured by the industry since it meets the demand of the Flowers Report and may legitimate further expansion. On the other hand, interim storage may improve safety both to workers and the future by enabling further research and development to be undertaken and by allowing cooling and some decay of wastes to occur. Opponents prefer this option since it maintains the visibility of a hazardous activity.

The uncertainty of risk calculations into the far future is an inescapable problem. The survival of the institutions producing and controlling nuclear waste cannot be guaranteed, even for a short period, especially if the industry continues to decline. Beyond 100 years the problem of surveillance will be increasingly difficult. Over the long time-scales during which some radionuclides continue to be harmful, it becomes impossible to predict the possibility of cataclysmic climatic and geological events which could lead to the release of radioactivity into the accessible environment. Whatever the notional statistical risk, it persists and can present a hazard threatening a vast area at any time. In Beck's doleful words, 'the least likely event will occur in the long run' (1995, p. 1). In so far as the risk can never be eliminated, permanent disposal represents a compromise with the principle of sustainable development.

A way forward?

Nuclear waste is not simply a technical problem for the industry to tackle, nor even a political problem for policy-makers. It is a social problem requiring a social solution. A solution must be feasible, credible, permanent and acceptable. What are the prospects for such a solution?

1 There must be agreement on the policy for the long-term. If deep disposal is to be accepted, it must first be demonstrably safe. This may require a sustained, rigorous and measured programme of research and experiment.
2 There needs to be a commitment to long-term management. The problem of nuclear waste extends beyond the short-term vagaries of markets or political horizons. There needs to be legislation to establish a process of management that is intended to survive political and economic change.
3 A consensus on the solution must be forged. At present nuclear policy is riven by conflict. A solution that serves to legitimate the continuation or expansion of the industry will not be acceptable to its opponents. Therefore, agreement on the phasing out of nuclear energy and the abandonment of reprocessing is a necessary, though not sufficient, condition for securing consensus.
4 The process for finding sites must be open and participative. This suggests that the ground rules must be agreed at the outset, and be clearly independent with all options fully considered. Selection confined to vulnerable nuclear oases will not be acceptable.

This will require a generic survey followed by a gradual elimination of unacceptable sites, leading to a comparative survey of sites from which one is selected.

5 The process must be manifestly fair. Once a site has been finally agreed, equity demands that a comprehensive package of benefits and investments must be provided. Furthermore, sufficient funds must be secured for the future management of the site up to the point when emplacement of wastes is completed and the site is relinquished.

6 The problem of radioactive wastes must be considered in its international context. Hitherto, radioactive waste management has been largely a matter for national determination. Yet, the trade in nuclear materials and the dangers of proliferation make it, inevitably, an international problem. Although in principle self-sufficiency is desirable, some countries, especially the smaller producers, will not be able to manage their nuclear waste in acceptable locations. Such wastes might be managed at a site in one of the larger producing nations. It would not be acceptable for an international solution to be found in poorer countries where there is little or no nuclear commitment. The problems of nuclear waste suggest that an international convention on management should be developed.

Even if all these precepts are agreed, there is no certainty that solutions will be found, at least in the short term. But, nuclear waste is a problem that must be managed in the long run if serious risks from proliferation, contamination and deterioration are to be avoided. The legacy is inevitable; it is incumbent on the present generation to protect the future as far as possible by finding a solution that is technically and socially acceptable.

References

Beck, U. (1995) *Ecological Politics in an Age of Risk* (Cambridge: Polity Press).

Blowers, A. and Leroy, P. (1994) 'Power, politics and environmental inequality: a theoretical and empirical analysis of the process of "peripheralisation"', *Environmental Politics*, 3, pp. 197–228.

Blowers, A., Lowry, D. and Solomon, B. (1991) *The International Politics of Nuclear Waste* (London: Macmillan).

International Atomic Energy Agency (IAEA) (1995) *The Principles of Radioactive Waste Management*, Safety Series, Safety Fundamentals, No. 111-F, Vienna.

Royal Commission on Environmental Pollution (RCEP) (1976) *Nuclear Power and the Environment*, Sixth Report, Cmnd 6618 (London: HMSO).

Royal Society, The (1998) *Management of Separated Plutonium* (London: Royal Society).

Source: Newly commissioned.

SECTION 3

Animal welfare and conservation: expanding the circle?

Introduction

The place of non-human animals in ecological discussion is a deeply contentious one. The existence of domesticated animals, as opposed to wild species, are part of the problem for some environmentalists (like the impact of cattle farming on the Brazilian forests). Yet for others, practices like hill farming with sheep in the English Lake District can be seen as contributing to the distinctive beauty of a managed landscape. A great deal of discussion within environmental movements has been oriented towards the concerns of animal welfare and these do not always fit neatly with the arguments over the conservation of wildlife. Much depends on whether the environmentalist or the ecologist in question is primarily concerned with an individual animal or with the whole species. In addition, a great deal follows from whether this concern is driven by instrumental motives or from a sense of the inherent value of non-human animals.

In Reading 3.1, from the cultural writings of Nick Fiddes, the way we regard animals is portrayed as historically and socially specific. In particular, Fiddes directs our attention to the symbolic associations, metaphors and myths which are associated with the uses of non-human animals in different times and places. These uses often only make sense and may only be acceptable when we think of them as part of social structures, each with a distinctive set of values about the welfare of animals and the conservation of wildlife. In Reading 3.2 you will encounter a series of four Hogarth prints from the seventeenth century. You will find these representations complex and open to different interpretations. Nevertheless, 'The Stages of Cruelty' is a story with an ethical code integrated into the narrative. In the two readings by Keith Thomas you will see how the shifting debates on the status of animals are closely connected to a broader shift in attitudes towards nature; the acceptance of 'Cruelty' (Reading 3.3a) gradually gave way to 'New Sensibilities' (Reading 3.3b) about appropriate treatment of non-human animals. Underlying all the readings so far is the recognition that we should acknowledge the ways in which the 'natural' is constructed within a broader fabric of ethical standards and rules.

In Reading 3.4, Peter Singer provides one of the most cogent cases for the application of the utilitarian approach to animal welfare. For Singer, just like racism and sexism, speciesism is portrayed as a prejudice waiting to be challenged through the rigorous application of human reason. He advocates an approach which weighs the pleasurable and painful consequences of our actions and practices, and applies this to human and non-human animals. For Singer, recognizing the suffering humankind creates for other species would transform the welfare of animals. This is challenged by Tom Regan (in Reading 3.5) for it allows the continuation of animal suffering where the benefits can be identified as outweighing the costs. For Regan, all higher animals are 'subjects of life' who possess an intrinsic value that should not be violated. For Regan, animals have rights. The response to the arguments of Singer and Regan is represented in this selection of extracts by Roger Scruton (Reading 3.6) who argues that non-human animals do not have the same standing as humans – that animals can neither comprehend nor participate as members of the moral community.

In the final writing by Robert Garner (Reading 3.7) we consider the earlier readings in the context of discussions on animal welfare and its associated ethical questions within the wider debates about the place of animals in their natural habitats. The impact of the conservation of wildlife on the way we view the natural world is considerable. Garner draws our attention to the differences between the arguments of philosophers (like Singer and Regan) who are concerned with individual animals and those concerned with the preservation of endangered species. The discussion of the place of non-human animals in terms of the moral community is a complex one. Much depends upon whether we value non-human animals because of their usefulness to humankind and how we place animals within the wider ecosystems they inhabit. Indeed, it is possible to place the interests of a species over and above the interests of trees (or their owners) which are damaged by them. Similarly, it is feasible to see animals as a renewable resource which should take second place to the conservation of finite natural resources. In each case, the solutions which are deemed appropriate are closely connected to the values of those who hold such views.

3.1

Nick Fiddes

More than a meal

All over the world food means much more than mere nutrition. Perhaps it is singled out for such significance because everybody, everywhere needs to eat. Perhaps it is because – along with only a few other similarly significant acts such as sex and defecation – eating breaches our normally sacrosanct bodily boundaries. Maybe it is important that when we consume we literally incorporate into our own bodies the physical material – and possibly the spiritual essence – of other animals and of the outside world in general. But whatever the reason, we routinely use food to express relationships: amongst ourselves and with our environment. The obtaining and sharing of food can be an eloquent statement of shared ideology and as such expresses group affiliation and apparent solidarity. W. Robertson Smith noted that 'those who eat and drink together are by this very act tied to one another by a bond of friendship and mutual obligation' (1889: 247); Radcliffe-Brown held that for the Andaman Islanders 'by far the most important social activity is the getting of food' (1922: 227); and Darlington (1969) suggests that commensality may be the most important basis of human associations. Food is a system of communication, a body of images, a protocol of usages, situations, and behaviour (Barthes 1975):

> Food is prestige, status and wealth . . . It is a means of communication and interpersonal relations, such as an 'apple for the teacher', or an expression of hospitality, friendship, affection, neighbourliness, comfort and sympathy in time of sadness or danger. It symbolises strength, athleticism, health and success. It is a means of pleasure and self-gratification and a relief from stress. It is feasts, ceremony, rituals, special days and nostalgia for home, family and the 'good old

days'. It is an expression of individuality and sophistication, a means of self-expression and a way of revolt. Most of all it is tradition, custom and security.

> Different foods satisfy these needs and beliefs of people in different cultures. Some foods are linked to the age and sex of the individual . . . There are Sunday foods and weekday foods, family foods and guest foods; foods with magical properties, and health and disease foods.
>
> (Todhunter 1973: 301)

> 'It sort of came as a final rejection of her cooking. It was actually at that level, when I was at home at the weekends. She was cooking lovely meals, and I was not eating the meat, which everyone else was enthusing over: it was "a beautiful piece of roast you've got this weekend", and "oh, it's lovely tender lamb for this time of year". And I was saying that I don't want it. And it was – it was like rejecting part of her . . .'

> *Meat is just a way of life for British families*
>
> Sir – You may not consider it very important, but I'd like to tell you a simple story about a piece of pork.
>
> Myself and my family bought it on Saturday for our Sunday lunch.
>
> It was an attractive joint and, despite some good Saturday night TV, the prospect of the meat remained in our minds and there was a hint of expectation on all our faces.
>
> For three hours on Sunday the smell of it cooking practically drove us wild. When we finally sat down at the table, all the troubles of the week seemed to drift away at the prospect of a delicious family lunch.
>
> The meat was wonderful and I thought how much the £5 joint had contributed to this typically British, family scene. My family left the table feeling well-fed and happy and the cold meat made a meal on Monday night as well.
>
> Meat is not just a meal, it is a way of life.
>
> T. Cook. Basildon, Essex.
>
> (*Meat Trades Journal* 1987: 2)

Conversely, those who diverge from community standards are commonly stigmatised since their dietary non-conformity is (correctly) taken to indicate broader differences in values. Consider, for example, the dismissive tone and marginalising vocabulary in an American sociologist's treatment of alternative diets:

> In studies of social movement and the formation of sects and dissident groups, the role of food cannot be underestimated. In adhering to some dietary rules, what to eat, when to eat, or when not to eat, groups maintain control over their members. They also require members to deviate from the general population when they venture outside their group. This behavior is one of the most effective ways of assuring adherence to special group codes. Vegetarianism, which has recently attracted a variety of individuals and been intimately connected with several modern movements . . . might serve as an example. It is hard to find a common denominator among these groups, except that they are all in some way

> intent on establishing a difference and attracting attention to it. A more recent phenomenon seems to be the fad for 'natural foods,' which often corresponds to some political and social dogma and thereby serves to bring adherents together.
>
> (Back 1977: 31–32)

Back seems desperate neatly to categorise all such individuals according to their culinary deviation from the mainstream. There are overtones of conspiracy theory with groups 'maintaining control' over members who must conform to 'some political or social dogma'. The disparaging use of terms such as fad and allusion to brainwashing in such value-loaded description implies the majority diet to be rational. But Back's dogmatic zeal obscures from him the diversity of ideas and meanings involved, not only in the minority groups he isolates (many of whom may indeed deliberately be rejecting what Pope called 'carnivoracity') but also within the dominant culture with which he evidently identifies. He assumes that vegetarianism ought to mean much the same thing to all vegetarians and is perplexed since the only common attribute he can advance is their 'difference' to which he weakly and unjustifiably ascribes a ubiquitous desire to attract attention. The seemingly bizarre food habits of exotic cultures or of minority groups may encourage us to look for extra meanings – of something communicative. But our own are just as expressive, and may seem equally strange to outsiders.

Foods do not intrinsically symbolise. They are used to symbolise. For example, when we are told that the cooked dinner of meat and two vegetables symbolises the woman's obligation as homemaker and her husband's as breadwinner in South Wales (Murcott 1982), clearly the food does not itself stand for homemaking and caring. The values are their respective gender roles, whilst the food is the medium through which that is communicated. This function of food is evident throughout our society.

So why is meat a 'Natural Symbol'? What does 'symbolic' in fact mean? The term is commonly used to imply communication that is distinct from direct, literal, discourse. Objects, ideas, or actions are called symbolic when they represent something beyond their obvious identities. A flag is symbolic because it is more than a piece of decorative cloth; a dove because, by common consent, it represents the concept of peace. As Anthony Cohen puts it, 'When guests compliment their host for having laid a good fire, they probably mean more than that he has achieved a commendable thermal output for a given quantity of fuel' (1986: 3).

Meat certainly qualifies as symbolic in these terms since its economic and social importance is frequently greater than might be anticipated from its purely nutritional value. But symbolism is more than a few ethereal associations which somehow disrupt our otherwise rational judgement. That is only the surface of an infinite system of thought that can be implicit or explicit, private or public, tacit or overt. Nutrition and economics, for example, are themselves symbolic (Sahlins 1976). As Douglas and Isherwood have argued, 'the problem goes so deep that nothing less profound than a corrected version of economic rationality is needed' (1980: 4). In fact all our goods are of communicative, at least as much as of utilitarian, value – they are good(s) to appreciate (and to be seen with by others who share our view of their value) as well as to consume.

When nutritionists or policy-makers discuss the energy, fat, or protein contents of foods, for example, and expect a willing public dutifully to adapt their habits, they are deceiving themselves in failing to accommodate the numerous other roles that foods play in people's lives. Indeed some 'authorities' seem to live in a fantastic science-fiction world, alien to those of us who simply enjoy buying, preparing, and eating food:

> Another transatlantic viewpoint was given by Dr Carl Unger, a meat consultant based in Georgia, USA, who projected a picture of computerised shopping in the mid-1990s.
>
> He predicted that the shopper would use a computer to work out daily dietary and nutritional needs. These would be translated into a number of dishes and meal components which would be ordered from the local supermarket.
>
> (*British Meat* 1987: 3)

We routinely stress meat's scientifically recognised function in terms of health and nutrition as the principal determinant of its status. From this, it is generally assumed, its value as an item of economic exchange is derived. Occasionally a certain symbolic importance may also be recognised, such as the macho steak which, like the macho car, may be purported to denote sexual prowess, or such as the view that its importance or prestige could be a sort of social relic from when we hunted to survive or from royal hunting in Days Of Olde.

But there is more to meat than protein. The Shorter Oxford English Dictionary, for example, defines meat as 'The flesh of animals used for food' and a few subsidiary and archaic variations; not even its common usage as a synonym for 'essence' or 'substance' is mentioned (e.g. 'The meat of the argument'). Whilst meat as food may be its principal accepted meaning, this falls far short of conveying the elusive depth of its signification. To one person the word *steak* might suggest a substance whereby affluence or culinary skill can be demonstrated; it might conjure up memories of celebratory dinners by candlelight; it might reassure about the consumption of good, body-building nutrition. To another, steak might stand for cruelty and nausea, with thoughts of horrific conditions in which animals are bred and slaughtered, and images of violence, blood, brutality. No two people will find quite the same meanings in the same word, and no such association can be called 'incorrect'.

The associations can be seen as relative, or 'symbolic truths', since it is upon belief and image that individuals base their actions, not absolute fact. This can help explain apparent contradictions in our behaviour. For example, amid increasing distrust of factory food production and of fatty meats, pork and chicken – in spite of being intensively farmed and of sometimes being no lower in fat than many 'red' meats – sell relatively well, partly because they are less bloody and have been marketed as white meat, and therefore less graphically meaty. Game and salmon are similarly favoured by many as 'wild' even though a high proportion of purchases are in fact of farmed stock.

Each meaning, and countless others, is true for the individuals concerned, extending the significance of the name of a particular meat, or of meat in general, far beyond its function as a foodstuff. It is the totality of these ideas which combine to form a language, and which constitute culture. We must look to the depth and diversity of meaning in all our ideas and actions. Thus economics and technology are symbolic in the very same sense as unicorns or religious icons, though they undoubtedly differ in both their form and content. And meat carries many meanings.

Sperber (1975) argues that symbolism deals with things which cannot be dealt with rationally (and so incorporated into everyday speech). Sometimes, however, symbols rather seem to concern things which simply *have not been* so classified and expressed, even though they might be. That a particular meaning associated with a symbol is not normally articulated may be because it is all the more persuasive for being unvoiced.

This is perhaps best illustrated by example. Advertisers and marketing executives are well aware that people buy more than utilitarian objects. Successful sales people offer 'lifestyles' and emphasise aspirations and associations rather than function. Much modern advertising, indeed, scarcely mentions the product. For example (excluding the purely image-building genre), a motor car is typically marketed in terms of comfort, economy, power ('to cope with emergencies'), and prestige, often with the implied promise of sexual success for the probable male purchaser, but these messages will be manifest in different forms. The most explicit messages will tend to concern the reputedly rational considerations that the potential buyer will happily acknowledge. Other more emotive meanings, perhaps concerning the social or business niche to which the target is characterised as aspiring, are unlikely to be spelled out in the text but are nonetheless clearly written in the situations portrayed.

Barthes (1975) identifies three themes that stand out in food advertising – the commemorative, eroticism, and health – arguing that this reflects the collective psychology more than it shapes it. But whether to sell the chicken or the egg, it is clear that much of advertising's symbolic power lies in its unvoiced suggestion. The message is altogether more compelling as part of the taken-for-granted context rather than as an overtly propagated claim. Indeed, an idea can have very different meanings depending upon whether it principally circulates explicitly or implicitly, perhaps because only when brought to awareness does it become open to criticism. The motorist who tacitly enjoys the endemic association of sports cars with sexual virility, for example, might feel considerably less comfortable when openly mocked by others for his 'penis substitute'. And were the *reasons* for the association between car and sex to be more precisely rendered, they would be open to discussion, demystification and dispute, and might rapidly lose their effect. The power of the ideas depends upon their being communicated without being rendered explicit, for their meaning can then be understood by all concerned, but at the level of assumption, common sense, and accepted fact.

That animals are killed for humans to eat meat is obvious to the point of banality. However, the inherent conquest is rarely discussed overtly in the context of food provision. Our willingness to eschew confronting certain aspects of meat's identity is more than a matter of preferring to sidestep that which might be unsavoury. The fact that most of us make little mention of the domination inherent in rearing animals for slaughter does not indicate that it is irrelevant. On the contrary, that which remains unsaid about meat conveys an added dimension of meaning which is particularly potent. It is the very taken-for-grantedness of values implicit in the meat system which makes the message so powerful, whilst rationalisations of meat's importance (such as nutrition and prestige) partly serve to obscure these values from our consciousness.

Paradoxically, this obscurity preserves and perpetuates the influence of these implicit meanings since, not being recognised, they can scarcely be challenged. Veal, for example, enjoyed high prestige for many years partly, I suggest, *because of* the extreme subjugation of the creatures intrinsic to its production. That, however, was seldom voiced; instead its value was explicitly attributed to such arbitrary qualities as delicate flavour and light colour. Only once its production methods were brought into the domain of explicit public consideration did they become intolerable, at which time this previously inherent meaning lost its positive power and instead became a negative influence on the meat's popularity.

What was true for veal yesterday could also become true for meat in general tomorrow. . . . [T]he unvoiced symbolic values which continue to underpin meat's popularity today principally concern our relationship with nature, as we perceive it. In this way changing attitudes to meat, as revealed by changing habits, may also be eloquent commentary on fundamental developments in society. Meat's signification, I suggest, principally relates to environmental control, and it has long held an unrivalled status amongst major foods on account of this meaning. But meat's stature is not inherent in its substance, but has been invested in it by successive generations who highly valued its meaning: who *liked* the notion of power over nature that it embodies. Its waning prestige – and outright rejection by many – may be indicative of more than changing tastes in food. Thus meat is more than just a meal; it also represents a way of life.

References

Back, K. (1977) 'Food, Sex and Theory', in T.K. Fitzgerald (ed.) *Nutrition and Anthropology in Action*: 24–34, Amsterdam: Van Gorcum.

Barthes, R. (1975) 'Towards a Psychosociology of Contemporary Food Consumption', in E. Forster and F. Forster (eds) *European Diet, from Pre-Industrial to Modern Times*: 47–59, New York: Harper and Row.

British Meat (1987) 'Minister congratulates British meat industry on record exports – opportunities for further expansion ahead', Milton Keynes: Meat and Livestock Commission, Summer: 2–3.

Cohen, A. (1986) 'Of Symbols and Boundaries, or, Does Ertie's Greatcoat Hold the Key?', in A. Cohen (ed.) *Symbolising Boundaries: Identity and Diversity in British Cultures*: 1–19, Manchester: Manchester University Press.

Darlington, C. (1969) *The Evolution of Man and Society*, New York: Simon & Schuster.

Douglas, M. and Isherwood, Baron (1980) *The World of Goods*, Harmondsworth: Penguin.

Meat Trades Journal (1987) 'Meat is just a way of life for British families', London: International Thomson, 2 July letter: 2.

Murcott, A. (1982) 'On the Social Significance of the "Cooked Dinner" in South Wales', *Social Science Information* 21, 4–5: 677–96.

Radcliffe-Brown, A.R. (1922) *The Andaman Islanders*, 1964 edn, New York: Free Press.

Robertson Smith, W. (1889) *The Religion of the Semites*, Edinburgh: Black.

Sahlins, M. (1976) *Culture and Practical Reason*, Chicago: University of Chicago Press.

Sperber, D. (1975) *Rethinking Symbolism*, Cambridge: Cambridge University Press.

Todhunter, E.N. (1973) 'Food Habits, Food Faddism and Nutrition', in M. Rechcigl (ed.) *Food, Nutrition and Health: World Review of Nutrition and Dietetics* 16: 186–317, Basel: Karger.

Source: Nick Fiddes (1991) *Meat: a natural symbol*, London: Routledge, pp. 38–45.

3.2

Mark J. Smith

Visual ethics: interpreting Hogarth's 'The Stages of Cruelty'

Moral messages can be communicated in many ways. Perhaps the most potent are those which demand that we use our visual interpretive skills. William Hogarth's engravings and prints were carefully constructed to highlight the ethical debates in seventeenth-century England, as well as to express his own judgements on such matters. The stories which Hogarth developed through his prints generated enormous interest throughout the ranks of English society. The use of symbols and well-known characters which mirrored his earlier stories or those popularized by the popular novels of Henry Fielding widened their appeal.

These four prints tell the story, or the 'progress', of Tom Nero, an unloved pauper orphan recognizable by the initials S.G. on his shoulder (signifying a child reared by the parish of St Giles). In the first plate, the focus is on the various forms of cruelty to animals inflicted by children without chaperones. Within this plate, the 'civilized' child (often taken as the young prince George) who attempts to remove an arrow from the impaled dog carries the message that we should learn from our betters. Starting the story in this way, with the child as much a victim as the animal he tortures, provides an opportunity for a more sympathetic reading. The second plate not only demonstrates how disregard for animals is carried into adulthood in the variety of ways depicted but also raises the effect of cruelty on those who perpetrate it. Beyond Tom Nero, who is exacting judgement on an old horse for collapsing and overturning a coach-load of barristers, are smaller scenes which highlight the way cruelty dehumanizes people. The barristers protest but they are also culpable for only being willing to pay the cheapest fare. The civilized boy is now a man who wishes to report the driver responsible for the accident, carrying the message that if such things are reported then some good may follow. In the background, the posters announce forthcoming events such as cockfighting and a variety of pugilistic sports.

In the third plate, entitled 'Cruelty in Perfection', Tom is discovered as the murderer of a maidservant, Ann Gill, whom he seduced to procure her mistress's plate. Her mutilated body forms the central focus while Tom's face displays a callous disregard for human life – that of the girl and of the child in her womb. The book of common prayer and 'God's Revenge Against Murder' have fallen from her chest, reinforcing the play on the phrase that 'we reap what we sow'. In the fourth plate, 'The Reward of Cruelty', the central message of Tom Nero's progress is driven home. Following his execution for murder, his body is designated for medical research and taken to the Surgeon's Hall. As Jenny Uglow argues: 'This is a moral as well as a physical dissection. Tom's cursing tongue has been torn out by its roots; his lusting eyes gouged out; his lusting heart thrown to a mangy dog. Skulls and bones boil over a fire, like a cannibal rite, a visceral pagan sacrifice' (Uglow, 1997, p. 505). The figure of the civilized man is harder to spot in the last two plates but is often seen as

indicating the final indignity for Tom; the shame of display as a skeleton in the chambers of the predatory surgeons.

On closer inspection you will find many other elements of narrative continuity between the plates. In every corner, you will find smaller scenes or artefacts which reinforce the central message of the narrative. However, as you work through the four plates of 'The Stages of Cruelty' (now some three hundred and more years later than Hogarth's intended audience), interpreting many of the symbols presents its own difficulties. We are situated in

a different culture with its own ways of articulating ethical values. We are also unfamiliar with the forms of representation which ran through Hogarth's prints, as well other caricatures within the same genre. The way we produce our own meanings to string the story together in a way which makes sense, helps us to recognize the intertextuality of cultural meaning (Smith, 1998, Ch. 6).

References

Smith, M. J. (1998) *Social Science in Question* (London: Sage).
Uglow, J. (1997) *Hogarth: A life and a world* (London: Faber and Faber).

3.3a

Keith Thomas

Cruelty

When English travellers went abroad in the late eighteenth century they were frequently shocked to see how foreigners treated animals. The Spanish bull-fight had long been notorious for what the first Earl of Clarendon called its 'rudeness and barbarity'. English tourists always went to see it, but usually only once. 'Fifteen or sixteen wretched bulls were massacred,' wrote the fastidious William Beckford after a Portuguese fight in 1787, adding on another occasion, 'I was highly disgusted with the spectacle. It set my nerves on edge and I seemed to feel cuts and slashes the rest of the evening.' It was 'a damnable sport', agreed Robert Southey.[1]

Continental methods of hunting were equally distasteful. When Sir Richard Colt Hoare went after wild boar with the King of Naples in 1786, he was appalled to discover that the boar, so far from being wild, came when whistled for, and that the hunters stuck it with spears when it was held fast by dogs. 'I was . . . thoroughly disgusted with this scene of slaughter and butchery . . . yet the King and his court seem[ed] to receive great pleasure from the acts of cruelty and to vie with each other in the expertness of doing them.'[2]

The treatment of domestic animals was also lamented. Tobias Smollett felt compassion for the wretched mules and donkeys in the south of France; Beau Brummell in exile was much upset by the way the Normans treated their horses; while Mrs Hervey, wife of the future Earl of Bristol *cum* Bishop of Derry, expressed the feelings of many subsequent Englishwomen when her coach stuck on a journey to Monte Cassino in 1766: 'What hurt me most was their barbarous treatment of the poor mules, whom they beat most unmercifully with their fists, feet, sticks and even stones. I walked about and begged them to be more gentle to them; they laughed at me.'[3]

These reactions reflect that growing concern about the treatment of animals which was one of the most distinctive features of late-eighteenth-century English middle-class culture. They also show the emergence of a belief which by Victorian times had become an entrenched conviction: that the unhappiest animals were those of the Latin countries of southern Europe, because it was there that the old Catholic doctrine that animals had no souls was still maintained.[4]

Yet previously it had been the English themselves who had been notorious among travellers for their cruelty to brutes. The staging of contests between animals was one of their most common forms of recreation. Bulls and bears were 'baited' by being tethered to a stake and then attacked by dogs, usually in succession, but sometimes all together. The dog would make for the bull's nose, often tearing off its ears or skin, while the bull would endeavour to toss the dog into the spectators. If the tethered animal broke loose, scenes of considerable violence ensued. Baiting of this kind was customarily regarded as an appropriate entertainment for royalty or foreign ambassadors. It also took place at country wakes and fairs and in the yards of ale-houses, where local dogs would be invited to challenge an itinerant bull or bear travelling round the country with its keeper. At Stamford and Tutbury there occurred an annual 'bull-running', when the animal, with its ears cropped, tail cut to a stump, body smeared with soap and nose blown full of pepper, was turned loose to see who could catch him in a general free-for-all. Badgers, apes, mules, and even horses might all be baited in similar fashion. Bull-baiting, wrote John Houghton in 1694, 'is a sport the English much delight in; and not only the baser sort, but the greatest ladies'.[5]

Cock-fighting had been equally popular since at least the twelfth century.[6] In the Stuart age it was a normal feature of fairs and race-meetings. The cock was brought up on a carefully chosen diet and specially trained for the fight. Its wings were clipped, its wattle and comb shorn off and its feet equipped with artificial spurs. Cock-fights were usually 'mains', that is contests between two rival teams paired off into a succession of individual combats, as in modern golf matches; a rougher version was the so-called 'Welsh main', which was a knockout competition. Most spectacular of all was the 'battle royal', when a large number of cocks were put into the same pit together, as at Lincoln in 1617, when James I was made 'very merry' by the spectacle.[7] The contests usually expressed regional rivalries, with different teams of cocks representing different villages or the 'gentlemen' of different counties. Meetings often lasted several days and were accompanied by heavy betting, with fresh wagers being laid at every stage of the fight. They involved the mingling of all social ranks, though only of men, for it was emphatically not a woman's sport. The refined Tudor humanist Roger Ascham was a passionate devotee; and when Pepys went to a cockpit in 1663 he saw everyone from 'parliament men' down to 'the poorest prentices, bakers, brewers, butchers, draymen and what not . . . all fellows one with another in swearing, cursing and betting'. The cocks themselves had a short life, even the best being unlikely to survive more than a dozen contests.[8]

In the countryside the pursuit and killing of wild animals for sport had been practised since time immemorial. . . . In the early modern period the prey was hunted either because it could be eaten, like the red and fallow deer, or because it was a pest, like the fox, or because its speed and agility made it an entertaining object of pursuit, as with the hare. Henry VIII's manner of hunting did not differ very much from that of the eighteenth-century King of Naples: he had two or three hundred deer rounded up and then loosed his greyhounds upon them.[9] Frequently, however, the methods of pursuit, capture and kill were highly stylized, and contemporary literature celebrated the majesty of the hunters, the nobility of the hounds and the music of the chase. Hunting, thought Gervase Markham, was 'compounded . . . of all the best parts of most refined pleasure'. No music, it was said, could be more 'ravishingly delightful' than the sound of a pack of dogs in full cry.[10] At the Inner Temple on St Stephen's Day it was customary to bring a fox and a cat into the hall and set hounds upon them. At Sheffield Park in the 1620s the Earl of Shrewsbury allowed his tenants to keep any buck they could kill, provided they used only their bare hands. At Smyrna in the late seventeenth century English merchants procured hounds and conducted a hunt, with the dogs following the hare by the scent, which the Turks regarded as 'a prodigious mystery'.[11] At Aleppo in 1716 they went in pursuit of antelopes with hawks, while later in the eighteenth century the Duke of Grafton planned to go to France with his horses and hounds to hunt wolves.[12] In all cases the climax of the hunt was the death of the hunted animal, for, as Montaigne observed, to hunt without killing was like having sexual intercourse without orgasm. . . .

Equally popular was the pursuit of wild birds, either with hawks or, increasingly, with guns. It was a splendid sight, thought a visitor to the fen country in 1635,

> to see a fleet of a hundred or two hundred sail of shell boats and . . . punts sailing . . . in the pursuit of a rout of fowl, driving them like sheep to their nets . . . sometimes they take a pretty feathered army prisoners, two or three thousand at one draught and give no quarter.[13]

Those who engaged in these sports were seldom inhibited by concern for the possible feelings of the animals themselves. Fishing involved the use of live bait, not just small fish, but also frogs.[14] Hawks were nourished on pigeons, hens and other birds. 'I once saw a gentleman,' recalled William Hinde in 1641, 'being about to feed his hawk, pull a live pigeon out of his falconer's bag, and taking her first by both wings, rent them with great violence from her body, and then taking hold of both legs, plucked them asunder in like manner, the body of the poor creature trembling in his hand, while his hawk was tiring upon the other parts, to his great contentment and delight upon his fist.' The *Gentleman's Recreation* (1674) recommended catching a hart in nets, cutting off one of his feet and letting him go to be pursued by young bloodhounds.[15]

The absence of any apparent moral consideration for the hunted animal is well revealed in the famous description of the entertainment provided for Queen Elizabeth when she visited Kenilworth in 1575. First she hunted the hart until it was killed by the hounds after it had taken to water. This, wrote a contemporary, Robert Laneham, was 'pastime delectable in so high a degree as for any person to take pleasure by most senses at once in mine opinion there can be none any way comparable to this'. A few days later a collection of mastiffs was let loose on to a group of thirteen bears. It was 'a sport very pleasant,' says Laneham, 'to see the bear . . . shake his ears twice or thrice with the blood; and the slaver about his physiognomy was a matter of a goodly relief.'[16] . . .

In addition to these stylized and highly formal methods of tormenting animals, there was an infinity of informal ones. Small boys were notorious for amusing themselves in the pursuit and torture of living creatures. In the grammar schools cock-throwing was a widely observed calendar ritual. On Shrove Tuesday the bird was tethered to a stake or buried in the ground up to its neck, while the pupils let fly at it until it was dead. ''Tis the bravest game,' wrote a seventeenth-century poet.[17] Outside school, children robbed birds' nests, hunted squirrels 'with drums, shouts and noises',[18] caught birds and put their eyes out, tied bottles or tin cans to the tails of dogs, killed toads by putting them on one end of a lever and hurling them into the air by striking the other end, dropped cats from great heights to see whether they would land on their feet, cut off pigs' tails as trophies and inflated the bodies of live frogs by blowing into them with a straw.[19] It was 'a familiar experiment among boys,' reported Thomas Willis in 1664, 'to thrust a needle through the head of a hen' to see how long it would survive the experience.[20] . . . No wonder that traditional nursery rhymes portray blind mice having their tails cut off with a carving-knife, blackbirds in a pie and pussy in the well. 'How full of mischief and cruelty are the sports of boys!' lamented the eighteenth-century Evangelical John Fletcher; and the refrain was echoed by scores of observers.[21]

Yet children merely reflected the standards of the adult world. The seventeenth century was an age when country gentlemen would entertain their visitors by putting their dogs to chase tame ducks or by throwing a goose or chicken into a pike-infested pond to watch its struggles.[22] At country fairs there were contests at biting off the heads of live chickens or sparrows.[23] . . . As the historian W. E. H. Lecky remarks, there were two kinds of cruelty: the cruelty which comes from carelessness or indifference; and the cruelty which comes from vindictiveness.[24] In the case of animals what was normally displayed in the early modern period was the cruelty of indifference. For most persons, the beasts were outside the terms of moral reference. Contemporaries resembled those 'primitive' peoples of whom a modern anthropologist writes that they neither seek to inflict pain on animals nor to avoid doing so: 'pain in human beings outside the social circle or in animals tends to be a matter of minimal interest.'[25] It was a world in which much of what would later be regarded as 'cruelty' had not yet been defined as such. A good example of how people were inured to the taking of animal life is provided by the diary kept by the schoolboy Thomas Isham, who grew up in Northamptonshire in the early 1670s. His little journal records much killing of cocks, slaughtering of oxen, drowning of puppies. It tells of coursing for hares, catching martens in traps, killing sparrows with stones and castrating bulls. None of these events evokes any special comment, and it is clear that the child was left emotionally unruffled.[26]

Notes

1 *The Miscellaneous Works of . . . Edward Earl of Clarendon* (2nd edn, 1751), 347; *The Journal of William Beckford in Portugal and Spain*, ed. Boyd Alexander (1954), 154, 127; *Letters of Robert Southey*, ed. Maurice H. Fitzgerald (1912), 33.

2 Kenneth Woodbridge, *Landscape and Antiquity* (Oxford, 1970), 84–5.

3 Tobias Smollett, *Travels through France and Italy* (1907 edn), 174; Captain Jesse, *The Life of George Brummell* (rev. edn, 1886), ii. 156; William S. Childe-Pemberton, *The Earl Bishop* (n.d.), i. 77.

4 See, e.g., *The Letters of Charles Dickens*, iv, ed. Kathleen Tillotson (Oxford, 1977), 272; Edward Maitland, *Anna Kingsford* (3rd edn, 1913), ii. 311–12.

5 John Houghton, *A Collection for Improvement of Husbandry and Trade*, v. 108 (24 Aug. 1694). On bear- and bull-baiting see E. K. Chambers, *The Elizabethan Stage* (Oxford, 1945 reprint), ii. 449–71, and Robert W. Malcolmson, *Popular Recreations in English Society, 1700–1850* (Cambridge, 1973), 45–6, 66–8. For baiting of other animals see, e.g., *HMC, Chequers*, 414; Charles Stevens and John Liebault, *Maison Rustique*, trans. Richard Surflet, ed. Gervase Markham (1616), 703; *Crime in England, 1550–1800*, ed. J. S. Cockburn (1977), 237; John Clare, 'Badger', *Poems*, ed. J. W. Tibble (1935), ii. 333–4 (badgers); *Lancashire Quarter Sessions Records*, ed. James Tait, i (Chetham Soc., 1917), 101 (apes); *Yorkshire Diaries*, ed. H. J. Morehouse (Surtees Soc., 1877), 307 (mules); Robert Surtees, *The History and Antiquities of the County Palatine of Durham* (1816–40), iv. 78n; *Notes & Queries*, 4th ser., xii (1873), 273 (horses).

6 *Materials for the History of Thomas Becket*, ed. James Craigie Robertson (Rolls Ser., 1875–85), iii. 9. For general accounts see *Games and Gamesters of the Restoration* (1930), 100–14; George Ryley Scott, *The History of Cockfighting* (n.d.); and Malcolmson, *Popular Recreations*, 49–50.

7 *HMC*, 14th rept., appendix, pt viii. 94.

8 Pepys, *Diary*, iv. 427–8; Laurence V. Ryan, *Roger Ascham* (1963), 229, 242; James Tyrrell, *A Brief Disquisition of the Law of Nature* (1692), 324.

9 *The Lisle Letters*, ed. Muriel St Clare Byrne (1981), vi. 177.

10 Gervase Markham, *Cavelarice* (1607), iii. 1; *Gentleman's Recreation*, 2.

11 William Dugdale, *Origines Juridiciales* (1666), 156; Evelyn Philip Shirley, *Some Account of English Deer Parks* (1867), 215–16; Roger North, *The Lives of Francis . . . Dudley . . . and . . . John North*, ed. Augustus Jessopp (1890), ii. 39.

12 *HMC, Portland*, ii. 258–9; Arthur Young, *Travels in France and Italy* (EL, 1915), 62.

13 *Camden Miscellany*, xvi (Camden ser., 1936), iii. 90.

14 See, e.g., Izaak Walton, *The Compleat Angler* (1653), i. 3; i. 8.

15 William Hinde, *A Faithfull Remonstrance of the Holy Life and Happy Death of Iohn Bruen* (1641), 33; *Gentleman's Recreation*, 11–12.

16 *Captain Cox, his Ballades and Books; or, Robert Laneham's Letter*, ed. Frederick J. Furnivall (Ballad Soc., 1871), 13–14, 17.
17 Martin Lluellyn, 'Cock-Throwing', in *Seventeenth-Century Lyrics*, ed. Norman Ault (1928), 191. See John Brand, *Observations on . . . Popular Antiquities*, ed. Sir Henry Ellis (new edn, 1849–55), i. 72–82; Jeffrey N. Boss in *Notes & Records of the Royal Society*, 32 (1977), 145.
18 Fabian Philipps, *Tenenda non Tollenda* (1660), sig. A4[v].
19 E.g. G. R. Owst, *The Destructorium Viciorum of Alexander Carpenter* (1952), 25–6; *Minor Poets of the Caroline Period*, ed. George Saintsbury (Oxford, 1921), iii. 342; *Diary of Thomas Isham*, trans. Norman Marlow, ed. Sir Gyles Isham (Farnborough, 1971), 119; James Orchard Halliwell, *A Dictionary of Archaic and Provincial Words* (5th edn, 1845), 'praaling', 'tail-piping'; Robert Holland, *A Glossary of Words used in the County of Chester* (EDS, 1884–6), 331; Ellis, vii(2). 104–5; *The Memoirs of James Stephen*, ed. Merle M. Bevington (1954), 62, 90–1; H. C. Maxwell Lyte, *A History of Eton College* (1875), 27.
20 *The Remaining Medical Works of . . . Thomas Willis*, trans. S. P[ordage] (1681), 93.
21 L. Tyerman, *Wesley's Designated Successor* (1882), 260. Cf. Jonathan Swift, *Gulliver's Travels* (1726), ii. 1; W. A. L. Vincent, *The State of School Education, 1640–1660* (1950), 36; Francis Coventry, *The History of Pompey the Little*, ed. Robert Adams Day (1974), 49; Lawrence, *Horses*, i. 137; Joseph Hunter, *The Hallamshire Glossary* (1829), 78.
22 James Edmund Harting, *The Ornithology of Shakespeare* (1871), 237; *The Art of Angling* (appended to *The Country Man's Recreation* (1654)), 11; Morton, *Northants.*, 422.
23 John Sykes, *Local Records* (new edn, Newcastle, 1833), i. 221; Halliwell, *Dictionary*, 'mumble-a-sparrow'.
24 Swift, *Gulliver's Travels*, ii. 1; William Edward Hartpole Lecky, *History of European Morals* (1913 edn), i. 134.
25 Raymond Firth, *Elements of Social Organization* (1951), 199–200.
26 *Diary of Thomas Isham*, *passim*.

Source: Keith Thomas (1984) *Man and the Natural World 1500–1800*, London: Penguin, pp. 143–8.

3.3b

Keith Thomas

New sensibilities

From the later seventeenth century onwards it had thus become an acceptable Christian doctrine that all members of God's creation were entitled to civil usage. Moreover, the area of moral concern had been widened to include many living beings which had been traditionally regarded as hateful or noxious. 'Even to the reptile,' wrote John Dyer in 1757, 'every cruel deed/Is high impiety.' Christopher Smart sang of the beetle, 'whose life is precious in the sight of God, tho' his appearance is against him', and the crocodile, 'which is pleasant and pure when he is interpreted, tho' his look is of terror and offence'. Parents should not let their children cause needless harm to any living thing, declared John Wesley, for the golden rule applied to all creatures – snakes, worms, toads and flies included. It was criminal, the Rev. James Granger told his rural congregation at Shiplake, Oxfordshire, in 1772, to destroy the 'meanest insect' without good reason. Worms, beetles, snails, earwigs and spiders all found their advocates; and naturalists began to seek for more humane methods of killing them.[1]

The eighteenth century abounds in these new susceptibilities. For if in literature we have Tristram Shandy's Uncle Toby, reluctant to kill the fly, then in actual life there was the Norwich doctor Sylas Neville, who in 1767 caught two mice in a trap, but then released them, being 'unwilling to kill the troublesome little vermin'. There were also the author William Melmoth, worried about the cruelty of destroying the snails on his garden peaches, and the Calvinist divine Augustus Toplady, deploring the cruelty of digging up ant-hills, and the writer William Chafin, lamenting the callow theft by schoolboys of nuts hoarded by industrious mice.[2]

Of course, spontaneous tender-heartedness, as such, was not new. [. . .] In sixteenth-century France Montaigne had denounced cruelty to animals, partly because they shared some qualities with men and partly because they were God's creatures worthy of respect, but also because such cruelty offended his innate sensibilities: 'If I see but a chicken's neck pulled off or a pig sticked, I cannot choose but grieve; and I cannot well endure a silly dew-bedabbled hare to groan when she is seized upon by the hounds.' It is not difficult to detect his influence upon Margaret Cavendish, who in 1667 described herself as 'tender-natured, for it troubles my conscience to kill a fly and the groans of a dying beast strike my soul'.[3] Shakespeare had also displayed explicit sympathy with hunted animals, trapped birds, tired horses, even flies, snails and 'the poor beetle that we tread upon'; and the Elizabethan John Stubbs (of *The Gaping Gulf*) alluded to the 'common compassion' which some men felt when they saw overburdened beasts.[4] The Seeker Thomas Taylor in 1661 contrasted the devotees of cruel animal sports with 'the tender nature of Christ and all Christians, truly so-called, who could never rejoice in any such things by reason of their tender, pitiful and merciful nature'. In his textbook on Greek history Francis Rous praised the Athenians because they showed themselves 'tender-hearted', not just to men, 'but even to brute beasts'. 'Tender compassion to the brutes,' agreed Richard Baxter, had been put by God into 'all good men.'[5]

These were not merely pious assertions, for there is no shortage of well-attested instances of such tender-heartedness at work. When in 1614 'a cur dog' was thrown into a London privy by 'an untoward lad . . . , taking delight in Knavish pastimes', and left to remain there 'starving and crying for food' for three days, it was eventually rescued by 'the good man of the house, who grieved to see a dumb beast so starved and for want of food thus to perish'. Among the Diggers in 1649, says Gerrard Winstanley, 'tender hearts' grieved to see their cows bruised and swollen after being beaten by the lord of the manor's bailiffs. Two decades later Samuel Pepys was maddened to see the son of Sir Heneage Finch beating a little dog to death and letting it lie in pain. At Florence in 1672 an experiment upon a live dog in the presence of some Englishmen was ruined when its agonies 'moved the compassion of one of the servants . . . out of an untimely charity to rid her from these torments by striking her on the head with a stick'; even the vivisectors Hooke and Boyle decline to experiment upon the same animal twice 'because of the torture of the creature'.[6] When Colonel Abraham Holmes, a supporter of Monmouth, was executed with some of his companions at Lyme Regis in 1685, the horses could not pull the sled carrying the condemned men to the scaffold. The attendants began to whip them furiously, whereupon Colonel Holmes, with one of those superb gestures of which the men of the seventeenth century were so frequently capable, got out to walk, saying, 'Come, gentlemen, don't let the poor creatures suffer on our account. I have often led you in the field. Let me lead you on in our way to Heaven.'[7]

All this was before 1700. In the eighteenth century the sensibilities were not different in kind, but they seem to have been much more widely dispersed, and they were much more explicitly backed up by the religious and philosophical teaching of the time. . . .

. . . [B]y the 1720s 'benevolence' and 'charity' had become the most favoured words in literary vocabulary. There was something in human nature, said William Wollaston, which made the pains of others obnoxious to us. 'It is grievous to see or hear (and almost to hear of) any man, or even any animal whatever, in torment.' The mid eighteenth century saw a cult of tender-heartedness, a vogue for weeping and a widespread acceptance by the middle classes of the principle that 'to communicate happiness is the characteristic of virtue'.[8] Kindliness and benevolence had become official ideals.

It was this mode of thought which gave rise to later utilitarianism, for the benevolent, as Cowper put it, wished 'all that are capable of pleasure pleased;[9] and, although its main implications were for the human species, whether slaves, children, the criminal or the insane, its relevance to animals was inescapable. . . .

What this new mode of thinking implied was that it was the *feelings* of the suffering object which mattered, not its intelligence or moral capacity. 'Pain is pain,' wrote Humphry Primatt in 1776, 'whether it be inflicted on man or beast.' As Rousseau had said twenty years earlier, neither animals nor men should be unnecessarily ill-treated; they were equally sentient beings.[10] Or, as Jeremy Bentham observed in 1789 in a famous passage, the question to be asked about animals was neither 'Can they *reason?*' nor 'Can they *talk?*', but 'Can they *suffer?*'[11] This was a new and altogether more secular mode of approach. It was now possible to attack cruelty to animals without invoking God's intentions at all. The ill-treatment of beasts was reprehensible on the purely utilitarian grounds that it diminished their happiness. Animals had feelings and those feelings ought to be respected. Whether animals had reason was irrelevant. After all, as one of John Wesley's correspondents observed, if pity was to be extended only to those with reason, then 'those would lose their claim to our compassion who stand in the greatest need of it, namely children, idiots and lunatics'.[12] Neither was it necessary to prove that animals had souls, for, if they did not, then their lack of future recompense was all the greater argument for treating them considerately in this world. The emphasis on sensation thus became basic to those who crusaded on behalf of animals. . . .

Inevitably the creatures which excited most sympathy were those who communicated their sense of pain in most recognizably human terms. As a commentator wrote in 1762, 'we are moved most by the distressful cries of those animals that have any similitude to the human voice, such as the fawn, and the hare when seized by dogs.'[13] Thomas Bewick's feelings of humanity to animals were first aroused when, as a boy, he caught a hare in his arms, while it was surrounded by the hunters and their dogs; 'the poor terrified creature screamed out so piteously, like a child, that I would have given any thing to save its life.' Yet despite his repugnance for blood sports, Bewick retained an enthusiasm for angling: 'I argued myself into a belief that fish had little sense, and scarcely any feeling.'[14] It was this uncertainty as to whether fish had sensation, since, as well as being virtually bloodless, they did not cry out or change expression, which enabled angling to retain its reputation as a philosophical, contemplative and innocent pastime, given impeccable ancestry by the New Testament and particularly suitable for clergymen. . . .

It was in keeping with the new emphasis upon sensation that the eighteenth century witnessed growing criticism of some forms of cruelty which had passed without much comment in earlier periods. Vivisection, which Isaac Barrow in 1654 had described as 'innocent' and 'easily excusable', was castigated a century later by Dr Johnson as the work of 'a race of men that have practised tortures without pity and related them without shame and are yet suffered to erect their heads among human beings'.[15] Many of his contemporaries shared Johnson's disgust. In 1816 a Dr Wilson Philip was rejected by the ballot of the Royal Society because of the great offence given to Fellows by the cruelty of his experiments upon animals.[16] Conventional methods of meat production also came under attack. 'What rapes are committed upon nature,' exclaimed Defoe, 'making the ewes bring lambs all the winter, fattening calves to a monstrous size, using cruelties and contrary diets to the poor brute, to whiten its flesh for the palates of the ladies!' The eighteenth century saw much protest against practices like that of crimping fish (i.e. cutting their live flesh to make it firmer) or plucking poultry before they were dead. Even William Cobbett, whose general attitude was nothing if not realistic, held that to cause pain to animals in order to heighten the pleasure of the palate was an abuse of man's God-given authority.[17] In the later part of the century methods of slaughter also came under critical scrutiny. The treatment of cattle at Smithfield market was put under statutory surveillance in 1781. In 1786 slaughterhouses had to be licensed and there was much discussion of humane-killers.[18] Meanwhile demands increased for legislative action against all kinds of cruelty to animals.[19] In the later eighteenth century some grammar schools introduced rules against the ill-treatment of animals;[20] and even before Parliament began to act there were prosecutions for cruelty brought under the head of trespass, nuisance or malicious damage.[21]

In such ways the notion that the feelings of all sentient beings should be regarded began to affect educated opinion. Of course, there were dissenters from the general trend, like William Whewell, the mid-nineteenth-century Master of Trinity College, Cambridge, who dismissed the idea that man's happiness should sometimes be sacrificed to increase the pleasure of animals as a *reductio ad absurdum* of Benthamite teaching; a view for which he was sharply reprimanded by John Stuart Mill, who, unlike Whewell, was quite certain that it was not only humanity which was entitled to humane treatment.[22]

At one point in the late eighteenth century there was a move to suggest that plants, no less than animals, were entitled to consideration, on the same utilitarian grounds. This was not an altogether novel idea. Montaigne had urged that trees and plants should be treated with humanity;[23] and various seventeenth-century English writers had embroidered upon the implications of the supposed antipathies and sympathies of vegetation. It was not hard to believe, thought Nathanael Homes in 1661,

> that plants and trees and herbs have their passions or affections; their love appearing in their sympathy as . . . in the ivy and oak, etc.; their hatred in their antipathy, as in the vine and colewort, that will not prosper if near each other; their sorrow in pining and withering; their joy in blossom and flowering.[24]

[. . .]

But most contemporaries dismissed such sentiments as poetical fancies and in the nineteenth century botany and poetry went their separate ways. It was animals, not plants, who gained most from the new disposition to found the rights of man upon his capacity for happiness. Indeed, if men had rights, so too did they. In a posthumous work published in 1755 the philosopher Francis Hutcheson declared that brutes 'have a *right* that no useless pain or misery should be inflicted on them'. Flies were as capable of pain as men, said Thomas Percival in 1775, and had no less a 'right' to life, liberty and enjoyment. 'The day may come,' wrote Jeremy Bentham in 1789, 'when the rest of the animal creation may acquire those rights which only human tyranny has withheld from them.' In 1798 John Lawrence proposed that the rights of beasts should be acknowledged by the state: 'the *ius animalium* . . . surely ought to form a part of the jurisprudence of every system founded on the principles of justice and humanity.' Cruelty now was not merely inhumane; it was unjust.[25]

There were also hints that the rights of animals extended to something more than mere protection from physical pain. Man's stewardship of creation, as Thomas Tryon had stressed, involved 'assisting those beasts to the obtaining of all the advantages their natures are by the great, beautiful and always beneficent creator made capable of'. They were capable of pleasures of the mind, Lord Monboddo would urge, of fellowship of the herd and affection for offspring.[26] Bernard Mandeville raised doubts about the ethics of castrating domestic animals as early as 1714. A hundred years later Shelley denounced the practice as 'unnatural and inhuman', while the physician William Lambe called it 'a shocking outrage on the common rights of nature'.[27] In the late Victorian age some humanitarians would urge the right of animals to 'self-realization'.[28]

So much then for the intellectual origins of the campaign against unnecessary cruelty to animals. It grew out of the (minority) Christian tradition that man should take care of God's creation. It was enhanced by the collapse of the old view that the world existed exclusively for humanity; and it was consolidated by a new emphasis on sensation and feeling as the true basis for a claim to moral consideration. In this way the anthropocentric tradition was, by a subtle dialectic, relentlessly adjusted to bring animals within the sphere of moral concern. The debate on animals thus furnishes yet another illustration of that shift to more secular modes of thinking which was characteristic of so much thought in the early modern period.

Notes

1 John Dyer, 'The Fleece', ii. lines 22–3; *Collected Poems of Christopher Smart*, i. 251, 252; *Child-Rearing. Historical Sources*, ed. Philip C. Greven (Itasca, Illinois, 1973), 66; James Granger, *An Apology for the Brute Creation* (1772), 17; Allen, *Naturalist*, 145–7. Cf. Cowper, 'The Task', vi. 560, 562–3.

2 Laurence Sterne, *The Life and Opinions of Tristram Shandy* (1759–67), bk ii, chap. 12; *The Diary of Sylas Neville*, ed. Basil Cozens-Hardy (1950), 25; [William Melmoth], *The Letters of Sir Thomas Fitzosborne* (1742; 1776 edn), letter 16; *The Works of August Toplady* (new edn, 1853), 526; Chafin, *Anecdotes and History of Cranbourn Chase*, 63–4.

3 *Essays of Montaigne*, trans. Florio, ii. 119; Margaret Cavendish, Duchess of Newcastle, *The Life of William Cavendish*, ed. C. H. Firth (2nd edn, n.d.), 175.

4 Shakespeare, *Measure for Measure*, iii. 1. line 79; *Titus Andronicus*, iii. 2. lines 54–65; and references cited in Caroline Spurgeon, *Shakespeare's Imagery* (Cambridge, 1935), 104–9; *John Stubbs's Gaping Gulf*, ed. Lloyd E. Berry (Charlottesville, 1968), 28.

5 *Truth's Innocency and Simplicity shining through . . . Thomas Taylor* (1697), 128–9; Francis Rous, *Archaeologiae Atticae* (8th edn, Oxford, 1675), 271; Frederick J. Powicke, 'The Reverend Richard Baxter's Last Treatise', *Bull. John Rylands Lib.*, x (1926), 197.

6 *Deeds against Nature, and Monsters by Kinde* (1614), sig. A4; *The Works of Gerrard Winstanley*, ed. George H. Sabine (Ithaca, N.Y., 1941), 329; Pepys, *Diary*, ix. 203; *Oldenburg*, ix. 187; Wallace Shugg, 'Humanitarian Attitudes in the Early Animal Experiments of the Royal Society', *Ann. Sci.*, 24 (1968), 231–2.

7 MS. note in copy in present writer's possession of William Turner, *A Compleat History of the Most Remarkable Providences* (1697), iii. 129.

8 John Hall of Richmond, *Of Government and Obedience* (1654), 300; Samuel Parker, *A Demonstration of the Divine Authority of the Law of Nature* (1681), 54–5; Cecil A. Moore, *Backgrounds of English Literature*, 1700–1760 (Minneapolis, 1953), chap. I; [William Wollaston), *The Religion of Nature* (5th edn, 1731), 139; *The Adventurer*, 37 (1753).

9 'The Task', vi. line 345.

10 Primatt, *Dissertation*, 7; *The Political Writings of Jean Jacques Rousseau*, ed. C. E. Vaughan (reprint, Oxford, 1962), i. 138.

11 Jeremy Bentham, *An Introduction to the Principles of Morals and Legislation*, ed. Wilfrid Harrison (Oxford, 1948), 412n; and a similar passage in Bentham MSS. in Univ. Coll. London cit. by Amnon Goldworth in *The New York Rev. of Books*, xx (20 Sep. 1973), 42. It is possible that Bentham subsequently reverted to the argument that cruelty to animals was bad because it led to cruelty to men; see David Baumgardt, *Bentham and the Ethics of Today* (Princeton, N.J., 1952), 338–9, 362–3.

12 *The Journal of John Wesley*, ed. Nehemiah Curnock (1909–16), iv. 176.

13 Thomas Sheridan, *A Course of Lectures on Elocution* (1762), 104.

14 *A Memoir of Thomas Bewick written by himself*, ed. Ian Bain (1979 edn), 6, 178.

15 *Johnson on Shakespeare*, ed. Walter Raleigh (1908), 181. Cf. *The Idler*, 17 (1758).

16 *The Banks Letters*, ed. Warren R. Dawson (1958), 506. Cf. *Gentleman's Mag.*, x (1740), 194; Pennant, *Zoology*, i. 165; [Robert Wallace], *Various Prospects* (1761), 17; [T. J. Mathias], *The Pursuits of Literature* (5th edn, 1798), 340–2; Joseph Spence, *Anecdotes*, ed. Samuel Weller Singer (1820), 293.

17 Daniel Defoe, *The Complete English Tradesman* (1841 edn), ii. 228–9; A Prebendary of York, *An Enquiry about the Lawfulness of Eating Blood* (1733), 10; Priscilla Wakefield, *Instinct Displayed* (1811), 289; Young, *Essay on Humanity to Animals*, chap. vi; William Cobbett, *Cottage Economy* (1926 edn), 126.

18 21 Geo. III c. 67 (1781); 26 Geo. III c. 71 (1786); Lawrence, *Horses*, i. 155–8; *Memoirs of Sir Samuel Romilly*, ii. 109–11.

19 E.g. Baker, *Peregrinations of the Mind*, 181; James Burgh, *An Account of the First Settlement . . . of the Cessares* (1764), 74–5; Cowper, 'The Task', vi. lines 432–3; *Works of Toplady*, 443; Primatt, *Dissertation*, 289; Lawrence, *Horses*, i. 123; Byng, iv. 151; *The Sporting Magazine*, iv (July 1794), 199.

20 E.g. L. Brettle, *A History of Queen Elizabeth's Grammar School for Boys, Mansfield* (Mansfield, n.d. [1961]), 43; W. T. Carless, *A Short History of Hereford School* (Hereford, 1914), 74.

21 E.g. *Bedfordshire County Records. Notes and Extracts from the . . . Quarter Sessions Rolls from 1714 to 1832* (Bedford, n.d.), i. 60; *The Sporting Magazine*, iv (July 1794), 188.

22 William Whewell, *Lectures on the History of Moral Philosophy* (1852), 223–5; John Stuart Mill, *Essays on Ethics, Religion and Society*, ed. J. M. Robson (1969), 185–7.

23 *Essays*, trans. Florio, ii. 126. Cf. Spink, *French Free-Thought*, 56, 61.

24 Nathaniel Homes, *The Resurrection-Revealed raised above Doubts and Difficulties* (1661), 244; J. W[orlidge], *Systema Horti-Culturae* (1677), 283; *The Autobiography of Edward, Lord Herbert of Cherbury*, ed. Sidney Lee (2nd edn, n.d.), 31.

25 Francis Hutcheson, *A System of Moral Philosophy* (1755), i. 314; Thomas Percival, *A Father's Instructions* (5th edn, 1781), 26–7; Bentham, *Introduction to Principles of Morals*, 412n; Lawrence, *Horses*, i. 123. Cf. Baker, *Peregrinations*, 182. Thomas Taylor's *A Vindication of the Rights of Brutes* (1792) cannot certainly be aligned with this school of thought. It is a curious work, in part satirical, in part serious, in part mildly pornographic.

26 Tryon, quoted above, p. 155; [James Burnet, Lord Monboddo], *Antient Metaphysics* (1779–99), ii. 103.

27 Mandeville, *Fable of the Bees*, i. 180; P. B. Shelley, *A Vindication of Natural Diet* (1813; 1884 edn), 13; William Lambe, *Additional Reports on the Effects of a Peculiar Regimen* (1815), 238.

28 E.g. Henry S. Salt, *Animals' Rights* (1892), 15.

Source: Keith Thomas (1984) *Man and the Natural World, 1500–1800*, London: Penguin, pp. 173–80.

3.4

Peter Singer

All animals are equal: the utilitarian case[1]

In recent years a number of oppressed groups have campaigned vigorously for equality. The classic instance is the Black Liberation movement, which demands an end to the prejudice and discrimination that has made blacks second-class citizens. The immediate appeal of the Black Liberation movement and its initial, if limited, success made it a model for other oppressed groups to follow. We became familiar with liberation movements for Spanish-Americans, gay people, and a variety of other minorities. When a majority group – women – began their campaign, some thought we had come to the end of the road. Discrimination on the basis of sex, it has been said, is the last universally accepted form of discrimination, practised without secrecy or pretence even in those liberal circles that have long prided themselves on their freedom from prejudice against racial minorities.

One should always be wary of talking of 'the last remaining form of discrimination'. If we have learnt anything from the liberation movements, we should have learnt how difficult it is to be aware of latent prejudice in our attitudes to particular groups until this prejudice is forcefully pointed out.

A liberation movement demands an expansion of our moral horizons and an extension or reinterpretation of the basic moral principle of equality. Practices that were previously regarded as natural and inevitable come to be seen as the result of an unjustifiable prejudice. Who can say with confidence that all his or her attitudes and practices are beyond criticism? If we wish to avoid being numbered amongst the oppressors, we must be prepared to re-think even our most fundamental attitudes. We need to consider them from the point of view of those most disadvantaged by our attitudes, and the practices that follow from these attitudes. If we can make this unaccustomed mental switch, we may discover a pattern in our attitudes and practices that consistently operates so as to benefit one group – usually the one to which we ourselves belong – at the expense of another. In this way we may come to see that there is a case for a new liberation movement. My aim is to advocate that we make this mental switch in respect of our attitudes and practices towards a very large group of beings: members of species other than our own – or, as we popularly though misleadingly call them, animals. In other words, I am urging that we extend to other species the basic principle of equality that most of us recognize should be extended to all members of our own species.

All this may sound a little far-fetched, more like a parody of other liberation movements than a serious objective. In fact, in the past the idea of 'The Rights of Animals' really has been used to parody the case for women's rights. When Mary Wollstonecraft, a forerunner of later feminists, published her *Vindication of the Rights of Women* in 1792, her ideas were widely regarded as absurd, and they were satirized in an anonymous publication entitled *A Vindication of the Rights of Brutes*. The author of this satire (actually Thomas Taylor, a distinguished Cambridge philosopher) tried to refute Wollstonecraft's reasonings by showing that they could be carried one stage further. If sound when applied to women, why should the arguments not be applied to dogs, cats, and horses? They seemed to hold equally well for these 'brutes'; yet to hold that brutes had rights was manifestly absurd; therefore the reasoning by which this conclusion had been reached must be unsound, and if unsound when applied to brutes, it must also be unsound when applied to women, since the very same arguments had been used in each case.

One way in which we might reply to this argument is by saying that the case for equality between men and women cannot validly be extended to non-human animals. Women have a right to vote, for instance, because they are just as capable of making rational decisions as men are; dogs, on the other hand, are incapable of understanding the significance of voting, so they cannot have the right to vote. There are many other obvious ways in which men and women resemble each other closely, while humans and other animals differ greatly. So, it might be said, men and women are similiar beings, and should have equal rights, while humans and non-humans are different and should not have equal rights.

The thought behind this reply to Taylor's analogy is correct up to a point, but it does not go far enough. There *are* important differences between humans and other animals, and these differences must give rise to *some* differences, in the rights that each have. Recognizing this obvious fact, however, is no barrier to the case for extending the basic principle of equality to non-human animals. The differences that exist between men and women are equally undeniable, and the supporters of Women's Liberation are aware that these differences may give rise to different rights. Many feminists hold that women have the right to an abortion on request. It does not follow that since these same people are campaigning for equality between men and women, they must support the right of men to have abortions too. Since a man cannot have an abortion, it is meaningless to talk of his right to have one. Since a pig can't vote, it is meaningless to talk of its right to vote. There is no reason why either Women's Liberation or Animal Liberation should get involved in such nonsense. The extension of the basic principle of equality from one group to another does not imply that we must treat both groups in exactly the same way, or grant exactly the same rights to both groups. Whether we should do so will depend on the nature of the members of the two groups. The basic principle of equality, I shall argue, is equality of consideration; and equal consideration for different beings may lead to different treatment and different rights.

So there is a different way of replying to Taylor's attempt to parody Wollstonecraft's arguments, a way which does not deny the differences between humans and non-humans, but goes more deeply into the question of equality, and concludes by finding nothing absurd in the idea that the basic principle of equality applies to so-called 'brutes'. I believe that we reach this conclusion if we examine the basis on which our opposition to discrimination on grounds of race or sex ultimately rests. We will then see that we would be on shaky ground if we were to demand equality for blacks, women, and other groups of oppressed humans while denying equal consideration to non-humans.

When we say that all human beings, whatever their race, creed, or sex, are equal, what is it that we are asserting? Those who wish to defend a hierarchical, inegalitarian society have often pointed out that by whatever test we choose, it simply is not true that all humans are equal. Like it or not, we must face the fact that humans come in different shapes and sizes; they come with differing moral capacities, differing intellectual abilities, differing amounts of benevolent feeling and sensitivity to the needs of others, differing abilities to communicate effectively, and differing capacities to experience pleasure and pain. In short, if the demand for equality were based on the actual equality of all human beings, we would have to stop demanding equality. It would be an unjustifiable demand.

[. . .]

. . . Equality is a moral ideal, not a simple assertion of fact. There is no logically compelling reason for assuming that a factual difference in ability between two people justifies any difference in the amount of consideration we give to satisfying their needs and interests. The principle of the equality of human beings is not a description of an alleged actual equality among humans: it is a prescription of how we should treat humans.

Jeremy Bentham incorporated the essential basis of moral equality into his utilitarian system of ethics in the formula: 'Each to count for one and none for more than one.' In other words, the interests of every being affected by an action are to be taken into account and given the same weight as the like interests of any other being. A later utilitarian, Henry Sidgwick, put the point in this way: 'The good of any one individual is of no more importance, from the point of view (if I may say so) of the Universe, than the good of any other.'[2] More recently, the leading figures in modern moral philosophy have shown a great deal of agreement in specifying as a fundamental presupposition of their moral theories some similar requirement which operates so as to give everyone's interests equal consideration – although they cannot agree on how this requirement is best formulated.[3]

It is an implication of this principle of equality that our concern for others ought not to depend on what they are like, or what abilities they possess – although precisely what this concern requires us to do may vary according to the characteristics of those affected by what we do. It is on this basis that the case against racism and the case against sexism must both ultimately rest; and it is in accordance with this principle that speciesism is also to be condemned. If possessing a higher degree of intelligence does not entitle one human to use another for his own ends, how can it entitle humans to exploit non-humans?

Many philosophers have proposed the principle of equal consideration of interests, in some form or other, as a basic moral principle; but, as we shall see in more detail shortly, not many of them have recognized that this principle applies to members of other species as well as to our own. Bentham was one of the few who did realize this. In a forward-looking passage, written at a time when black slaves in the British dominions were still being treated much as we now treat non-human animals, Bentham wrote:

> The day *may* come when the rest of the animal creation may acquire those rights which never could have been witholden from them but by the hand of tyranny. The French have already discovered that the blackness of the skin is no reason why a human being should be abandoned without redress to the caprice of a tormentor. It may one day come to be recognized that the number of the legs, the villosity of the skin, or the termination of the *os sacrum*, are reasons equally insufficient for abandoning a sensitive being to the same fate. What else is it that should trace the insuperable line? Is it the faculty of reason, or perhaps the faculty of discourse? But a full-grown horse or dog is beyond comparison a more rational, as well as a more conversable animal, than an infant of a day, or a week, or even a month, old. But suppose they were otherwise, what would it avail? The question is not, Can they *reason*? nor Can they *talk*? but, *Can they suffer*?[4]

In this passage Bentham points to the capacity for suffering as the vital characteristic that gives a being the right to equal consideration. The capacity for suffering – or more strictly, for suffering and/or enjoyment or happiness – is not just another characteristic like the capacity for language, or for higher mathematics. Bentham is not saying that those who try to mark 'the insuperable line' that determines whether the interests of a being should be considered happen to have selected the wrong characteristic. The capacity for suffering and enjoying things is a pre-requisite for having interests at all, a condition that must be satisfied before we can speak of interests in any meaningful way. It would be nonsense to say that it was not in the interests of a stone to be kicked along the road by a schoolboy. A stone does not have interests because it cannot suffer. Nothing that we can do to it could possibly make any difference to its welfare. A mouse, on the other hand, does have an interest in not being tormented, because it will suffer if it is.

If a being suffers, there can be no moral justification for refusing to take that suffering into consideration. No matter what the nature of the being, the principle of equality requires that its suffering be counted equally with the like suffering – in so far as rough comparisons can be made – of any other being. If a being is not capable of suffering, or of experiencing enjoyment or happiness, there is nothing to be taken into account. This is why the limit of sentience (using the term as a convenient, if not strictly accurate, shorthand for the capacity to suffer or experience enjoyment or happiness) is the only defensible boundary of concern for the interests of others. To mark this boundary by some characteristic like intelligence or rationality would be to mark it in an arbitrary way. Why not choose some other characteristic, like skin colour?

The racist violates the principle of equality by giving greater weight to the interests of members of his own race, when there is a clash between their interests and the interests of those of another race. Similarly the speciesist allows the interests of his own species to override the greater interests of members of other species.[5] The pattern is the same in each case. Most human beings are speciesists. I shall now very briefly describe some of the practices that show this.

For the great majority of human beings, especially in urban, industrialized societies, the most direct form of contact with members of other species is at meal-times: we eat them. In doing so we treat them purely as means to our ends. We regard their life and well-being as subordinate to our taste for a particular kind of dish. I say 'taste' deliberately – this is purely a matter of pleasing our palate. There can be no defence of eating flesh in terms of satisfying nutritional needs, since it has been established beyond doubt that we could satisfy our need for protein and other essential nutrients far more efficiently with a diet that replaced animal flesh by soy beans, or products derived from soy beans, and other high-protein vegetable products.[6]

It is not merely the act of killing that indicates what we are ready to do to other species in order to gratify our tastes. The suffering we inflict on the animals while they are alive is perhaps an even clearer indication of our speciesism than the fact that we are prepared to kill them.[7] In order to have meat on the table at a price that people can afford, our society tolerates methods of meat production that confine sentient animals in cramped, unsuitable conditions for the entire duration of their lives. Animals are treated like machines that convert fodder into flesh, and any innovation that results in a higher 'conversion ratio' is liable to be adopted. As one authority on the subject has said, 'cruelty is acknowledged only when profitability ceases'.[8]

Since, as I have said, none of these practices cater for anything more than our pleasures of taste, our practice of rearing and killing other animals in order to eat them is a clear instance of the sacrifice of the most important interests of other beings in order to satisfy trivial interests of our own. To avoid speciesism we must stop this practice, and each of us has a moral obligation to cease supporting the practice. Our custom is all the support that the meat industry needs. The decision to cease giving it that support may be difficult, but it is no more difficult than it would have been for a white Southerner to go against the traditions of his society and free his slaves: if we do not change our dietary habits, how can we censure those slave-holders who would not change their own way of living?

The same form of discrimination may be observed in the widespread practice of experimenting on other species in order to see if certain substances are safe for human beings, or to test some psychological theory about the effect of severe punishment on learning, or to try out various new compounds just in case something turns up . . .

In the past, argument about vivisection has often missed this point, because it has been put in absolutist terms: Would the abolitionist be prepared to let thousands die if they could be saved by experimenting on a single animal? The way to reply to this purely hypothetical question is to pose another: Would the experimenter be prepared to perform his experiment on an orphaned human infant, if that were the only way to save many lives? (I say 'orphan' to avoid the complication of parental feelings, although in doing so I am being over-fair to the experimenter, since the non-human subjects of experiments are not orphans.) If the experimenter is not prepared to use an orphaned human infant, then his readiness to use non-humans is simple discrimination, since adult apes, cats, mice, and other mammals are more aware of what is happening to them, more self-directing and, so far as we can tell, at least as sensitive to pain, as any human infant. There seems to be no relevant characteristic that human infants possess that adult mammals do not have to the same or a higher degree. (Someone might try to argue that what makes it wrong to experiment on a human infant is that the infant will, in time and if left alone, develop into more than the non-human, but one would then, to be consistent, have to oppose abortion, since the foetus has the same potential as the infant – indeed, even contraception and abstinence might be wrong on this ground, since the egg and sperm, considered jointly, also have the same potential. In any case, this argument still gives us no reason for selecting a non-human, rather than a human with severe and irreversible brain damage, as the subject for our experiments.)

The experimenter, then, shows a bias in favour of his own species whenever he carries out an experiment on a non-human for a purpose that he would not think justified him in using a human being at an equal or lower level of sentience, awareness, ability to be self-directing, etc. No one familiar with the kind of results yielded by most experiments on animals can have the slightest doubt that if this bias were eliminated the number of experiments performed would be a minute fraction of the number performed today.

Notes

1 Part of this essay appeared in the *New York Review of Books* (5 Apr. 1973), and is reprinted by permission of the Editor. This version is an abridged form of an essay which was first published in *Philosophic Exchange* vol. 1, no. 5 (Summer 1974).

2 *The Methods of Ethics* (7th edn.), p. 382.

3 For example, R.M. Hare, *Freedom and Reason* (Oxford, 1963) and J. Rawls, *A Theory of Justice* (Harvard, 1972); for a brief account of the essential agreement on this issue between these and

other positions, see R.M. Hare, 'Rules of War and Moral Reasoning', *Philosophy and Public Affairs*, vol. I, no. 2 (1972).

4 *Introduction to the Principles of Morals and Legislation*, ch. XVII.

5 I owe the term 'speciesism' to Richard Ryder.

6 In order to produce 1 lb. of protein in the form of beef or veal, we must feed 21 lb. of protein to the animal. Other forms of livestock are slightly less inefficient, but the average ratio in the US is still 1:8. It has been estimated that the amount of protein lost to humans in this way is equivalent to 90 per cent of the annual world protein deficit. For a brief account, see Frances Moore Lappé, *Diet for a Small Planet* (Friends of The Earth/Ballantine, New York, 1971), pp. 4–11.

7 Although one might think that killing a being is obviously the ultimate wrong one can do to it, I think that the infliction of suffering is a clearer indication of speciesism because it might be argued that at least part of what is wrong with killing a human is that most humans are conscious of their existence over time, and have desires and purposes that extend into the future – see, for instance, M. Tooley, 'Abortion and Infanticide', *Philosophy and Public Affairs*, vol. 2, no. 1 (1972). Of course, if one took this view one would have to hold – as Tooley does – that killing a human infant or mental defective is not in itself wrong, and is less serious than killing certain higher mammals that probably do have a sense of their own existence over time.

8 Ruth Harrison, *Animal Machines* (London, 1964). For an account of farming conditions, see my *Animal Liberation* (New York, 1975).

Source: Peter Singer (1986) 'All animals are equal', in Peter Singer (ed.) *In Defence of Animals*, Oxford: Blackwell, pp. 13–24.

3.5

Tom Regan

Animal rights: the Kantian case

I regard myself as an advocate of animal rights – as a part of the animal rights movement. That movement, as I conceive it, is committed to a number of goals, including:

- the total abolition of the use of animals in science;
- the total dissolution of commercial animal agriculture;
- the total elimination of commercial and sport hunting and trapping.

There are, I know, people who profess to believe in animal rights but do not avow these goals. Factory farming, they say, is wrong – it violates animals' rights – but traditional animal agriculture is all right. Toxicity tests of cosmetics on animals violates their rights, but important medical research – cancer research, for example – does not. The clubbing of baby seals is abhorrent, but not the harvesting of adult seals. I used to think I understood this reasoning. Not any more. You don't change unjust institutions by tidying them up.

What's wrong – fundamentally wrong – with the way animals are treated isn't the details that vary from case to case. It's the whole system. The forlornness of the veal calf is pathetic, heart wrenching; the pulsing pain of the chimp with electrodes planted deep in her brain is repulsive; the slow, tortuous death of the racoon caught in the leg-hold trap is agonizing. But what is wrong isn't the pain, isn't the suffering, isn't the deprivation. These compound what's wrong. Sometimes – often – they make it much, much worse. But they are not the fundamental wrong.

The fundamental wrong is the system that allows us to view animals as *our resources*, here for *us* – to be eaten, or surgically manipulated, or exploited for sport or money. Once we accept this view of animals – as our resources – the rest is as predictable as it is regrettable. Why worry about their loneliness, their pain, their death? Since animals exist for us, to benefit us in one way or another, what harms them really doesn't matter – or matters only if it starts to bother us, makes us feel a trifle uneasy when we eat our veal escalope, for example. So, yes, let us get veal calves out of solitary confinement, give them more space, a little straw, a few companions. But let us keep our veal escalope.

But a little straw, more space and a few companions won't eliminate – won't even touch – the basic wrong that attaches to our viewing and treating these animals as our resources. A veal calf killed to be eaten after living in close confinement is viewed and treated in this way: but so, too, is another who is raised (as they say) 'more humanely'. To right the wrong of our treatment of farm animals requires more than making rearing methods 'more humane'; it requires the total dissolution of commercial animal agriculture.

How we do this, whether we do it or, as in the case of animals in science, whether and how we abolish their use – these are to a large extent political questions. People must change their beliefs before they change their habits. Enough people, especially those elected to public office, must believe in change – must want it – before we will have laws that protect the rights of animals. This process of change is very complicated, very demanding, very exhausting, calling for the efforts of many hands in education, publicity, political organization and activity, down to the licking of envelopes and stamps. As a trained and practising philosopher, the sort of contribution I can make is limited but, I like to think, important. The currency of philosophy is ideas – their meaning and rational foundation – not the nuts and bolts of the legislative process, say, or the mechanics of community organization. That's what I have been exploring over the past ten years or so in my essays and talks and, most recently, in my book, *The Case for Animal Rights*. I believe the major conclusions I reach in the book are true because they are supported by the weight of the best arguments. I believe the idea of animal rights has reason, not just emotion, on its side.

In the space I have at my disposal here I can only sketch, in the barest outline, some of the main features of the book. Its main themes – and we should not be surprised by this – involve asking and answering deep, foundational moral questions about what morality is, how it should be understood and what is the best moral theory, all considered. I hope I can convey something of the shape I think this theory takes. The attempt to do this will be (to use a word a friendly critic once used to describe my work) cerebral, perhaps too cerebral. But this is misleading. My feelings about how animals are sometimes treated run just as deep and just as strong as those of my more volatile compatriots. Philosophers do – to use the jargon of the day – have a right side to their brains. If it's the left side we contribute (or mainly should), that's because what talents we have reside there.

How to proceed? We begin by asking how the moral status of animals has been understood by thinkers who deny that animals have rights. Then we test the mettle of their ideas by seeing how well they stand up under the heat of fair criticism. If we start our thinking in this way, we soon find that some people believe that we have no duties directly to animals, that we owe nothing to them, that we can do nothing that wrongs them. Rather, we can do wrong acts that involve animals, and so we have duties regarding them, though none to them. Such views may be called indirect duty views. By way of illustration: suppose your neighbour kicks your dog. Then your neighbour has done something wrong. But not to your dog. The wrong that has been done is a wrong to you. After all, it is wrong to upset people, and your neighbour's kicking your dog upsets you. So you are the one who is wronged, not your dog. Or again: by kicking your dog your neighbour damages your property. And since it is wrong to damage another person's property, your neighbour has done something wrong – to you, of course, not to your dog. Your neighbour no more wrongs your dog than your car would be wronged if the windshield were smashed. Your neighbour's duties involving your dog are indirect duties to you. More generally, all of our duties regarding animals are indirect duties to one another – to humanity.

How could someone try to justify such a view? Someone might say that your dog doesn't feel anything and so isn't hurt by your neighbour's kick, doesn't care about the pain since none is felt, is as unaware of anything as is your windshield. Someone might say this, but no rational person will, since, among other considerations, such a view will commit anyone who holds it to the position that no human being feels pain either – that human beings also don't care about what happens to them. A second possibility is that though both humans and your dog are hurt when kicked, it is only human pain that matters. But, again, no rational person can believe this. Pain is pain wherever it occurs. If your neighbour's causing you pain is wrong because of the pain that is caused, we cannot rationally ignore or dismiss the moral relevance of the pain that your dog feels.

Philosophers who hold indirect duty views – and many still do – have come to understand that they must avoid the two defects just noted: that is, both the view that animals don't feel anything as well as the idea that only human pain can be morally relevant. Among such thinkers the sort of view now favoured is one or other form of what is called *contractarianism*.

Here, very crudely, is the root idea: morality consists of a set of rules that individuals voluntarily agree to abide by, as we do when we sign a contract (hence the name contractarianism). Those who understand and accept the terms of the contract are covered directly; they have rights created and recognized by, and protected in, the contract. And these contractors can also have protection spelled out for others who, though they lack the ability to understand morality and so cannot sign the contract themselves, are loved or cherished by those who can. Thus young children, for example, are unable to sign contracts and lack rights. But they are protected by the contract none the less because of the sentimental interests of others, most notably their parents. So we have, then, duties involving these children, duties regarding them, but no duties to them. Our duties in their case are indirect duties to other human beings, usually their parents.

As for animals, since they cannot understand contracts, they obviously cannot sign; and since they cannot sign, they have no rights. Like children, however, some animals are the objects of the sentimental interest of others. You, for example, love your dog or cat. So those animals that enough people care about (companion animals, whales, baby seals, the American bald eagle), though they lack rights themselves, will be protected because of the sentimental interests of people. I have, then, according to contractarianism, no duty directly to your dog or any other animal, not even the duty not to cause them pain or suffering; my duty not to hurt them is a duty I have to those people who care about what happens to them.

As for other animals, where no or little sentimental interest is present – in the case of farm animals, for example, or laboratory rats – what duties we have grow weaker and weaker, perhaps to vanishing point. The pain and death they endure, though real, are not wrong if no one cares about them.

When it comes to the moral status of animals, contractarianism could be a hard view to refute if it were an adequate theoretical approach to the moral status of human beings. It is not adequate in this latter respect, however, which makes the question of its adequacy in the former case, regarding animals, utterly moot. For consider: morality, according to the (crude) contractarian position before us, consists of rules that people agree to abide by. What people? Well, enough to make a difference – enough, that is, *collectively* to have the power to enforce the rules that are drawn up in the contract. That is very well and good for the signatories but not so good for anyone who is not asked to sign. And there is nothing in contractarianism of the sort we are discussing that guarantees or requires that everyone will have a chance to participate equally in framing the rules of morality. The result is that this approach to ethics could sanction the most blatant forms of social, economic, moral and political injustice, ranging from a repressive caste system to systematic racial or sexual discrimination. Might, according to this theory, does make right. Let those who are the victims of injustice suffer as they will. It matters not so long as no one else – no contractor, or too few of them – cares about it. Such a theory takes one's moral breath away . . . as if, for example, there would be nothing wrong with apartheid in South Africa if few white South Africans were upset by it. A theory with so little to recommend it at the level of the ethics of our treatment of our fellow humans cannot have anything more to recommend it when it comes to the ethics of how we treat our fellow animals.

The version of contractarianism just examined is, as I have noted, a crude variety, and in fairness to those of a contractarian persuasion it must be noted that much more refined, subtle and ingenious varieties are possible. For example, John Rawls, in his *A Theory of Justice*, sets forth a version of contractarianism that forces contractors to ignore the accidental features of being a human being – for example, whether one is white or black, male or female, a genius or of modest intellect. Only by ignoring such features, Rawls believes, can we ensure that the principles of justice that contractors would agree upon are not based on bias or prejudice. Despite the improvement a view such as Rawls's represents over the cruder forms of contractarianism, it remains deficient: it systematically denies that we have direct duties to those human beings who do not have a sense of justice – young children, for instance, and many mentally retarded humans. And yet it seems reasonably certain that, were we to torture a young child or a retarded elder, we would be doing something that wronged him or her, not something that would be wrong if (and only if) other humans with a sense of justice were upset. And since this is true in the case of these humans, we cannot rationally deny the same in the case of animals.

Indirect duty views, then, including the best among them, fail to command our rational assent. Whatever ethical theory we should accept rationally, therefore, it must at least recognize that we have some duties directly to animals, just as we have some duties directly to each other. The next two theories I'll sketch attempt to meet this requirement.

The first I call the cruelty–kindness view. Simply stated, this says that we have a direct duty to be kind to animals and a direct duty not to be cruel to them. Despite the familiar, reassuring ring of these ideas, I do not believe that this view offers an adequate theory. To make this clearer, consider kindness. A kind person acts from a certain kind of motive

– compassion or concern, for example. And that is a virtue. But there is no guarantee that a kind act is a right act. If I am a generous racist, for example, I will be inclined to act kindly towards members of my own race, favouring their interests above those of others. My kindness would be real and, so far as it goes, good. But I trust it is too obvious to require argument that my kind acts may not be above moral reproach – may, in fact, be positively wrong because rooted in injustice. So kindness, notwithstanding its status as a virtue to be encouraged, simply will not carry the weight of a theory of right action.

Cruelty fares no better. People or their acts are cruel if they display either a lack of sympathy for or, worse, the presence of enjoyment in another's suffering. Cruelty in all its guises is a bad thing, a tragic human failing. But just as a person's being motivated by kindness does not guarantee that he or she does what is right, so the absence of cruelty does not ensure that he or she avoids doing what is wrong. Many people who perform abortions, for example, are not cruel, sadistic people. But that fact alone does not settle the terribly difficult question of the morality of abortion. The case is no different when we examine the ethics of our treatment of animals. So, yes, let us be for kindness and against cruelty. But let us not suppose that being for the one and against the other answers questions about moral right and wrong.

Some people think that the theory we are looking for is utilitarianism. A utilitarian accepts two moral principles. The first is that of equality: everyone's interests count, and similar interests must be counted as having similar weight or importance. White or black, American or Iranian, human or animal – everyone's pain or frustration matter, and matter just as much as the equivalent pain or frustration of anyone else. The second principle a utilitarian accepts is that of utility: do the act that will bring about the best balance between satisfaction and frustration for everyone affected by the outcome.

As a utilitarian, then, here is how I am to approach the task of deciding what I morally ought to do: I must ask who will be affected if I choose to do one thing rather than another, how much each individual will be affected, and where the best results are most likely to lie – which option, in other words, is most likely to bring about the best results, the best balance between satisfaction and frustration. That option, whatever it may be, is the one I ought to choose. That is where my moral duty lies.

The great appeal of utilitarianism rests with its uncompromising *egalitarianism*: everyone's interests count and count as much as the like interests of everyone else. The kind of odious discrimination that some forms of contractarianism can justify – discrimination based on race or sex, for example – seems disallowed in principle by utilitarianism, as is speciesism, systematic discrimination based on species membership.

The equality we find in utilitarianism, however, is not the sort an advocate of animal or human rights should have in mind. Utilitarianism has no room for the equal moral rights of different individuals because it has no room for their equal inherent value or worth. What has value for the utilitarian is the satisfaction of an individual's interests, not the individual whose interests they are. A universe in which you satisfy your desire for water, food and warmth is, other things being equal, better than a universe in which these desires are frustrated. And the same is true in the case of an animal with similar desires. But neither you nor the animal have any value in your own right. Only your feelings do.

Here is an analogy to help make the philosophical point clearer: a cup contains different liquids, sometimes sweet, sometimes bitter, sometimes a mix of the two. What has value are the liquids: the sweeter the better, the bitterer the worse. The cup, the container, has no value. It is what goes into it, not what they go into, that has value. For the utilitarian, you and I are like the cup; we have no value as individuals and thus no equal value. What has value is what goes into us, what we serve as receptacles for; our feelings of satisfaction have positive value, our feelings of frustration negative value.

Serious problems arise for utilitarianism when we remind ourselves that it enjoins us to bring about the best consequences. What does this mean? It doesn't mean the best consequences for me alone, or for my family or friends, or any other person taken individually. No, what we must do is, roughly, as follows: we must add up (somehow!) the separate satisfactions and frustrations of everyone likely to be affected by our choice, the satisfactions in one column, the frustrations in the other. We must total each column for each of the options before us. That is what it means to say the theory is aggregative. And then we must choose that option which is most likely to bring about the best balance of totalled satisfactions over totalled frustrations. Whatever act would lead to this outcome is the one we ought morally to perform – it is where our moral duty lies. And that act quite clearly might not be the same one that would bring about the best results for me personally, or for my family or friends, or for a lab animal. The best aggregated consequences for everyone concerned are not necessarily the best for each individual.

That utilitarianism is an aggregative theory – different individuals' satisfactions or frustrations are added, or summed, or totalled – is the key objection to this theory. My Aunt Bea is old, inactive, a cranky, sour person, though not physically ill. She prefers to go on living. She is also rather rich. I could make a fortune if I could get my hands on her money, money she intends to give me in any event, after she dies, but which she refuses to give me now. In order to avoid a huge tax bite, I plan to donate a handsome sum of my profits to a local children's hospital. Many, many children will benefit from my generosity, and much joy will be brought to their parents, relatives and friends. If I don't get the money rather soon, all these ambitions will come to naught. The once-in-a-lifetime opportunity to make a real killing will be gone. Why, then, not kill my Aunt Bea? Oh, of course I *might* get caught. But I'm no fool and, besides, her doctor can be counted on to co-operate (he has an eye for the same investment and I happen to know a good deal about his shady past). The deed can be done . . . professionally, shall we say. There is *very* little chance of getting caught. And as for my conscience being guilt-ridden, I am a resourceful sort of fellow and will take more than sufficient comfort – as I lie on the beach at Acapulco – in contemplating the joy and health I have brought to so many others.

Suppose Aunt Bea is killed and the rest of the story comes out as told. Would I have done anything wrong? Anything immoral? One would have thought that I had. Not according to utilitarianism. Since what I have done has brought about the best balance between totalled satisfaction and frustration for all those affected by the outcome, my action is not wrong. Indeed, in killing Aunt Bea the physician and I did what duty required.

This same kind of argument can be repeated in all sorts of cases, illustrating, time after time, how the utilitarian's position leads to results that impartial people find morally callous. It *is* wrong to kill my Aunt Bea in the name of bringing about the best results for others. A good end does not justify an evil means. Any adequate moral theory will have to explain why this is so. Utilitarianism fails in this respect and so cannot be the theory we seek.

What to do? Where to begin anew? The place to begin, I think, is with the utilitarian's view of the value of the individual – or, rather, lack of value. In its place, suppose we consider that you and I, for example, do have value as individuals – what we'll call *inherent value*. To say we have such value is to say that we are something more than, something different from, mere receptacles. Moreover, to ensure that we do not pave the way for such injustices as slavery or sexual discrimination, we must believe that all who have inherent value have it equally, regardless of their sex, race, religion, birthplace and so on. Similarly to be discarded as irrelevant are one's talents or skills, intelligence and wealth, personality or pathology, whether one is loved and admired or despised and loathed. The genius and the retarded child, the prince and the pauper, the brain surgeon and the fruit vendor, Mother Teresa and the most unscrupulous used-car salesman – all have inherent value, all possess it equally, and all have an equal right to be treated with respect, to be treated in ways that do not reduce them to the status of things, as if they existed as resources for others. My value as an individual is independent of my usefulness to you. Yours is not dependent on your usefulness to me. For either of us to treat the other in ways that fail to show respect for the other's independent value is to act immorally, to violate the individual's rights.

Some of the rational virtues of this view – what I call the rights view – should be evident. Unlike (crude) contractarianism, for example, the rights view *in principle* denies the moral tolerability of any and all forms of racial, sexual or social discrimination; and unlike utilitarianism, this view *in principle* denies that we can justify good results by using evil means that violate an individual's rights – denies, for example, that it could be moral to kill my Aunt Bea to harvest beneficial consequences for others. That would be to sanction the disrespectful treatment of the individual in the name of the social good, something the rights view will not – categorically will not – ever allow.

The rights view, I believe, is rationally the most satisfactory moral theory. It surpasses all other theories in the degree to which it illuminates and explains the foundation of our duties to one another – the domain of human morality. On this score it has the best reasons, the best arguments, on its side. Of course, if it were possible to show that only human beings are included within its scope, then a person like myself, who believes in animal rights, would be obliged to look elsewhere.

But attempts to limit its scope to humans only can be shown to be rationally defective. Animals, it is true, lack many of the abilities humans possess. They can't read, do higher mathematics, build a bookcase or make *baba ghanoush*. Neither can many human beings, however, and yet we don't (and shouldn't) say that they (these humans) therefore have less inherent value, less of a right to be treated with respect, than do others. It is the *similarities* between those human beings who most clearly, most non-controversially have such value (the people reading this, for example), not our differences, that matter most. And the really crucial, the basic similarity is simply this: we are each of us the experiencing subject of a life, a conscious creature having an individual welfare that has importance to us whatever our usefulness to others. We want and prefer things, believe and feel things, recall and expect things. And all these dimensions of our life, including our pleasure and pain, our enjoyment and suffering, our satisfaction and frustration, our continued existence or our untimely death – all make a difference to the quality of our life as lived, as experienced, by us as individuals. As the same is true of those animals that concern us (the ones that are eaten and trapped, for example), they too must be viewed as the experiencing subjects of a life, with inherent value of their own.

Some there are who resist the idea that animals have inherent value. 'Only humans have such value,' they profess. How might this narrow view be defended? Shall we say that only humans have the requisite intelligence, or autonomy, or reason? But there are many, many humans who fail to meet these standards and yet are reasonably viewed as having value above and beyond their usefulness to others. Shall we claim that only humans belong to the right species, the species *Homo sapiens*? But this is blatant speciesism. Will it be said, then, that all – and only – humans have immortal souls? Then our opponents have their work cut out for them. I am myself not ill-disposed to the proposition that there are immortal souls. Personally, I profoundly hope I have one. But I would not want to rest my position on a controversial ethical issue on the even more controversial question about who or what has an immortal soul. That is to dig one's hole deeper, not to climb out. Rationally, it is better to resolve moral issues without making more controversial assumptions than are needed. The question of who has inherent value is such a question, one that is resolved more rationally without the introduction of the idea of immortal souls than by its use.

Well, perhaps some will say that animals have some inherent value, only less than we have. Once again, however, attempts to defend this view can be shown to lack rational justification. What could be the basis of our having more inherent value than animals? Their lack of reason, or autonomy, or intellect? Only if we are willing to make the same judgement in the case of humans who are similarly deficient. But it is not true that such humans – the retarded child, for example, or the mentally deranged – have less inherent value than you or I. Neither, then, can we rationally sustain the view that animals like them in being the experiencing subjects of a life have less inherent value. *All* who have inherent value have it *equally*, whether they be human animals or not.

Inherent value, then, belongs equally to those who are the experiencing subjects of a life. Whether it belongs to others – to rocks and rivers, trees and glaciers, for example – we do not know and may never know. But neither do we need to know, if we are to make the case for animal rights. We do not need to know, for example, how many people are eligible to vote in the next presidential election before we can know whether I am. Similarly, we do not need to know how many individuals have inherent value before we can know that some do. When it comes to the case for animal rights, then, what we need to know is whether the animals that, in our culture, are routinely eaten, hunted and used in our laboratories, for example, are like us in being subjects of a life. And we do know this. We do know that many – literally, billions and billions – of these animals are the subjects of a life in the sense explained and so have inherent value if we do. And since, in order to arrive at the best theory of our duties to one another, we must recognize our equal inherent value as individuals, reason – not sentiment, not emotion – reason compels us to recognize the equal inherent value of these animals and, with this, their equal right to be treated with respect.

Source: Tom Regan (1984) 'The case for animal rights', in Peter Singer (ed.) *Applied Ethics*, Oxford: Oxford University Press.

3.6

Roger Scruton

The moral status of animals

Non-moral beings

. . . If there are non-human animals who are rational and self-conscious, then they, like us, are persons, and should be described and treated accordingly. If *all* animals are persons, then there is no longer a problem as to how we should treat them. They would be full members of the moral community, with rights and duties like the rest of us. But it is precisely because there are animals who are not persons that the moral problem exists. And to treat these non-personal animals as persons is not to grant to them a privilege nor to raise their chances of contentment. It is to ignore what they essentially are, and so to fall out of relation with them altogether.

The concept of the person belongs to the ongoing dialogue which binds the moral community. Creatures who are by nature incapable of entering into this dialogue have neither rights nor duties nor personality. If animals had rights, then we should require their consent before taking them into captivity, training them, domesticating them or in any way putting them to our uses. But there is no conceivable process whereby this consent could be delivered or withheld. Furthermore, a creature with rights is duty-bound to respect the rights of others. The fox would be duty-bound to respect the right to life of the chicken, and whole species would be condemned out of hand as criminal by nature. Any law which compelled persons to respect the rights of non-human species would weigh so heavily on the predators as to drive them to extinction in a short while. Any morality which *really* attributed rights to animals would therefore constitute a gross and callous abuse of them.

Those considerations are obvious, but by no means trivial. For they point to a deep difficulty in the path of any attempt to treat animals as our equals. By ascribing rights to animals, and so promoting them to full membership of the moral community, we tie them in obligations that they can neither fulfil nor comprehend. Not only is this senseless cruelty in itself; it effectively destroys all possibility of cordial and beneficial relations between us and them. Only by refraining from personalising animals do we behave towards them in ways that they can understand. And even the most sentimental animal lovers know this, and confer 'rights' on their favourites in a manner so selective and arbitrary as to show that they are not really dealing with the ordinary moral concept. When a dog savages a sheep no-one believes that the dog, rather than its owner, should be sued for damages. Sei Shonagon, in *The Pillow Book*, tells of a dog breaching some rule of court etiquette and being horribly beaten, as the law requires. The scene is most disturbing to the modern reader. yet surely, if dogs have rights, punishment is what they must expect when they disregard their duties.

But the point does not concern rights only. It concerns the deep and impassable difference between personal relations, founded on dialogue, criticism and the sense of justice, and animal relations, founded on affections and needs. The moral problem of animals arises because they cannot enter into relations of the first kind, while we are so much bound by those relations that they seem to tie us even to creatures who cannot themselves be bound by them.

Defenders of 'animal liberation' have made much of the fact that animals suffer as we do: they feel pain, hunger, cold and fear and therefore, as Singer puts it, have 'interests' which form, or ought to form, part of the moral equation. While this is true, it is only part of the truth. There is more to morality than the avoidance of suffering: to live by no other standard than this one is to avoid life, to forgo risk and adventure, and to sink into a state of cringing morbidity. Moreover, while our sympathies ought to be, and unavoidably will be, extended to the animals, they should not be indiscriminate. Although animals have no rights, we still have duties and responsibilities towards them, or towards some of them, and these will cut across the utilitarian equation, distinguishing the animals who are close to us and who have a claim on our protection from those towards whom our duties fall under the broader rule of charity.

This is important for two reasons. Firstly, we relate to animals in three distinct situations, which define three distinct kinds of responsibility: as pets, as domestic animals reared for human purposes and as wild creatures. Secondly, the situation of animals is radically and often irreversibly changed as soon as human beings take an interest in them. Pets and other domestic animals are usually entirely dependent on human care for their survival and well-being; and wild animals too are increasingly dependent on human measures to protect their food supplies and habitats.

Some shadow version of the moral law therefore emerges in our dealings with animals. I cannot blithely count the interests of my dog as on a par with the interests of any other dog, wild or domesticated, even though they have an equal capacity for suffering and an equal need for help. My dog has a special claim on me, not wholly dissimilar from the claim of my child. I caused it to be dependent on me, precisely by leading it to expect that I would cater for its needs.

The situation is further complicated by the distinction between species. Dogs form life-long attachments, and a dog brought up by one person may be incapable of living comfortably with another. A horse may be bought or sold many times, with little or no distress, provided it is properly cared for by each of its owners. Sheep maintained in flocks are every bit as dependent on human care as dogs and horses; but they do not notice it, and regard their shepherds and guardians as little more than aspects of the environment, which rise like the sun in the morning and depart like the sun at night.

For these reasons we must consider our duties towards animals under three separate heads: pets, animals reared for our purposes and creatures of the wild.

Pets

A pet is an honorary member of the moral community, though one relieved of the burden of duty which that status normally requires. Our duties towards these creatures in whom, as Rilke puts it, we have 'raised a soul', resemble the general duties of care upon which households depend. A man who sacrificed his child or a parent for the sake of his pet would be acting wrongly; but so too would a man who sacrificed his pet for the sake of a wild animal towards which he has had no personal responsibility – say by feeding it to a lion. As in the human case, moral judgement depends upon a prior assignment of responsibilities. I do not release myself from guilt by showing that my pet starved to death only because I neglected it in order to take food to hungry strays; for my pet, unlike those strays, depended completely on *me* for its well-being.

In this area our moral judgements derive not only from ideas of responsibility, but also from our conception of human virtue. We judge callous people adversely not merely on account of the suffering that they cause, but also, and especially, for their thoughtlessness. Even if they are calculating for the long-term good of all sentient creatures, we are critical of them precisely for the fact that they are *calculating*, in a situation where some other creature has a direct claim on their compassion. The fanatical utilitarian, like Lenin, who acts always with the long-term goal in view, loses sight of what is near at hand and what should most concern him, and may be led thereby, like Lenin, into unimaginable cruelties. Virtuous people are precisely those whose sympathies keep them alert and responsive to those who are near to them, dependent on their support and most nearly affected by their heartlessness.

If morality were no more than a device for minimising suffering, it would be enough to maintain our pets in a state of pampered somnolence, awakening them from time to time with a plate of their favourite tit-bits. But we have a conception of the fulfilled animal life which reflects, however distantly, our conception of human happiness. Animals must flourish according to their nature: they need exercise, interests and activities to stimulate desire. Our pets depend upon *us* to provide these things – and not to shirk the risks involved in doing so.

Pets also have other, and more artificial, needs, arising from their honorary membership of the moral community. They need to ingratiate themselves with humans, and therefore to acquire their own equivalent of the social virtues. Hence they must be elaborately trained and disciplined. If this need is neglected, then they will be a constant irritation to the human beings upon whose good will they depend. This thought is obvious to anyone who keeps a dog or a horse. But its implications are not always appreciated. For it imposes on us an obligation to deal strictly with our pets, to punish their vices, to constrain their desires and to shape their characters. In so far as punishment is necessary for the education of children, we regard it as justified: parents who spoil their children produce defective moral beings. This is not merely a wrong towards the community; it is a wrong towards the children themselves, who depend for their happiness on the readiness of others to accept them. Pets must likewise be educated to the standards required by the human community in which their life, for better or worse, is to be led.

Furthermore, we must remember the ways in which pets enhance the virtues and vices of their owners. By drooling over a captive animal, the misanthrope is able to dispense more easily with those charitable acts and emotions which morality requires. The sentimentalising and 'kitschification' of pets may seem to many to be the epitome of kind-heartedness. In fact it is very often the opposite: a way of enjoying the luxury of warm emotions without the usual cost of feeling them, a way of targeting an innocent victim with simulated love that it lacks the understanding to reject or criticise, and of confirming thereby a habit of heartlessness. To this observation I shall return.

Pets are part of a complex human practice, and it is important also to consider the nature of this practice and its contribution to the well-being of the participants. Even if we fulfil all our obligations to the animals whom we have made dependent, and even if we show no vicious motives in doing so, the question remains whether the net result of this is positive or negative for the humans and the animals concerned. There are those who believe that the effect on the animals is so negative, that they ought to be 'liberated' from human control. This dubious policy exposes the animals to risks for which they are ill-prepared; it also shows a remarkable indifference to the *human* suffering that ensues. People depend upon their pets, and for many people a pet may be their only object of affection. Pets may suffer from their domestication, as do dogs pent up in a city flat. Nevertheless, the morality of the practice could be assessed only when the balance of joy and suffering is properly drawn up. In this respect the utilitarians are right: we have no way of estimating the value of a practice or an institution except through its contribution to the total good of those involved. If it could be shown that, in the stressful conditions of modern life, human beings could as well face the prospect of loneliness without pets as with them, then it would be easier to condemn a practice which, as it stands, seems to make an indisputable contribution to the sum of human happiness, without adding sensibly to that of animal pain.

We should also take note of the fact that most pets exist only *because* they are pets. The alternative, for them, is not another and freer kind of existence, but no existence at all. No utilitarian could really condemn the practice of keeping pets therefore, unless he believed that the animals in question suffer so much that their lives are not worthwhile.

This point touches on many of our modern concerns. We recognise the increasing dependence of animals on human decisions. Like it or not, we must accept that a great many of the animals with which we are in daily contact are there only because of a human choice. In such circumstances we should not hasten to criticise practices which renew the supply of animals, while at the same time imposing upon us clear duties to look after them.

Animals for human use and exploitation

The most urgent moral questions concern not pets, but animals which are used for specific purposes – including those which are reared for food. There are five principal classes of such animals:

- beasts of burden, notably horses, used to ride or drive;
- animals used in sporting events – for example, in horse-racing, dog-racing, bull-fighting and so on;
- animals kept in zoos or as specimens;
- animals reared for animal products: milk, furs, skins, meat, etc;
- animals used in research and experimentation.

No person can be used in any of those five ways; but it does not follow that an animal who is so used will suffer. To shut a horse in a stable is not the same act as to imprison a free agent. It would normally be regarded as conclusive justification for shutting up the horse, that it is better off in the stable than elsewhere, regardless of its own views in the matter. Such a justification is relevant in the second case only if the victim has either forfeited freedom through crime or lost it through insanity.

The first two uses of animals often involve training them to perform activities that are not natural to them, but which exploit their natural powers. Two questions need to be addressed. First, does the training involve an unacceptable measure of suffering? Second, does the activity allow for a fulfilled animal life? These questions are empirical, and cannot be answered without detailed knowledge of what goes on. However, there is little doubt in the mind of anyone who has worked with horses, for example, that they are willing to learn, require only light punishment and are, when properly trained, the objects of such care and affection as to provide them with ample reward. It should be added that we have one reliable criterion of enjoyment, which is the excitement and eagerness with which an animal approaches its work. By this criterion there is no doubt that greyhounds enjoy racing, that horses enjoy hunting, team-chasing and cross-country events in which they can run with the herd and release their energies, and even that terriers enjoy, however strange this seems to us, those dangerous adventures underground in search of rats and rabbits.

But this should not blind us to the fact that sporting animals are exposed to real and unnatural dangers. Many people are exercised by this fact, and particularly by the conduct of sports like horse-racing and polo, in which animals are faced with hazards from which they would normally shy away, and which may lead to painful and often fatal accidents. Ought we to place animals in such predicaments?

To answer such a question we should first compare the case of human danger. Many of our occupations involve unnatural danger and extreme risk – soldiering being the obvious example. People willingly accept the risk, in return for the excitement, status or material reward which attends it. This is a normal calculation that we make on our own behalf and also on behalf of our children when choosing a career. In making this calculation we are motivated not only by utilitarian considerations, but also by a conception of virtue. There are qualities which we admire in others, and would wish for in ourselves and our children. Courage, self-discipline, and practical wisdom are promoted by careers in which risk is paramount; and this is a strong reason for choosing those careers.

Now animals do not freely choose a career, since long-term choices lie beyond their mental repertoire. Nevertheless, a career may be chosen *for* them; and, since the well-being of a domesticated animal depends upon the attitude of those who care for it, its career must be one in which humans have an interest and which leads them to take proper responsibility for its health and exercise. The ensuing calculation may be no different from the calculation undertaken in connection with a human career. The risks attached to horse-racing, for example, are offset, in many people's minds, by the excitement, abundant feed and exercise and constant occupation which are the horse's daily reward, and by the human admiration and affection which a bold and willing horse may win, and which have made national heroes of several privileged animals, like Red Rum and Desert Orchid.

But this brings us to an interesting point. Because animals cannot deliberate and take no responsibility for themselves and others, human beings find no moral obstacle to breeding them with their future use in mind. Almost all the domestic species that surround us have been shaped by human decisions, bred over many generations to perform by instinct a task which for us is part of a conscious plan. This is especially true of dogs, cats and horses, and true for a different reason of the animals which we rear for food. Many people feel that it would be morally objectionable to treat humans in this way. There is something deeply disturbing in the thought that a human being should be bred for a certain purpose, or that genetic engineering might be practised on the human foetus in order to secure some desired social result. The picture painted by Aldous Huxley in *Brave New World* has haunted his readers ever since, with a vision of human society engineered for happiness, and yet deeply repugnant to every human ideal. It is not that the planned person, once grown to maturity, is any less free than the normal human accident. Nevertheless, we cannot accept the kind of manipulation that produced him, precisely because it seems to disrespect his nature as a moral being and to assume a control over his destiny to which we have no right. This feeling is an offshoot of piety and has no real ground either in sympathy or in the moral law.

Pious feelings also forbid the more presumptuous kind of genetic engineering in the case of animals. There is a deep-down horror of the artificially-created monster which, should it ever be lost, would be lost to our peril. Yet the conscious breeding of dogs, for instance, seems to most eyes wholly innocent. Indeed, it is a way of incorporating dogs more fully into human plans and projects, and so expressing and enhancing our love for them. And there are breeds of dog which have been designed precisely for risky enterprises, like the terrier, the husky and the St Bernard, just as there are horses bred for racing. Such creatures, deprived of their intended career, are in a certain measure unfulfilled, and we may find ourselves bound, if we can, to give them a crack at it. Given our position, after several millennia in which animals have been bred for our purposes, we have no choice but to accept that many breeds of animal have needs which our own ancestors planted in them.

Once we have understood the complex interaction between sporting animals and the human race, it seems clear that the same moral considerations apply here as in the case of pets. Provided the utilitarian balance is (in normal circumstances) in the animal's favour, and provided the responsibilities of owners and trainers are properly fulfilled, there can be no objection to the use of animals in competitive sports. Moreover, we must again consider the human values that have grown around this use of animals. In Britain, for example, the horse race is an immensely important social occasion: a spectacle which does not merely generate great excitement and provide a cathartic climax, but which is a focus of elaborate social practices and feelings. For many people a day at the races is a high point of life, a day when they exist as eager and affectionate members of an inclusive society. And animals are an indispensable part of the fun – imparting to the human congress some of the uncomplicated excitement and prowess upon which the spectators, long severed from their own instinctive emotions, draw for their heightened sense of life.

Indeed, history has brought people and animals together in activities which are occasions of individual pleasure and social renewal. Take away horse-racing, and you remove a cornerstone of ordinary human happiness. This fact must surely provide ample justification for the risks involved. It does not follow that horse-racing can be conducted anyhow, and there are serious question to be raised about the racing of very young horses who, when so abused, are unlikely to enjoy a full adult life thereafter. But, provided the victims of accidents are humanely treated, such sports cannot be dismissed as immoral. Indeed, we have a duty to encourage them as occasions of cheerful association between strangers.

Source: Roger Scruton (1996) *Animal Rights and Wrongs*, London: Demos, pp. 66–76.

3.7

Robert Garner

Conserving wildlife

Wildlife conservation is a confused issue. In particular, the motives behind it are so varied that it appeals to a wide constituency from animal welfare and rights advocates on the one hand to those who emphasise the benefits to humans of biological diversity on the other.[1] It is, though, the dominant anthropocentrism – which comes in various guises – that explains why the problems facing many wild animals through human exploitation and neglect have generated, as we shall see, more public concern and inter-governmental action than those relating to the domesticated species we have examined so far. What this anthropocentric outlook also reveals, however, is the difficulty of reconciling much of the theory and practice of conservation, as articulated by some pressure groups as well as public authorities, with the moral theories discussed in relation to companion, farm and laboratory animals. . . . [The] application of these moral theories to the plight of wild animals can reveal very different conclusions from those reached by conservationists concerned primarily with human interests.

Conservation and the moral status of animals

Why conserve wild animals? Clearly, there are many answers to this question but a crucial dimension concerns the interests served by conservation. We might want to keep animals around because it serves our interests to do so and the fact that the interests of some animals are promoted as a result is merely an indirect consequence of the furtherance of human interests. According to this view, the natural world has no value in itself but is merely a resource for humans to manage. As N. W. Moore, a British conservationist, explains:

> We should never forget that the objective of conservation is no less than to maintain the living resources of the world so that each generation can use and enjoy them.[2]

Alternatively, we might want to conserve animals for their own sakes because they have interests which must be promoted directly, irrespective of any benefits we may derive from so doing.

The gulf between these two approaches should be appreciated. The latter, of course, is consistent with the various moral theories we considered in relation to the issues discussed in the preceding three chapters. All of these theories – from the moral orthodoxy to the rights view – recognise that animals can be harmed and that we owe duties directly to them. The emphasis is on the suffering and/or death of individual animals, since this matters to them directly. By contrast, anthropocentric conservationism – the dominant form at present – is more concerned with the protection of species than the suffering of individual animals and more concerned with the protection of species that are of use to humans than those which are not.

This dichotomy is hugely important in the debate about wild animals. It explains why there is little co-operation between conservation groups and the rest of the animal protection movement in addition to providing the essence of the conflict within the conservation movement itself. Further, the human-centred nature of much conservationism explains why it has a much higher public profile (with the possible exception of companion animals) than . . . other issues relating to animals. . . . In order to illustrate this we will look at the way wild animals are treated in national and international law in addition to the debates surrounding wildlife conservation.

International wildlife treaties

Co-operation between sovereign states in the field of nature conservation is not a new phenomenon. With the widespread incidence of virtually uncontrolled hunting in the nineteenth century, for instance, both hunters and naturalists became concerned about the increasing number of endangered species. It is estimated that in British East Africa some 10,000 animals were shot by hunting parties annually.[3] As a result, seven countries including Britain signed one of the first treaties – the Convention for the Preservation of Animals, Birds and Fish in Africa – in 1900.[4] In recent years, with the growing threat to the world's flora and fauna and the rise of a greater environmental awareness, though, a greater urgency has been provoked.

Since the seventeenth century around 350 species and sub-species of animals – including the sea mink and Steller's sea cow – have become extinct and one authority has estimated that half of the species which are known to have disappeared during the past 2,000 years have been lost since 1900.[5] The exact figure will never be known, not least because man has been aware of only a fraction of the species which have lived, and do live, on the earth. Excluding plants and vegetation, estimates of the total number of living species vary between five and thirty million and yet only about one and a half million have so far been identified.[6] It is a humbling thought to recognise that many species may have disappeared recently without humans ever realising they existed. Perhaps the most potent symbol of the slaughter was the American passenger pigeon. Once one of the most prolific species in North America with tens of millions existing in the eighteenth and early nineteenth centuries, it was, due to widespread shooting and destruction of habitat, completely wiped out by 1914 when the last one (poignantly named Martha) died in Cincinnati Zoo.[7]

Wildlife conservation has taken on an international dimension for obvious reasons.[8] In the first place, animals do not respect national boundaries. Members of an endangered species may exist in a number of countries and effective protection involves united action. In particular, it is impossible effectively to conserve birds at the national level since the vast majority migrate. There is little point in having strong regulations protecting birds in one country if a neighbouring state allows them to be indiscriminately slaughtered when they pass over its territory. Secondly, live animals, and the various products that can be derived from them, form a significant part of international trade. It is much more difficult to protect animals in one country if others do not seek to prevent them being imported. Thirdly, although some animals may only exist in one country, others may have an interest in protecting them and participation in an inter-governmental organisation is an effective means of offering their assistance.

Following Lyster, we can distinguish between three types of treaties.[9] Firstly, there are those designed to protect either a single species or a group of species. Here, most notably, is the International Convention for the Regulation of Whaling (ICRW) although treaties also exist to protect seals, polar bears and birds. Secondly, there are regional nature conservation treaties such as the Convention of Nature Protection and Wildlife Preservation in the Western Hemisphere and thirdly there are the 'big four' wildlife treaties all concluded in the 1970s and all open for most countries to join. These four treaties which provide 'the centrepiece of international wildlife law' are the Convention on Wetlands of International Importance Especially as Waterfowl Habitat (known as Ramsar), the Convention Concerning the Protection of the World Cultural and Natural Heritage, the Convention on the Conservation of Migratory Species of Wild Animals (the Bonn Convention) and finally the Convention on International Trade in Endangered Species of Wild Fauna and Flora (CITES).

These examples of international co-operation did not occur in a vacuum. They have come about as the result of an increasingly complex network of permanent inter-governmental and non-governmental organisations. Thus, the Council of Europe, the European Community, UNESCO and the UN Environment Programme have all been involved in either initiating, utilising or administering conservation agreements between nations. Mention should also be made of IUCN which, as an organisation containing governments and government agencies as members as well as pressure groups, has been particularly influential in the conservation field doing much of the groundwork, for instance, which led to the signing of the CITES treaty in 1973.[10]

It is not possible here to consider all these treaties in detail. Instead, two treaties – those concerned with whaling and the trade in endangered species – will be examined since these enable us both to identify some of the common characteristics of international agreements and to highlight major conservation issues. Two more general points, though, can be made here. In the first place, mainly as a result of the influence of ecology within the conservation movement, it is now widely accepted that the major threat to most species is not man's deliberate exploitation of animals but the indirect consequence of the destruction of their habitat – usually for agricultural purposes. Given this, there is a need for a world-wide habitat treaty to complement the regional agreements that already exist, although the failure to conclude such a treaty is not surprising given the threat to human interests it would pose.[11]

Secondly, it is striking how often one comes across anthropocentric justifications in the wording of treaties. Some, such as the whaling convention and the Convention for the Protection of Birds Useful to Agriculture (signed in 1902), were unmistakably set up with economic interests in mind. With others, the motive soon becomes clear. Thus, a treaty to protect birds signed by the USA and Japan in 1972 refers to their 'aesthetic' and 'scientific' qualities whilst the preamble to a bilateral treaty concluded by the USA and the Soviet Union in 1976 states that 'migratory birds are a natural source of great scientific, economic, aesthetic, cultural, educational, recreational and ecological value'. Likewise, the Ramsar treaty explains that wetlands 'constitute a resource of great economic, cultural, scientific and recreational value' whilst the Bonn Convention insists that 'wild animals . . . must be conserved for the good of mankind.'[12] All of this might be true but there is no mention here of the argument that birds and mammals should be protected because they themselves have an interest in not suffering or not being killed.

[. . .]

Conservation and the animal protection movement

It might be asked at this point why it is that so much emphasis is focused upon endangered or threatened species. Some, particularly within the animal protection movement, seek to protect them because they recognise their right to exist. This is certainly one explanation for the opposition towards a strategy which involves a limited amount of sustainable utilisation of such species. Indeed, the attempt to protect African game (symbolised by the CITES ban on the ivory trade), even when this is achieved at the expense of the interests of local indigenous populations coupled with the authority given to the wardens of some national parks to kill poachers, would seem to represent a victory for animal rights advocates.

It is true that mainstream animal welfare and rights views are becoming increasingly influential within the conservation movement. This is seen not only in the hardening of attitudes towards the protection of endangered species but also in a greater willingness to campaign for all wild animals. The group Elefriends (an offshoot from Zoo Check), for instance, has been fundamentally opposed to a limited resumption of the ivory trade as a conservation strategy.[13] Other recently founded groups – such as the People's Trust for Endangered Species and the IFAW – also emphasise compassionate grounds for conservation and recognise the duties we owe directly to animals irrespective of the status given to them by humans. Thus, the IFAW's campaign against the killing of seal pups in Canada based on the cruelty involved and not their endangered status, caught the popular imagination and, more recently, both the RSPCA and RSPB have initiated a campaign against the bird trade – a trade which causes the death, and often atrocious suffering, of an estimated one million birds a year. Significantly, the campaign is directed at the legal trade in relatively common birds as much as the illegal transportation of protected species.[14]

Conflict between groups and individuals emphasising the animal protection approach and those emphasising a human-centred conservationism has been endemic. In Britain, both the LACS and the HSA, for instance, have had run-ins with the RSPB because of the latter's unwillingness to oppose the shooting of grouse on the grounds that wildlife habitats are managed effectively as a result. Likewise, animal protectionists have been critical of the priority given to humans by FoE and, particularly in its earlier days, by Greenpeace. FoE did not, for instance, support the whaling moratorium since it regarded the ethnic rights of Inuit as superior to the protection of whales. Similarly, Greenpeace decided in the 1980s to drop their campaign against the fur trade when it came to their attention that fur provided a crucial source of income for some Canadian and American peoples. This decision so annoyed some Greenpeace activists that, under the leadership of Mark Glover, they split away in 1985 to form the anti-fur group Lynx, which holds an uncompromising anti-fur position.[15]

The animal protection emphasis has also become more noticeable within the membership of the WWF, if not yet the leadership, and this has caused conflict. This explains the furore that has occurred as a result of revelations that the WWF sells its logo for use by commercial concerns. The criticism, which resulted in the resignation of the WWF's senior corporate fund-raiser, centred particularly on the logo's appearance on Natrel, a deodorant manufactured by Gillette, which it was claimed has been tested on animals.[16] The WWF's long-term sponsorship deal with the household products giant Procter and Gamble is also unlikely to have endeared it to radicals in the animal protection movement. In addition, as we have seen, the WWF's move towards support for a 'sustainable use' strategy has provoked profound disagreement. This arose again at the 1990 IUCN General Assembly in Australia when a proposal to allow the Fur Council of Canada to join as part of a new business sector membership category was rejected by the membership. For much of the leadership of IUCN and the WWF, the proposal was quite consistent with their anthropocentric conservationism. After all, Eugene Lapointe, the deposed CITES general secretary, said after his forced resignation that the 'international fur trade association has done more for conservation than conservationists'. For the new brand of animal protectionists, however, this move was clearly anathema to their perceived demarcation between 'animal protectors' and 'animal abusers'.[17]

One final point here is that we should be careful that we do not simplify the divisions within the conservation movement. In particular, the theory of sustainable use can be seen as both an ideology and a strategy. As an ideology, it serves to emphasise the view that wildlife conservation, as a matter of principle, should never take precedence over the interests of humans so that, in the event of a clash between the interests of wildlife and the interests of humans, the latter should never be sacrificed for the former. As a strategy, on the other hand, it serves to emphasise the view that, whatever the personal views of conservationists, it is recognised that without stressing the human benefits of conservation it is, in practice, unlikely to gain enough support to succeed. Clearly, both strands of opinion are held within the WW/IUCN leadership. Martin Holdgate, the director general of IUCN, for instance, seems to lean towards the second view. As he wrote in defence of his organisation's position:

> The real debate in IUCN over controlled use of wildlife is not between fundamentalist 'greenies' and game hunters. It is about ensuring the survival of economically important species . . . in a world that is bound to exploit them.[18]

Eco-imperialism

The logic of animal rights philosophy militates against any priority being given to endangered species. As Regan points out, if the case for animal rights is successful, it applies to animals which are common as well as those which belong to endangered species. Indeed, for Regan, if we had to choose between saving the last remaining member of a particular species or an animal from a common species, we should choose the latter option if death would cause a greater harm to that individual. In normal circumstances, then, we should save members of endangered species, not because they are endangered but because they have rights which we violate by harming them.[19]

Few, though, who support the protection of endangered species would accept the implications of this consistent application of animal rights philosophy. What are the grounds, then, for treating such animals differently in practice? The answer, as we have intimated throughout this chapter, is that they are treated differently because it is in the commercial, scientific, aesthetic, ecological or sporting interests of humans to do so. Thus, for example, whales are protected, at least in the eyes of the whaling nations, because of the need to revive stocks before whaling can begin again. For many others, though, whales, and many other species, should be protected because aesthetically the world would be a poorer place without them – although poorer, it should be added, for humans. This explains why, even amongst the many species that are threatened with extinction, only some – such as the African elephant, the polar bear, the whale and many primates – have been subject to widespread public attention. It is easy to see why such majestic creatures catch the public imagination, but their plight is no worse than other, less attractive, endangered species and individual members of many of these latter species have an equal capacity to experience suffering.

The result of the excessive anthropocentric attention devoted to endangered wild animals is a clear case of double standards which amounts to a kind of 'eco-imperialism'. Thus, those, such as whalers and Third World governments and peoples, who stand to gain from the exploitation of endangered species (and, indeed, other species of wild animals to whom the developed nations attach some status) and who stand to lose by allowing the protection of such animals to come before development projects which will benefit their populations, are expected to sacrifice their interests whilst the West continues to exploit animals by the millions in factory farms and laboratories. If what we have described as the moral orthodoxy is applied to endangered species (as it is applied to domesticated animals in the West), interesting conclusions follow. According to this orthodoxy, whether a species is endangered or not is irrelevant since what matters is whether or not any suffering inflicted is necessary. Killing animals painlessly is not a problem since *whether* animals die or not is not a welfare issue. It is only *how* they lived and died that matters.

What this reveals is not, then, a conflict between human interests and animal interests. Rather, it is a conflict between *competing human interests*. Seen as such, the case for allowing poor Third World countries to utilise animals, even if they are listed as endangered, becomes much stronger. At the very least, Western conservationists should demonstrate how local people will benefit from the protection of their country's wildlife or else show why killing endangered species inflicts unnecessary suffering on them. If neither stands up to analysis (as it often does not), then there is no justification for preventing the utilisation of endangered species *even if* this results in extinction for some of them. The only way of logically avoiding this conclusion is to argue that the animals should be protected from human exploitation because they have a right to be treated with respect. But once this is accepted, then, as we have seen, nothing would stand in the way of granting this status to other wild and domesticated animals. The choice for Western conservationists (and governments) is therefore clear. Either accept that there is nothing in principle wrong with killing threatened species (painlessly) even to the point of extinction or accept that the killing of animals for food or in pursuit of scientific progress is morally wrong.

Of course, in practice, wild animals do suffer at the hands of humans, in the same ways and sometimes worse than domesticated species. One can, for instance, criticise the methods used to trap fur-bearing animals or some of the practices involved in animal trading on the grounds that they are inhumane. Whaling too would seem to fit into this category. Leaving aside the more contentious issue as to whether killing a mammal with a brain larger than humans and with an average lifespan of seventy years is morally wrong, the method used to catch whales (a harpoon carrying an explosive charge which is detonated inside the body) is almost certainly inhumane and it is difficult to see how an improved system of killing them could be devised which does not cause suffering to these intelligent sentient creatures.[20] To their credit, the British government has changed its tactics within the IWC to argue, consistently with its position regarding domestic animals, that whaling should not be resumed until a humane way of killing them is found.[21]

Given the suffering inflicted on wild animals, the onus is on trappers, traders and hunters and those who support their activities to show that this suffering is necessary. Since whaling is now more concerned with issues of national pride rather than economic or dietary necessity, the case of the whaling nations is weak.[22] The one exception here might be the Alaskan Inuit, who do rely on their 'take' of (bowhead) whales much more than the big rich whaling nations, although even their dependence on this catch has declined. The key point, though, is that once we get involved in arguments about what constitutes unnecessary suffering, then it applies universally to all animals, irrespective of whether they are part of an endangered species. On these terms, unless we introduce the right to life principle, we would be more justified in painlessly killing an animal from an endangered species than painfully and unnecessarily killing an animal from a species that was common. That this is not the conventional approach is a reflection of, on the one hand, the anthropocentric nature of our approach to animals and, on the other, the way in which the peoples of the developing nations are perceived by the rest of the world. Here, one can answer Richard North's question in the affirmative: 'Isn't it true . . . that for most of us, the animals and wildlife of the Third World seem glorious, and their peoples an embarrassment?'[23]

[. . .]

Blood sports

[. . .]

Given that suffering is clearly inflicted in blood sports, we need to ask how far it is justified. If the only justification for hunting, shooting and fishing is that it provides entertainment for the participants then, unless we have an almost Cartesian regard for animals, these practices should be banned particularly as, in the case of hunting, there is an alternative in the form of drag hunting. We do not, though, have such a low regard for animals. Public opinion is against hunting presumably on the grounds that it is regarded as unnecessary. Furthermore, if we accept the entertainment logic for hunting and killing foxes then why not for domesticated animals too? Yet we do not accept this as there are uncontentious laws protecting domestic animals from gratuitous cruelty.

The hunting fraternity recognises that they have to provide more weighty justifications for their 'sport'. The first shot in their armoury is that the fox is a pest and needs to be controlled. This argument was also accepted for a long time by the RSPCA until the 1970s when the radicals began to change the Society's direction. The view that foxes are pests is based in particular on the belief that they cause immense damage to agriculture by taking poultry and lambs. This is clearly exaggerated. Foxes will kill poultry but since the vast majority are kept indoors in battery cages, they are hardly likely to be greatly affected. In addition, it is difficult to see why free-range poultry cannot be protected against foxes at night, when the animals are active. There is now strong evidence too that foxes are not a threat to lambs. The impression that they are has probably come about because the remains of dead lambs are found in fox earths but this does not prove that they killed them as opposed to removing those that had already died. A research project undertaken by Ray Hewson from Aberdeen University provides the best evidence yet that foxes are not a threat to farmers. Over a three-year period, Hewson, whose work was financed by the LACS [League Against Cruel Sports], studied the effects, on a site in Scotland, of an uncontrolled fox population. Significantly, he found that fox numbers did not increase and that the losses of lambs were small. Ironically, it may even be the case that foxes are useful to farmers since the report showed that foxes preferred to prey on rabbits, voles, carrions, slugs and beetles, all of whom can be extremely destructive.[24] If this is true, maybe someone should explain it to the Scottish Office who, according to Barry Kew, fund twenty-nine fox destruction clubs which kill approximately 10,000 foxes a year.[25]

Blood sports enthusiasts have increasingly (and cleverly) turned their attention to conservation justifications for hunting, fishing and shooting.[26] Thus, it is argued that foxes and deer survive because they and their habitats are protected by farmers who enjoy hunting them. If hunting was banned, foxes would be killed as pests and their habitat destroyed. Anglers, similarly, maintain that they are the best guarantee of unpolluted rivers since they have a vested interest in keeping them clean, just as game bird shooting is justified on the grounds that it provides an economic incentive for preserving habitats which would otherwise disappear. Note here that hares are not pests and there is no conservation justification for hunting or coursing them.

What should we make of these arguments? Well, in the first place, the protection of a particular species is not an animal welfare issue. Indeed, if the choice was between the humane slaughter of all foxes, deer and game birds and the continuation of suffering through hunting, then the former would have to be accepted. Leaving that aside (and the inconsistency between the pest and the conservation angle), there would seem to be some logic in the conservation case. It is true that farmers would not necessarily kill foxes as pests. A recent opinion poll conducted by NOP for the LACS indicated that 70 per cent of all farmers questioned did not consider foxes to be significantly harmful to their interests.[27] Nevertheless, farmers deprived of hunting would be more likely to destroy fox habitats.[28] But then this would not necessarily involve any greater suffering to individual foxes (and probably far less) than hunting them, bearing in mind that the survival of the species or the reduction in numbers (which would be the consequence of a reduction in suitable breeding sites) are not animal welfare issues. If deer were no longer hunted, there would probably have to be culling. Again, though, the killing of animals by experienced marksmen has to be preferable to the fear and often violence involved in the hunt.

We can apply similar arguments to fishing and shooting. It is undeniably true that pressure from anglers helps to keep rivers clean but the question we have to ask is whether this benefit to fish and other wildlife outweighs the suffering caused by fishing. Certainly, there would seem to be a strong case that this is so. The implications are, though, that if it were found that anglers ceased to have an impact on the condition of rivers, either because, despite their efforts, the condition of rivers deteriorated or because they remained free of pollution in any case, then fishing would not be justified. This serves to emphasise that, given the suffering inflicted, fishing for pleasure alone is not justified even in terms of the moral orthodoxy. Shooting animals is a more complex matter because of the diversity of activities covered by the term. Killing animals instantly and painlessly is not a welfare problem but, particularly given the lack of expertise prevalent in game bird shoots where anyone who pays enough can participate, suffering is inflicted. Given that it is not necessary to shoot at living creatures, there is little justification for it to continue. It probably is true that habitat is preserved as a result of the need to breed and raise game birds but many other wild animals who are a threat to the game (and to the success of the shoot) are, often illegally, killed by gamekeepers as a result.[29]

Thus, there would seem to be a strong case for the banning of at least some blood sports and the widespread public opposition to hunting and hare coursing, coupled with the commitment of some political parties to abolish them, could well mean that their days are numbered. Remember that the case against them is based on the moral orthodoxy. If one adopts the alternative rights view, then all forms of blood sports are clearly illegitimate since to kill some, even if the consequence is to lessen the total amount of harm, violates the rights of individuals.[30]

The fur trade

Killing animals for their fur is another issue where conservation can come into conflict with animal welfare. Thus, the anthropocentric version of the former can justify conserving fur-bearing species on the grounds that extinction would damage the fur trade or that we should not allow such animals to become extinct because they are aesthetically pleasing to us. Animal welfare, on the other hand, is more concerned with the level of suffering involved for individual animals (irrespective of whether or not they are part of an endangered species) and whether or not that suffering is justified. Again, as with blood sports, the rights view would automatically prohibit killing animals for their fur as a violation of their rights.

Seen in animal welfare terms, the case for ending the fur trade would seem to be strong. The level of suffering inflicted on wild-caught animals is intense. The steel jawed (or gin) trap is still widely used throughout the world (although it was banned in Britain over thirty years ago) and animals (including some so-called 'trash' animals who were not the intended victims) may be left in agony for several days until they die, gnaw off a trapped limb and escape, or are put out of their misery when the trapper returns and kills them, either by standing on them until they suffocate or by bludgeoning. Given this, we require a very substantial benefit to accrue from it and yet wearing fur is not necessary, being, in the West at least, essentially an item of fashion. Of course, there are economic interests involved but society must ask itself whether, in this case at least (even though we are accepting here that animals have a fairly minimal moral status), the suffering inflicted really does justify the production of an essentially trivial item where alternatives readily exist.

These alternatives, of course, often involve products – such as leather – derived from other animals and defenders of fur are entitled to ask whether it is inconsistent to attack the farming of fur-bearing animals (the method which provides over 90 per cent of British fur from, in 1991, thirty-one mink farms and two Arctic fox farms) whilst continuing to accept the farming of other animals for human benefits. There is a case to answer here. It should be noted firstly that those who challenge the moral orthodoxy would reject the farming of any animals and are therefore immune from the inconsistency charge. From the standpoint of the moral orthodoxy it is, I think, valid to say that there are fewer objections to fur farming than to the cruel trapping of wild animals. Here we must apply the same criteria as we did to farm animals in general and ask how much suffering it involves. In terms of slaughter, as long as the methods are painless, there is not a problem and the killing of fur farm animals by injection, gassing or electrocution, as long as the procedures are carried out correctly, would not appear to be a problem – at least as far as the moral orthodoxy is concerned. Where there is a difference is in the husbandry of fur-bearing animals. Minks and Arctic foxes are used to ranging across wide areas in the wild and yet are kept in small wire cages. If one then objects that other farmed animals, such as veal calves and battery hens, are just as much deprived of performing their natural behaviour patterns, then the moral orthodoxy would concur and argue that all such practices should be prohibited.

In Britain, the fur trade has gone into a steep decline. Department of Trade and Industry figures reveal that the biggest fur traders in Britain sold only £11 million worth of goods in the first half of 1989 compared to £47 million in 1987 and £80 million in 1984.[31] Few fur retailers are left. Harrods, for instance, announcing the closure of its famed fur department after 140 years of trading in 1990, reported a 40 per cent drop in sales over eight years. Even Oxfam decided at around the same time to ban the sale of fur in its 800 or so shops.[32] In so far as this reflects public recognition of the cruelties involved, as opposed to, say, a change in fashion or even the onset of milder winters (as the fur trade often claims), then public opinion would seem to be ahead of legislative action for it is perfectly legal to catch most animals for their fur (although certain methods are illegal), and perfectly legal to sell fur which has been imported from countries which do not have any restrictions on the methods used to catch animals.

In other countries, the fur trade has not declined to the same degree. World-wide, some thirty million animals are trapped in the wild for their skins and a further forty million are raised and killed on fur farms.[33] America, in particular, would seem to be particularly guilty of operating the double standards which derive from focusing on anthropocentric conservationism. For, although the US has some of the strongest conservation programmes and legislation in the world, it also catches more animals for their fur than any other country in the world and sales of fur in that country account for about one-third of the world market. Even in the US (and, perhaps more surprisingly, Canada as well), though, there is evidence to suggest that the fur trade is slipping.[34]

An international ban on the trade of furs, of course, would be the most appropriate mechanism for halting the fur trade but it is unlikely to happen. CITES does prohibit trade in endangered animals such as the leopard, tiger, jaguar and ocelot, but the fur trappers simply turned to more common species such as the margay and the lynx. Given that the sustainable use strategy of IUCN and the WWF positively encourages the continuation of the fur trade, animal welfare, let alone animal rights, views would seem to have been defeated, at least for the present, internationally if not in Britain.

. . . [W]ildlife conservation is a problematic issue. Some wild animals are protected by a considerable bulk of national and international law and, whilst there is still much to be done to prevent the further killing and suffering of these animals, it is heartening to see that an increasing number of governments support the attempts to keep them alive in their natural habitats. Here, though, lies the problem. For this protection is afforded only to those animals to whom humans attach a value, and this value is equated with the promotion of human interests. It is not generally equated with a recognition that we have duties to animals because they are sentient beings who have an interest in not suffering and even in not being killed.

As a consequence, inconsistencies abound. Thus, those wild animals that humans regard as being more important alive are treated much more favourably than common domesticated animals on farms and in laboratories. Worse still, many unfavoured wild animals are not, in Britain as elsewhere, protected at all despite the fact that their moral status – and, therefore, what we are entitled to do to them – is not altered by their fecundity or their value to humans. This does not mean, of course, that we should protect all wild animals. According to the moral orthodoxy, for instance, significant human interests should take precedence. But even if this minimal moral principle is applied, then all wild animals are entitled to some protection from unnecessary suffering. Significantly, as a result of the value applied by Western conservationists and the public at large, it tends to be the interests of endangered species in the Third World which are (indirectly) promoted ahead of many Third World people. Not only has this proved to be of dubious validity as a strategy, it is also morally repugnant since it involves sacrificing the interests of often very poor native populations without any corresponding sacrifice on the part of the developed world.

Of course, it is recognised that in order to 'sell' conservation, it is necessary to stress its value to humans. This, though, is a dangerous strategy since the implication is that once conservation ceases to be of value, then wildlife is dispensable. Now, in many cases, conservation is very much in the interests of humans, but it is not always so and conflict between human and animal interests is inevitable. Thus, a more enduring basis for conservation is the approach which holds that wild animals (and even the whole of nature if one thinks this is philosophically sound[35]) should be treated as ends in themselves with interests that must be taken into account. The significant point is that this must apply to all animals whether endangered or not, whether in the Third World or in rich Western nations, and whether useful to humanity or not. The challenges to the moral orthodoxy argue that animal and human interests should be given equal consideration. If one holds this view, then a considerable number of human interests will have to be sacrificed but the riposte is that justice is very rarely painless.

Notes

1 See P. Lowe, 'Values and institutions in the history of British nature conservation', in A. Warren and F. B. Goldsmith (eds), *Conservation in Perspective*, Chichester, 1983, pp. 329–52.
2 N. W. Moore, *The Bird of Time: The Science and Politics of Nature Conservation*, Cambridge, 1987, p. 257. See also J. Passmore, *Man's Responsibility for Nature*, London, 1974, for the classic statement of this view.
3 R. Boardman, *International Organisations and the Conservation of Nature*, London, 1981, p. 144.
4 R. Ryder, *Animal Revolution: Changing Attitudes towards Speciesism*, Oxford, 1989, p. 215.
5 Ibid., pp. 214–15; L. Regenstein, 'Animal rights, endangered species and human survival', in P. Singer, *In Defence of Animals*, Oxford, 1985, pp. 118–32.
6 Regenstein, 'Endangered species', p. 119.
7 P. and A. Ehrlich, 'Extinction', in T. Regan and P. Singer, *Animal Rights and Human Obligations* (2nd edition), Englewood Cliffs, NJ, Prentice Hall.
8 Boardman, *International Organisations*, p. 4.
9 S. Lyster, *International Wildlife Law*, Cambridge, 1985, p. xxii.
10 Boardman, *International Organisations*, pp. 88–91.
11 Lyster, *Wildlife Law*, p. 303.
12 Ibid., pp. 75–6, 180.
13 I. Guest, 'Agony and ivory', *The Guardian*, 12 April 1991.

14 *The Guardian*, 20 May 1991. The main development in this campaign to date is the now substantial number of airlines who are refusing to transport wild birds.
15 Ryder, *Animal Revolution*, pp. 219, 234, 236.
16 *The Sunday Times*, 3 March 1991.
17 Guest, 'Agony and ivory' *The Guardian*, 12 April 1991; D. Lavigne, 'Slipping into the marketplace', *BBC Wildlife Magazine*, February 1991, p. 128.
18 B. Holdgate, 'That old-time utilisation', *BBC Wildlife Magazine*, May 1991, p. 374.
19 Regan, *The Case for Animal Rights*, New York, 1982, pp. 359–60.
20 The oldest whale to be caught was calculated to be 114 years old. For a detailed examination of the capacities of whales see Cherfas, *Hunting of the Whale*, London, 1988, pp. 15–56.
21 *The Guardian*, 27 May 1991.
22 See T. Regan, 'Why whaling is wrong', in *All that Dwell Therein*, Berkeley, 1982, pp. 104–7.
23 R. North, 'The way ahead for a crowded planet', *The Sunday Times*, 30 June 1991.
24 *BBC Wildlife Magazine*, January 1991, p. 61. The report has been published by the LACS as *Victim of Myth*.
25 B. Kew, *The Pocketbook of Animal Facts and Figures*, London, 1991, p. 21.
26 See, for instance, *Report on the British Field Sports and Conservation Conference*, British Field Sports Society, London, 1988.
27 'Hunting and Public Opinion', the LACS leaflet 1974.
28 J. Bryant, *Fettered Kingdoms*, Winchester, 1990, p. 62.
29 Kew, *Animal Facts and Figures*, p. 19.
30 Regan, *Animal Rights*, pp. 353–6.
31 *The Guardian*, 11 June 1990; see also 22 February 1990.
32 Ibid., 15 February 1990; 6 February 1990.
33 Ryder, *Animal Revolution*, p. 235.
34 See *The Sunday Times*, 14 January 1990; for Canadian evidence see *BBC Wildlife Magazine*, April 1991, p. 290, and for an account of American fur farming see *The Animals' Agenda*, November 1991, pp. 12–15.
35 See T. Regan, 'The nature and possibility of an environmental ethic', in *All That Dwell Therein*, pp. 192–8.

Source: Robert Garner (1993) *Animals, Politics and Morality*, Manchester: Manchester University Press, ch. 6.

SECTION 4

Values and obligations: rethinking nature

Introduction

The central concern of this section is to focus on the interventions in ecological thought which have attempted to think through what it would mean to take streams, rivers, trees, mountains, and ecosystems more seriously within the moral community. Some of the readings push back the ethical horizons of much environmental discussion – they are concerned with identifying the intrinsic value of natural things rather than seeing them as having a value purely in terms of the instrumental needs of people (as resources to be consumed in the pursuit of human ends). This debate emerged in the discussions over the preservation and conservation of wild places. In Reading 4.1, from the naturalist travelogues of John Muir, we witness an approach which combines an aesthetic appreciation of the wide open spaces of the wilderness with a sense of desolation at the avarice of humankind. This is mixed with an appeal to the American people to put some of the wild aside for the future before it is too late (which resulted in the American National Park system). By contrast, Gifford Pinchot (Reading 4.2) presents a case for conservation through the scientific management of natural resources for the welfare of the Union. Whilst Muir wished to leave the wild to its own devices, Pinchot sought to establish the principle of careful stewardship, so that natural resources were used intelligently and rationally as well as efficiently distributed. The squandering of natural resources in wasteful ways was a technical issue rather than a moral one.

A more explicit defence of the natural world can be seen in Aldo Leopold's advocacy of the 'land ethic' (Reading 4.3). Leopold's case is conveyed through a travelogue where vivid descriptions of natural flora and fauna contain carefully crafted ethical claims. Leopold presents a case for transforming the meaning of the moral community. This approach is concerned with the 'integrity, stability and beauty of the biotic community'. In the classic statement on deep ecology by Arne Naess (Reading 4.4) we can see how the division between preservation and conservation has hardened into a normatively driven distinction between deep ecologists and the shallow ecologists whose main concern is the maintenance of the health and welfare of the populations of Western societies. For deep ecologists, however, all life forms should be venerated – they have a right to 'live and blossom'. These themes are reinforced in Reading 4.5, where Bill Devall and George Sessions use the deep ecology approach to challenge the conservationist account of the stewardship of natural resources. In Reading 4.6 Holmes Rolston III provides a contemporary restatement of the land ethic. Rolston places a special emphasis upon nature as a community (a parallel with the biospherical egalitarianism of deep ecologists) rather than as a commodity – that the land ethic could transform human experience so that we no longer live inferior lives.

In the next two readings, Christopher D. Stone explores the ways in which the legal system could be transformed if the interests of natural things were granted legal standing. In Reading 4.7 Stone focuses on the grounds for including or excluding trees from the moral community, constructs a case for recognizing their legal standing through the roles of guardians, and explores how common law has denied legal standing for natural things. In Reading 4.8 he uses the legal definition of rights to construct a case for establishing the legal considerateness of forests, oceans, lakes and so on. More specifically, Stone provides ways of defining the interests, intactness and preferences of 'non-persons' in order to argue for the award of compensation for existing damage and to protect natural things from human activities. Stone provides a wealth of evidence and argumentation to make a case for their legal recognition, in the same way that corporate lawyers act on behalf of companies and other organizations or as parents act for their children.

In Reading 4.9 Luke Martell demonstrates how it is possible to clarify the approaches considered in this section. He offers three ways of thinking through the ethical choices developed so far. First, he considers whether the values attached to a natural object are intrinsic to the animal, tree or ecosystem in question or exist by virtue of their uses by humankind. Second, he explores the different ways in which moral standing can be attributed – sentiency, flourishing, diversity, species preservation or membership of the moral community. Finally, Martell offers a classification system defining entities based on whether they are human or non-human, sentient or non-sentient and living or non-living.

4.1

John Muir

Preserving the wilderness

Wild parks of the West

[. . .] Only thirty years ago, the great Central Valley of California, five hundred miles long and fifty miles wide, was one bed of golden and purple flowers. Now it is ploughed and pastured out of existence, gone forever – scarce a memory of it left in fence corners and along the bluffs of the streams. The gardens of the Sierra, also, and the noble forests in both the reserved and unreserved portions are sadly hacked and trampled, notwithstanding the ruggedness of the topography – all excepting those of the parks guarded by a few soldiers. In the noblest forests of the world, the ground, once divinely beautiful, is desolate and repulsive, like a face ravaged by disease. This is true also of many other Pacific Coast and Rocky Mountain valleys and forests. The same fate, sooner or later, is awaiting them all, unless awakening public opinion comes forward to stop it. Even the great deserts in Arizona, Nevada, Utah, and New Mexico, which offer so little to attract settlers, and which a few years ago pioneers were afraid of, as places of desolation and death, are now taken as pastures at the rate of one or two square miles per cow, and of course their plant treasures are passing away – the delicate abronias, phloxes, gilias, etc. Only a few of the bitter, thorny, unbitable shrubs are left, and the sturdy cactuses that defend themselves with bayonets and spears.

Most of the wild plant wealth of the East also has vanished – gone into dusty history. Only vestiges of its glorious prairie and woodland wealth remain to bless humanity in boggy, rocky, unploughable places. Fortunately, some of these are purely wild, and go far to keep Nature's love visible. White water-lilies, with rootstocks deep and safe in mud, still send up every summer a Milky Way of starry, fragrant flowers around a thousand lakes, and many a tuft of wild grass waves its panicles on mossy rocks, beyond reach of trampling feet, in company with saxifrages, bluebells, and ferns. Even in the midst of farmers' fields, precious sphagnum bogs, too soft for the feet of cattle, are preserved with their charming plants unchanged – chiogenes, Andromeda, Kalmia, Linnæa, Arethusa, etc.

[. . .]

The Yellowstone National Park

[. . .] Camp out among the grass and gentians of glacier meadows, in craggy garden nooks full of Nature's darlings. Climb the mountains and get their good tidings. Nature's peace will flow into you as sunshine flows into trees. The winds will blow their own freshness into you, and the storms their energy, while cares will drop off like autumn leaves. As age comes on, one source of enjoyment after another is closed, but Nature's sources never fail. Like a generous host, she offers here brimming cups in endless variety, served in a grand hall, the sky its ceiling, the mountains its walls, decorated with glorious paintings and enlivened with bands of music ever playing. The petty discomforts that beset the awkward guest, the unskilled camper, are quickly forgotten, while all that is precious remains. Fears vanish as soon as one is fairly free in the wilderness.

Most of the dangers that haunt the unseasoned citizen are imaginary; the real ones are perhaps too few rather than too many for his good. The bears that always seem to spring up thick as trees, in fighting, devouring attitudes before the frightened tourist whenever a camping trip is proposed, are gentle now, finding they are no longer likely to be shot; and rattlesnakes, the other big irrational dread of over-civilized people, are scarce here, for most of the park lies above the snake-line. Poor creatures, loved only by their Maker, they are timid and bashful, as mountaineers know; and though perhaps not possessed of much of that charity that suffers long and is kind, seldom, either by mistake or by mishap, do harm to any one. Certainly they cause not the hundredth part of the pain and death that follow the footsteps of the admired Rocky Mountain trapper. Nevertheless, again and again, in season and out of season, the question comes up, "What are rattlesnakes good for?" As if nothing that does not obviously make for the benefit of man had any right to exist; as if our ways were God's ways. Long ago, an Indian to whom a French traveler put this old question replied that their tails were good for toothache, and their heads for fever. Anyhow, they are all, head and tail, good for themselves, and we need not begrudge them their share of life.

Fear nothing. No town park you have been accustomed to saunter in is so free from danger as the Yellowstone. It is a hard place to leave.

[. . .]

The American Forests

Under the timber and stone act of 1878, which might well have been called the "dust and ashes act," any citizen of the United States could take up one hundred and sixty acres of timber land, and by paying two dollars and a half an acre for it obtain title. There was some virtuous effort made with a view to limit the operations of the act by requiring that the purchaser should make affidavit that he was entering the land exclusively for his own use, and by not allowing any association to enter more than one hundred and sixty acres. Nevertheless, under this act wealthy corporations have fraudulently obtained title to from ten thousand to twenty thousand acres or more. The plan was usually as follows: A mill company, desirous of getting title to a large body of redwood or sugar-pine land, first blurred the eyes and ears of the land agents, and then hired men to enter the land they wanted, and immediately deed it to the company after a nominal compliance with the law; false swearing in the wilderness against the government being held of no account. In one case which came under the observation of Mr. Bowers, it was the practice of a lumber company to hire the entire crew of every vessel which might happen to touch at any port in the redwood belt, to enter one hundred and sixty acres each and immediately deed the land to the company, in consideration of the company's paying all expenses and giving the jolly sailors fifty dollars apiece for their trouble.

By such methods have our magnificent redwoods and much of the sugar-pine forests of the Sierra Nevada been absorbed by foreign and resident capitalists. Uncle Sam is not often called a fool in business matters, yet he has sold millions of acres of timber land at two dollars and a half an acre on which a single tree was worth more than a hundred dollars. But this priceless land has been patented, and nothing can be done now about the crazy bargain. According to the everlasting law of righteousness, even the fraudulent buyers at less than one per cent of its value are making little or nothing, on account of fierce competition. The trees are felled, and about half of each giant is left on the ground to be converted into smoke and ashes; the better half is sawed into choice lumber and sold to citizens of the United States or to foreigners: thus robbing the country of its glory and impoverishing it without right benefit to anybody – a bad, black business from beginning to end.

The redwood is one of the few conifers that sprout from the stump and roots, and it declares itself willing to begin immediately to repair the damage of the lumberman and also that of the forest-burner. As soon as a redwood is cut down or burned it sends up a crowd of eager, hopeful shoots, which, if allowed to grow, would in a few decades attain a height of a hundred feet, and the strongest of them would finally become giants as great as the original tree. Gigantic second and third growth trees are found in the redwoods, forming magnificent temple-like circles around charred ruins more than a thousand years old. But not one denuded acre in a hundred is allowed to raise a new forest growth. On the contrary, all the brains, religion, and superstition of the neighborhood are brought into play to prevent a new growth. The sprouts from the roots and stumps are cut off again and again, with zealous concern as to the best time and method of making death sure. In the clearings of one of the largest mills on the coast we found thirty men at work, last summer, cutting off redwood shoots "in the dark of the moon," claiming that all the stumps and roots cleared at this auspicious time would send up no more shoots. Anyhow, these vigorous, almost immortal trees are killed at last, and black stumps are now their only monuments over most of the chopped and burned areas.

The redwood is the glory of the Coast Range. It extends along the western slope, in a nearly continuous belt about ten miles wide, from beyond the Oregon boundary to the south of Santa Cruz, a distance of nearly four hundred miles, and in massive, sustained grandeur and closeness of growth surpasses all the other timber woods of the world. Trees from ten to fifteen feet in diameter and three hundred feet high are not uncommon, and a few attain a height of three hundred and fifty feet or even four hundred, with a diameter at the base of

Figure 10 Young Big Tree felled for shingles

fifteen to twenty feet or more, while the ground beneath them is a garden of fresh, exuberant ferns, lilies, gaultheria, and rhododendron. This grand tree, Sequoia sempervirens, is surpassed in size only by its near relative, Sequoia gigantea, or Big Tree, of the Sierra Nevada, if, indeed, it is surpassed. The sempervirens is certainly the taller of the two. The gigantea attains a greater girth, and is heavier, more noble in port, and more sublimely beautiful. These two Sequoias are all that are known to exist in the world, though in former geological times the genus was common and had many species. The redwood is restricted to the Coast Range, and the Big Tree to the Sierra.

As timber the redwood is too good to live.

Source: John Muir (1901) *Our National Parks*, Boston: Houghton Mifflin, pp. 5–6, 56–8, 347–50.

4.2

Gifford Pinchot

Conservation and human welfare

The prodigal squandering of our mineral fuels proceeds unchecked in the face of the fact that such resources as these, once used or wasted, can never be replaced. If waste like this were not chiefly thoughtless, it might well be characterized as the deliberate destruction of the nation's future.

Many fields of iron ore have already been exhausted, and in still more, as in the coal mines, only the higher grades have been taken from the mines, leaving the least valuable beds to be exploited at increased cost or not at all. Similar waste in the case of other minerals is less serious only because they are less indispensable to our civilization than coal and iron. Mention should be made of the annual loss of millions of dollars worth of by-products from coke, blast, and other furnaces now thrown into the air, often not merely without benefit but to the serious injury of the community. In other countries these by-products are saved and used.

We are in the habit of speaking of the solid earth and the eternal hills as though they, at least, were free from the vicissitudes of time and certain to furnish perpetual support for prosperous human life. This conclusion is as false as the term "inexhaustible" applied to other natural resources. The waste of soil is among the most dangerous of all wastes now in progress in the United States. In 1896, Professor Shaler, than whom no one has spoken with greater authority on this subject, estimated that in the upland regions of the states south of Pennsylvania three thousand square miles of soil had been destroyed as the result of forest denudation, and that destruction was then proceeding at the rate of one hundred square miles of fertile soil per year. No seeing man can travel through the United States without being struck with the enormous and unnecessary loss of fertility by easily preventable soil wash. The soil so lost, as in the case of many other wastes, becomes itself a source of damage and expense, and must be removed from the channels of our navigable streams at an enormous annual cost. The Mississippi River alone is estimated to transport yearly four hundred million tons of sediment, or about twice the amount of material to be excavated from the Panama Canal. This material is the most fertile portion of our richest fields, transformed from a blessing to a curse by unrestricted erosion.

The destruction of forage plants by overgrazing has resulted, in the opinion of men most capable of judging, in reducing the grazing value of the public lands by one-half. This enormous loss of forage, serious though it be in itself, is not the only result of wrong methods of pasturage. The destruction of forage plants is accompanied by loss of surface soil through erosion; by forest destruction; by corresponding deterioration in the water supply; and by a serious decrease in the quality and weight of animals grown on overgrazed lands. These sources of loss from failure to conserve the range are felt to-day. They are accompanied by the certainty of a future loss not less important, for range lands once badly overgrazed can be restored to their former value but slowly or not at all. The obvious and certain remedy is for the Government to hold and control the public range until it can pass into the hands of settlers who will make their homes upon it. As methods of agriculture improve and new dry-land crops are introduced, vast areas once considered unavailable for cultivation are being made into prosperous homes; and this movement has only begun.

[. . .]

It is well to remember that there is no foreign source from which we can draw cheap and abundant supplies of timber to meet a demand per capita so large as to be without parallel in the world, and that the suffering which will result from the progressive failure of our timber has been but faintly foreshadowed by temporary scarcities of coal.

What will happen when the forests fail? In the first place, the business of lumbering will disappear. It is now the fourth greatest industry in the United States. All forms of building industries will suffer with it, and the occupants of houses, offices, and stores must pay the added cost. Mining will become vastly more expensive; and with the rise in the cost of mining there must follow a corresponding rise in the price of coal, iron, and other minerals. The railways, which have as yet failed entirely to develop a satisfactory substitute for the wooden tie (and must, in the opinion of their best engineers, continue to fail), will be profoundly affected, and the cost of transportation will suffer a corresponding increase. Water power for lighting, manufacturing, and transportation, and the movement of freight and passengers by inland waterways, will be affected still more directly than the steam railways. The cultivation of the soil, with or without irrigation, will be hampered by the increased cost of agricultural tools, fencing, and the wood needed for other purposes about the farm. Irrigated agriculture will suffer most of all, for the destruction of the forests means the loss of the waters as surely as night follows day. With the rise in the cost of producing food, the cost of food itself will rise. Commerce in general will necessarily be affected by the difficulties of the primary industries upon which it depends. In a word, when the forests fail, the daily life of the average citizen will inevitably feel the pinch on every side. And the forests have already begun to fail, as the direct result of the suicidal policy of forest destruction which the people of the United States have allowed themselves to pursue.

[. . .]

Principles of conservation

The first great fact about conservation is that it stands for development. There has been a fundamental misconception that conservation means nothing but the husbanding of resources for future generations. There could be no more serious mistake. Conservation does mean provision for the future, but it means also and first of all the recognition of the right of the present generation to the fullest necessary use of all the resources with which this country is so abundantly blessed. Conservation demands the welfare of this generation first, and afterward the welfare of the generations to follow.

The first principle of conservation is development, the use of the natural resources now existing on this continent for the benefit of the people who live here now. There may be just as much waste in neglecting the development and use of certain natural resources as there is in their destruction. We have a limited supply of coal, and only a limited supply. Whether it is to last for a hundred or a hundred and fifty or a thousand years, the coal is limited in amount; unless through geological changes which we shall not live to see, there will never be any more of it than there is now. But coal is in a sense the vital essence of our civilization.

If it can be preserved, if the life of the mines can be extended, if by preventing waste there can be more coal left in this country after we of this generation have made every needed use of this source of power, then we shall have deserved well of our descendants.

Conservation stands emphatically for the development and use of water-power now, without delay. It stands for the immediate construction of navigable waterways under a broad and comprehensive plan as assistants to the railroads. More coal and more iron are required to move a ton of freight by rail than by water, three to one. In every case and in every direction the conservation movement has development for its first principle, and at the very beginning of its work. The development of our natural resources and the fullest use of them for the present generation is the first duty of this generation. So much for development.

In the second place conservation stands for the prevention of waste. There has come gradually in this country an understanding that waste is not a good thing and that the attack on waste is an industrial necessity. I recall very well indeed how, in the early days of forest fires, they were considered simply and solely as acts of God, against which any opposition was hopeless and any attempt to control them not merely hopeless but childish. It was assumed that they came in the natural order of things, as inevitably as the seasons or the rising and setting of the sun. To-day we understand that forest fires are wholly within the control of men. So we are coming in like manner to understand that the prevention of waste in all other directions is a simple matter of good business. The first duty of the human race is to control the earth it lives upon.

[. . .]

In addition to the principles of development and preservation of our resources there is a third principle. It is this: The natural resources must be developed and preserved for the benefit of the many, and not merely for the profit of a few. We are coming to understand in this country that public action for public benefit has a very much wider field to cover and a much larger part to play than was the case when there were resources enough for every one, and before certain constitutional provisions had given so tremendously strong a position to vested rights and property in general.

[. . .]

. . . Conservation means the greatest good to the greatest number for the longest time. One of its great contributions is just this, that it has added to the worn and well-known phrase, "the greatest good to the greatest number," the additional words "for the longest time," thus recognizing that this nation of ours must be made to endure as the best possible home for all its people.

Conservation advocates the use of foresight, prudence, thrift, and intelligence in dealing with public matters, for the same reasons and in the same way that we each use foresight, prudence, thrift, and intelligence in dealing with our own private affairs. It proclaims the right and duty of the people to act for the benefit of the people.

[. . .]

The moral issue

The central thing for which Conservation stands is to make this country the best possible place to live in, both for us and for our descendants. It stands against the waste of the natural resources which cannot be renewed, such as coal and iron; it stands for the perpetuation of the resources which can be renewed, such as the food-producing soils and the forests; and most of all it stands for an equal opportunity for every American citizen to get his fair share of benefit from these resources, both now and hereafter.

Conservation stands for the same kind of practical common-sense management of this country by the people that every business man stands for in the handling of his own business. It believes in prudence and foresight instead of reckless blindness; it holds that resources now public property should not become the basis for oppressive private monopoly; and it demands the complete and orderly development of all our resources for the benefit of all the people, instead of the partial exploitation of them for the benefit of a few. It recognizes fully the right of the present generation to use what it needs and all it needs of the natural resources now available, but it recognizes equally our obligation so to use what we need that our descendants shall not be deprived of what they need.

Conservation has much to do with the welfare of the average man of to-day. It proposes to secure a continuous and abundant supply of the necessaries of life, which means a reasonable cost of living and business stability. It advocates fairness in the distribution of the benefits which flow from the natural resources. It will matter very little to the average citizen, when scarcity comes and prices rise, whether he can not get what he needs because there is none left or because he can not afford to pay for it. In both cases the essential fact is that he can not get what he needs. Conservation holds that it is about as important to see that the people in general get the benefit of our natural resources as to see that there shall be natural resources left.

Conservation is the most democratic movement this country has known for a generation. It holds that the people have not only the right, but the duty to control the use of the natural resources, which are the great sources of prosperity. And it regards the absorption of these resources by the special interests, unless their operations are under effective public control, as a moral wrong. Conservation is the application of common-sense to the common problems for the common good, and I believe it stands nearer to the desires, aspirations, and purposes of the average man than any other policy now before the American people.

The danger to the Conservation policies is that the privileges of the few may continue to obstruct the rights of the many, especially in the matter of water power and coal.

Source: Gifford Pinchot (1901) *The Fight for Conservation*, New York: Harcourt Brace, pp. 8–17, 42–8, 79–82.

4.3a

Aldo Leopold

The chit-chat of the woods

Why is the shovel regarded as a symbol of drudgery? Perhaps because most shovels are dull. Certainly all drudges have dull shovels, but I am uncertain which of these two facts is cause and which effect. I only know that a good file, vigorously wielded, makes my shovel sing as it slices the mellow loam. I am told there is music in the sharp plane, the sharp chisel, and the sharp scalpel, but I hear it best in my shovel; it hums in my wrists as I plant a pine. I suspect that the fellow who tried so hard to strike one clear note upon the harp of time chose too difficult an instrument.

It is well that the planting season comes only in spring, for moderation is best in all things, even shovels. During the other months you may watch the process of becoming a pine.

The pine's new year begins in May, when the terminal bud becomes 'the candle.' Whoever coined that name for the new growth had subtlety in his soul. 'The candle' sounds like a platitudinous reference to obvious facts: the new shoot is waxy, upright, brittle. But he who lives with pines knows that candle has a deeper meaning, for at its tip burns the eternal flame that lights a path into the future. May after May my pines follow their candles skyward, each headed straight for the zenith, and each meaning to get there if only there be years enough before the last trumpet blows. It is a very old pine who at last forgets which of his many candles is the most important, and thus flattens his crown against the sky. You may forget, but no pine of your own planting will do so in your lifetime.

If you are thriftily inclined, you will find pines congenial company, for, unlike the hand-to-mouth hardwoods, they never pay current bills out of current earnings; they live solely on their savings of the year before. In fact every pine carries an open bankbook, in which his cash balance is recorded by 30 June of each year. If, on that date, his completed candle has developed a terminal cluster of ten or twelve buds, it means that he has salted away enough rain and sun for a two-foot or even a three-foot thrust skyward next spring. If there are only four or six buds, his thrust will be a lesser one, but he will nevertheless wear that peculiar air that goes with solvency.

Hard years, of course, come to pines as they do to men, and these are recorded as shorter thrusts, i.e. shorter spaces between the successive whorls of branches. These spaces, then, are an autobiography that he who walks with trees may read at will. In order to date a hard year correctly, you must always subtract one from the year of lesser growth. Thus the 1937 growth was short in all pines; this records the universal drouth of 1936. On the other hand the 1941 growth was long in all pines; perhaps they saw the shadow of things to come, and made a special effort to show the world that pines still know where they are going, even though men do not.

When one pine shows a short year but his neighbors do not, you may safely interpolate some purely local or individual adversity: a fire scar, a gnawing meadowmouse, a windburn, or some local bottleneck in that dark laboratory we call the soil.

There is much small-talk and neighborhood gossip among pines. By paying heed to this chatter, I learn what has transpired during the week when I am absent in town. Thus in March, when the deer frequently browse white pines, the height of the browsings tells me how hungry they are. A deer full of corn is too lazy to nip branches more than four feet above the ground; a really hungry deer rises on his hind legs and nips as high as eight feet. Thus I learn the gastronomic status of the deer without seeing them, and I learn, without visiting his field, whether my neighbor has hauled in his cornshocks.

In May, when the new candle is tender and brittle as an asparagus shoot, a bird alighting on it will often break it off. Every spring I find a few such decapitated trees, each with its wilted candle lying in the grass. It is easy to infer what has happened, but in a decade of watching I have never once *seen* a bird break a candle. It is an object lesson: one need not doubt the unseen.

In June of each year a few white pines suddenly show wilted candles, which shortly thereafter turn brown and die. A pine weevil has bored into the terminal bud cluster and deposited eggs; the grubs, when hatched, bore down along the pith and kill the shoot. Such a leaderless pine is doomed to frustration, for the surviving branches disagree among themselves who is to head the skyward march. They all do, and as a consequence the tree remains a bush.

It is a curious circumstance that only pines in full sunlight are bitten by weevils; shaded pines are ignored. Such are the hidden uses of adversity.

In October my pines tell me, by their rubbed-off bark, when the bucks are beginning to 'feel their oats.' A jackpine about eight feet high, and standing alone, seems especially to incite in a buck the idea that the world needs prodding. Such a tree must perforce turn the other cheek also, and emerges much the worse for wear. The only element of justice in such combats is that the more the tree is punished, the more pitch the buck carries away on his not-so-shiny antlers.

The chit-chat of the woods is sometimes hard to translate. Once in midwinter I found in the droppings under a grouse roost some half-digested structures that I could not identify. They resembled miniature corncobs about half an inch long. I examined samples of every local grouse food I could think of, but without finding any clue to the origin of the 'cobs.' Finally I cut open the terminal bud of a jackpine, and in its core I found the answer. The grouse had eaten the buds, digested the pitch, rubbed off the scales in his gizzard, and left the cob, which was, in effect, the forthcoming candle. One might say that this grouse had been speculating in jackpine 'futures.'

Source: Aldo Leopold (1949) *A Sand County Almanac – and Sketches Here and There*, Oxford: Oxford University Press, pp. 82–5.

4.3b

Aldo Leopold

The land ethic

There is as yet no ethic dealing with man's relation to land and to the animals and plants which grow upon it. Land, like Odysseus' slave-girls, is still property. The land-relation is still strictly economic, entailing privileges but not obligations.

The extension of ethics to this third element in human environment is, if I read the evidence correctly, an evolutionary possibility and an ecological necessity. It is the third step in a sequence. The first two have already been taken. Individual thinkers since the days of Ezekiel and Isaiah have asserted that the despoliation of land is not only inexpedient but wrong. Society, however, has not yet affirmed their belief. I regard the present conservation movement as the embryo of such an affirmation.

An ethic may be regarded as a mode of guidance for meeting ecological situations so new or intricate, or involving such deferred reactions, that the path of social expediency is not discernible to the average individual. Animal instincts are modes of guidance for the individual in meeting such situations. Ethics are possibly a kind of community instinct in-the-making.

The community concept

All ethics so far evolved rest upon a single premise: that the individual is a member of a community of interdependent parts. His instincts prompt him to compete for his place in that community, but his ethics prompt him also to co-operate (perhaps in order that there may be a place to compete for).

The land ethic simply enlarges the boundaries of the community to include soils, waters, plants, and animals, or collectively: the land.

This sounds simple: do we not already sing our love for and obligation to the land of the free and the home of the brave? Yes, but just what and whom do we love? Certainly not the soil, which we are sending helter-skelter downriver. Certainly not the waters, which we assume have no function except to turn turbines, float barges, and carry off sewage. Certainly not the plants, of which we exterminate whole communities without batting an eye. Certainly not the animals, of which we have already extirpated many of the largest and most beautiful species. A land ethic of course cannot prevent the alteration, management, and use of these 'resources,' but it does affirm their right to continued existence, and, at least in spots, their continued existence in a natural state.

In short, a land ethic changes the role of *Homo sapiens* from conqueror of the land-community to plain member and citizen of it. It implies respect for his fellow-members, and also respect for the community as such.

In human history, we have learned (I hope) that the conqueror role is eventually self-defeating. Why? Because it is implicit in such a role that the conqueror knows, *ex cathedra,* just what makes the community clock tick, and just what and who is valuable, and what and who is worthless, in community life. It always turns out that he knows neither, and this is why his conquests eventually defeat themselves.

[. . .]

Substitutes for a land ethic

When the logic of history hungers for bread and we hand out a stone, we are at pains to explain how much the stone resembles bread. I now describe some of the stones which serve in lieu of a land ethic.

One basic weakness in a conservation system based wholly on economic motives is that most members of the land community have no economic value. Wildflowers and songbirds are examples. Of the 22,000 higher plants and animals native to Wisconsin, it is doubtful whether more than 5 per cent can be sold, fed, eaten, or otherwise put to economic use. Yet these creatures are members of the biotic community, and if (as I believe) its stability depends on its integrity, they are entitled to continuance.

When one of these non-economic categories is threatened, and if we happen to love it, we invent subterfuges to give it economic importance. At the beginning of the century songbirds were supposed to be disappearing. Ornithologists jumped to the rescue with some distinctly shaky evidence to the effect that insects would eat us up if birds failed to control them. The evidence had to be economic in order to be valid.

It is painful to read these circumlocutions today. We have no land ethic yet, but we have at least drawn nearer the point of admitting that birds should continue as a matter of biotic right, regardless of the presence or absence of economic advantage to us.

A parallel situation exists in respect of predatory mammals, raptorial birds, and fish-eating birds. Time was when biologists somewhat overworked the evidence that these creatures preserve the health of game by killing weaklings, or that they control rodents for the farmer, or that they prey only on 'worthless' species. Here again, the evidence had to be economic in order to be valid. It is only in recent years that we hear the more honest argument that predators are members of the community, and that no special interest has the right to exterminate them for the sake of a benefit, real or fancied, to itself. Unfortunately this

enlightened view is still in the talk stage. In the field the extermination of predators goes merrily on: witness the impending erasure of the timber wolf by fiat of Congress, the Conservation Bureaus, and many state legislatures.

Some species of trees have been 'read out of the party' by economics-minded foresters because they grow too slowly, or have too low a sale value to pay as timber crops: white cedar, tamarack, cypress, beech, and hemlock are examples. In Europe, where forestry is ecologically more advanced, the non-commercial tree species are recognized as members of the native forest community, to be preserved as such, within reason. Moreover some (like beech) have been found to have a valuable function in building up soil fertility. The interdependence of the forest and its constituent tree species, ground flora, and fauna is taken for granted.

Lack of economic value is sometimes a character not only of species or groups, but of entire biotic communities: marshes, bogs, dunes, and 'deserts' are examples. Our formula in such cases is to relegate their conservation to government as refuges, monuments, or parks. The difficulty is that these communities are usually interspersed with more valuable private lands; the government cannot possibly own or control such scattered parcels. The net effect is that we have relegated some of them to ultimate extinction over large areas. If the private owner were ecologically minded, he would be proud to be the custodian of a reasonable proportion of such areas, which add diversity and beauty to his farm and to his community.

In some instances, the assumed lack of profit in these 'waste' areas has proved to be wrong, but only after most of them had been done away with. The present scramble to reflood muskrat marshes is a case in point.

[. . .]

The outlook

It is inconceivable to me that an ethical relation to land can exist without love, respect, and admiration for land, and a high regard for its value. By value, I of course mean something far broader than mere economic value; I mean value in the philosophical sense.

Perhaps the most serious obstacle impeding the evolution of a land ethic is the fact that our educational and economic system is headed away from, rather than toward, an intense consciousness of land. Your true modern is separated from the land by many middlemen, and by innumerable physical gadgets. He has no vital relation to it; to him it is the space between cities on which crops grow. Turn him loose for a day on the land, and if the spot does not happen to be a golf links or a 'scenic' area, he is bored stiff. If crops could be raised by hydroponics instead of farming, it would suit him very well. Synthetic substitutes for wood, leather, wool, and other natural land products suit him better than the originals. In short, land is something he has 'outgrown.'

Almost equally serious as an obstacle to a land ethic is the attitude of the farmer for whom the land is still an adversary, or a taskmaster that keeps him in slavery. Theoretically, the mechanization of farming ought to cut the farmer's chains, but whether it really does is debatable.

One of the requisites for an ecological comprehension of land is an understanding of ecology, and this is by no means co-extensive with 'education'; in fact, much higher education seems deliberately to avoid ecological concepts. An understanding of ecology does not necessarily originate in courses bearing ecological labels; it is quite as likely to be labeled geography, botany, agronomy, history, or economics. This is as it should be, but whatever the label, ecological training is scarce.

The case for a land ethic would appear hopeless but for the minority which is in obvious revolt against these 'modern' trends.

The 'key-log' which must be moved to release the evolutionary process for an ethic is simply this: quit thinking about decent land-use as solely an economic problem. Examine each question in terms of what is ethically and esthetically right, as well as what is economically expedient. A thing is right when it tends to preserve the integrity, stability, and beauty of the biotic community. It is wrong when it tends otherwise.

It of course goes without saying that economic feasibility limits the tether of what can or cannot be done for land. It always has and it always will. The fallacy the economic determinists have tied around our collective neck, and which we now need to cast off, is the belief that economics determines *all* land-use. This is simply not true. An innumerable host of actions and attitudes, comprising perhaps the bulk of all land relations, is determined by the land-users' tastes and predilections, rather than by his purse. The bulk of all land relations hinges on investments of time, forethought, skill, and faith rather than on investments of cash. As a land-user thinketh, so is he.

I have purposely presented the land ethic as a product of social evolution because nothing so important as an ethic is ever 'written.' Only the most superficial student of history supposes that Moses 'wrote' the Decalogue; it evolved in the minds of a thinking community, and Moses wrote a tentative summary of it for a 'seminar.' I say tentative because evolution never stops.

The evolution of a land ethic is an intellectual as well as emotional process. Conservation is paved with good intentions which prove to be futile, or even dangerous, because they are devoid of critical understanding either of the land, or of economic land-use. I think it is a truism that as the ethical frontier advances from the individual to the community, its intellectual content increases.

The mechanism of operation is the same for any ethic: social approbation for right actions: social disapproval for wrong actions.

By and large, our present problem is one of attitudes and implements. We are remodeling the Alhambra with a steam-shovel, and we are proud of our yardage. We shall hardly relinquish the shovel, which after all has many good points, but we are in need of gentler and more objective criteria for its successful use.

Source: Aldo Leopold (1949) *A Sand County Almanac – and Sketches Here and There*, Oxford: Oxford University Press, pp. 201–4, 210–12, 223–6.

4.4

Arne Naess

The shallow and the deep

> Ecologically responsible policies are concerned only in part with pollution and resource depletion. There are deeper concerns which touch upon principles of diversity, complexity, autonomy, decentralization, symbiosis, egalitarianism, and classlessness.

The emergence of ecologists from their former relative obscurity marks a turning-point in our scientific communities. But their message is twisted and misused. A shallow, but presently rather powerful movement, and a deep, but less influential movement, compete for our attention. I shall make an effort to characterize the two.

(a) *The Shallow Ecology movement:*
Fight against pollution and resource depletion. Central objective: the health and affluence of people in the developed countries.

(b) *The Deep Ecology movement:*
1 Rejection of the man-in-environment image in favour of *the relational, total-field image*. Organisms as knots in the biospherical net or field of intrinsic relations. An intrinsic relation between two things *A* and *B* is such that the relation belongs to the definitions or basic constitutions of *A* and *B*, so that without the relation, *A* and *B* are no longer the same things. The total-field model dissolves not only the man-in-environment concept, but every compact thing-in-milieu concept – except when talking at a superficial or preliminary level of communication.

2 *Biospherical egalitarianism* – in principle. The 'in principle' clause is inserted because any realistic praxis necessitates some killing, exploitation, and suppression. The ecological field-worker acquires a deep-seated respect, or even veneration, for ways and forms of life. He reaches an understanding from within, a kind of understanding that others reserve for fellow men and for a narrow section of ways and forms of life. To the ecological field-worker, *the equal right to live and blossom* is an intuitively clear and obvious value axiom. Its restriction to humans is an anthropocentrism with detrimental effects upon the life quality of humans themselves. This quality depends in part upon the deep pleasure and satisfaction we receive from close partnership with other forms of life. The attempt to ignore our dependence and to establish a master–slave role has contributed to the alienation of man from himself.

Ecological egalitarianism implies the reinterpretation of the future-research variable, 'level of crowding', so that *general* mammalian crowding and loss of life-equality is taken seriously, not only human crowding. (Research on the high requirements of free space of certain mammals has, incidentally, suggested that theorists of human urbanism have largely underestimated human life-space requirements. Behavioural crowding symptoms [neuroses, aggressiveness, loss of traditions. . .] are largely the same among mammals.)

3 *Principles of diversity and of symbiosis*. Diversity enhances the potentialities of survival, the chances of new modes of life, the richness of forms. And the so-called struggle of life, and survival of the fittest, should be interpreted in the sense of ability to coexist and cooperate in complex relationships, rather than ability to kill, exploit, and suppress. 'Live and let live' is a more powerful ecological principle than 'Either you or me'.

The latter tends to reduce the multiplicity of kinds of forms of life, and also to create destruction within the communities of the same species. Ecologically inspired attitudes therefore favour diversity of human ways of life, of cultures, of occupations, of economies. They support the fight against economic and cultural, as much as military, invasion and domination, and they are opposed to the annihilation of seals and whales as much as to that of human tribes or cultures.

4 *Anti-class posture*. Diversity of human ways of life is in part due to (intended or unintended) exploitation and suppression on the part of certain groups. The exploiter lives differently from the exploited, but both are adversely affected in their potentialities of self-realization. The principle of diversity does not cover differences due merely to certain attitudes or behaviours forcibly blocked or restrained. The principles of ecological egalitarianism and of symbiosis support the same anti-class posture. The ecological attitude favours the extension of all three principles to any group conflicts, including those of today between

developing and developed nations. The three principles also favour extreme caution towards any overall plans for the future, except those consistent with wide and widening classless diversity.

5 Fight against *pollution and resource depletion*. In this fight ecologists have found powerful supporters, but sometimes to the detriment of their total stand. This happens when attention is focused on pollution and resource depletion rather than on the other points, or when projects are implemented which reduce pollution but increase evils of the other kinds. Thus, if prices of life necessities increase because of the installation of anti-pollution devices, class differences increase too. An ethics of responsibility implies that ecologists do not serve the shallow, but the deep ecological movement. That is, not only point (5), but all seven points must be considered together.

Ecologists are irreplaceable informants in any society, whatever their political colour. If well organized, they have the power to reject jobs in which they submit themselves to institutions or to planners with limited ecological perspectives. As it is now, ecologists sometimes serve masters who deliberately ignore the wider perspectives.

6 *Complexity, not complication*. The theory of ecosystems contains an important distinction between what is complicated without any Gestalt or unifying principles – we may think of finding our way through a chaotic city – and what is complex. A multiplicity of more or less lawful, interacting factors may operate together to form a unity, a system. We make a shoe or use a map or integrate a variety of activities into a workaday pattern. Organisms, ways of life, and interactions in the biosphere in general, exhibit complexity of such an astoundingly high level as to colour the general outlook of ecologists. Such complexity makes thinking in terms of vast systems inevitable. It also makes for a keen, steady perception of the profound *human ignorance* of biospherical relationships and therefore of the effect of disturbances.

Applied to humans, the complexity-not-complication principle favours division of labour, *not fragmentation of labour*. It favours integrated actions in which the whole person is active, not mere reactions. It favours complex economies, an integrated variety of means of living. (Combinations of industrial and agricultural activity, of intellectual and manual work, of specialized and non-specialized occupations, of urban and non-urban activity, of work in city and recreation in nature with recreation in city and work in nature . . .)

It favours soft technique and 'soft future-research', less prognosis, more clarification of possibilities. More sensitivity towards continuity and live traditions, and – most importantly – towards our state of ignorance.

The implementation of ecologically responsible policies requires in this century an exponential growth of technical skill and invention – but in new directions, directions which today are not consistently and liberally supported by the research policy organs of our nation-states.

7 *Local autonomy and decentralization*. The vulnerability of a form of life is roughly proportional to the weight of influences from afar, from outside the local region in which that form has obtained an ecological equilibrium. This lends support to our efforts to strengthen local self-government and material and mental self-sufficiency. But these efforts presuppose an impetus towards decentralization. Pollution problems, including those of thermal pollution

and recirculation of materials, also lead us in this direction, because increased local autonomy, if we are able to keep other factors constant, reduces energy consumption. . . .

Summing up, then, it should, first of all, be borne in mind that the norms and tendencies of the Deep Ecology movement are not derived from ecology by logic or induction. Ecological knowledge and the life-style of the ecological field-worker have *suggested, inspired, and fortified* the perspectives of the Deep Ecology movement. Many of the formulations in the above seven-point survey are rather vague generalizations, only tenable if made more precise in certain directions. But all over the world the inspiration from ecology has shown remarkable convergencies. The survey does not pretend to be more than one of the possible condensed codifications of these convergencies.

Secondly, it should be fully appreciated that the significant tenets of the Deep Ecology movement are clearly and forcefully *normative*. They express a value priority system only in part based on results (or lack of results, cf. point [6]) of scientific research. Today, ecologists try to influence policy-making bodies largely through threats, through predictions concerning pollutants and resource depletion, knowing that policy-makers accept at least certain minimum *norms* concerning health and just distribution. But it is clear that there is a vast number of people in all countries, and even a considerable number of people in power, who accept as valid the wider norms and values characteristic of the Deep Ecology movement. There are political potentials in this movement which should not be overlooked and which have little to do with pollution and resource depletion. In plotting possible futures, the norms should be freely used and elaborated.

Thirdly, in so far as ecology movements deserve our attention, they are *ecophilosophical* rather than ecological. Ecology is a *limited* science which makes *use* of scientific methods. Philosophy is the most general forum of debate on fundamentals, descriptive as well as prescriptive, and political philosophy is one of its subsections. By an *ecosophy* I mean a philosophy of ecological harmony or equilibrium. A philosophy as a kind of *sofia* wisdom, is openly normative, it contains *both* norms, rules, postulates, value priority announcements *and* hypotheses concerning the state of affairs in our universe. Wisdom is policy wisdom, prescription, not only scientific description and prediction.

The details of an ecosophy will show many variations due to significant differences concerning not only 'facts' of pollution, resources, population, etc., but also value priorities. Today, however, the seven points listed provide one unified framework for ecosophical systems.

Select Bibliography

Commoner, B., *The Closing Circle: Nature, Man, and Technology*, Alfred A. Knopf, New York 1971.

Ehrlich, P.R. and A. H., *Population, Resources, Environment: Issues in Human Ecology*, 2nd ed, W.H. Freeman & Co., San Francisco 1972.

Ellul, J., *The Technological Society*, English edn, Alfred A. Knopf, New York 1964.

Glacken, C. J., *Traces on the Rhodian Shore. Nature and Culture in Western Thought*, University of California Press, Berkeley 1967.

Kato, H., 'The Effects of Crowding', Quality of Life Conference, Oberhausen, April 1972.

McHarg, Ian L., *Design with Nature*, 1969. Paperback 1971, Doubleday & Co., New York.

Meynaud, J., *Technocracy*, English edn, Free Press of Glencoe, Chicago 1969.

Mishan, E. J., *Technology and Growth: The Price We Pay*, Frederick A. Praeger, New York 1970.

Odum, E. P., *Fundamentals of Ecology*, 3rd edn, W.E. Saunders Co., Philadelphia 1971.
Shepard, Paul, *Man in the Landscape*, A.A. Knopf, New York.

Source: A. Naess (1973) 'The shallow and the deep, long-range ecology movement', *Inquiry*, 16, pp. 95–100.

4.5

Bill Devall and George Sessions

Deep ecology

Ecological consciousness and deep ecology are in sharp contrast with the dominant worldview of technocratic-industrial societies which regards humans as isolated and fundamentally separate from the rest of Nature, as superior to, and in charge of, the rest of creation. But the view of humans as separate and superior to the rest of Nature is only part of larger cultural patterns. For thousands of years, Western culture has become increasingly obsessed with the idea of *dominance:* with dominance of humans over nonhuman Nature, masculine over the feminine, wealthy and powerful over the poor, with the dominance of the West over non-Western cultures. Deep ecological consciousness allows us to see through these erroneous and dangerous illusions.

For deep ecology, the study of our place in the Earth household includes the study of ourselves as part of the organic whole. Going beyond a narrowly materialist scientific understanding of reality, the spiritual and the material aspects of reality fuse together. While the leading intellectuals of the dominant worldview have tended to view religion as "just superstition," and have looked upon ancient spiritual practice and enlightenment, such as found in Zen Buddhism, as essentially subjective, the search for deep ecological consciousness is the search for a more objective consciousness and state of being through an active deep questioning and meditative process and way of life.

Many people have asked these deeper questions and cultivated ecological consciousness within the context of different spiritual traditions – Christianity, Taoism, Buddhism, and Native American rituals, for example. While differing greatly in other regards, many in these traditions agree with the basic principles of deep ecology.

Warwick Fox, an Australian philosopher, has succinctly expressed the central intuition of deep ecology: "It is the idea that we can make no firm ontological divide in the field of existence: That there is no bifurcation in reality between the human and the non-human realms . . . to the extent that we perceive boundaries, we fall short of deep ecological consciousness."[1]

From this most basic insight or characteristic of deep ecological consciousness, Arne Naess has developed two *ultimate norms* or intuitions which are themselves not derivable from other principles or intuitions. They are arrived at by the deep questioning process and reveal the importance of moving to the philosophical and religious level of wisdom. They cannot be validated, of course, by the methodology of modern science based on its usual mechanistic assumptions and its very narrow definition of data. These ultimate norms are *self-realization* and *biocentric equality.*

Self-realization

In keeping with the spiritual traditions of many of the world's religions, the deep ecology norm of self-realization goes beyond the modern Western *self* which is defined as an isolated ego striving primarily for hedonistic gratification or for a narrow sense of individual salvation in this life or the next. This socially programmed sense of the narrow self or social self dislocates us, and leaves us prey to whatever fad or fashion is prevalent in our society or social reference group. We are thus robbed of beginning the search for our unique spiritual/biological personhood. Spiritual growth, or unfolding, begins when we cease to understand or see ourselves as isolated and narrow competing egos and begin to identify with other humans from our family and friends to, eventually, our species. But the deep ecology sense of self requires a further maturity and growth, an identification which goes beyond humanity to include the nonhuman world. We must see beyond our narrow contemporary cultural assumptions and values, and the conventional wisdom of our time and place, and this is best achieved by the meditative deep questioning process. Only in this way can we hope to attain full mature personhood and uniqueness.

A nurturing nondominating society can help in the "real work" of becoming a whole person. The "real work" can be summarized symbolically as the realization of "self-in-Self" where "Self" stands for organic wholeness. This process of the full unfolding of the self can also be summarized by the phrase, "No one is saved until we are all saved," where the phrase "one" includes not only me, an individual human, but all humans, whales, grizzly bears, whole rain forest ecosystems, mountains and rivers, the tiniest microbes in the soil, and so on.

Biocentric equality

The intuition of biocentric equality is that all things in the biosphere have an equal right to live and blossom and to reach their own individual forms of unfolding and self-realization within the larger Self-realization. This basic intuition is that all organisms and entities in the ecosphere, as parts of the interrelated whole, are equal in intrinsic worth. Naess suggests that biocentric equality as an intuition is true in principle, although in the process of living, all species use each other as food, shelter, etc. Mutual predation is a biological fact of life, and many of the world's religions have struggled with the spiritual implications of this. Some animal liberationists who attempt to side-step this problem by advocating vegetarianism are forced to say that the entire plant kingdom including rain forests have no right to their own existence. This evasion flies in the face of the basic intuition of equality.[2] Aldo Leopold expressed this intuition when he said humans are "plain citizens" of the biotic community, not lord and master over all other species.

Biocentric equality is intimately related to the all-inclusive Self-realization in the sense that if we harm the rest of Nature then we are harming ourselves. There are no boundaries and everything is interrelated. But insofar as we perceive things as individual organisms or entities, the insight draws us to respect all human and non-human individuals in their own right as parts of the whole without feeling the need to set up hierarchies of species with humans at the top.

The practical implications of this intuition or norm suggest that we should live with minimum rather than maximum impact on other species and on the Earth in general. Thus we see another aspect of our guiding principle: "simple in means, rich in ends."....

A fuller discussion of the biocentric norm as it unfolds itself in practice begins with the realization that we, as individual humans, and as communities of humans, have vital needs which go beyond such basics as food, water, and shelter to include love, play, creative expression, intimate relationships with a particular landscape (or Nature taken in its entirety) as well as intimate relationships with other humans, and the vital need for spiritual growth, for becoming a mature human being.

Our vital material needs are probably more simple than many realize. In technocratic-industrial societies there is overwhelming propaganda and advertising which encourages false needs and destructive desires designed to foster increased production and consumption of goods. Most of this actually diverts us from facing reality in an objective way and from beginning the "real work" of spiritual growth and maturity.

Many people who do not see themselves as supporters of deep ecology nevertheless recognize an overriding vital human need for a healthy and high-quality natural environment for humans, if not for all life, with minimum intrusion of toxic waste, nuclear radiation from human enterprises, minimum acid rain and smog, and enough free flowing wilderness so humans can get in touch with their sources, the natural rhythms and the flow of time and place.

Drawing from the minority tradition and from the wisdom of many who have offered the insight of interconnectedness, we recognize that deep ecologists can offer suggestions for gaining maturity and encouraging the processes of harmony with Nature, but that there is no grand solution which is guaranteed to save us from ourselves.

The ultimate norms of deep ecology suggest a view of the nature of reality and our place as an individual (many in the one) in the larger scheme of things. They cannot be fully grasped intellectually but are ultimately experiential....

As a brief summary of our position ... Table 3 summarizes the contrast between the dominant worldview and deep ecology.

Basic principles of deep ecology

In April 1984, during the advent of spring and John Muir's birthday, George Sessions and Arne Naess summarized fifteen years of thinking on the principles of deep ecology while camping in Death Valley, California. In this great and special place, they articulated these principles in a literal, somewhat neutral way, hoping that they would be understood and accepted by persons coming from different philosophical and religious positions.

Readers are encouraged to elaborate their own versions of deep ecology, clarify key concepts and think through the consequences of acting from these principles.

Table 3 Summary of the contrast between the dominant worldview and deep ecology

Dominant worldview	*Deep ecology*
Dominance over nature	Harmony with nature
Natural environment as resource for humans	All Nature has intrinsic worth/biospecies equality
Material/economic growth for growing human population	Elegantly simple material needs (material goals serving the larger goal of self-realization)
Belief in ample resource reserves	Earth "supplies" limited
High technological progress and solutions	Appropriate technology; nondominating science
Consumerism	Doing with enough/recycling
National/centralized community	Minority tradition/bioregion

Basic principles

1 The well-being and flourishing of human and nonhuman Life on Earth have value in themselves (synonyms: intrinsic value, inherent value). These values are independent of the usefulness of the nonhuman world for human purposes.
2 Richness and diversity of life forms contribute to the realization of these values and are also values in themselves.
3 Humans have no right to reduce this richness and diversity except to satisfy *vital* needs.
4 The flourishing of human life and cultures is compatible with a substantial decrease of the human population. The flourishing of nonhuman life requires such a decrease.
5 Present human interference with the nonhuman world is excessive, and the situation is rapidly worsening.
6 Policies must therefore be changed. These policies affect basic economic, technological, and ideological structures. The resulting state of affairs will be deeply different from the present.
7 The ideological change is mainly that of appreciating *life quality* (dwelling in situations of inherent value) rather than adhering to an increasingly higher standard of living. There will be a profound awareness of the difference between big and great.
8 Those who subscribe to the foregoing points have an obligation directly or indirectly to try to implement the necessary changes.

[. . .]

Stewardship in practice

1 A brief history of Resource Conservation and Development (RCD)

After spending three nights with President Theodore Roosevelt under the oaks and pine trees of Yosemite National Park in 1903, John Muir proclaimed in his journal, "Now Ho! for righteous management." Muir was hopeful at the beginning of the twentieth century that under the leadership of wise managers, the national parks and forests would be left essentially wild, preserved as watershed and wildlife habitat. The national parks would remain largely wilderness. Utilitarian uses of the national forests would respect the ongoing, healthy functioning of ecosystems. But his hopes for the wise management of the nation's forests and wild lands were soon to be destroyed.

The story of professionalized, scientific management of natural resources and public land by new experts working within the framework of centralized corporations and national legislation in the United States begins with Gifford Pinchot. Pinchot was trained in Germany in forest management and fought under the label of the "conservation movement" to change the then-prevalent attitude, especially in the western United States, that all land was open for taking minerals, grazing on open range, cutting timber, plowing fields, and appropriating water without planning for the future or considering what the economists were to call "externalities" – air and water pollution, for example.[3]

While Muir was striving to protect large areas of land from the machines of technocratic-industrial society, first through the institution of national forests and then through the institution of national parks, Pinchot was striving to develop a professional cadre of managers to develop resources and encourage legislation which would institutionalize scientific management of renewable resources.

Pinchot's ideology was adopted in law and through the actions of public agencies and private organizations. Conservation became a way of allocating natural resources more efficiently through scientific management and manipulation of natural systems on an ever-larger scale. The "wise use" and multiple use of natural resources meant management for development and economic growth.

As Pinchot said:

> The first great fact about conservation is that it stands for development. There has been a fundamental misconception that conservation means nothing but the husbanding of resources for future generations. There could be no more serious mistake. . . . The first principle of conservation is the use of the natural resources now existing on this continent for the benefit of the people who live here now.[4]

It now seems obvious that we are in the midst of an environmental and spiritual crisis more severe than the one that sent Muir to his grave defending wild Nature in Yosemite. For now the whole planet is threatened by the possible holocaust of nuclear war and by the continued "peaceful" development of natural resources in the tropical rain forests and in the oceans. "Balanced use of resources," "wise use," "scientific management," and "genetic improvement" of forests are all central concepts of the management ideology based upon the assumption that humans are the central figures and actors in history, together with the idea that the whole of Nature is to be understood as resources for humans and thus is open for unlimited human manipulation.

In the United States these assumptions were enacted in land use laws passed by most county governments and federal agencies such as the U.S. Forest Service and Bureau of Reclamation as well as the Tennessee Valley Authority, and form the dominant ideology taught in professional schools of forestry, wildlife management, water resources management, range management and agriculture.

Various types of natural resources were given to special recreation managers, soil scientists, foresters, range managers, environmental engineers, and energy managers, for example. Some reform environmental groups developed their own professionals specializing in these fields. Historian Stephen Fox in his history of the conservation movement argues that radical amateurs arose time and again to revitalize reform groups such as the Audubon Society and Sierra Club, but the experts continue to dominate the normal decision-making processes.[5]

The experts found a congenial home in colleges and universities, which were interested in keeping student enrollment up by training hordes of these experts. In some ways the modern university is like a sponge, sopping up new professions. The university has been called the citadel of expertise. It is not surprising that some of the leading theorists of reform environmentalism have been university professors who wish to appear progressive and professional.

"Expert Testimony" became something of a growth industry. Frequently, in congressional committee hearings or in administrative hearings, or in court cases concerning some natural resource issue, expert was pitted against expert. Theodore Roszak calls this strategy one of "countervailing expertise" and contends that it is a shallow practice because:

> while undeniably well-intentioned and capable of stopgap success on specific political issues, it leaves wholly untouched the great cultural question of our times. It does not challenge the universally presumed rightness of the urban-industrial order of life. Therefore it cannot address itself to the possibility that high industrial society, due to its scale, pace, and complexity, is *inherently* technocratic, and so inherently undemocratic. At most, it leaves us with the hope that the bastardized technocracies of our day might be converted into ideal technocracies.[6]

2 *Resource conservation and development ideology*

Among resource managers there seems to be some awareness of the philosophical assumptions underlying the anthropocentric resource ideology. But generally the problems that arise in this kind of management are perceived to be technical, economic, or political issues. Many people trained in this ideology see themselves as being "value-free" and beyond politics in their decisions. In keeping with this ideology, when environmentalists try to discuss forestry management with public agencies such as the Forest Service, their positions and arguments are viewed from the subjective standpoint of a "special interest" group. The generally nonreflective position of RCD managers makes it almost impossible to discuss issues on a deeper philosophical level. The anthropocentric versus biocentric worldviews of land use managers and environmentalists generally mean that they share little common ground and, as a result, they talk past each other. The basic philosophical differences tend to be obscured or deflected into discussions of technical issues. For example, those who oppose aerial spraying of herbicides on forests are trapped into arguing over the research data of very technical studies of dispersion rates, the effects on pregnant women, and so on. But the chains of interrelationships in an ecosystem are so complex that the results of such studies are usually tentative and inconclusive. And if the burden of proof is on those opposed to the spraying to demonstrate its harmful effects, then the spraying will continue. Those with a philosophically biocentric perspective that respects all of Nature and its processes would most likely arrive at a contrary conclusion.

The usual rhetoric of "conservation," "stewardship," and "wise use" in the contemporary version of RCD now means in practice the development of resources as quickly as is technically possible with the available capital to serve human "needs." The whole of Nature and nonhuman species are not seen as having value in themselves and the right to follow out their own evolutionary destinies. In the ideology of RCD, humans are not understood or experienced to be an integral part of natural processes, but rather as rightfully dominating and controlling the rest of Nature based on principles of scientific management. This means altering Nature to produce more or "better" commodities for human consumption and directing Nature to do the bidding of humans on the utilitarian principle of the "greatest good for the greatest number" of humans.[7]

The ultimate foundations of RCD scientific management of Nature appear to be a profound faith, almost a religion of management. A commissioner of the Bureau of Reclamation, in a speech defending the Reagan administration's management policies, said:

> Research, engineering, and resource management all have a role in solving our water resource problems. We need better water resource technology, especially for ground water resources. Although the Bureau of Reclamation is recognized internationally for its engineering excellence, we're constantly working to improve our engineering. *And the quest for better management is practically a religion in America.* [Emphasis added.][8]

The value-free managers, the experts and technocrats, even if they do not espouse some extreme version of Christian stewardship and domination, do profess a secular religion of faith and hope – a faith in never-ending technological progress and a hope that what they do will work. Even technological failure on an alarmingly regular basis, as in the case of nuclear power, only seems to generate more faith, more hope, and a stronger belief that we need more and more managers and technical experts to solve the problems. Case after case of technological fixes that produce even greater environmental backlash seem not to daunt them in the least.[9]

The metaphor of the Earth as just natural resources to be exploited and consumed by humans remains the dominant image embedded in the psyches of modern RCD managers. As sociologist William Burch wrote, "Though the conversion of all the world into a commodity is periodically challenged and even modified, it remains the basic metaphor in high energy societies both communist and capitalist."[10]

Modern managers can rationalize that they are only serving the needs of the people because of their commitment to unlimited growth and ever-expanding markets. As Karl Polanyi points out, the modern consumer society in which everything is marketable and is assigned an economic value is a completely new social form of society that involves "no less a transformation than that of the natural and human substance of society into commodities. Yet labor, land, and money are *not* commodities. Labor is simply another name for human activity. But on the basis of the fiction that labor, land and money are commodities, markets are organized."[11] The trinity of beliefs underlying the ethics of RCD is the metaphor of the market, the Earth as a collection of human resources or commodities, and the Earth as a machine or spaceship. The dominant quest for better management in modern industrial societies, however, is *not* righteous management as envisioned by John Muir.

The RCD position easily translates into the economizing of forests, rivers and anything defined within the specific human economy as a natural resource. When this degeneration of the RCD position is coupled with the anthropocentric assumptions which are its underpinning, and when it is believed to be natural and desirable for human populations to increase indefinitely together with the assumption that it is desirable for humans to continue expanding their demands and wants, then there is very little room to consider any rights of dolphins, spotted owls, or California condors to their own habitats. Indeed, the logical outcome is to consider other species as just genetic resources whose DNA can be frozen and stored in gene banks for manipulation by scientist-technologists at the command of corporations or government agencies.

Within the assumptions of the dominant worldview, the basic challenge of the forester, water resource manager, range manager, fisheries manager, etc., is to produce more and more commodities in shorter and shorter periods of time.

Nature and its processes are too slow and inefficient in terms of the economizing model. Indeed, "efficiency of production," virtually without regard for the larger ecological context, is the major slogan of managers who take a homocentric rather than biocentric position.

For example, the rotation cycle, the number of years between cutting a stand of timber and its recutting after regrowth, has been progressively reduced from perhaps 120 years to eighty, sixty, or forty. One official of a major corporation in the western United States asked his scientific managers and technologists to develop and plant "genetically enhanced" trees which could be "harvested" in twenty years. "Trees are just a crop, like corn," say many commercial foresters.

Relatively small natural areas and stretches of free-flowing rivers are allowed to exist in the context of RCD, but only if they do not intrude upon basic resource production. As one forest products industry official said, "Maybe 100 acres of old-growth redwood would be enough for our grandchildren to see." "Nonproductive" land may be left as wilderness, but the borders get smaller and smaller and environmentalists find themselves arguing with forestry officials over fifty-foot buffer strips along streams where logging will not be allowed. At best, under RCD, the "environmental effects" of proposed development projects will be studied and *some* effort will be made to "mitigate" the known negative environmental impacts.

Recent Forest Service efforts to develop a "decision procedure" for differing recreational uses on forest lands have fared little better than economic analyses of forest values, and continue to highlight the failure of land use managers to recognize more objective ecological criteria. For example, in a recent study of "scenic preferences" as baseline data for decisions concerning "scenic management," a sample of persons was shown photos of various kinds of scenes ranging from clear-cut forests to super freeways. They were asked to rate their preferences on a subjective scale and then the averages were tabulated. The average preference ratings were then considered to be public opinion concerning the types of landscape to be valued for recreational purposes.[12]

The older imagery of RCD sees humans as happy gardeners and stewards weeding the Earth of "undesirable pests" and predators. Biologist René Dubos has presented this image in his book, *The Wooing of the Earth*, along with the claim that humans are simply "bringing out the potential" of the planet. But even Dubos admits that "The belief that we can manage the Earth and improve on Nature is probably the ultimate expression of human conceit, but it has deep roots in the past and is almost universal."[13]

Notes

1 Warwick Fox, "Deep Ecology: A New Philosophy of Our Time?" *The Ecologist*, v. 14, 5–6, 1984, pp. 194–200. Arnie Naess replies, "Intuition, Intrinsic Value and Deep Ecology," *The Ecologist*, v. 14, 5–6, 1984, pp. 201–204.
2 Tom Regan, *The Case for Animal Rights* (New York: Random House, 1983). For excellent critiques of the animal rights movement, see John Rodman, "The Liberation of Nature?" *Inquiry* 20 (Oslo, 1977). J. Baird Callicott, "Animal Liberation," *Environmental Ethics* 2, 4, (1980); see also John Rodman, "Four Forms of Ecological Consciousness Reconsidered" in T. Attig and D. Scherer, eds., *Ethics and the Environment* (Englewood Cliffs, N.J.: Prentice-Hall, 1983).
3 Gifford Pinchot, *Breaking New Ground* (New York: Harcourt, Brace and Co., 1947).
4 Ibid., p. 261.
5 Stephen Fox, *John Muir and His Legacy*, (Boston: Little, Brown and Co., 1981), chapter four.
6 Theodore Roszak, *Where the Wasteland Ends* (New York: Anchor, 1972), pp. 26–67.
7 John Rodman, "Resource Conservation: Economics and Beyond" (Unpublished paper, Claremont, Ca.: Pitzer College, 1976).
8 Remarks by Robert Broadbent, Commissioner of Reclamation, U.S. Department of the Interior, National Cotton Outlook Conference, South Padre Island, Texas (16 June 1982).
9 See J. P. Milton and M. T. Favor, *The Careless Technology* (New York: Natural History Press, 1971).
10 William Burch, Jr., *Daydreams and Nightmares* (New York: Harper & Row, 1971), p. 154.
11 Karl Polyani, *The Great Transformation* (Boston: Beacon, 1944), p. 72.
12 Terry Daniel and Ron Boster, "Measuring Landscape Esthetics: The Scenic Beauty Estimation Method," USFS Research Paper RM-167 (Rocky Mountain Experimental Station, May 1976).
13 René Dubos, *The Wooing of the Earth* (New York: Scribner's, 1980), p. 79.

Source: Bill Devall and George Sessions (1985) *Deep Ecology*, Salt Lake City: Peregrine Smith Books, pp. 65–70, 132–8.

4.6

Holmes Rolston III

Valuing the environment

Ethics in ecosystems

In an environmental ethic, what humans want to value is not compassion, charity, rights, personality, justice, fairness or even pleasure and the pursuit of happiness. Those values belong in interhuman ethics – in culture, not nature – and to look for them here is to make a category mistake. What humans value in nature is an ecology, a pregnant Earth, a projective and prolife system in which (considering biology alone, not culture) individuals can prosper but are also sacrificed indifferently to their pains and pleasures, individual well-being a lofty but passing role in a storied natural history. From the perspective of individuals there is violence, struggle, death; but from a systems perspective, there is also harmony, interdependence, and ever-continuing life.

The beauty, integrity, stability of an ecosystem can put constraints on appropriate human conduct in both small and larger ways. This is what is wrong, at the deepest ethical level, with such seemingly trivial behavior as putting soap in geysers, carving names on trees and boulders, carving mountains into monuments to human pride, tossing toilet paper off the Mount of the Holy Cross, or bulldozing a giant fire off the lip of Glacier Point. Mere sport hunting is wrong on this count alone, even if those killed feel no pain. To make of nature a mere plaything is to profane it, just as to make playthings of persons is to misunderstand them. We humans ought to have our parks and pleasuring places, but we ought to check the types of enjoyments there with an appropriate appreciation of those places.

An Indiana Dunes National Lakeshore poster depicts a clump of marram grass, sand, a lake, with the injunction *"Let it be!"*[1] The Indiana sand dunes are adjacent to Chicago on the shores of Lake Michigan; they are valuable industrial and residential property as well as ecosystems of biological and historical interest. The science of ecology was, in significant part, founded by Henry Cowles as a result of his studies there, discovering dune succession. Preserving areas of the dunes was achieved only after one of the longest, bitterest fights in environmental conservation, a century of struggle against powerful industrial and development interests.[2] A major reason was to preserve a playground for Chicagoans. But in the imperative "Let it be!" there seems an element also of respect for ecosystems; at least some token of the ecosystem type that natural history placed on the Lake Michigan shore ought to remain in the midst of the Chicago culture.

When humans make their living off the land, this ethic asks a gentle presence rather than a domineering and thoughtless one. Humans are permitted to make many wild areas rural and some areas urban, and they do rebuild the environment dramatically. But this ethic requires that rural places be kept as full of nature as is consistent with their being agricultural places as well. It thinks of nature as a community, not just a commodity. It limits road-building to minimize the impact on wildlife. It likes brushy fencerows and dislikes clean (and barren) ones. It protects all species, not just the "game." It leaves the hardwoods along the stream courses when converting the uplands to pinewoods for paper pulp. It appreciates a forest, not just board feet of timber. It sees water first as the lifeblood of an ecosystem, secondly as acre-feet in a reservoir. It sees humans as biotic citizens (if also kings) who belong to the land, not man as conqueror to whom the land belongs. This ethic urges multiple appreciations of the landscape, not just multiple uses. It says that humans ought to let sand dune ecosystems be – sometimes, at least. Humans are not free to make whatever uses of nature suit their fancy, amusement, need, or profit.

It has been necessary in the course of human history to sacrifice most of the wildlands, converting them to rural and urban settlements, and this is both good and ecological; nevertheless, when humans prey on nature to build culture and make the land yield its wealth, these "moral predators" – who can have a view of the whole and a conscience about their presence – have some duties to the ecosystems of which they are part. This demands that places of especially striking site integrity be left untrammeled by humans. These are named national parks, wildernesses, wildlife refuges. But not only do the highly distinctive places present values that count morally. Representatives of once common ecosystems (hemlock forests, tall-grass prairies, sand dunes) also have their integrity, now threatened by advancing agriculture and culture, and ought to be preserved.

Perhaps it was necessary for the plainsmen to reduce the buffalo herds so that they could put cattle on the range; the plains states could hardly have been settled any other way. But it was not right to destroy the bison; respect for those bison in their plains ecosystems should have preserved ample grasslands wilderness and national parks in every plains state. Alas, the United States has not a single grasslands national park or grasslands wilderness with free-ranging buffalo.

The extension of an ethic to the land gives humans a comprehensive situated fitness in the global ecosystems. Such fitness, more than predatory success, makes the human behavior here right – right because humans respectfully appreciate the integrity of the places they inhabit.

Consider the following argument for the preservation of wilderness.

1 Seeking the unchecked domination of others is self-defeating.
2 Persons who (only) dominate other persons are not free to appreciate them.
3 Persons who (only) dominate other persons lead inferior lives. Slaveowners, for example, lead better lives after the slaves are freed, when human relationships are more just and generous.
4 Persons who (only) dominate nature are not free to appreciate nature.
5 Persons who (only) dominate nature lead inferior lives.
6 Ours is a society that (only) seeks to dominate nature.
7 Ours is an inferior society – to the extent that it seeks unchecked domination of nature.
8 Wilderness is a region undominated by persons. A wilderness is "an area where the earth and its community of life are untrammeled by man, where man himself is a visitor who does not remain."[3]
9 Designating wilderness by deliberate resolve checks the human domination of nature. (Congress and government agencies resolve to "let it be!").
10 Maintaining wilderness by citizen cooperation checks the human domination of nature. (Visitors observe wilderness regulations, walking gently, leaving little or no trace. They "let it be!").
11 In wilderness, nature may be appreciated for what it is in itself. Wilderness serves as a living symbol or representative of pristine nature.
12 Persons who appreciate nature live better lives than those who seek (only) to dominate nature. Be a better person! Let wilderness be! Is this a self-serving argument?. . .
13 Wilderness is essential to a better society. "In Wildness is the preservation of the World" (Thoreau).[4]

Civilization needs to be tamed as well as nature. American society in earlier centuries tamed nature, but in this century civilization needs to tame itself and recognize the integrity of wild places.

Would it be better, as a symbolic gesture of nondomination, to leave some wilderness areas unmapped, at least on small scales? Is mapping the last acre of wilderness necessary for, or might it prevent, positive human appreciation of wild nature?

The land ethic rests upon the discovery of certain values – integrity, projective creativity, life support, community – already present in ecosystems, and it imposes an obligation to act so as to maintain these. This is not, we have repeatedly warned, an ethic concerning culture, not an interhuman ethic. We will continue to need the Ten Commandments, categorical imperatives, the Golden Rule, concepts of justice, and the utilitarian calculus. But we are developing an extension of ethics into environmental attitudes, a new commandment, about landscapes and ecosystems.

[. . .]

Our duties to persons in culture will at times bring us into conflict with this land ethic, and we will have to adjudicate such conflicts. We may even take a clue from the sorts of values here defended. What if humans simplify the native ecosystems of Iowa and Kansas, planting monocultures of corn and wheat as Americans have done, in order to feed a growing nation? What if the plainsmen reduce the bison herds so that they can put cattle on the range? They will have done harm to the original ecosystems in the sense that farming and ranching has reduced natural values there, even though this may be justified by the increased value produced in society. Beauty, integrity, stability, community (which, when found in culture, moralists usually call justice and utility) are increased in American society when persons are better fed; but the increase is purchased by the sacrifice of beauty, integrity, stability, community in the native grasslands.

Notes

1 U.S. Government Printing Office, Poster 751–269 (1981).
2 J. Ronald Engel, *Sacred Sands* (Middletown, Conn.: Wesleyan University Press, 1983).
3 Wilderness Act of 1964, sec. 2 (c) Public Law 88–577, 78 Stat. 891.
4 Henry David Thoreau, "Walking," in *The Portable Thoreau* (New York: Penguin Books, 1947, 1980), pp. 592–630, citation on p. 609.

Source: Holmes Royston, III (1988) *Environmental Ethics*, Philadelphia: Temple University Press, pp. 225–9.

4.7

Christopher D. Stone

Should trees have standing?

[. . .]

Throughout legal history, each successive extension of rights to some new entity has been, theretofore, a bit unthinkable. We are inclined to suppose the rightlessness of rightless "things" to be a decree of Nature, not a legal convention acting in support of some status quo. It is thus that we defer considering the choices involved in all their moral, social, and economic dimensions. And so the United States Supreme Court could straight-facedly tell us in *Dred Scott* that Blacks had been denied the rights of citizenship "as a subordinate and inferior class of beings, who had been subjugated by the dominant race. . . ."[1]

[. . .]

. . .The fact is, that each time there is a movement to confer rights onto some new "entity," the proposal is bound to sound odd or frightening or laughable.[2] This is partly because until the rightless thing receives its rights, we cannot see it as anything but a *thing* for the use of "us" – those who are holding rights at the time.[3]. . .

The reason for this little discourse on the unthinkable, the reader must know by now, if only from the title of the paper. I am quite seriously proposing that we give legal rights to forests, oceans, rivers and other so-called "natural objects" in the environment – indeed, to the natural environment as a whole.[4]

As strange as such a notion may sound, it is neither fanciful nor devoid of operational content. In fact, I do not think it would be a misdescription of recent developments in the law to say that we are already on the verge of assigning some such rights, although we have not faced up to what we are doing in those particular terms.[5] We should do so now, and begin to explore the implications such a notion would hold.

Toward rights for the environment

Now, to say that the natural environment should have rights is not to say anything as silly as that no one should be allowed to cut down a tree. We say human beings have rights, but – at least as of the time of this writing – they can be executed.[6] Corporations have rights, but they cannot plead the fifth amendment;[7] *In re Gault* gave 15-year-olds certain rights in juvenile proceedings, but it did not give them the right to vote. Thus, to say that the environment should have rights is not to say that it should have every right we can imagine, or even the same body of rights as human beings have. Nor is it to say that everything in the environment should have the same rights as every other thing in the environment.

What the granting of rights does involve has two sides to it. The first involves what might be called the legal-operational aspects; the second, the psychic and socio-psychic aspects. I shall deal with these aspects in turn.

The legal-operational aspects

What it means to be a holder of legal rights

There is, so far as I know, no generally accepted standard for how one ought to use the term "legal rights." Let me indicate how I shall be using it in this piece.

First and most obviously, if the term is to have any content at all, an entity cannot be said to hold a legal right unless and until *some public authoritative body* is prepared to give *some amount of review* to actions that are colorably inconsistent with that "right." For example, if a student can be expelled from a university and cannot get any public official, even a judge or administrative agent at the lowest level, either (i) to require the university to justify its actions (if only to the extent of filling out an affidavit alleging that the expulsion "was not wholly arbitrary and capricious") or (ii) to compel the university to accord the student some procedural safeguards (a hearing, right to counsel, right to have notice of charges), then the minimum requirements for saying that the student has a legal right to his education do not exist.[8]

But for a thing to be *a holder of legal rights*, something more is needed than that some authoritative body will review the actions and processes of those who threaten it. As I shall use the term, "holder of legal rights," each of three additional criteria must be satisfied. All three, one will observe, go towards making a thing *count* jurally – to have a legally recognized worth and dignity in its own right, and not merely to serve as a means to benefit "us" (whoever the contemporary group of rights-holders may be). They are, first, that the thing can institute legal actions *at its behest*; second, that in determining the granting of legal relief, the court must take *injury to it* into account; and, third, that relief must run to the *benefit of it*.

[. . .]

The rightlessness of natural objects at common law

Consider, for example, the common law's posture toward the pollution of a stream. True, courts have always been able, in some circumstances, to issue orders that will stop the pollution. . . . But the stream itself is fundamentally rightless, with implications that deserve careful reconsideration.

The first sense in which the stream is not a rights-holder has to do with standing. The stream itself has none. So far as the common law is concerned, there is in general no way to challenge the polluter's actions save at the behest of a lower riparian – another human being – able to show an invasion of *his* rights. This conception of the riparian as the holder of the right to bring suit has more than theoretical interest. The lower riparians may simply not care about the pollution. They themselves may be polluting, and not wish to stir up legal waters. They may be economically dependent on their polluting neighbor.[9] And, of course, when they discount the value of winning by the costs of bringing suit and the chances of success, the action may not seem worth undertaking. Consider, for example, that while the polluter might be injuring 100 downstream riparians $10,000 a year *in the aggregate*, each riparian separately might be suffering injury only to the extent of $100 – possibly not enough for any one of them to want to press suit by himself, or even to go to the trouble and cost of securing co-plaintiffs to make it worth everyone's while. This hesitance will be especially likely when the potential plaintiffs consider the burdens the law puts in their way:[10] proving, e.g., specific damages, the "unreasonableness" of defendant's use of the water, the fact that practicable means of abatement exist, and overcoming difficulties raised by issues such as joint causality, right to pollute by prescription, and so forth. Even in states which, like California, sought to overcome these difficulties by empowering the attorney-general to sue for abatement of pollution in limited instances, the power has been sparingly invoked and, when invoked, narrowly construed by the courts.[11]

The second sense in which the common law denies "rights" to natural objects has to do with the way in which the merits are decided in those cases in which someone is competent and willing to establish standing. At its more primitive levels, the system protected the "rights" of the property owning human with minimal weighing of any values: "*Cujus est solum, ejus est usque ad coelum et ad infernos*."[12] Today we have come more and more to make balances – but only such as will adjust the economic best interests of identifiable humans. For example, continuing with the case of streams, there are commentators who speak of a "general rule" that "a riparian owner is legally entitled to have the stream flow by his land with its quality unimpaired" and observe that "an upper owner has, prima facie, no right to pollute the water."[13] Such a doctrine, if strictly invoked, would protect the stream absolutely whenever a suit was brought; but obviously, to look around us, the law does not work that way. Almost everywhere there are doctrinal qualifications on riparian "rights" to an unpolluted stream.[14] Although these rules vary from jurisdiction to jurisdiction, and upon whether one is suing for an equitable injunction or for damages, what they all have in common is some sort of balancing. Whether under language of "reasonable use," "reasonable methods of use," "balance of convenience" or "the public interest doctrine,"[15] what the courts are balancing, with varying degrees of directness, are the economic hardships on the upper riparian (or dependent community) of abating the pollution vis-à-vis the economic hardships of continued pollution on the lower riparians. What does not weigh in the balance is the damage to the stream, its fish and turtles and "lower" life. So long as the natural environment itself is rightless, these are not matters for judicial cognizance. Thus, we find the highest court of Pennsylvania refusing to stop a coal company from discharging polluted mine water into a tributary of the Lackawana River because a plaintiff's "grievance is for a mere personal inconvenience; and . . . mere private personal inconveniences . . . must yield to the necessities of a great public industry, which although in the hands of a private corporation, subserves a great public interest."[16] The stream itself is lost sight of in "a quantitative compromise between *two* conflicting interests."[17]

The third way in which the common law makes natural objects rightless has to do with who is regarded as the beneficiary of a favorable judgment. Here, too, it makes a considerable difference that it is not the natural object that counts in its own right. To illustrate this point, let me begin by observing that it makes perfectly good sense to speak of, and ascertain, the legal damage to a natural object, if only in the sense of "making it whole" with respect to the most obvious factors.[18] The costs of making a forest whole, for example, would include the costs of reseeding, repairing watersheds, restocking wildlife – the sorts of costs the Forest Service undergoes after a fire. Making a polluted stream whole would include the costs of restocking with fish, water-fowl, and other animal and vegetable life, dredging, washing out impurities, establishing natural and/or artificial aerating agents, and so forth. Now, what is important to note is that, under our present system, even if a plaintiff riparian wins a water pollution suit for damages, no money goes to the benefit of the stream itself to repair *its* damages.[19] This omission has the further effect that, at most, the law confronts a polluter with what it takes to make the plaintiff riparians whole; this may be far less than the damages to the stream,[20] but not so much as to force the polluter to desist. For example, it is easy to imagine a polluter whose activities damage a stream to the extent of $10,000 annually, although the aggregate damage to all the riparian plaintiffs who come into the suit is only $3,000. If $3,000 is less than the cost to the polluter of shutting down, or making the requisite technological changes, he might prefer to pay off the damages (i.e., the legally cognizable damages) and continue to pollute the stream. Similarly, even if the jurisdiction issues an injunction at the plaintiffs' behest (rather than to order payment of damages), there is nothing to stop the plaintiffs from "selling out" the stream, i.e., agreeing to dissolve or not enforce the injunction at some price (in the example above, somewhere between plaintiffs' damages – $3,000 – and defendant's next best economic alternative). Indeed, I take it this is exactly what Learned Hand had in mind in an opinion in which, after issuing an anti-pollution injunction, he suggests that the defendant "make its peace with the plaintiff as best it can."[21] What is meant is a peace between *them*, and not amongst them and the river.

I ought to make clear at this point that the common law as it affects streams and rivers, which I have been using as an example so far, is not exactly the same as the law affecting other environmental objects. Indeed, one would be hard pressed to say that there was a "typical" environmental object, so far as its treatment at the hands of the law is concerned. There are some differences in the law applicable to all the various resources that are held in common: rivers, lakes, oceans, dunes, air, streams (surface and subterranean), beaches, and so forth.[22] And there is an even greater difference as between these traditional communal resources on the one hand, and natural objects on traditionally private land, e.g., the pond on the farmer's field, or the stand of trees on the suburbanite's lawn.

On the other hand, although there be these differences which would make it fatuous to generalize about a law of the natural environment, most of these differences simply underscore the points made in the instance of rivers and streams. None of the natural objects, whether held in common or situated on private land, has any of the three criteria of a rightsholder. They have no standing in their own right; their unique damages do not count in determining outcome; and they are not the beneficiaries of awards. In such fashion, these objects have traditionally been regarded by the common law, and even by all but the most recent legislation, as objects for man to conquer and master and use . . . Even where special measures have been taken to conserve them, as by seasons on game and limits on timber cutting, the dominant motive has been to conserve them *for us* – for the greatest good of the greatest number of human beings. Conservationists, so far as I am aware, are generally reluctant to maintain otherwise.[23] As the name implies, they want to conserve and guarantee *our* consumption and *our* enjoyment of these other living things. In their own right, natural objects have counted for little, in law as in popular movements.

As I mentioned at the outset, however, the rightlessness of the natural environment can and should change; it already shows some signs of doing so.

Toward having standing in its own right

It is not inevitable, nor is it wise, that natural objects should have no rights to seek redress in their own behalf. It is no answer to say that streams and forests cannot have standing because streams and forests cannot speak. Corporations cannot speak either; nor can states, estates, infants, incompetents, muncipalities or universities. Lawyers speak for them, as they customarily do for the ordinary citizen with legal problems. One ought, I think, to handle the legal problems of natural objects as one does the problems of legal incompetents – human beings who have become vegetable. If a human being shows signs of becoming senile and has affairs that he is de jure incompetent to manage, those concerned with his well being make such a showing to the court, and someone is designated by the court with the authority to manage the incompetent's affairs. The guardian[24] (or "conservator"[25] or "committee"[26] – the terminology varies) then represents the incompetent in his legal affairs. Courts make similar appointments when a corporation has become "incompetent" – they appoint a trustee in bankruptcy or reorganization to oversee its affairs and speak for it in court when that becomes necessary.

On a parity of reasoning, we should have a system in which, when a friend of a natural object perceives it to be endangered, he can apply to a court for the creation of a guardianship.[27] Perhaps we already have the machinery to do so. California law, for example, defines an incompetent as "any person, whether insane or not, who by reason of old age, disease, weakness of mind, or other cause, is unable, unassisted, properly to manage and take care of himself or his property, and by reason thereof is likely to be deceived or imposed upon by artful or designing persons."[28] Of course, to urge a court that an endangered river is "a person" under this provision will call for lawyers as bold and imaginative as those who convinced the Supreme Court that a railroad corporation was a "person" under the fourteenth amendment, a constitutional provision theretofore generally thought of as designed to secure the rights of freedmen.[29] . . . If such an argument based on present statutes should fail, special environmental legislation could be enacted along traditional guardianship lines. Such provisions could provide for guardianship both in the instance of public natural objects and also, perhaps with slightly different standards, in the instance of natural objects on "private" land.[30]

The potential "friends" that such a statutory scheme would require will hardly be lacking. The Sierra Club, Environmental Defense Fund, Friends of the Earth, Natural Resources Defense Counsel, and the Izaak Walton League are just some of the many groups which have manifested unflagging dedication to the environment and which are becoming increasingly capable of marshalling the requisite technical experts and lawyers. If, for example, the Environmental Defense Fund should have reason to believe that some company's strip mining operations might be irreparably destroying the ecological balance of large tracts of land, it could, under this procedure, apply to the court in which the lands were situated to be appointed guardian.[31] As guardian, it might be given rights of inspection (or visitation) to determine and bring to the court's attention a fuller finding on the land's condition. If there were indications that under the substantive law some redress might be available on the land's behalf, then the guardian would be entitled to raise the land's rights in the land's name, i.e., without having to make the roundabout and often unavailing demonstration, discussed below, that the "rights" of the club's members were being invaded. Guardians would also be looked to for a host of other protective tasks, e.g., monitoring effluents (and/or monitoring the monitors), and representing their "wards" at legislative and administrative hearings on such matters as the setting of state water quality standards. Procedures exist, and can be strengthened, to move a court for the removal and substitution of guardians, for conflicts of interest or for other reasons,[32] as well as for the termination of the guardianship.[33]

[. . .]

Unlike the liberalized standing approach, the guardianship approach would secure an effective voice for the environment even where federal administrative action and public-lands and waters were not involved. It would also allay one of the fears courts – such as the Ninth Circuit – have about the extended standing concept: if any ad hoc group can spring up overnight, invoke some "right" as universally claimable as the esthetic and recreational interests of its members and thereby get into court, how can a flood of litigation be prevented?[34] If an ad hoc committee loses a suit brought *sub nom.* Committee to Preserve our Trees, what happens when its very same members reorganize two years later and sue *sub nom.* the Massapequa Sylvan Protection League? Is the new group bound by res judicata? Class action law may be capable of ameliorating some of the more obvious problems. But even so, court economy might be better served by simply designating the guardian de jure representative of the natural object, with rights of discretionary intervention by others, but with the understanding that the natural object is "bound" by an adverse judgment.[35] The guardian concept, too, would provide the endangered natural object with what the trustee in bankruptcy provides the endangered corporation: a continuous supervision over a period of time, with a consequent deeper understanding of a broad range of the ward's problems, not just the problems present in one particular piece of litigation. It would thus assure the courts that the plaintiff has the expertise and genuine adversity in pressing a claim which are the prerequisites of a true "case or controversy."

The guardianship approach, however, is apt to raise two objections, neither of which seems to me to have much force. The first is that a committee or guardian could not judge the needs of the river or forest in its charge; indeed, the very concept of "needs," it might be said, could be used here only in the most metaphorical way. The second objection is that such a system would not be much different from what we now have: is not the Department of Interior already such a guardian for public lands, and do not most states have legislation empowering their attorneys general to seek relief – in a sort of *parens patriae* way – for such injuries as a guardian might concern himself with?

As for the first objection, natural objects *can* communicate their wants (needs) to us, and in ways that are not terribly ambiguous. I am sure I can judge with more certainty and meaningfulness whether and when my lawn wants (needs) water, than the Attorney General can judge whether and when the United States wants (needs) to take an appeal from an adverse judgment by a lower court. The lawn tells me that it wants water by a certain dryness of the blades and soil – immediately obvious to the touch – the appearance of bald spots, yellowing, and a lack of springiness after being walked on; how does "the United States" communicate to the Attorney General? For similar reasons, the guardian-attorney for a smog-endangered stand of pines could venture with more confidence that his client wants the smog stopped, than the directors of a corporation can assert that "the corporation" wants dividends declared. We make decisions on behalf of, and in the purported interests of, others every day; these "others" are often creatures whose wants are far less verifiable, and even far more metaphysical in conception, than the wants of rivers, trees, and land.[36]

As for the second objection, one can indeed find evidence that the Department of Interior was conceived as a sort of guardian of the public lands.[37] But there are two points to keep in mind. First, insofar as the Department already is an adequate guardian it is only with respect to the federal public lands as per Article IV, section 3 of the Constitution.[38] Its guardianship includes neither local public lands nor private lands. Second, to judge from the environmentalist literature and from the cases environmental action groups have been bringing, the Department is itself one of the bogeys of the environmental movement. (One thinks of the uneasy peace between the Indians and the Bureau of Indian Affairs.) Whether the various charges be right or wrong, one cannot help but observe that the Department has been charged with several institutional goals (never an easy burden), and is currently looked to for action by quite a variety of interest groups, only one of which is the environmentalists. In this context, a guardian outside the institution becomes especially valuable. Besides, what a person wants, fully to secure his rights, is the ability to retain independent counsel even when, and perhaps especially when, the government is acting "for him" in a beneficent way. I have no reason to doubt, for example, that the Social Security System is being managed "for me"; but I would not want to abdicate my right to challenge its actions as they affect me, should the need arise.[39] I would not ask more trust of national forests, vis-à-vis the Department of Interior. The same considerations apply in the instance of local agencies, such as regional water pollution boards, whose members' expertise in pollution matters is often all too credible.[40]

The objection regarding the availability of attorneys-general as protectors of the environment within the existing structure is somewhat the same. Their statutory powers are limited and sometimes unclear. As political creatures, they must exercise the discretion they have with an eye toward advancing and reconciling a broad variety of important social goals, from preserving morality to increasing their jurisdiction's tax base. The present state of our environment, and the history of cautious application and development of environmental protection laws long on the books,[41] testifies that the burdens of an attorney-general's broad responsibility have apparently not left much manpower for the protection of nature. . . . No doubt, strengthening interest in the environment will increase the zest of public attorneys even where, as will often be the case, well-represented corporate polluters are the quarry. Indeed, the United States Attorney General has stepped up anti-pollution activity, and ought to be further encouraged in this direction.[42] The statutory powers of the attorneys-general should be enlarged, and they should be armed with criminal penalties made at least commensurate with the likely economic benefits of violating the law.[43] On the other hand, one cannot ignore the fact that there is increased pressure on public law-enforcement offices to give more attention to a host of other problems, from crime "on the streets" (why don't we say "in the rivers"?) to consumerism and school bussing. If the environment is not to get lost in the shuffle, we would do well, I think, to adopt the guardianship approach as an additional safeguard, conceptualizing major natural objects as holders of their own rights, raisable by the court-appointed guardian.

Notes

1 Dred Scott v. Sandford, 60 U.S. (19 How.) 396, 404–5 (1856). In Bailey v. Poindexter's Ex'r, 56 Va. (14 Gratt.) 132, 142–43 (1858) a provision in a will that testator's slaves could choose between emancipation and public sale was held void on the ground that slaves have no legal capacity to choose:

> These decisions are legal conclusions flowing naturally and necessarily from the one clear, simple, fundamental idea of chattel slavery. That fundamental idea is, that, in the eye of the law, so far certainly as civil rights and relations are concerned, the slave is not a person, but a thing. The investiture of a chattel with civil rights or legal capacity is indeed a legal solecism and absurdity. The attribution of legal personality to a chattel slave, – legal conscience, legal intellect, legal freedom, or liberty and power of free choice and action, and corresponding legal obligations growing out of such qualities, faculties and action – implies a palpable contradiction in terms.

2 Recently, a group of prison inmates in Suffolk County tamed a mouse that they discovered, giving him the name Morris. Discovering Morris, a jailer flushed him down the toilet. The prisoners brought a proceeding against the Warden complaining, *inter alia*, that Morris was subjected to discriminatory discharge and was otherwise unequally treated. The action was unsuccessful, on grounds that the inmates themselves were "guilty of imprisoning Morris without a charge, without a trial, and without bail," and that other mice at the prison were not treated more favorably. "As to the true victim the Court can only offer again the sympathy first proffered to his ancestors by Robert Burns. . . ." The Judge proceeded to quote from Burns' "To a Mouse." Morabito v. Cyrta, 9 CRIM. L. REP. 2472 (N.Y. Sup. Ct. Suffolk Co. Aug. 26, 1971).

The whole matter seems humorous, of course. But what we need to know more of is the function of humor in the unfolding of a culture, and the ways in which it is involved with the social growing pains to which it is testimony. Why do people make jokes about the Women's Liberation Movement? Is it not on account of – rather than in spite of – the underlying validity of the protests, and the uneasy awareness that a recognition of them is inevitable? A. Koestler rightly begins his study of the human mind, *Act of Creation* (1964), with an analysis of humor, entitled "The Logic of Laughter." And cf. Freud, *Jokes and the Unconscious*, 8 Standard Edition of the Complete Psychological Works of Sigmund Freud (J. Strachey transl. 1905). (Query too: what is the relationship between the conferring of proper *names*, e.g., Morris, and the conferring of social and legal *rights?*)

3 Thus it was that the Founding Fathers could speak of the inalienable rights of all men, and yet maintain a society that was, by modern standards, without the most basic rights for Blacks,

Indians, children and women. There was no hypocrisy; emotionally, no one *felt* that these other things were men.

4 In this article I essentially limit myself to a discussion of non-animal but natural objects. I trust that the reader will be able to discern where the analysis is appropriate to advancing our understanding of what would be involved in giving "rights" to other objects not presently endowed with rights – for example, not only animals (some of which already have rights in some senses) but also humanoids, computers, and so forth. *Cf.* the National Register for Historic Places, 16 U.S.C. §470 (1970), discussed in Ely v. Velde, 321 F. Supp. 1088 (E.D. Va. 1971).

As the reader will discover, there are large problems involved in defining the boundaries of the "natural object." For example, from time to time one will wish to speak of that portion of a river that runs through a recognized jurisdiction; at other times, one may be concerned with the entire river, or the hydrologic cycle – or the whole of nature. One's ontological choices will have a strong influence on the shape of the legal system, and the choices involved are not easy. See notes 49, 73 and accompanying text *infra*.

On the other hand, the problems of selecting an appropriate ontology are problems of all language – not merely of the language of legal concepts, but of ordinary language as well. Consider, for example, the concept of a "person" in legal or in everyday speech. Is each *person* a fixed bundle of relationships, persisting unaltered through time? Do our molecules and cells not change at every moment? Our hypostatizations always have a pragmatic quality to them. See D. Hume, *Of Personal Identity*, in *Treatise of Human Nature* bk. 1, pt. IV, §VI, in *The Philosophical Works of David Hume* 310–18, 324 (1854); T. Murti, *The Central Philosophy of Buddhism* 70–73 (1955). In *Loves Body* 146–47 (1966) Norman O. Brown observes:

> The existence of the "let's pretend" boundary does not prevent the continuance of the real traffic across it. Projection and introjection, the process whereby the self as distinct from the other is constituted, is not past history, an event in childhood, but a present process of continuous creation. The dualism of self and external world is built up by a constant process of reciprocal exchange between the two. The self as a stable substance enduring through time, an identity, is maintained by constantly absorbing good parts (or people) from the outside world and expelling bad parts from the inner world. "There is a continual 'unconscious' wandering of other personalities into ourselves."
>
> Every person, then, is many persons; a multitude made into one person; a corporate body; incorporated, a corporation. A "corporation sole"; every man a parson-person. The unity of the person is as real, or unreal, as the unity of the corporation.

See generally, W. Bishin & C. Stone, *Law, Language and Ethics*, Ch. 5 (1972).

In different legal systems at different times, there have been many shifts in the entity deemed "responsible" for harmful acts: an entire clan was held responsible for a crime before the notion of individual responsibility emerged; in some societies the offending hand, rather than an entire body, may be "responsible." Even today, we treat father and son as separate jural entities for some purposes, but as a single jural entity for others. I do not see why, in principle, the task of working out a legal ontology of natural objects (and "qualities," e.g., climatic warmth) should be any more unmanageable. Perhaps someday all mankind shall be, for some purposes, one jurally recognized "natural object."

5 The statement in text is not quite true; *cf.* Murphy, *Has Nature Any Right to Life?*, 22 Hast. L.J. 467 (1971). An Irish court, passing upon the validity of a testamentary trust to the benefit of someone's dogs, observed in dictum that "'lives' means lives of human beings, not of animals or trees in California." Kelly v. Dillon, 1932 Ir. R. 255, 261. (The intended gift over on the death of the last surviving dog was held void for remoteness, the court refusing "to enter into the question of a dog's expectation of life," although prepared to observe that "in point of fact neighbor's [*sic*] dogs and cats are unpleasantly long-lived. . . ." *Id.* at 260–61).

6 Four cases dealing with the Constitutionality of the death penalty under the eighth and fourteenth amendments are pending before the United States Supreme Court. Branch v. Texas, 447 S.W.2d 932 (Tex. 1969), *cert. granted*, 91 S. Ct. 2287 (1970); Aikens v. California, 70 Cal. 2d 369, 74 Cal. Rptr. 882, 450 P.2d 258 (1969), *cert. granted*, 91 S. Ct. 2280 (1970); Furman v. Georgia, 225 Ga. 253, 167 S.E.2d

628 (1969), *cert. granted*, 91 S. Ct. 2282 (1970); Jackson v. Georgia, 225 Ga. 790, 171 S.E.2d 501 (1969), *cert. granted*, 91 S. Ct. 2287 (1970).

7 See George Campbell Painting Corp. v. Reid, 392 U.S. 286 (1968); Oklahoma Press Pub. Co. v. Walling, 327 U.S. 186 (1946); Baltimore & O.R.R. v. ICC, 221 U.S. 612 (1911); Wilson v. United States, 221 U.S. 361 (1911); Hale v. Henkel, 201 U.S. 43 (1906).

8 See Dixon v. Alabama State Bd. of Educ., 294 F.2d 150 (5th Cir.), *cert. denied*, 368 U.S. 930 (1961).

9 For example, see People *ex rel.* Ricks Water Co. v. Elk River Mill & Lumber Co., 107 Cal. 221, 40 Pac. 531 (1895) (refusing to enjoin pollution by a upper riparian at the instance of the Attorney General on the grounds that the lower riparian owners, most of whom were dependent on the lumbering business of the polluting mill, did not complain).

10 The law in a suit for injunctive relief is commonly easier on the plaintiff than in a suit for damages. See J. Gould, LAW OF WATERS § 206 (1883).

11 However, in 1970 California amended its Water Quality Act to make it easier for the Attorney General to obtain relief, *e.g.*, one must no longer allege irreparable injury in a suit for an injunction. CAL. WATER CODE § 13350(b) (West 1971).

12 To whomsoever the soil belongs, he owns also to the sky and to the depths. See W. BLACKSTONE, 2 COMMENTARIES •18.

At early common law, the owner of land could use all that was found under his land "at his free will and pleasure" without regard to any "inconvenience to his neighbour." Acton v. Blundell, 12 Meeson & Welsburg 324, 354, 152 Eng. Rep. 1223, 1235 (1843). "He [the landowner] may waste or despoil the land as he pleases. . . ." R. MOGARRY & H. WADE. THE LAW OF REAL PROPERTY 70 (3d edn 1966). *See* R. POWELL, 5 THE LAW OF REAL PROPERTY ¶725 (1971).

13 See Note, *Statutory Treatment of Industrial Stream Pollution*, 24 GEO. WASH. L. REV. 302, 306 (1955); H. FARNHAM, 2 LAW OF WATERS AND WATER RIGHTS § 461 (1904); GOULD, *supra* note 10, at § 204.

14 For example, courts have upheld a right to pollute by prescription, Mississippi Mills Co. v. Smith, 69 Miss. 299, 11 So. 26 (1882), and by easement, Luama v. Bunker Hill & Sullivan Mining & Concentrating Co., 41 F.2d 358 (9th Cir. 1930).

15 See Red River Roller Mills v. Wright, 30 Minn. 249, 15 N.W. 167 (1883) (enjoyment of stream by riparian may be modified or abrogated by reasonable use of stream by others); Townsend v. Bell, 167 N.Y. 462, 60 N.E. 757 (1901) (riparian owner not entitled to maintain action for pollution of stream by factory where he could not show use of water was unreasonable); Smith v. Staso Milling Co., 18 F.2d 736 (2d Cir. 1927) (in suit for injunction, right on which injured lower riparian stands is a quantitative compromise between two conflicting interests); Clifton Iron Co. v. Dye, 87 Ala. 468, 6 So. 192 (1889) (in determining whether to grant injunction to lower riparian, court must weigh interest of public as against injury to one or the other party). See also Montgomery Limestone Co. v. Bearder, 256 Ala. 269, 54 So. 2d 571 (1951).

16 Pennsylvania Coal Co. v. Sanderson, 113 Pa. 126, 149, 6 A. 453, 459 (1886).

17 Hand, J. in Smith v. Staso Milling Co., 18 F.2d 736, 738 (2d Cir. 1927) (emphasis added). See also Harrisonville v. Dickey Clay Co., 289 U.S. 334 (1933) (Brandeis. J.).

18 Measuring plantiff's damages by "making him whole" has several limitations . . .

19 Here, again, an analogy to corporation law might be profitable. Suppose that in the instance of negligent corporate management by the directors, there were no institution of the stockholder derivative suit to force the directors to make *the corporation* whole, and the only actions provided for were direct actions by stockholders to collect for damages *to themselves qua* stockholders. Theoretically and practically, the damages might come out differently in the two cases, and not merely because the creditors' losses are not aggregated in the stockholders' direct actions.

20 And even far less than the damages to all human economic interests derivately through the stream . . .

21 Smith v. Staso, 18 F.2d 736, 738 (2d Cir. 1927).

22 Some of these public properties are subject to the "public trust doctrine," which, while ill-defined, might be developed in such fashion as to achieve fairly broad-ranging environmental protection. See Gould v. Greylock Reservation Comm'n, 350 Mass. 410, 215 N.E.2d 114 (1966), discussed in Sax, *The Public Trust Doctrine in Natural Resource Law: Effective Judicial Intervention*, 68 MICH. L. REV. 471, 492–509 (1970).

23 By contrast, for example, with humane societies.

24 See, e.g., CAL. PROB CODE §§ 1460–62 (West Supp. 1971).
25 CAL. PROB CODE § 1751 (West Supp. 1971) provides for the appointment of a "conservator."
26 In New York the Supreme Court and county courts outside New York City have jurisdiction to appoint a committee of the person and/or a committee of the property for a person "incompetent to manage himself or his affairs." N.Y. MENTAL HYGIENE LAW § 100 (McKinney 1971).
27 This is a situation in which the ontological problems discussed in note 4 *supra* become acute. One can conceive a situation in which a guardian would be appointed by a county court with respect to a stream, bring a suit against alleged polluters, and lose. Suppose now that a federal court were to appoint a guardian with respect to the larger river system of which the stream were a part, and that the federally appointed guardian subsequently were to bring suit against the same defendants in state court, now on behalf of the river, rather than the stream. (Is it possible to bring a still subsequent suit, if the one above fails, on behalf of the entire hydrologic cycle, by a guardian appointed by an international court?)

While such problems are difficult, they are not impossible to solve. For one thing, pre-trial hearings and rights of intervention can go far toward their amelioration. Further, courts have been dealing with the matter of potentially inconsistent judgments for years, as when one state appears on the verge of handing down a divorce decree inconsistent with the judgment of another state's courts. Kempson v. Kempson, 58 N.J. Eg. 94, 43 A. 97 (Ch. Ct. 1899). Courts could, and of course would, retain some natural objects in the res nullius classification to help stave off the problem. Then, too, where (as is always the case) several "objects" are interrelated, several guardians could all be involved, with procedures for removal to the appropriate court – probably that of the guardian of the most encompassing "ward" to be acutely threatened. And in some cases subsequent suit by the guardian of more encompassing ward, not guilty of laches, might be appropriate. The problems are at least no more complex than the corresponding problems that the law has dealt with for years in the class action area.

28 CAL. PROB. CODE § 1460 (West Supp. 1971). The N.Y. MENTAL HYGIENE LAW (McKinney 1971) provides for jurisdiction "over the custody of a person and his property if he is incompetent to manage himself or his affairs by reason of age, drunkenness, mental illness or other cause. . . ."
29 Santa Clara County v. Southern Pac. R.R., 118 U.S. 394 (1886). Justice Black would have denied corporations the rights of "persons" under the fourteenth amendment. See Connecticut Gen. Life Ins. Co. v. Johnson, 303 U.S. 77, 87 (1938) (Black, J. dissenting): "Corporations have neither race nor color."
30 The laws regarding the various communal resources had to develop along their own lines, not only because so many different persons' "rights" to consumption and usage were continually and contemporaneously involved, but also because no one had to bear the costs of his consumption of public resources in the way in which the owner of resources on private land has to bear the costs of what he does. For example, if the landowner strips his land of trees, and puts nothing in their stead, he confronts the costs of what he has done in the form of reduced value of his land; but the river polluter's actions are costless, so far as he is concerned – except insofar as the legal system can somehow force him to internalize them. The result has been that the private landowner's power over natural objects on his land is far less restrained by law (as opposed to economics) than his power over the public resources that he can get his hands on. If this state of affairs is to be changed, the standard for interceding in the interests of natural objects on traditionally recognized "private" land might well parallel the rules that guide courts in the matter of people's children whose upbringing (or lack thereof) poses social threat. The courts can, for example, make a child "a dependent of the court" where the child's "home is an unfit place for him by reason of neglect, cruelty, or depravity of either of his parents. . . . " CAL. WELF. & INST. CODE § 600(b) (West 1966). See also id at § 601: any child "who from any cause is in danger of leading an idle, dissolute, lewd, or immoral life [may be adjudged] a ward of the court."
31 See note 30 *supra*. The present way of handling such problems on "private" property is to try to enact legislation of general application under the police power, see Pennsylvania Coal Co. v. Mahon, 260 U.S. 393 (1922), rather than to institute civil litigation which, though a piecemeal process, can be tailored to individual situations.

32 Cal. Prob. Code § 1580 (West Supp. 1971) lists specific causes for which a guardian may, after notice and a hearing, be removed.

Despite these protections, the problem of overseeing the guardian is particularly acute where, as here, there are no immediately identifiable human beneficiaries whose self-interests will encourage them to keep a close watch on the guardian. To ameliorate this problem, a page might well be borrowed from the law of ordinary charitable trusts, which are commonly placed under the supervision of the Attorney General. See Cal. Corp. Code §§ 9505, 10207 (West 1955).

33 See Cal. Prob. Code § 1472, 1590 (West 1956 and Supp. 1971).

34 Concern over an anticipated flood of litigation initiated by environmental organizations is evident in Judge Trask's opinion in Alameda Conservation Ass'n v. California, 437 F.2d 1087 (9th Cir.), *cert. denied*, Leslie Salt Co. v. Alameda Conservation Ass'n, 402 U.S. 908 (1971), where a non-profit corporation having as a primary purpose protection of the public's interest in San Francisco Bay was denied standing to seek an injunction prohibiting a land exchange that would allegedly destroy wildlife, fisheries and the Bay's unique flushing characteristics.

> Standing is not established by suit initiated by this association simply because it has as one of its purposes the protection of the "public interest" in the waters of the San Francisco Bay. However well intentioned the members may be, they may not by uniting create for themselves a super-administrative agency or a *parens patriae* official status with the capability of over-seeing and of challenging the action of the appointed and elected officials of the state government. Although recent decisions have considerably broadened the concept of standing, we do not find that they go this far. [Citation.]
>
> Were it otherwise the various clubs, political, economic and social now or yet to be organized, could wreak havoc with the administration of government, both federal and state. There are other forums where their voices and their views may be effectively presented, but to have standing to submit a "case or controversy" to a federal court, something more must be shown.

437 F.2d at 1090.

35 See note 27 *supra*.

36 Here, too, we are dogged by the ontological problem discussed in note 4 *supra*. It is easier to say that the smog-endangered stand of pines "wants" the smog stopped (assuming that to be a jurally significant entity) then it is to venture that the mountain, or the planet earth, or the cosmos, is concerned about whether the pines stand or fall. The more encompassing the entity of concern, the less certain we can be in venturing judgments as to the "wants" of any particular substance, quality, or species within the universe. Does the cosmos care if we humans persist or not? "Heaven and earth . . . regard all things as insignificant, as though they were playthings made of straw." Lao-Tzu, *Tao The King* 13 (D. Goddard transl. 1919).

37 See Knight v. United States Land Ass'n, 142 U.S. 161, 181 (1891).

38 Clause 2 gives Congress the power "to dispose of and make all needful Rules and Regulations respecting the Territory or other Property belonging to the United States."

39 See Flemming v. Nestor, 363 U.S. 603 (1960).

40 See the *L.A. Times* editorial "Water: Public vs. Polluters" criticizing:

> the ridiculous built-in conflict of interests on Regional Water Quality Control Board. By law, five of the seven seats are given to spokesmen for industrial, governmental, agricultural or utility users. Only one representative of the public at large is authorized, along with a delegate from fish and game interests.
>
> (Feb. 12, 1969, Part II, at 8, cols. 1–2)

41 The Federal Refuse Act is over 70 years old. Refuse Act of 1899, 33 U.S.C. §407 (1970).

42 *See* Hall, *Refuse Act of 1899 and the Permit Program*, 1 Nat'l Res. Defense Council Newsletter i (1971).

43 To be effective as a deterrent, the sanction ought to be high enough to bring about an internal reorganization of the corporate structure which minimizes the chances of future violations.

Because the corporation is not necessarily a profit-maximizing "rationally economic man," there is no reason to believe that setting the fine as high as – but no higher than – anticipated profits from the violation of the law, will bring the illegal behavior to an end.

Source: Christopher D. Stone (1972) 'Should trees have standing? Toward legal rights for natural objects', *Southern California Law Review*, vol. 45, no. 2, pp. 453–67, 450–73.

4.8

Christopher D. Stone

Legal considerateness: trees and the moral community

The nature of a thing's *legal status*, and, in particular, the relationship between its legal status and its moral status, can be clarified if we confront at the start two popular misconceptions. The first mistake is to suppose that all questions of *legal status* can be conflated into questions of *legal rights*. That is, people tend to identify the question "Can some entity be accorded legal recognition?" with "Can (or does) the entity have a legal right?" In fact, the motive for recognizing something as a legal person may have nothing to do with *its* legal rights. For example, the courts may give a stillborn fetus the status of "person" in order to fulfill a technical prerequisite for the parents to file a malpractice case against the doctors.[1] Giving the fetus its independent legal status is designed to secure the legal rights of the parents, not those of the fetus.

This first misconception is commonly compounded by a second error. This is the view that the only basis on which we can support according a thing a legal right (or other legal recognition) is if we can show it has something like its own moral right, underneath. That, too, is a misconception . . .

Let me begin here by placing *legal rights-holding* in perspective. The short of it is that having a legal right is one way to provide something a concern-manifesting legal recognition. But it is not the only way. When the law criminalizes dog beating it institutionalizes concern for dogs. It does so, however, by creating a prospective liability for the dog beater, who is made answerable to public prosecutors at their discretion. In no accepted sense does such a statute create a "right" in the dog. The same principle is at work in legislation establishing animal sanctuaries, and laws that compel cattle transporters to provide minimally "humane" standards at the risk of losing their certificate. The law is enlisted in an effort to protect Nonpersons, but legal rights are not required. The federal government recently issued regulations requiring fishermen who accidentally land sea turtles on their decks to give them artificial respiration. The technique is set out in detail. But the term *turtle's rights* is never mentioned. Nor need it be. The turtles are provided a measure of legal protection by creating enforceable legal duties *in regard to* them, duties enforceable by others, presumably the Commerce Department and Coast Guard. . . .

This should remind us that while allocating rights is a fundamental way of operationalizing legal concern, and I myself emphasized it in *Trees*, such concern can be implemented through a broad range of arrangements, not all of which can be forced into the classic "rights-holding," or even "duties-bearing," mold. Therefore, let me introduce as the more comprehensive notion *legal considerateness*. A terse operational definition would look like this. Consider a lake. The lake is considerate within a legal system if the system's rules have as their immediate object to affect (as to preserve) some condition of the lake. The law's operation would turn on proof that the lake is not in the condition that the law requires, without any further need to demonstrate anyone else's interests in or claims touching the lake.

We can illustrate by reference to a lake which is being polluted by a factory. Under conventional law, the pollution of the lake can be restrained at some human's behest (call him Jones) if Jones can show that he has a legally protectable interest in the lake. Jones might prove that the lake is on his land, or that, as owner of lakeside property, he has a right to the water in a condition suited for his domestic or agricultural use. Changes in the state of the lake – for example, its degradation – are relevant to Jones's proof, but his case cannot stop with the proof of such changes. As plaintiff, he has to prove that he has some right in respect of the lake's condition, which right the factory is infringing. Damages, if any, go to Jones. If the court awards him an injunction, it is *his* injunction; Jones has the liberty to sit down with the factory owners and negotiate an agreement not to enforce it, to let the pollution restart – if the factory owners make an offer that is *to Jones's advantage*.

Contrast that system to one in which the rules empower a suit to be brought against the factory in the name not of Jones, but of the lake, through a guardian or trustee. The factory's liability is established on the showing that without justification, it degraded the lake from one condition, which is lawful, to another, which is not.

In the first system described, the lake is not legally considerate. It has no protection that is not wholly parasitic upon the rights of some person ready and willing to assert them. In the latter system, it is the lake that is considerate, not the person empowered to assert claims on its behalf.

Now, if this were intended as a jurisprudence text, we might proceed to unpack legal considerateness into the whole array of terms with which legal philosophers deal: rights, duties, privileges, immunities, no-rights, and all the rest. For our purposes, however, we can simply conceptualize considerateness into two broad categories: *legal advantage* (typified by holding a legal right) and *legal disadvantage* (typified by bearing a legal duty). The question I will pursue, therefore, is to what extent and in what senses we can coherently situate Nonpersons in positions of legal advantage and legal disadvantage (hereinafter Advantage and Disadvantage), respectively.

Are interests a requirement of legal considerateness?

At this point, someone is bound to object that all this talk about "positioning" a Nonperson in a position of Advantage or of Disadvantage simply slides over the most serious problem. In regard to legislation touching some Nonpersons – as when we criminalize dog beating, for example – few would deny we are acting for the dog's benefit, *really* advantaging it. But one cannot so plausibly characterize an arrangement that nestles the lake in the protective custody of a guardian as being "for the lake's 'benefit' " in any familiar sense of "benefit." We have simply plugged the lake into the legal system roughly in the place that a person might occupy, to a person's advantage; but it is a place that can be of no advantage to the lake, only to people. The same may be said of whatever we might do concerning many Nonpersons, from ants and aquifers to zoophytes and zygotes. Because they have no self-conscious interests in their own fates (as distinct from our interests in them), it is unclear how notions of Advantage and Disadvantage apply.

How do we respond to this challenge? Inasmuch as we cannot examine all Nonpersons at the same time and with equal emphasis, I will set aside for the moment the most sympathetic cases for the Nonperson advocate to represent – those, for example, of distant persons and higher animals. To examine the general problem, we best concentrate on the most implausible of my Nonperson clients, thereby confronting the most fundamental objections. What space is there in law and morals, what toehold even, for the subset of Nonpersons I will call "Things" – utterly disinterested entities devoid of feelings or interests except in the most impoverished or even metaphorical sense?

There are various classes of these disinterested Things. There are man-made inanimates exemplified by artworks and artificial intelligence. There are Things that were not always so disinterested – corpses – and Things with the potential to be otherwise – embryos and fetuses. There are trees and algae which, while without self-consciousness and preferences, are nonetheless living organisms with biological requirements. There are what we discern as functional systems, but not systems that conform to the boundaries of any organism; for example, the hydrologic cycle. Habitats are of this sort. That is, while the habitat may include higher animals, we may find ourselves wishing to speak for some value not reducible to the sum of the values of the habitat's parts, the various things that the habitat sustains in relation. There are several sorts of natural and conventional membership sets already mentioned: species, tribes, nations, corporations. We might for completeness' sake keep in mind such "things" as in the happenstance of our language are rendered as qualities, for example, the quality of the light in the Arizona desert at sunset. Finally, there are events that might be of legal and moral concern, such as the flooding of the Nile.

Whatever might motivate and justify the lawmakers to try to make some such Thing legally considerate, how can we coherently account for an entity that has no welfare? We can imagine our interlocutor putting it this way. "All right," he says, "let us suppose that somewhere in the text that follows, you will be able to demonstrate at least some rational basis that could motivate us to devise legal rules in which a Thing – a lake, say – is made legally considerate, as you use the term. We will even agree to provide it a court-appointed guardian empowered to get up in court and 'take on' polluters in the lake's name. But now comes the tough part. The lake itself being utterly indifferent to whether it is clear and full of fish or muddy and lifeless, when the guardian for the river gets up to speak, *what is he or she supposed to say?*"

My answer will sound, I am afraid, a bit anticlimactic. As in any situation in which a legal guardian is empowered to speak for a ward, what the guardian says will depend upon what the legal rules touching on the ward provide. By definition a Thing can neither be benefited nor detrimented in the ordinary sense, so that the rules cannot orient to its best interests in the way some rules of child custody enlist "the best interests of the child." But what is implied? Not that no legal rule is imaginable. It only means that the state of the Thing for the preservation or attainment of which the guardian speaks will have to be some state the law decrees without reference to what the choiceless Thing would choose. What we should be asking, then, is this: What states of a disinterested entity (a Thing), of necessity unrooted in its own interests or welfare or preferences, are available for the law to embody in legal rules, and what would be the implications of so embodying them?

Intactness as a legal advantage

To examine what sort of legally defined Advantages might be conferred on a Thing, and with what implications, let us begin by considering a system that took as its target preserving the lake's intactness. The lawmakers could provide stiff criminal penalties for anyone who polluted the lake in the least degree. They could fortify this "advantaging" by assimilating the lake into the civil-liability rules in a way that approximated constituting the lake a rights holder with guardian. Specifically, the law could provide that in case someone violated established effluent standards, altering the state of the lake, a complaint could be instituted in the name of the lake, as party plaintiff, against the polluter. As I spelled out in *Trees*, this suit could be initiated and maintained – as suits are maintained as a matter of course for infants, the senile, and corporations – by a lawyer authorized to represent it, by ad hoc court appointment, or otherwise. Assuming that the guardian had to show damages, the law could simply provide that the lake's legal damages were to be measured by the costs of making the lake "whole" in the sense of restoring the lake to the condition it would have been in had its "legal right" to intactness not been violated. That is, if the defendant's liability were established (if the upstream plant were found to have violated the applicable standards), it would have to pay into a trust fund, for the repair of the river, funds adequate to cover such items as aeration, restocking with fish and aquatic plants, filtering, and dredging.

[. . .]

Commonwealth of Puerto Rico v. *SS Zue Colocotroni*[2] is a striking illustration. In *Zue Colocotroni*, an oil tanker by that name was allowed by its owners to deteriorate into an unseaworthy condition, to be launched without proper charts, and to be manned, as the courts put it, by a "hopelessly lost" and "incompetent crew." The ship ran aground off Puerto Rico, spilling thousands of tons of crude oil. Puerto Rico, as trustee for its resources, submitted an estimate of damage to an area around a twenty-acre mangrove swamp. The major item was for a decline of 4,605,486 organisms per acre. "This means," the district court said, "92,109,720 marine animals were killed," largely sand crabs, segmented worms, and the like. Trying to get a handle on damages (the court's own can of segmented worms), the judge wrote:

> The uncontradicted evidence establishes that there is a ready market with reference to biological supply laboratories, thus allowing a reliable calculation of the cost of replacing these organisms. The lowest possible replacement cost figure is \$.06 per animal, with many species selling from \$1.00 to \$4.50 per individual. Accepting the lowest replacement cost, and attaching damages only to the lost marine animals in the West Mangrove area, we find the damages caused by Defendants to amount to \$5,526,584.20.

There was not a lot of precedent, as the reader can imagine. But the court of appeals affirmed the award, filing an opinion clearly sympathetic to deterring damage to nature – keeping it intact – without too finicky a regard either for human economic value or for the lower court's sympathetic but wondrously coarse technique for approximating the "replacement costs."

[. . .]

What the *Zue Colocotroni* litigation suggests is that an intactness standard may "make sense," in the sense that it can be operationalized coherently. But it may not always "make sense" when evaluated as policy. Certainly the policy of making the environment whole will be resisted when the costs of restoration or preservation are vastly out of line with the resource's consumption and use value, and the injury arises from a conscientious miscalculation rather than, as in *Zue Colocotroni*, from gross, almost willful, negligence. That is, even if one recognizes a moral claim to expend for the resource *some* sum in excess of its beneficial social value (as we do in undertaking the rescue of a trapped miner), one is rightly queasy about committing society to a higher price-tag than any moral theory we can devise will warrant. The issue of overprotection is not peculiar to entities that lack, or take no interest in their well-being. The problem arises from our uncertainty about *preferences*. It is thus one of the common "general theory" problems that has to be faced in connection not only with lakes and forests, but also with some "higher" Nonpersons such as whales, primates, and mental defectives, entities that, while possessing interests and even preferences, are at best restricted in their capacities to express them.[3]

To understand the difficulty introduced by the want of preferences, we must consider for a moment the significance of preferences in the legal system in its ordinary operations, that is, as it affects dealings among Persons. As the law develops, a body of rules establishes or confirms certain entitlements, such as the right of a homeowner to his home, or of a car owner to her car. In the familiar interpersonal situations, the existence of legally enforceable claims does not freeze the social ordering. They only establish ground rules for the operation of a primarily consensual, mutually beneficial system.

[. . .]

Imputing preferences

First, while Things have (or "take") no interests by definition, the same concession is not required of all Nonpersons, for example, future and spatially remote humans. We cannot know their preferences with certainty, but we can make some good guesses within reason. Even regarding "higher" animals, we cannot dismiss the feasibility of correctly identifying and imputing at least some preferences. Note that the issue being raised here is not the conventional query of the animal rights literature – whether higher animals have properties essential for possessing moral rights. All we are asking here is whether we can project a Nonperson's preferences confidently enough to allocate it some Advantage, without locking ourselves into a position of utter inflexibility.

To illustrate how a Nonperson's preferences, where available, may be enlisted to support the entity's legal position, let me return to the case of the bowhead whales. I am not certain how clear a picture we have of what whales like, of what welfare economists might call their "preference profile." Presumably a whale's preferences are less richly detailed than a

Person's, if only because the richness of many alternatives available to humans – to go bowling or sit in a movie – do not present themselves to animals for reasons both physical and intellectual.[4] On the other hand, we could say much the same for infants and the mentally disabled, on whose behalf the law constructs rough preferences fairly routinely. Surely, unless one is prepared to deny that whales have intelligence and are capable of exercising choice, it is fair to infer that they prefer their present route through the Beaufort Sea to any other. (If the path is not exhaustively prescribed by instinct, why do they select the variations they do?) True, we cannot be certain *why* and *how much* they would find an alternate route less preferable. Perhaps a more northerly path would take them into less comfortable waters in terms of temperature. Another route would involve a less familiar, more bewildering (anxiety-ridden), path; the food at the greater distances from shoreline, even if comparable nutritionally, might be less palatable.

We are unclear on all the details. But we do know some things about their preferences – and we could learn more. For example, I presume that marine biologists know the whales' favorite foods, and it does not seem beyond our capabilities to determine experimentally what meals they would prefer to whatever they can locate on the northernmost edge of their current journey-route. And let us simplify matters by supposing that none of their favorite fare (krill, plankton, or – God forbid! – snail darters) are recognized as legal persons in their own right. The point is, the more we are able to approximate such interests, the more freely we can allocate Advantages to Nonpersons with the assurance that if we need to modify them, we can "compensate" the way we do with Persons. This considerably mitigates the inflexibility problem referred to earlier.

To illustrate how such compensation would work, let us suppose that we do regard the whales as having established an Advantage to their traditional migratory route – something like what the law would call an easement by prior occupancy or prescription. Even if we elevated such an Advantage to equivalent rank of a human's property right, it would still be subject to condemnation, just as the ordinary person's easements (such as a long-established access path across a neighbor's property) can be condemned. To carry out this line of thought, suppose that the government had a higher public use for the whale's easement. The Treasury might realize, say, $10 billion in selling the oil rights. If so, mankind could well stand to proceed with the oil sale and still "pay off" the whales with a trust fund of $1 million for making their new course more comfortable. This could be accomplished by, say, "chumming" the alternate, northerly route with whatever foodstuff whale research indicated was high on the whales' preferences. As an ideal, some such solution would be better for everyone (would constitute, in the academic lingo, an interspecies-pareto improvement). The U.S. citizens would be better off through a reduced tax burden. The oil companies and their customers would be better off through the prospect of new domestic oil reserves. And the whales would be no worse off, tided over by a trust fund expended in a way as to compensate them – and help steer them clear of dangers at the drilling site.

[. . .]

Boundaries and ideals

Unfortunately, the feasibility of such a preference-enlisting solution exists only in regard to Nonpersons at the higher end of the intelligence scale. In the case of many Nonpersons, and of all Things, that tactic has no place. They have no preferences. What, then, could comprise a workable solution?

One way to avoid inflexibility is to build some threshold conditions into the rules.[5] Such boundary-sensitive Advantages would not be out of character with ordinary human "rights." To take an obvious example, people have a right to a trial in certain federal civil cases if, but only if, the amount in controversy exceeds (depending on the controversy) twenty dollars or ten thousand dollars. By analogy, we could assign some monetary value to a Thing we chose to protect, say, $25 million for snail darters. This would mean that we were committed to forgo up to that amount to preserve the species. But if the tangible social benefits of some proposed snail-darter-jeopardizing action exceeded $25 million, then the action could proceed under the condition that that sum would be applied either to mitigate the risk, or else be allocated to some other part of the environment, perhaps to preserve some endangered, closely related species in a nearby biotic community.

. . . A polluter might be required to return the lake to its pre-pollution condition unless the amount required to restore it exceeded a fixed figure. There is no other response when we are dealing with mishaps that are technologically impossible to repair at any price. Instead of demanding an infinite sum in damages, we have to develop feasible alternative remedies. Something like this occurred in the wake of the notorious discharges of the pesticide Kepone into the James River in 1975. A full dredging of the river bottom would not only have been extravagant in relation to the dangers; many believe that stirring up toxins from the bottom was riskier than just leaving them lie. As a consequence, Allied Chemical agreed to fund an $8 million trust fund with the mandate to mitigate the damage to the river and to foster general environmental-health research in the area.[6]

Moreover, we do not have to express boundary parameters in monetary terms. We can employ physical definitions. If, say, a lake's level of dissolved oxygen should fall below so many parts per million (a common measure of biological degradation), or if the aggregate biomass it supports should decline below so many specified tons, then the guardian would be empowered to invoke some legal remedy, in whatever way specified. A comparable approach is written into the Marine Mammal Protection Act. Individual endangered sea mammals, such as porpoises, are not protected as such. But if a human activity, such as seining for tuna, threatens the "optimum sustainable population" of their habitat, then the courts are authorized to intervene. Indeed, at that point, as the courts have interpreted the law, "balancing of interests between the commercial fishing fleet and the porpoise is irrelevant; the porpoise must prevail."[7]

Notes

1 See Summerfield v. Superior Court of the State of Arizona, Supreme Court of Arizona, April 24, 1985. And see MacDonald v. Time, Inc., 554 F.Supp. 1053 (D.C. N.J. 1983) (libel suit not mooted by plaintiff's death).

2 456 F.Supp. 1327 (D.Puerto Rico 1978), *aff'd* 628 F.2d 652 (1st Cir. 1979).

3 See, however, David Lamb, "Animal Rights and Liberation Movements," *Environmental Ethics* 4 (1982): 215, 231.
4 On the other fin, a whale swimming in the ocean may routinely mull alternatives that would not occur to a person swimming alongside.
5 Even without express threshold conditions, judicial institutions are prone to dismiss complaints on traditionally available grounds when any harm the plaintiff suffered can be regarded as de minimis – beneath the threshold of what the law is prepared to recognize. See United States v. Chevron, 583 F.2d 1357 (5th Cir. 1978). The company was found to have discharged oil into a navigable waterway, which caused a presumption of violation of law; but the only consequence was a "sheen" on the water. The company was allowed to rebut the presumption of illegality by demonstrating that the sheen-causing discharge was "less than harmful."
6 See Christopher D. Stone, "A Slap on the Wrist for the Kepone Mob," *Business and Society Review* 22 (Summer 1977): 4–11.
7 Committee for Humane Legislation v. Richardson, 540 F.2d 1141, 1151, n. 39 (C.A. D.C. 1976).

Source: Christopher D. Stone (1987) *Earth and Other Ethics: The Case for Moral Pluralism*, New York: Harper & Row, pp. 43–62.

4.9

Luke Martell

On values and obligations to the environment

Ecologists argue that what is distinctive about environmental ethics is that it extends rights and obligations beyond humans to other entities in the wider environment: animals and other living and non-living non-sentient beings. Anthropocentric arguments justify protection of parts of the environment – resources, animals, wilderness, bio-diversity and such like – for the practical or aesthetic value they have for humans. Many environmentalists argue that such parts of the environment should command obligations in themselves. They should be protected regardless of, and in cases where they do not have, value for humans.[1]

Coming to conclusions on where value resides or obligations are due has implications for which parts of the environment we protect. It may mean, for instance, protecting parts of the environment which have little value for humans but have value in themselves. Intrinsic value in nature broadens our policy responsibilities.

I wish to discuss here arguments for extending obligations to non-humans, why we should do so and to what range of entities. Some environmentalists want to include animals. Others want to include living non-sentient entities like plants or even non-living things like rocks and stones. I will be making three sets of distinctions between (1) different sorts of value; (2) different bases for attributing value and moral standing; and (3) different sorts of being to which value should be attached.

The first distinction is between intrinsic and extrinsic value. Intrinsic value is in something itself regardless of its value for other things. Humans could be said to have a value in themselves in their capacity to experience pleasure or flourish or in their nature as conscious intelligent beings. Humans have a value in such properties regardless of their use or value for other things. A spanner has extrinsic value. It has a value which derives from its objective properties but it is not intrinsic in the spanner itself but in the use it has for humans by virtue of its functions. Its value comes from its objective properties but is a value for something else.

Another distinction is between the differing bases on which non-humans have value or moral standing. These are listed under 2 in Table 4. Finally, there are different entities in the world, as under 3 in Table 4. I wish to discuss to which of these entities in 3 the reasons in 2 suggest value and concern should be extended and whether this value is intrinsic or extrinsic as distinguished in 1.

Using the distinction between the two different sorts of value, let me proceed to category 2: different arguments for attributing value and being concerned for, or holding obligations to, things. Which arguments are favoured determines which entities are attributed value or moral standing.

1 *Sentience* I will start with sentience – having the power to experience a sense of well-being or suffering.[2] We should extend obligations to entities in the world that have such a capacity. It is wrong to cause suffering to a being or curtail its ability to experience well-being. Using sentience as a basis for extending obligations incorporates animals alongside humans as a group to whom these are due. Animals, like humans, have the capacity to feel pain and pleasure and so on sentient criteria should also command obligations. At present we keep many animals in conditions that cause pain, distress or discomfort or we curtail their ability to lead a pleasurable life by killing them for sport or food. As such, animals are often not given the moral respect sentient arguments say they are due.[3]

Other living beings like plants or non-living things like rocks and stones do not, as far as we know, have the capacity to experience pleasure or pain. On sentient criteria, therefore, they do not have value in themselves and cannot command moral concern or obligations.

Table 4 Value in and obligations to the non-human environment

1 Sorts of value
- (a) intrinsic
- (b) extrinsic

2 Bases for attributing value or moral standing
- (a) sentience
- (b) capacity to flourish and develop
- (c) preservation of diversity
- (d) preservation of species and systems
- (e) membership of community

3 Entities to which value or moral standing attached
- (a) humans
- (b) non-human sentient living beings, i.e. animals
- (c) non-human non-sentient living beings, e.g. plants
- (d) non-human non-sentient non-living beings, e.g. rocks

They do have a value but that value is for beings who can experience well-being or suffering from the existence or flourishing of plants or stones. They should be preserved for their value to such beings. But they cannot have a value or command obligations in abstraction from sentient experience which comes only in their relation to other sentient groups.

For many of us, our emotional feelings and intuitions are that there *is* an intrinsic value in the being, life or development of plants or rocks. But, as I argue below, we should not trust our intuitions. It is hard to see a value in just being, living or growing. Value is in the experience of these. Plants and rocks do not have the capacity to experience being or growing or gain well-being from them. But experience or well-being, which *are* of intrinsic value, can be felt by sentient beings – humans and animals – and it is in them that intrinsic value lies.

2 *Flourishing* The debate with sentience is based not so much on a rejection of sentient arguments (although this sometimes features as I will discuss below) as on the argument that they are not enough. It is argued that sentience is part but not all of what gives a being value and a claim to rights and obligations. There are beings who do not have sentience, plants for example, but have a claim to rights and obligations because of other capacities they have which can pin down such claims – the capacity to grow, develop and flourish, for example. We should respect the rights of, and hold obligations to, anything which can flourish and develop and should restrain from actions which interfere with such capacities.[4]

A problem here is that arguments for the capacity for flourishing as a criterion on which obligations are due distinguish too sharply between it and sentience. What makes it of value is the joy of flourishing, not just flourishing by itself. Where it brings suffering it is not of value and we may not want to give rights and obligations to entities if their growth has ill effects. Think of locusts or plants that strangle other plants. We should judge flourishing according to the experiences it is wrapped up in. It is they which are of value.

The value in non-sentient flourishing beings is not intrinsic, as it is in its implications for things other than flourishing itself. Values and obligations of an extrinsic sort can be extended to non-sentient flourishing beings. They have a value but in the well-being which derives from their capacities rather than in those capacities themselves. They evoke well-being not in themselves because they are non-sentient but in other sentient beings. This is why their value is extrinsic (for other things). Intrinsic value is located in sentient rather than non-sentient beings because it is sentience which is of value in itself. Intrinsic value is not divorced from flourishing because sentient experience is wrapped up in it. But it is experience which is the locus of value and not flourishing independently of the experience it is associated with.

We should, in sum, be responsive to the capacities of flourishing beings to flourish. But obligations go to sentient beings because it is the sentient experience involved in their own or others' flourishing which is of value. Flourishing itself cannot be a basis for commanding respect and obligations. First, it does not by itself have the weight commanding of respect that it has when wrapped up with sentient experience. Second, it could involve flourishing with ill effects to which we would not want to give value, respect or obligations.

3 *Diversity* So far I have considered two characteristics of entities which might make them of value and deserving of respect or obligations: sentience and the capacity to flourish. I have argued for the former. Let me now turn to three other arguments in environmental

ethics for giving value, obligations or respect to entities in the world. The arguments I want to consider now do not, on the face of it, turn so much on the characteristics of individual beings as on structures or principles which are seen to be of value: diversity; species or systems; and community obligations.

Diversity can be seen as a value in itself. It is of positive value and it is because it is good that we should value and extend respect, rights and obligations to diverse things in the world.[5] We should respect the place of all things in the world not so much for the sake of those things but because diversity is desirable. Plurality rather than the entities of which there are plural instances is what should make us want to respect them. In ecology diversity has a special ring to it because diversity and interdependence are said to be functional for the smooth running of ecosystems.[6]

There are a number of problems here. The first is on the functions of diversity for the system. If it is this that is desirable then it is the system, rather than diversity which is of intrinsic value and which we should want to protect. Diversity only has extrinsic value and we would not want to respect it where it fails to fulfil its systemic functions. It is not diversity itself which is of intrinsic value. The argument made for it here is better covered under the valuing of systems which I will discuss below. We should not extend rights on the basis of a respect for diversity if it is the system which is of value and not diversity, which could potentially be of disservice.

On the other hand, if it is a concern for the individuals in the system which makes the functions of diversity for the system valuable then the environmental ethic is concerned about individuals rather than diversity. The value of individuals is covered by the discussions above on sentience and flourishing or by properties of value such as consciousness, intelligence, control or autonomy which individuals have.

Functions apart, one of the things which makes diversity of value is the fulfilment that living in a diverse world brings to beings with the sensory capacities to experience it. It is that diversity has such consequences rather than just the existence of diversity in abstraction that is behind our convictions when we say the world is better for being pluralistic. It is not diversity which is of value but the benefits it brings. What is of value is the experience facilitated by it. Gaining this experience is based on having a capacity for sensory experience – sentience. In itself it is difficult to see why diversity – just having lots of kinds of things – is good on its own. What is good about diversity is in its connection with the experience it contributes to and it is of value where it does so positively but may be a principle which we do not wish to respect or value where it does so negatively.

4 Species and systems In much environmental thinking value is put on the preservation of collective entities like species or ecosystems. These are said to have an intrinsic value in themselves. The death of the last member of a species is worse than the death of a member of a not endangered species. A species is seen to have a value in itself over and above the value of its individual members.[7]

One of the arguments on species comes from the case for diversity just discussed. We should preserve species because if one is lost there is a loss of a type of thing and a loss, therefore, to the diversity of things in the world. However, I have already explained why I think arguments on diversity are weak in abstraction from arguments for individual well-being. They are strengthened by being linked to well-being but then become based on the value of well-being rather than diversity.

In my view, it is difficult to see how arguments on species can work independently and without resort to other arguments on which they ultimately rest. They do not stand on the intrinsic value of having species alone but come down to arguments on the sentient well-being of members of the species or of other individuals who suffer as a result of the loss of a species. Loss of a species can be a loss because it involves losing its individual members. It is a loss of individuals rather than the collective entity they make up. Or it is a loss because a particular type of thing is no longer around. This does not make sense as a loss unless it is linked to a lessening of well-being of members of the world as a result. In abstraction from a diminution of well-being the loss of diversity of species remains statistical. It is difficult to see why there should be just more and more categories of things except if linked to the life of members of the species or the well-being of individuals from other species who benefit from the richness of life in a world of natural diversity or from the special value of a species.

The loss of species is bad. But it is so because of the loss of individual members or a diminution of the well-being of members of other species, rather than just the loss of a category itself in abstraction from such other considerations. Species have a value but it is not intrinsic. It is a value for members of the species or other beings in the world who benefit from its existence. Individuals of a species or the individuals of others may have a case on which to call for moral consideration from us. But abstract categories of species cannot make good claims for rights or value in themselves.

Another argument in which value is put on collective entities in the environment is on preserving ecosystems.[8] Leopold (1968: 224) argues that 'a thing is right when it tends to preserve the integrity, stability and beauty of the biotic community'. This suggests that value resides in the biotic community and that actions should be judged according to their contribution to the good of the community. The whole itself has an intrinsic value and characteristics worthy of respect and accommodation.

According to this view, our respecting and valuing of nature should be for it as a whole entity rather than, or as well as, for its parts because nature has an identity and functions as a whole. This can go further to a strong fetishizing naturalism. Nature knows best and we should not interfere with it as a system because this goes against what is natural and best for the survival of life. Nature is a whole, we should respect the 'natural' and we should practise non-interference with regard to it.[9]

In my view, there are a number of problems in the arguments tangled up in this 'holist' perspective. First, there is a question mark hanging over the scientific validity of what is claimed. Brennan (1988), by no means an opponent of a more relational and environmental ethics, argues that there is not a factual or scientific basis for the holism that greens aspire to. Greens tend to argue that we should respect ecosystems because we are bound up in them and because it is according to holist systemic principles that nature works. However, on Brennan's analysis it is not clear that ecosystems do actually function according to principles of holism and interdependence. The fact of holism should be analysed rather than assumed. . . .

Second, there is a problem with the view that we should respect the 'natural'.[10] It is not clear what it is about being natural that means we should respect it. To say we should respect something because it is natural is not enough. This fetishizes 'nature'. It needs to be said what it is about being natural that makes it worthy of respect.

Third, the very dichotomy between the natural and social needs to be challenged. What is it about humans that makes our behaviour not natural and in need of being accommodated to what is? It could be said that humans are just as natural as anything else. We have natural capacities and live within and in relation to nature. What reason is there to define our actions and capacities, development of social organization and technology and our purposive transformation of our surroundings as not natural or not taking place as part of nature? If humans *are* natural then accommodating to nature does not involve changing our patterns of behaviour to fit in with other principles. (On such issues see Dickens 1992.)

Fourth, what nature is is open to question. What goes on in nature is contradictory and often downright undesirable. Nature exhibits both toleration and killing, diversity and extinction, equality and exploitation. There is no apparent general design, guide, intention or rationale in this to show what is the preferred way of nature. It is not clear that there is something which is nature – distinctive or coherent characteristics which are identifiable and can be followed and given respect and value.

There is a fifth problem on interference and non-interference.[11] To defer to nature, not interfere with it and act in accordance with its principles can be a recipe for not doing what seems the best thing in the light of ethical consideration and the perceived best consequences. Further, it can inhibit actions which might seem to be the best for nature itself. Human interference may have played a large part in contributing to environmental problems but it is part of the solution as well. Yet interference in nature to protect it – building dams to protect natural habitats or killing members of species (e.g. locusts or strangling plants) to protect others, for example – is ruled out by deference to nature. We may need greater restraint but on the basis that it is good for the environment rather than because it is 'natural' and not to the exclusion of intervention in 'natural' processes to protect the environment.

Sixth and last, there is a problem with value residing in systems. To say that a system has intrinsic value means that the value is in the system rather than the individuals who make it up. I would argue that there cannot be intrinsic value in an ecosystem. A system's value and claim to respect rest in the value it has for its individual members. This is not to say that value is purely a perception of individuals and not in the objective properties of the system itself. The value may be a result of properties of the system irrespective of whether individuals recognize it or not. My point is that it is a value *for* individuals who make up the system and not of the system itself. The system has no value in itself divorced from the well-being of the individuals it contributes to.

Giving value to systems has dangerous implications. It means we can value systems over individuals and individuals can be sacrificed for the sake of an impersonal structure. Making the ecosystem of intrinsic value creates a conflict between its interests and the interests of the individuals who make it up. Yet it is the latter who matter and the former which should serve them. If the system gains value in itself over and above individuals this can be very dangerous for them.

It ought to be mentioned that I am not arguing for epistemological, ontological or methodological individualisms. It is not my claim that individuals are the source of knowledge or value, or the basic building blocks in natural or social life or the unit on which explanatory analysis should focus. On epistemology, for example, my argument is that value is in objective properties of the environment and not just in the eye of the beholder. But it is a value *for* individuals if not one just dreamt up by them. I am arguing for an ethical individualism and within this for a particular variant of it. My argument in ethical individualism is not for individual liberty (although autonomy is an important *part* of the good of individuals) or for atomistic or egoistic individualism. A scheme within which the well-being of individuals is the end may be collectivist or one in which rules restrict the uninhibited pursuit of self-interests. The well-being of individuals is the end with which my ethical individualism here is concerned.

5 *Community* Value, rights or obligations may be extended to non-human entities on the basis that they are part of the same community as humans. This is connected to the argument on systems and holism because it suggests that as members of the same whole different entities have obligations to one another. Humans have rights and obligations to non-human entities because they are part of the interdependent whole to which we all belong. Different entities have mutual obligations which come from interdependence, participation and membership in the same community.[12]

I am not going to dwell on whether entities in the environment *are* interdependent or members of the same community (see Brennan 1988). In my view, the argument on community falls down earlier than this – on the idea that ethics should be based on shared community in the first place. Why should we have obligations to someone because they are members of the same community? And why should we not have obligations to someone because they are not? I have already argued, regarding future generations, that we should have obligations to people and other beings who are strangers and not members of our community and with whom we are not in a position of interdependence.

We have obligations to the present-day third world poor because they are needy and we can help them. Even were we not responsible for their circumstances or not dependent on them (neither of which is the case) we would still have obligations to them for these reasons. It would be irresponsible for us not to help suffering beings when we can, regardless of the status of any other connections we may or may not have with them. The same goes for future generations. Because we can both adversely and positively affect their circumstances, we are obliged to at least not do the former. This should be incumbent upon us whether or not we are in a relation of mutual dependency or shared community with them.

Shared community and mutual dependence as the basis for obligations depend on ideas of contract and self-interest. We are said to owe obligations to others because of the mutual contract involved in joining a society with them or because we depend on one another. We agree to hold obligations to others because we wish to take part in the community with them, depend on them or want them to do likewise for us. We have obligations to members of our own community rather than to non-community justifications that claimants from outside it could make for our attentions.

My argument, however, is that there are beings in the world who have the capacity to experience well-being and suffering. If we have it in our power to help them without sacrificing our own prospects we have an obligation to do so, as long as they are not needy because of injustices or lack of effort on their part. Obligations extend beyond boundaries of community and such boundaries as the basis of obligations can prevent us from fulfilling obligations to those outside our community to whom we owe them. Community is not only too exclusive in this way but also too inclusive. It incorporates among those to whom we owe obligations people who can make claims on us on the grounds of shared membership of the same community but who have no claims on grounds of needs or well-being.

Notes

1 For influential classic statements by deep ecologists on intrinsic value in nature see Naess (1973) and Leopold (1968). Also see the discussion of value in the environment in Goodin (1992: ch. 2).

2 Griffin (1986) discusses issues such as these using the term 'well-being'. I will return in more depth to sentience and the literature on this issue later in the chapter.

3 For recent discussions of animal rights see Tester (1991), Benton (1993) and Garner (1993). For influential 'classic' statements see Bentham (1960), Salt (1980), Singer (1976), Clark (1977), Regan (1988) and Midgeley (1983). Also the collections edited by Singer (1985), Regan and Singer (1976) and Miller and Williams (1983). Different theorists argue for obligations to animals on different grounds and by no means all do so on the sentient grounds that Singer (1976), for example, and I favour.

4 Clark (1977) argues on flourishing in relation to animals. See also Attfield (1983: 151–4) and Taylor (1986).

5 See Naess (1973), Dobson (1990: 121–2), Sale (1984), Norton (1987) and Attfield (1983: 149–50) on the intrinsic value of diversity.

6 Brennan (1988) suspects deep ecological appeals to scientific claims about diversity in nature do not hold up – see pp. 43–4, 119, 122–3. Further, he argues that if diversity *is* an ecological reality this is not a sufficient basis for it to be of value – see pp. 152 and 164 [in Martell, L. (1994) *Ecology and Society*, Polity Press, Cambridge].

7 Naess (1984) and Norton (1986 and 1987) propose that species have an intrinsic value. See also Eckersley (1992: 46–7) and Feinberg (1980: 171–3, 204–5). Attfield (1983: 150–1, 155–6) is a critic of the idea.

8 On wholes or systems as having a value in themselves see Goodpaster (1978), Rodman (1977) and Callicott (1980). A scientific basis for ethical claims on holism is often made; see Lovelock (1979), Capra (1985) and Callicott (1985). See the discussion in Attfield (1983: 156–60, 179–82). Brennan (1988) argues that the scientific basis claimed by ethical holists and their ethical claims themselves are faulty. In this chapter I reject ethical holism. In chapter 6 [of source work] I reject ontological or explanatory holism which fetishizes the natural.

9 The idea that the system as a whole provides conditions optimal for life comes through strongly in the influential 'Gaia' thesis advanced by Lovelock (1979). For an accessible introduction to 'Gaia' see Dobson (1990: 42–7) and Dobson (1991: 264–8). Again, I discuss in this chapter why I think holist Gaia-type ideas are ethically dangerous. In chapter 6 [of source work] I explain why I think they are flawed as explanations of society–nature relations.

10 See Dobson (1990: 24–8), Sale (1984 and 1985), Bookchin (1982).

11 For an argument for non-interference see Regan's (1981) 'preservation principle' rejected, rightly in my view, by Brennan (1988: 198).

12 On community and obligations in environmental ethics see Leopold (1968: 203), Callicott (1979), Attfield (1983: 157–8).

References

Attfield, Robin 1983: *The Ethics of Environmental Concern*. Oxford: Blackwell.

Bentham, Jeremy 1960: *An Introduction to the Principles of Morals and Legislation*. Oxford: Blackwell.

Benton, Ted 1993: *Natural Relations: Ecology, Animal Rights and Social Justice*. London: Verso.

Bookchin, Murray 1982: *The Ecology of Freedom*. Palo Alto: Cheshire Books.

Brennan, Andrew 1988: *Thinking about Nature*. London: Routledge.

Callicott, J. Baird 1979: 'Elements of an Environmental Ethic: Moral Considerability and the Biotic Community'. *Environmental Ethics*, 1, 71–81.

Callicott, J. Baird 1980: 'Animal Liberation: A Triangular Affair'. *Environmental Ethics*, 2, 311–38.

Callicott, J. Baird 1985: 'Intrinsic Value, Quantum Theory and Environmental Ethics'. *Environmental Ethics*, 7, 257–75.

Capra, Fritjof 1985: *The Turning Point: Science, Society and the Rising Culture*. London: Flamingo.

Clark, Stephen R. L. 1977: *The Moral Status of Animals*. Oxford: Clarendon Press.
Dickens, Peter 1992: *Society and Nature: Towards a Green Social Theory*. Hemel Hempstead: Harvester Wheatsheaf.
Dobson, Andrew 1990: *Green Political Thought*. London: Andre Deutsch.
Dobson, Andrew (ed.) 1991 *The Green Reader*. London: Andre Deutsch.
Eckersley, Robyn 1992: *Environmentalism and Political Theory*. London: U.C.L. Press.
Feinberg, J. 1980: *Rights, Justice and the Bounds of Liberty*. Princeton: Princeton University Press.
Garner, Robert 1993: *Animals, Politics and Morality*. Manchester: Manchester University Press.
Goodin, Robert E. 1992: *Green Political Theory*. Cambridge: Polity Press.
Goodpaster, Kenneth 1978: On Being Morally Considerable. *Journal of Philosophy*, 75, 308–25.
Griffin, J. 1986: *Well-Being: Its Meaning, Measurement and Moral Importance*. Oxford: Oxford University Press.
Leopold, Aldo 1968: *A Sand County Almanac*. Oxford: Oxford University Press.
Lovelock, James 1979: *Gaia: A New Look at Life on Earth*. Oxford: Oxford University Press.
Midgeley, Mary 1983: *Animals and Why They Matter*. Harmondsworth: Penguin.
Miller, H. B. and Williams, W. (eds) 1983: *Ethics and Animals*. Clifton, N.J.: Humana Press.
Naess, Arne 1973: 'The Shallow and the Deep, Long-Range Ecology Movement: A Summary'. *Inquiry*, 16, 95–100.
Naess, Arne 1984: 'Intuition, Intrinsic Value and Deep Ecology'. *The Ecologist*, 14, 5–6.
Norton, Bryan G. (ed.) 1986: *The Preservation of Species: The Value of Biological Diversity*. Princeton: Princeton University Press.
Norton, Bryan G. 1987: *Why Preserve Natural Variety?* Princeton: Princeton University Press.
Regan, Tom 1981: 'The Nature and Possibility of an Environmental Ethic'. *Environmental Ethics*, 3, 16–31.
Regan, Tom 1988: *The Case for Animal Rights*. London: Routledge.
Regan, Tom and Singer, Peter (eds) 1976: *Animal Rights and Human Obligations*. Englewood Cliffs, N.J.: Prentice-Hall.
Rodman, John 1977: 'The Liberation of Nature'. *Inquiry*, 20, 83–145.
Sale, Kirkpatrick 1984: 'Bioregionalism – a New Way to Treat the Land'. *The Ecologist* 14, 167–73.
Sale, Kirkpatrick 1985: *Dwellers in the Land: The Bioregional Vision*. San Francisco: Sierra Club Books.
Salt, Henry S. 1980: *Animal Rights Considered in Relation to Social Progress*. London: Centaur.
Singer, Peter 1976: *Animal Liberation*. London: Cape.
Singer, Peter (ed.) 1985: *In Defence of Animals*. Oxford: Blackwell.
Taylor, P. 1986: *Respect For Nature: A Theory of Environmental Ethics*. Princeton: Princeton University Press.
Tester, Keith 1991: *Animals and Society: The Humanity of Animal Rights*. London: Routledge.

Source: Luke Martell (1995) *Ecology and Society*, Cambridge: Polity Press, pp. 86–94, 205–7, 213–25.

SECTION 5

Ecology, order and individualism

Introduction

The readings in this section are concerned with the ways in which environmental issues and ecological thinking have had an impact upon the mainstream traditions of Western social and political thought, liberalism and conservatism. Like all labels, each conceals considerable variety, so it is also open to question whether a branch of each tradition is more or less well equipped to respond to ecological concerns – such as the differences between consequentialist or utilitarian accounts of the outcomes of human activities and the deontological approaches which focus upon the emergence of sets of rules which organize the social and political order. Indeed, these differences have already been outlined in Section 3 in the readings by Singer and Regan. These differences are outlined and discussed in Reading 5.1 where Mark Sagoff explores how they have interacted with environmentalism (which Sagoff defines broadly to include ecological approaches as well as human-centred forms of environmentalism). More specifically, Sagoff raises questions about whether the aims of environmentalists are compatible with a liberal social and political order.

In Reading 5.2 a classic statement of the causes of the 'tragedy of the commons', Garrett Hardin provides a basis for clarifying some of the central problems which liberals have to confront when trying to find a way of resolving environmental problems. Hardin uses Game Theory to problematize many conventional liberal assumptions based upon utilitarian arguments and the blind belief in markets as a solution for all problems. To remedy the partial deficits in liberal thinking, he highlights the ways in which the tragedy of the commons can be avoided through property ownership. When looking at pollution problems, he advocates a solution which carries the price of sacrificing liberal freedoms – most significantly the freedom to reproduce. As a result, Hardin is a controversial figure in these debates for arguing that it is morally justifiable to use coercion to curb population growth. This theme is developed in Reading 5.3, where Robert Young assesses the different strategies for limiting or restraining the size of humankind. Not only does this address how many people are likely to exist in the future, it also considers the differential impact of social practices on the environment in different societies – population growth now takes place beyond the Western industrial societies which have a disproportionate impact on the environment. The sense of urgency in Hardin's and Young's accounts is driven by a fear that members of developing societies will be just as materialistic as the members of industrial societies. Young's conclusion that the failure of non-coercive measures justifies the consideration of more radical steps is challenged in Reading 5.4 by Gary Malinas, who responds by proposing a more flexible approach. Malinas suggests that we should take greater account of the complexities of social relations if we are to develop a workable population management strategy which involves a patchwork of measures configured in ways suitable for different societies. Malinas also highlights how Young's argument occupies a halfway house between consequentialist and deontological liberalisms.

The last two readings focus more on the impact of market institutions on the environment. Ecologists have consistently attacked the consequences of the capitalist growth machine, the role of profit-making over environmental concerns, and the way that companies have been able to dodge responsibility for the industrial waste they produce. You have the opportunity, in Reading 5.5, to explore a strident defence of the capitalist system. Peter

Saunders offers the technocentric neo-liberal response which flourished in Western societies in the 1980s. Saunders draws from Hardin's defence of private property to suggest that ecologically based criticisms of capitalism fail to acknowledge the significant benefits of the market system and the evidence that it can help solve environmental problems (that the arguments of Greens and Reds are closely connected). This market-led technofix approach is challenged by John Gray (Reading 5.6), who draws upon conservative values and assumptions. In particular, Gray draws out the affinities between ecological thinking and conservative philosophy – that they both adopt a multi-generational perspective, the primacy of the common life and have a distrust of technological innovation and social change for its own sake. This approach also argues in favour of property ownership as a way of conserving the environment but also recognizes the socially disorganizing effects of market institutions.

5.1

Mark Sagoff

Can environmentalists be liberals?

Classical liberalism, as Brian Barry notes, comprises many ideas, but one is "certainly the idea that the state is an instrument for satisfying the wants that men happen to have rather than a means of making good men (e.g., cultivating desirable wants or dispositions in its citizens)."[1] The state, on this view, seeks to ensure that all its citizens will be able to pursue personal interests and private preferences under conditions that are convenient and equitable to all. "The state, on the liberal view," Barry summarizes, "must be capable of fulfilling the same self-effacing function as a policeman on point duty, who facilitates the motorists' getting to their several destinations without bumping into one another but does not have any power to influence those destinations."[2]

Once liberalism is defined in this way, as an individualism, it merges easily with the value premise on which many economists base the cost-benefit or efficiency criterion in public policy. "The value premise," as Kneese and Bower state it, "is that the personal wants of the individuals in the society should guide the use of resources in production, distribution, and exchange, and that these personal wants can most efficiently be met through the seeking of maximum profits by all producers."[3]

Liberal political theory, likewise, may construe values as "personal wants of the individuals in the society"; thus it may regard public values as a peculiar kind of personal desire. In that case, political theory may dismiss idealistic, impersonal, or community values as illegitimate meddling in other people's affairs, or it may treat them as a weird sort of "intangible" that deserves a surrogate market price. "What underlies this view," as Brian Barry explains, "is a rejection of any suggestion that an ideal-regarding judgement should be treated as anything other than a peculiar kind of want."[4]

Those who support a cost-benefit approach to social regulation, as we have seen, consider the welfare of the individual to be the major desideratum of public policy. They often appeal for support to individualistic concepts that are central to the institutions of a liberal society, such as private property, personal freedom, and individual choice. Environmentalists, as I have argued, would base social regulation largely on shared or public values, which may express not our wants and preferences as individuals but our identity, character, and aspirations as a community. Environmentalism may seem, then, to involve a sort of communitarianism that is inconsistent with principles traditionally associated with a liberal state.

On the one hand, environmentalists (e.g., the Greens in Germany) apparently belong to the political left. On the other hand, they cannot (as I have argued) derive their policies simply from considerations of efficiency or equality, interests or rights. Where, then, do environmentalists fit into the political spectrum? Are the policies they propose consistent with the concepts and principles on which the institutions of a liberal society are based?

Two kinds of environmentalism

"Conservation," Aldo Leopold wrote, "is a state of harmony between men and land."[5] Leopold supposed that natural communities possess an order, integrity, and life that command our love and admiration and which, therefore, we should seek to protect for their sake and not simply to increase our own welfare. . . .

This [Leopoldian] "ethic" contrasts with the economic approach to environmental policy advocated by early conservationists like Gifford Pinchot. "The first great fact about conservation," Pinchot wrote, "is that it stands for development."[6] He added, "Conservation demands the welfare of this generation first, and afterward the welfare of the generations to follow."[7]

The difference between the positions of Leopold and Pinchot may be summarized as follows. Both recognize that only human beings (so far as we know) have values; in other words, only human beings make judgments of the kind: "This is valuable" or "This is good." Leopold and Pinchot agree, then, that human values and only human values count in resource policy. They disagree, however, over which values are important. In that sense, they disagree about *what is valuable.*

Leopold argued that land use and environmental policy ought to respond to the love, admiration, and respect many of us feel for the natural world. Love, admiration, and respect are human values, of course, but they do not necessarily involve human welfare. Rather, these values (although they arise in human beings) may be directed to the well-being and integrity of the rest of nature. Values such as these engender a widely shared attitude of aesthetic contemplation and moral altruism, for example, toward other species, for love typically seeks benefits not for itself only but also for its object.[8] Thus, the values Leopold emphasized, although they are human values, are directed toward the good of nature, not toward Leopold's own good or the good of humanity.

Pinchot, on the other hand, apparently believed that resource policy should serve the good of humanity and therefore should attempt to maximize social welfare as this is understood in economic theory. Pinchot assumed that only human welfare – and therefore nothing else in nature – can be valued for its own sake or have intrinsic worth. On this view, the reverence and respect people feel for nature do not endow it with intrinsic value; rather, these attitudes simply represent preferences the satisfaction of which will contribute to human "satisfaction."

Thus, Leopold and Pinchot agree that only human beings have values; only humans, so far as we know, value things. Those in the tradition of Pinchot, however, assert that the only object that can have intrinsic value or worth – the only goal that can be considered an end in itself – is human welfare. This differs from the view of Leopold and his followers, who assert that nature, as an object of reverence, love, and respect, itself has a moral worth and therefore should be protected for its own sake and not simply for the "satisfactions" or "benefits" it offers human beings.

I shall be concerned . . . with environmentalism as a movement that follows Leopold in espousing on ethical grounds the political goal of maintaining harmony between people and their environment. This movement asserts the importance of the cultural, historical, aesthetic, and religious values [previously described]: . . . it attempts – at times successfully – to embody these values in legislation. This sort of environmentalism rejects the individualistic view that society is essentially an "assemblage associated by a common acknowledgment of right and community of interest."[9] Instead, it visualizes society as a nation or people, which is, in Augustine's phrase, "an assemblage of reasonable beings bound together by the objects of their love."[10]

The tradition of classical liberalism, in emphasizing the importance of the individual, may support Pinchot's view that individual welfare is what matters in policy choices. It is easy, for example, to show how Locke's conception of property might justify the idea that perfectly competitive markets define the best or most valuable uses of land . . . I need only refer to the kinship many commentators have noted between traditional statements of liberal political theory, for example, in John Locke and Adam Smith, and classical economic theory. "The classical liberal view of individuality merged easily with economic rationality, and together these two ideologies spoke against any intervention" by the government except to ensure the fair and efficient functioning of markets.[11]

Today, many liberal political theorists emphasize the importance of state neutrality among the competing goals, values, or ends individuals may seek to achieve.[12] . . .

Liberals strive to prevent "moral" majorities from imposing ethical views and religious beliefs on minorities, for example, with respect to abortion, homosexuality, and school prayer. Environmentalists, however, may be said to constitute a moral lobby, if not a moral majority, of a sort, insofar as they advocate laws that embody ethical and perhaps even religious ideals concerning the way we ought to treat our natural surroundings.[13] If the laws and policies supported by the environmental lobby are not neutral among ethical, aesthetic, and religious ideals but express a moral conception of people's appropriate relation to nature, can environmentalists be liberals? May liberals support environmental laws even when these conflict with the utilitarian and egalitarian goals we usually associate with liberalism?

Two kinds of liberalism

Let me begin to answer these questions by presenting a view of what liberalism is. Liberalism is the political theory that holds that many conflicting and even incommensurable conceptions of the good may be fully compatible with free, autonomous, and rational action.

Liberals contend, therefore, that political and social institutions should be structured to allow free and equal individuals the widest opportunities, consistent with the like opportunities of others, to plan their own lives and to live the lives they plan.

Liberals differ in this respect from conservatives, who believe that social institutions should reward virtue and punish vice, as these are conceived within a particular cultural or religious tradition, and that these institutions therefore should not be neutral among the ways people may choose to live.[14] The conservative will favor the conception of the good life associated with the religion and culture of his or her community, for example, with respect to prayer, pornography, and sexual behavior, and he or she may wish to enforce that conception with the steel of the law.

Socialists differ from liberals because they, like conservatives, subscribe to a conception of virtue they would oblige citizens to practice. Socialists would officially discourage a hedonic or bourgeois life-style, for example, in the classless society they expect to flourish after the Marxist revolution. The socialist derives his or her conception of virtue and vice, however, from a priori arguments and philosophical theories, of the sort known to a political vanguard. In this the socialist differs greatly from the conservative, whose view of the good life is much less esoteric and rests in familiar religious and cultural traditions.[15]

Liberalism has been understood historically in terms of a distinction between two imaginary entities: civil society and the state.[16] According to this picture, individuals are joined in civil society to pursue their own interests, whatever they may be, by cooperating and, if necessary, by competing with one another within a system of rights that is fair to all. Individuals are joined as citizens in the state strictly for the purpose of enforcing those rights. The liberal state does not dictate the moral goals its citizens are to achieve; it simply referees the means they use to satisfy their own preferences. It respects the right of each person to pursue his or her own conception of the good life as long as his or her actions do not infringe on the same right of others.

It is common nowadays to sort liberal political theories under two headings: deontological (or "Kantian") and utilitarian. These theories differ essentially in the way they construe the relationship between the *right* and the *good*. *Rightness* is a quality that attaches to actions, for example, insofar as those actions are just or meet some other ethical criterion. *Goodness* attaches primarily to the consequences of actions, for example, insofar as these consequences increase happiness, satisfy preferences, or achieve some other goal assumed to be worthwhile.

Deontological approaches to liberalism, which I shall discuss presently, hold that a legal or political decision is right insofar as it is just and fair and respects the fundamental equality of persons. For the deontological liberal the principles of justice are established independently of social interests and preferences, and "against these principles neither the intensity of feeling nor its being shared by the majority counts for anything."[17] Deontological liberals argue, therefore, that policies that advance justice, fairness, and social equality "trump" claims that may be made on behalf of the general welfare.[18]

Utilitarian political theories argue, on the contrary, that a policy or decision is right not independently of its effect on social welfare but precisely because of it. The utilitarian liberal may argue, indeed, that rights themselves are justified only because they maximize overall welfare when consistently enforced over the long run. A utilitarian may concede, then, that the rights secured by a theory of justice "trump" the claims of social welfare in specific cases. Nevertheless, the utilitarian will argue that, at a higher level of analysis, the principles of justice are themselves to be justified in relation to their consequences for social welfare.[19]

Utilitarian liberalism differs from deontological liberalism, then, primarily because it takes the right to be subservient to the good. By this I mean that utilitarians consider an action or a decision to have the moral quality of rightness to the extent that it leads to (or is derived from rules that lead to) the maximization of good consequences, conceived in terms of social welfare or utility, over the long run.

Deontological liberals may agree with a conception of the good that ties it to social welfare, wealth maximization, or utility, insofar as such a conception remains arguably neutral among the values that preferences, desires, or satisfactions express. The deontological liberal insists, however, that the rightness, fairness, or justice of decisions cannot be analyzed at any level in terms of the satisfaction of preferences or the maximization of utility. In that sense the deontological liberal takes the right to be prior to the good.

[. . .]

Earlier this century, utilitarian liberals joined conservationist movements to advocate the prudent use and wise exploitation of natural resources. As one commentator observes, conservationist movements

> were mostly concerned with making sure that natural resources and environments were used in a fashion that reflected their true worth to man. This resulted in a utilitarian conception of environments and in the adoption of means to partially preserve them – for example, cost-benefit analysis and policies of multiple use on federal lands.[20]

The environmental, or "ecology," movement that arose in the 1960s and 1970s differs from conservationism in defending a nonutilitarian conception of man's relationship to nature. Environmentalists often refer to a dictum of Aldo Leopold's to describe this relationship. "A thing is right when it tends to preserve the integrity, stability, and beauty of the biotic community. It is wrong otherwise."[21] Speaking of actions insofar as they affect the environment, commentators add that "the good of the biotic *community* is the ultimate measure of the moral value, the rightness or wrongness, of actions,"[22] and that "the effect on ecological systems is the decisive factor in the determination of the ethical quality of actions."[23]

If environmentalists take a moral position about environmental policy, as they seem to do, and if, therefore, they would not regard preference satisfaction or welfare as the desideratum of social choice, can they be liberals? To answer this question, we shall next consider deontological liberalism and its relation to environmentalism.

[. . .]

[Liberalism and environmentalism]

Deontological, or "Kantian," liberalism may best be understood as a reaction to liberal political theories associated with utilitarianism.[24] Deontological liberals typically argue that utilitarianism fails to respect the boundaries between individuals and the fact of their separate existences; they claim that utilitarianism replaces persons with their pleasures or preferences, all of which it then combines, in a fungible way, into a single social aggregate.[25]

Utilitarians treat persons with equal respect and concern, so this criticism goes, by treating them with no respect or concern but only as locations where pleasures may be produced and preferences may be found.[26]

The deontological approach, on the contrary, recognizes that justice, equality, and autonomy are the irreducible conditions under which freedom is possible, and persons may be said to choose and not merely to channel their preferences and desires.[27] A utilitarian state, its critics further contend, fails to treat its citizens as ends in themselves but regards them merely as means to be dedicated to the maximization of social welfare or utility. And thus a utilitarian government will sacrifice the interests of some individuals unfairly in order to confer greater benefits on others or on society as a whole.[28]

Utilitarian liberals are by now familiar with this criticism, and many respond that intuitions about justice and equality are, indeed, important; therefore, a trade-off or balance must be struck between equity and efficiency.[29] As deontological liberals are quick to argue, however, goals like "allocatory efficiency," "preference satisfaction," and "wealth maximization" are not to be considered as independent ideals, to be weighed or balanced against other ideals, namely, distributional justice and equality, which unfortunately conflict with them.[30] Rather, an efficient allocation of resources, insofar as it differs from an equitable one, has no value to begin with and therefore has no moral claim against which to balance the claims of equity.[31]

This argument goes back to Kant, who considered wants, desires, and preferences to be mere "inclinations," which may be arbitrary or contingent from a moral point of view, and thus the satisfaction of which per se has no value or moral significance.[32] As John Rawls puts this point: "The satisfaction of these feelings has no value that can be put in the scales against the claims of equal liberty."[33]

Many environmentalists, agreeing with this critique of utilitarianism, have tried to make common cause with deontological liberalism. They have attempted to do this in two ways. First, environmentalists have appealed to the rights of future generations as reasons to protect wilderness and other natural areas.[34] This appeal fails, however, because it amounts to no more than the conservationist principle that we should exploit environmental resources wisely to maximize the long-run benefits nature offers humankind. Utilitarianism itself may treat present and future interests, pleasures, and preferences on an equitable basis, moreover, by insisting that cost-benefit analyses employ a social discount rate that balances the welfare of future individuals fairly with our own.[35]

Second, some environmentalists, seeking deontological arguments for preserving the natural environment, have appealed to the rights and interests of animals and other natural things. Indeed, a few scholars have explored the possibility that natural objects, like animals and trees, might have rights of the sort that give them legal standing or, failing that, interests that might be entered into the cost-benefit analyses on which social regulations may be based.[36]

These suggestions proved futile, however, in part because only individuals, that is, particular plants or animals, could possess rights or interests, but it is collections, such as species, communities, and ecosystems, that environmentalists are concerned to protect. As Joel Feinberg observes, species cannot be a proper object of moral concern in the context of a theory of rights, fairness, or justice. "A whole collection, as such, cannot have beliefs, expectations, wants or desires. . . . Individual elephants can have interests, but the species elephant cannot."[37]

To protect a few members of one species, it may be necessary to seal the fate of many more of another, for example, the millions of krill eaten by a single whale. To preserve the healthy functioning and integrity of an ecosystem, it might be necessary again to let many individual creatures perish – deer, for example – that might easily be saved from starvation by human intervention or might even prosper in a managed environment.

Although the animal rights movement correctly emphasizes the important truth that man ought not to be cruel to animals and thus has insisted, quite properly, on humane conditions for pets and livestock, it is unclear how the rights or interests of animals and other natural objects can be systematically connected with the goals and values of environmentalism. Accordingly, the rights and interests of animals, although important in the domestic context, will not allow environmentalism to hitch its wagon to the star of either deontological or utilitarian liberalism.[38]

From the point of view of the environmentalist, indeed, there may be little to choose between utilitarian and deontological liberalism, for all the controversy between them. The controversy comes down to this: The utilitarian allows certain trade-offs the deontological liberal refuses to permit.

[. . .]

Can environmentalists be liberals? . . . [we] may conclude that liberalism defines every policy question as one of maximizing utility or enforcing rights. If liberalism makes these assumptions – which, perhaps, it need not – then it is plainly incompatible with environmentalism.

[. . .]

Liberalism and public policy

Liberalism, as I understand it, relies on two distinctions, the first of which . . . divides the state from civil society. We need not interpret this distinction as drawing a sharp division, however, between rights and preferences or between the rules that govern competition and the interests that motivate it.

[. . .]

. . . Environmental decisions, by and large, have to do with what goes on out of doors not indoors; they concern the character and quality of the public household not of the private home. Environmental policies, in general, restrict what corporations and municipalities may do with their investments and effluents – not what individuals may do with their lovers, co-worshipers, or friends. Thus the content of environmental policy rarely becomes relevant to the kind of neutrality essential to liberalism.

Accordingly, the distinction between civil society and the state, at least as I interpret it, need not prevent environmentalists from being liberals. This distinction, moreover, need not prevent liberals from endorsing even those environmental policies that are based on particular ethical, cultural, or aesthetic convictions. These convictions must not infringe on the right of every citizen to make his or her own intimate decisions, for example, with respect to choices of friends, religion, and sexual relationships. I cannot think of any environmental statute that restricts these personal choices and beliefs.

The second distinction on which an understanding of liberalism depends divides between the basic structure of institutions and the social policies that emerge from those institutions. Liberal political theory concerns only the former, that is, the basic structure of social arrangements. At this level, liberals insist on structures that are fair among the individuals who participate in them. These arrangements must be neutral among conceptions of the good and treat individuals as equals independently of their race, sex, color, preferences, principles, or beliefs. Thus, liberal theory as a Comprehensive View applies at the level of social structure, not at the level of social policy.

This is not to say that liberals, as liberals, have no view of the good society and no particular conception of what social policy should be. What I suggest is simply that liberal social policy cannot be inferred from liberal political theory. Instead, liberals endorse, for a variety of reasons, social policies that provide a lively, diverse, and hospitable environment in which people can develop their own values and exercise their talent and imagination. No theory, of course, tells liberals what kind of environment this is. Liberals depend, at the level of policy . . . on aesthetic judgment, moral intuition, human compassion, honesty, intelligence, and common sense.

[. . .]

Notes

1 Brian Barry, *Political Argument* (London: Routledge & Kegan Paul, 1965), p. 66.
2 Ibid., p. 74.
3 Allen Kneese and Blair Bower, *Environmental Quality and Residuals Management* (Baltimore: Johns Hopkins University Press, 1979), pp. 4–5.
4 Barry, *Political Argument*, p. 71.
5 Aldo Leopold, "The Land Ethic," in *A Sand County Almanac* (New York: Oxford University Press, 1966), p. 222.
6 Gifford Pinchot, *The Fight for Conservation* (Seattle: University of Washington Press, 1910), p. 42.
7 Ibid., p. 43.
8 Aristotle, *Nichomachean Ethics* 1115a–1157b.
9 Saint Augustine ascribes this view to Cicero. See Saint Augustine, *The City of God*, trans. Marcus Dods (New York: Random House, Modern Library, 1950), pp. 61–62.
10 Ibid., p. 706.
11 Andred Dobelstein, *Politics, Economics, and Public Welfare* (Englewood Cliffs, N.J.: Prentice-Hall, 1980), p. 109.
12 John Rawls summarizes: "Systems of ends are not ranked in value." *A Theory of Justice* (Cambridge, Mass.: Harvard University Press, 1971), p. 19.
13 For a statement of these ideals, see, for example, John Muir, *The Wilderness World of John Muir* (Boston: Houghton Mifflin, 1976). Muir writes (p. 317), "Why should man value himself as more than a small part of the one great unit of creation? And what creature of all the Lord has taken the pains to make is not essential to the completeness of that unit – the cosmos? The universe would be incomplete without the smallest transmicroscopic creature that dwells beyond our conceitful eyes and knowledge."

For further development of similar themes, see Leopold, *A Sand County Almanac*; Marjorie Hope Nicolson, *Mountain Gloom and Mountain Glory: The Development of the Aesthetic of the Infinite* (New York: Norton, 1963); John Passmore, *Man's Responsibility for Nature: Ecological Problems and Western Traditions* (New York: Scribner, 1974); and Joseph Sax, *Mountains without Handrails: Reflections on the National Parks* (Ann Arbor: University of Michigan Press, 1980).

14 See, for example Patrick Devlin, *The Enforcement of Morals* (New York: Oxford University Press, 1965). Devlin writes on pp. 13–14, "Society is justified in taking the same steps to preserve its moral code as it does to preserve its government and other essential institutions. The suppression of vice is as much the law's business as the suppression of subversive activities; it is no more possible to define a sphere of private morality than it is to define a sphere of private

subversive activity. It is wrong to talk of private morality or of the law not being concerned with immorality as such or to try to set rigid bounds to the part which the law may play in the suppression of vice."

For a subtle defense of Devlin's general position, see Roger Scruton, *The Meaning of Conservatism* (Totowa, N.J.: Barnes & Noble, 1980), esp. pp. 71–93. For a liberal reply, see Ronald Dworkin, *Taking Rights Seriously* (Cambridge, Mass.: Harvard University Press, 1977), chap. 10, and H. L. A. Hart, *Law, Liberty, and Morality* (New York: Random House, Vintage Books, 1966).

15 Ronald Dworkin characterizes conservatism and various forms of socialism or Marxism as adopting the thesis "that the treatment government owes its citizens is at least partly determined by some conception of the good life." Marxism and conservatism differ, of course, in the conception of the good life they endorse. Dworkin, "Liberalism," in Stuart Hampshire, ed., *Public and Private Morality* (Cambridge: Cambridge University Press, 1978), pp. 113–43; quotation on p. 128.

16 The distinction between civil society and the state is defined by Hegel in T. M. Knox, ed., *Hegel's Philosophy of Right* (New York: Oxford University Press, 1952), esp. sec. 258, p. 156. For discussion of the distinction, see Shlomo Avineri, *Hegel's Theory of the Modern State* (Cambridge: Cambridge University Press, 1972), pp. 141–54.

17 John Rawls, *A Theory of Justice* (Cambridge, Mass.: Harvard University Press, 1971), p. 450.

18 Dworkin, "Liberalism," p. 136.

19 For discussion, see Dworkin, *Taking Rights Seriously*, pp. 94–100.

20 Martin Krieger, "What's Wrong with Plastic Trees?" *Science* 179 (1973): 446–80; quotation on p. 446.

21 Aldo Leopold, "The Land Ethic," p. 240.

22 J. Baird Callicott, "Animal Liberation: A Triangular Affair," *Environmental Ethics* 2 (1980): 320.

23 Ibid. See also Don Marietta, Jr., "The Interrelationship of Ecological Science and Environmental Ethics," *Environmental Ethics* 1 (1979): 195–207. "The basic concept behind an ecological ethic is that morally acceptable treatment of the environment is that which does not upset the integrity of the ecosystem as it is seen in a diversity of life forms existing in a dynamic and complex but stable interdependency" (p. 197).

24 For a critical study of deontological liberalism and its relation to Kantian moral theory, see Michael J. Sandel, *Liberalism and the Limits of Justice* (Cambridge: Cambridge University Press, 1982), esp. pp. 1–14.

25 See Amartya Sen and Bernard Williams, "Introduction," in *Utilitarianism and Beyond* (Cambridge: Cambridge University Press, 1982), p. 4: "Essentially, utilitarianism sees persons as locations of their respective utilities – as the sites at which such activities as desiring and having pleasure and pain take place. . . . Utilitarianism is the combination . . . of welfarism, sum ranking and consequentialism, and each of these components contribute to this narrow view of a person."

26 For discussion, see H. L. A. Hart, "Between Utility and Rights," in A. Ryan, ed., *The Idea of Freedom* (New York: Oxford University Press, 1979), pp. 77–98.

27 Fried, *Right and Wrong*, (Cambridge, Mass.: Harvard university Press, 1978), pp. 7–17.

28 Nozick, *Anarchy, State, and Utopia*, (New York: Basic Books, 1974), pp. 71–84.

29 Arthur Okun, *Equality and Efficiency: The Big Tradeoff* (Washington, D.C.: Brookings Institution, 1975).

30 Ronald Dworkin, "Why Efficiency?," *Hofstra Law Review* 8 (1980): 563–90.

31 Rawls, *A Theory of Justice*, p. 31.

32 Immanuel Kant, *Critique of Practical Reason*, trans. L. W. Beck (Indianapolis, Ind.: Bobbs-Merrill, 1956), esp. pp. 18–20.

33 Rawls, *A Theory of Justice*, p. 450.

34 See, for example, Bryan Norton, "Environmental Ethics and the Rights of Future Generations," *Environmental Ethics* 4 (1982): 319–37. For good anthologies collecting relevant essays, see Douglas MacLean and Peter Brown, eds., *Energy and the Future* (Totowa, N.J.: Rowman & Littlefield, 1983); Ernest Partridge, ed., *Responsibilities to Future Generations* (Buffalo, N.Y.: Prometheus Press, 1980); and Richard Sikora and Brian Barry, eds., *Obligations to Future Generations* (Philadelphia: Temple University Press, 1978). For an excellent review of the issues, see Annette Baier, "For the Sake of Future Generations," in Tom Regan, ed., *Earthbound: New Introductory Essays in Environmental Ethics* (New York: Random House, 1984).

35 For discussion of the social discount rate, see Talbot Page, "Intergenerational Justice as Opportunity," in MacLean and Brown, *Energy and the Future*, pp. 38–58, and sources cited therein.
36 See, for example, Christopher Stone, *Should Trees Have Standing? Toward Legal Rights for Natural Objects* (Los Altos, Calif.: Kaufmann, 1974), and Laurence Tribe, "Ways Not to Think about Plastic Trees: New Foundations for Environmental Law," *Yale Law Journal* 83 (1974): 1315–48. I have commented on this literature . . . in Mark Sagoff, "Animal Liberation and Environmental Ethics: Bad Marriage, Quick Divorce," *Osgoode Hall Law Journal* 22 (1984): 297–307.
37 Joel Feinberg, "The Rights of Animals and Unborn Generations," in William T. Blackstone, ed., *Philosophy and Environmental Crisis* (Athens: University of Georgia Press, 1974), pp. 55–56.
38 In reaching this conclusion, I have drawn upon a large literature, including Callicott, "Animal Liberation: A Triangular Affair"; Bryan Norton, "Environmental Ethics and Nonhuman Rights," *Environmental Ethics* 4 (1982): 17–36; Eric Katz, "Is There a Place for Animals in the Moral Consideration of Nature?," *Ethics and Animals* 4 (1983): 74–85. For arguments for an opposing conclusion, see Tom Regan, *The Case for Animal Rights* (Berkeley: University of California Press, 1983), pp. 361–63; and Edward Johnson, "Treating the Dirt: Environmental Ethics and Moral Theory," in Regan, ed., *Earthbound*, pp. 336–65, esp. pp. 351–54.

Source: Mark Sagoff (1988) *The Economy of the Earth: Philosophy, Law and the Environment*, Cambridge: Cambridge University Press, pp. 146–57, 162, 165–7, 246–50.

5.2

Garrett Hardin

The tragedy of the commons

At the end of a thoughtful article on the future of nuclear war, Wiesner and York[1] concluded that: "Both sides in the arms race are . . . confronted by the dilemma of steadily increasing military power and steadily decreasing national security. *It is our considered professional judgment that this dilemma has no technical solution.* If the great powers continue to look for solutions in the area of science and technology only, the result will be to worsen the situation."

I would like to focus your attention not on the subject of the article (national security in a nuclear world) but on the kind of conclusion they reached, namely that there is no technical solution to the problem. An implicit and almost universal assumption of discussions published in professional and semipopular scientific journals is that the problem under discussion has a technical solution. A technical solution may be defined as one that requires a change only in the techniques of the natural sciences, demanding little or nothing in the way of change in human values or ideas of morality.

[. . .]

. . . Recall the game of tick-tack-toe. Consider the problem, "How can I win the game of tick-tack-toe?" It is well known that I cannot, if I assume (in keeping with the conventions of game theory) that my opponent understands the game perfectly. Put another way, there is no "technical solution" to the problem. I can win only by giving a radical meaning to the word "win." I can hit my opponent over the head; or I can drug him; or I can falsify the records. Every way in which I "win" involves, in some sense, an abandonment of the game . . .

The class of "No technical solution problems" has members. My thesis is that the "population problem," as conventionally conceived, is a member of this class. How it is conventionally conceived needs some comment. It is fair to say that most people who anguish over the population problem are trying to find a way to avoid the evils of over-population without relinquishing any of the privileges they now enjoy. They think that farming the seas or developing new strains of wheat will solve the problem – technologically. I try to show here that the solution they seek cannot be found. The population problem cannot be solved in a technical way, any more than can the problem of winning the game of tick-tack-toe.

What shall we maximize?

Population, as Malthus said, naturally tends to grow "geometrically," or, as we would now say, exponentially. In a finite world this means that the per capita share of the world's goods must steadily decrease. Is ours a finite world?

A fair defense can be put forward for the view that the world is infinite; or that we do not know that it is not. But, in terms of the practical problems that we must face in the next few generations with the foreseeable technology, it is clear that we will greatly increase human misery if we do not, during the immediate future, assume that the world available to the terrestrial human population is finite. "Space" is no escape.[2]

A finite world can support only a finite population; therefore, population growth must eventually equal zero. . . . When this condition is met, what will be the situation of mankind? Specifically, can Bentham's goal of "the greatest good for the greatest number" be realized?

No – for two reasons, each sufficient by itself. The first is a theoretical one. It is not mathematically possible to maximize for two (or more) variables at the same time. This was clearly stated by von Neumann and Morgenstern[3]. . .

The second reason springs directly from biological facts. To live, any organism must have a source of energy (for example, food). This energy is utilized for two purposes: mere maintenance and work. For man, maintenance of life requires about 1600 kilocalories a day ("maintenance calories"). Anything that he does over and above merely staying alive will be defined as work, and is supported by "work calories" which he takes in. Work calories are used not only for what we call work in common speech; they are also required for all forms of enjoyment, from swimming and automobile racing to playing music and writing poetry. If our goal is to maximize population it is obvious what we must do: We must make the work calories per person approach as close to zero as possible. No gourmet meals, no vacations, no sports, no music, no literature, no art. . . . I think that everyone will grant, without argument or proof, that maximizing population does not maximize goods. Bentham's goal is impossible.

[. . .]

We want the maximum good per person; but what is good? To one person it is wilderness, to another it is ski lodges for thousands. To one it is estuaries to nourish ducks for hunters to shoot; to another it is factory land. Comparing one good with another is, we usually say, impossible because goods are incommensurable. Incommensurables cannot be compared.

Theoretically this may be true; but in real life incommensurables *are* commensurable. Only a criterion of judgment and a system of weighting are needed.

[. . .]

We can make little progress in working toward optimum poulation size until we explicitly exorcize the spirit of Adam Smith in the field of practical demography. In economic affairs, *The Wealth of Nations* (1776) popularized the "invisible hand," the idea that an individual who "intends only his own gain," is, as it were, "led by an invisible hand to promote . . . the public interest."[4] Adam Smith did not assert that this was invariably true, and perhaps neither did any of his followers. But he contributed to a dominant tendency of thought that has ever since interfered with positive action based on rational analysis, namely, the tendency to assume that decisions reached individually will, in fact, be the best decisions for an entire society. If this assumption is correct, it justifies the continuance of our present policy of *laissez-faire* in reproduction. If it is correct we can assume that men [*sic*] will control their individual fecundity so as to produce the optimum population. If the assumption is not correct, we need to reexamine our individual freedoms to see which ones are defensible.

Tragedy of freedom in a commons

The rebuttal to the invisible hand in population control is to be found in a scenario first sketched in a little-known pamphlet[5] in 1833 by a mathematical amateur named William Forster Lloyd (1794–1852). We may well call it "the tragedy of the commons," using the word "tragedy" as the philosopher Whitehead used it: "The essence of dramatic tragedy is not unhappiness. It resides in the solemnity of the remorseless working of things."[6] He then goes on to say, "This inevitableness of destiny can only be illustrated in terms of human life by incidents which in fact involve unhappiness. For it is only by them that the futility of escape can be made evident in the drama."

The tragedy of the commons develops in this way. Picture a pasture open to all. It is to be expected that each herdsman will try to keep as many cattle as possible on the commons. Such an arrangement may work reasonably satisfactorily for centuries because tribal wars, poaching, and disease keep the numbers of both man and beast well below the carrying capacity of the land. Finally, however, comes the day of reckoning, that is, the day when the long-desired goal of social stability becomes a reality. At this point, the inherent logic of the commons remorselessly generates tragedy.

As a rational being, each herdsman seeks to maximize his gain. Explicitly or implicitly, more or less consciously, he asks, "What is the utility *to me* of adding one more animal to my herd?" This utility has one negative and one positive component.

1 The positive component is a function of the increment of one animal. Since the herdsman receives all the proceeds from the sale of the additional animal, the positive utility is nearly +1.
2 The negative component is a function of the additional overgrazing created by one more animal. Since, however, the effects of overgrazing are shared by all the herdsmen, the negative utility for any particular decision-making herdsman is only a fraction of −1.

Adding together the component partial utilities, the rational herdsman concludes that the only sensible course for him to pursue is to add another animal to his herd. And another; and another.... But this is the conclusion reached by each and every rational herdsman sharing a commons. Therein is the tragedy. Each man is locked into a system that compels him to increase his herd without limit – in a world that is limited. Ruin is the destination toward which all men rush, each pursuing his own best interest in a society that believes in the freedom of the commons. Freedom in a commons brings ruin to all.

[. . .]

In an approximate way, the logic of the commons has been understood for a long time, perhaps since the discovery of agriculture or the invention of private property in real estate. But it is understood mostly only in special cases which are not sufficiently generalized. Even at this late date, cattlemen leasing national land on the western ranges demonstrate no more than an ambivalent understanding, in constantly pressuring federal authorities to increase the head count to the point where overgrazing produces erosion and weed-dominance. Likewise, the oceans of the world continue to suffer from the survival of the philosophy of the commons. Maritime nations still respond automatically to the shibboleth of the "freedom of the seas." Professing to believe in the "inexhaustible resources of the oceans," they bring species after species of fish and whales closer to extinction.[7]

The National Parks present another instance of the working out of the tragedy of the commons. At present, they are open to all, without limit. The parks themselves are limited in extent – there is only one Yosemite Valley – whereas population seems to grow without limit. The values that visitors seek in the parks are steadily eroded. Plainly, we must soon cease to treat the parks as commons or they will be of no value to anyone.

What shall we do? We have several options. We might sell them off as private property. We might keep them as public property, but allocate the right to enter them. The allocation might be on the basis of wealth, by the use of an auction system. It might be on the basis of merit, as defined by some agreed-upon standards. It might be by lottery. Or it might be on a first-come, first-served basis, administered to long queues. These, I think, are all the reasonable possibilities. They are all objectionable. But we must choose – or acquiesce in the destruction of the commons that we call our National Parks.

Pollution

In a reverse way, the tragedy of the commons reappears in problems of pollution. Here it is not a question of taking something out of the commons, but of putting something in – sewage, or chemical, radioactive, and heat wastes into water; noxious and dangerous fumes into the air; and distracting and unpleasant advertising signs into the line of sight. The calculations of utility are much the same as before. The rational man finds that his share of the cost of the wastes he discharges into the commons is less than the cost of purifying his wastes before releasing them. Since this is true for everyone, we are locked into a system of "fouling our own nest," so long as we behave only as independent, rational, free-enterprisers.

The tragedy of the commons as a food basket is averted by private property, or something formally like it. But the air and waters surrounding us cannot readily be fenced, and so the tragedy of the commons as a cesspool must be prevented by different means, by coercive laws or taxing devices that make it cheaper for the polluter to treat his pollutants than to discharge them untreated. We have not progressed as far with the solution of this problem as we have with the first. Indeed, our particular concept of private property, which deters us from exhausting the positive resources of the earth, favors pollution. The owner of a factory on the bank of a stream – whose property extends to the middle of the stream – often has difficulty seeing why it is not his natural right to muddy the waters flowing past his door. The law, always behind the times, requires elaborate stitching and fitting to adapt it to this newly perceived aspect of the commons.

The pollution problem is a consequence of population. It did not much matter how a lonely American frontiersman disposed of his waste. "Flowing water purifies itself every 10 miles," my grandfather used to say, and the myth was near enough to the truth when he was a boy, for there were not too many people. But as population became denser, the natural chemical and biological recycling processes became overloaded, calling for a redefinition of property rights.

How to legislate temperance?

Analysis of the pollution problem as a function of population density uncovers a not generally recognized principle of morality, namely: *the morality of an act is a function of the state of the system at the time it is performed.*[8] Using the commons as a cesspool does not harm the general public under frontier conditions, because there is no public; the same behavior in a metropolis is unbearable. A hundred and fifty years ago a plainsman could kill an American bison, cut out only the tongue for his dinner, and discard the rest of the animal. He was not in any important sense being wasteful. Today, with only a few thousand bison left, we would be appalled at such behavior.

[. . .]

That morality is system-sensitive escaped the attention of most codifiers of ethics in the past. "Thou shalt not . . ." is the form of traditional ethical directives which make no allowance for particular circumstances. The laws of our society follow the pattern of ancient ethics, and therefore are poorly suited to governing a complex, crowded, changeable world. Our epicyclic solution is to augment statutory law with administrative law. Since it is practically impossible to spell out all the conditions under which it is safe to burn trash in the back yard or to run an automobile without smog-control, by law we delegate the details to bureaus. The result is administrative law, which is rightly feared for an ancient reason – *Quis custodiet ipsos custodes?* – "Who shall watch the watchers themselves?". . .

Prohibition is easy to legislate (though not necessarily to enforce); but how do we legislate temperance? Experience indicates that it can be accomplished best through the mediation of administrative law. . . . The great challenge facing us now is to invent the corrective feedbacks that are needed to keep custodians honest. We must find ways to legitimate the needed authority of both the custodians and the corrective feedbacks.

Freedom to breed is intolerable

The tragedy of the commons is involved in population problems in another way. In a world governed solely by the principle of "dog eat dog" – if indeed there ever was such a world – how many children a family had would not be a matter of public concern. Parents who bred too exuberantly would leave fewer descendants, not more, because they would be unable to care adequately for their children. David Lack and others have found that such a negative feedback demonstrably controls the fecundity of birds.[9] But men are not birds, and have not acted like them for millennia, at least.

If each human family were dependent only on its own resources; *if* the children of improvident parents starved to death; *if*, thus, overbreeding brought its own "punishment" to the germ line – *then* there would be no public interest in controlling the breeding of families. But our society is deeply committed to the welfare state,[10] and hence is confronted with another aspect of the tragedy of the commons.

In a welfare state, how shall we deal with the family, the religion, the race, or the class (or indeed any distinguishable and cohesive group) that adopts overbreeding as a policy to secure its own aggrandizement?[11] To couple the concept of freedom to breed with the belief that everyone born has an equal right to the commons is to lock the world into a tragic course of action.

Unfortunately this is just the course of action that is being pursued by the United Nations. In late 1967, some 30 nations agreed to the following:

> The Universal Declaration of Human Rights describes the family as the natural and fundamental unit of society. It follows that any choice and decision with regard to the size of the family must irrevocably rest with the family itself, and cannot be made by anyone else.[12]

It is painful to have to deny categorically the validity of this right; denying it, one feels as uncomfortable as a resident of Salem, Massachusetts, who denied the reality of witches in the seventeenth century. At the present time, in liberal quarters, something like a taboo acts to inhibit criticism of the United Nations. There is a feeling that the United Nations is "our last and best hope," that we shouldn't find fault with it; we shouldn't play into the hands of the archconservatives. However, let us not forget what Robert Louis Stevenson said: "The truth that is suppressed by friends is the readiest weapon of the enemy." If we love the truth we must openly deny the validity of the Universal Declaration of Human Rights, even though it is promoted by the United Nations. . . .

Conscience is self-eliminating

It is a mistake to think that we can control the breeding of mankind in the long run by an appeal to conscience. Charles Galton Darwin made this point when he spoke on the centennial of the publication of his grandfather's great book. The argument is straightforward and Darwinian.

People vary. Confronted with appeals to limit breeding, some people will undoubtedly respond to the plea more than others. Those who have more children will produce a larger fraction of the next generation than those with more susceptible consciences. The difference will be accentuated, generation by generation.

In C. G. Darwin's words: "It may well be that it would take hundreds of generations for the progenitive instinct to develop in this way, but if it should do so, nature would have taken her revenge, and the variety *Homo contracipiens* would become extinct and would be replaced by the variety *Homo progenitivus*."[13]

[. . .]

Pathogenic effects of conscience

The long-term disadvantage of an appeal to conscience should be enough to condemn it; but has serious short-term disadvantages as well. If we ask a man who is exploiting a commons to desist "in the name of conscience," what are we saying to him? What does he hear? . . . Sooner or later, consciously or subconsciously, he senses that he has received two communications, and that they are contradictory: (i) (intended communication) "If you don't do as we ask, we will openly condemn you for not acting like a responsible citizen"; (ii) (the unintended communication) "If you *do* behave as we ask, we will secretly condemn you for a simpleton who can be shamed into standing aside while the rest of us exploit the commons."

[. . .]

To conjure up a conscience in others is tempting to anyone who wishes to extend his control beyond the legal limits. Leaders at the highest level succumb to this temptation. Has any President during the past generation failed to call on labor unions to moderate voluntarily their demands for higher wages, or to steel companies to honor voluntary guidelines on prices? I can recall none. The rhetoric used on such occasions is designed to produce feelings of guilt in noncooperators.

[. . .]

. . .We hear much talk these days of responsible parenthood; the coupled words are incorporated into the titles of some organizations devoted to birth control. Some people have proposed massive propaganda campaigns to instill responsibility into the nation's (or the world's) breeders. But what is the meaning of the word responsibility in this context? Is it not merely a synonym for the word conscience? When we use the word responsibility in the absence of substantial sanctions, are we not trying to browbeat a free man in a commons into acting against his own interest? Responsibility is a verbal counterfeit for a substantial *quid pro quo*. It is an attempt to get something for nothing.

[. . .]

Mutual coercion mutually agreed upon

The social arrangements that produce responsibility are arrangements that create coercion, of some sort. Consider bank-robbing. The man who takes money from a bank acts as if the bank were a commons. How do we prevent such action? Certainly not by trying to control his behavior solely by a verbal appeal to his sense of responsibility. Rather than rely on propaganda we . . . insist that a bank is not a commons; we seek the definite social arrangements that will keep it from becoming a commons. That we thereby infringe on the freedom of would-be robbers we neither deny nor regret.

The morality of bank-robbing is particularly easy to understand because we accept complete prohibition of this activity. We are willing to say "Thou shalt not rob banks," without providing for exceptions. But temperance also can be created by coercion. Taxing is a good coercive device. To keep downtown shoppers temperate in their use of parking space, we introduce parking meters for short periods, and traffic fines for longer ones. We need not actually forbid a citizen to park as long as he wants to; we need merely make it increasingly expensive for him to do so. Not prohibition, but carefully biased options are what we offer him. A Madison Avenue man might call this persuasion; I prefer the greater candor of the word coercion.

Coercion is a dirty word to most liberals now, but it need not forever be so. As with the four-letter words, its dirtiness can be cleansed away by exposure to the light, by saying it over and over without apology or embarrassment. To many, the word coercion implies arbitrary decisions of distant and irresponsible bureaucrats; but this is not a necessary part of its meaning. The only kind of coercion I recommend is mutual coercion, mutually agreed upon by the majority of the people affected.

To say that we mutually agree to coercion is not to say that we are required to enjoy it, or even to pretend we enjoy it. Who enjoys taxes? We all grumble about them. But we accept compulsory taxes because we recognize that voluntary taxes would favor the conscienceless. We institute and (grumblingly) support taxes and other coercive devices to escape the horror of the commons.

An alternative to the commons need not be perfectly just to be preferable. With real estate and other material goods, the alternative we have chosen is the institution of private property coupled with legal inheritance. Is this system perfectly just? As a genetically trained biologist I deny that it is. It seems to me that, if there are to be differences in individual inheritance, legal possession should be perfectly correlated with biological inheritance – that those who are biologically more fit to be the custodians of property and power should legally inherit more. But genetic recombination continually makes a mockery of the doctrine of "like father, like son" implicit in our laws of legal inheritance. An idiot can inherit millions, and a trust fund can keep his estate intact. We must admit that our legal system of private property plus inheritance is unjust – but we put up with it because we are not convinced, at the moment, that anyone has invented a better system. The alternative of the commons is too horrifying to contemplate. Injustice is preferable to total ruin.

[. . .]

Recognition of necessity

Perhaps the simplest summary of this analysis of man's population problems is this: the commons, if justifiable at all, is justifiable only under conditions of low-population density. As the human population has increased, the commons has had to be abandoned in one aspect after another.

First we abandoned the commons in food gathering, enclosing farm land and restricting pastures and hunting and fishing areas. These restrictions are still not complete throughout the world.

Somewhat later we saw that the commons as a place for waste disposal would also have to be abandoned. Restrictions on the disposal of domestic sewage are widely accepted in the Western world; we are still struggling to close the commons to pollution by automobiles, factories, insecticide sprayers, fertilizing operations, and atomic energy installations.

[. . .]

Every new enclosure of the commons involves the infringement of somebody's personal liberty. Infringements made in the distant past are accepted because no contemporary complains of a loss. It is the newly proposed infringements that we vigorously oppose; cries of "rights" and "freedom" fill the air. But what does "freedom" mean? When men mutually agreed to pass laws against robbing, mankind became more free, not less so. Individuals locked into the logic of the commons are free only to bring on universal ruin; once they see the necessity of mutual coercion, they become free to pursue other goals. I believe it was Hegel who said, "Freedom is the recognition of necessity."

The most important aspect of necessity that we must now recognize, is the necessity of abandoning the commons in breeding. No technical solution can rescue us from the misery of overpopulation. Freedom to breed will bring ruin to all. At the moment, to avoid hard decisions many of us are tempted to propagandize for conscience and responsible parenthood. The temptation must be resisted, because an appeal to independently acting consciences selects for the disappearance of all conscience in the long run, and an increase in anxiety in the short.

The only way we can preserve and nurture other and more precious freedoms is by relinquishing the freedom to breed, and that very soon. "Freedom is the recognition of necessity" – and it is the role of education to reveal to all the necessity of abandoning the freedom to breed. Only so, can we put an end to this aspect of the tragedy of the commons.

References

1 J. B. Wiesner and H. F. York, *Sci. Amer.* 211 (No. 4), 27 (1964).
2 G. Hardin, *J. Hered.* 50, 68 (1959); S. von Hoernor, *Science* 137, 18 (1962).
3 J. von Neumann and O. Morgenstern, *Theory of Games and Economic Behavior* (Princeton Univ. Press, Princeton, N.J., 1947), p. 11.
4 A. Smith, *The Wealth of Nations* (Modern Library, New York, 1937), p. 423.
5 W. F. Lloyd, *Two Lectures on the Checks to Population* (Oxford Univ. Press, Oxford, England, 1833), reprinted (in part) in *Population, Evolution, and Birth Control*, G. Hardin, Ed. (Freeman, San Francisco, 1964), p. 37.
6 A. N. Whitehead, *Science and the Modern World* (Mentor, New York, 1948), p. 17.
7 S. McVay, *Sci. Amer.* 216 (No. 8), 13 (1966).
8 J. Fletcher, *Situation Ethics* (Westminster, Philadelphia, 1966).
9 D. Lack, *The Natural Regulation of Animal Numbers* (Clarendon Press, Oxford, 1954).
10 H. Girvetz, *From Wealth to Welfare* (Stanford Univ. Press, Stanford, Calif., 1950).
11 G. Hardin, *Perspec. Biol. Med.* 6 366 (1963).
12 U. Thant, *Int. Planned Parenthood News*, No. 168 (February 1968), p. 3.
13 S. Tax, ed., *Evolution after Darwin* (Univ. of Chicago Press, Chicago, 1960), vol. 2, p. 469.

Source: Garrett Hardin (1968) 'The tragedy of the commons', *Science*, 162, pp. 1243–8.

5.3

Robert Young

Population policies, coercion and morality

Despite the fact that we in the affluent countries have been shielded (by our geographical isolation from the developing world and by our very affluence) from obtaining a full appreciation of the structural as against the merely visible ingredients in the threat to our delicate relationship with our natural environment, there has been an increasing awareness of the ways in which that relationship has been put at risk. We have learned that excessive population growth, ever-growing consumerism, and the resort to faulty technologies which make difficult the absorption of wastes and over-utilize scarce natural resources, are causally interlocked and have not just an additive but a multiplicative impact.[1]

[. . .]

. . . One response to the threat has been that we should urgently try to influence population levels. I shall argue that even though compulsory methods of population control can be shown to be morally justifiable, an alternative strategy is available for curbing excessive population growth. Such an alternative should be preferred, I argue, because it more fairly locates the responsibility for corrective action with those of us in the affluent world rather than with those in the third world.

I

The world's population has been growing *exponentially*, or, if you like, *at compound interest*. When growth is exponential each addition becomes a contributor of new additions. One quick way to grasp the significance of this point is in terms of 'doubling time'. Doubling time roughly equals 70 years divided by the annual percentage increase. While the world's human population continues to grow at an annual rate of about 2 per cent it will take 35 years to double itself. Energy resources have been much in people's minds in recent years. As it happens, consumption of energy is a useful index of resource consumption as a whole and of impact on the environment. Now energy consumption has been growing world-wide at about 5 per cent p.a. (about 2 per cent due to population growth, the rest to rising consumption per head). Thus, energy consumption will double in 14 years at the current rate of growth.

Exponential growth is carrying mankind *at an accelerating rate* toward the numerical limit which the resources of this planet can sustain. Hence, it is no accident that certain environmental problems have materialized so suddenly in the last few decades, for it is a characteristic of exponential growth that limits are approached with surprising suddenness. The likelihood of *over-shooting* the limit imposed by available resources is increased by delays between the operation of causes and the appearance of their effects in the ecosystem, and by the fact that some kinds of damage are irreversible by the time they are visible.

. . . We are, however, not in sight of the end of the exponential phase. There are two considerations which warrant mentioning here in this connection:

1 The skewed age distribution – today's population is heavily weighted with young people (whose reproductive years are ahead of them). Nearly two-fifths of the world's population at the present time is under 15 years of age;
2 Attitudes towards child-bearing have deep biological and cultural roots. Such considerations make it obvious that population levels are in the main resistant to *rapid* reduction short of war, famine or wholesale slaughter. The issue should, therefore, be couched in terms of the resistance of population levels to *change as such*. If the claim that we are moving at an accelerating rate toward the limit of our environment's capacity to sustain the world's population is correct, as the best available evidence suggests, what options are open to us to achieve an optimum population level?

I presume that the only humane means of achieving any change in population levels on a global scale is by reducing the birth rate. The alternative – permitting premature deaths – must be rejected out of hand. . . . There is very little agreement as to how far and how fast limitation should proceed, and, for our purposes more importantly still, very little agreement as to the means for achieving whatever goals are finally agreed on as being desirable.

At least three distinct aims have been mooted in proposals for the regulation of population levels:

1 A reduction in the *rate of growth* (although there would still be a positive rate);
2 Stabilization of the *absolute size* of the population by means of a zero rate of growth;
3 A reduction in the absolute size of the population by way of a negative 'rate of growth'.

[. . .]

I shall proceed . . . to the more clearly moral questions about determining what measures might justifiably be adopted to achieve any such goal as is agreed on. Four measures have been advocated which warrant consideration. The first three might be bracketed as supposedly relying on self-motivated human effort. To that extent they are mutually compatible. The four measures are:

1 Family planning policies (whatever one's view of abortion, realism demands that we envisage these policies as including a liberalization of abortion laws);
2 Employment of socio-economic 'incentives';
3 Reliance on the 'demographic transition';
4 Compulsory fertility control.

I shall consider consecutively the defensibility of each.

II

The most immediately noticeable point about family planning is that it operates on an individual rather than on a social plane. Precisely because it provides for the maximum freedom of choice for individuals, it involves no conscious consideration of the optimum population size for a society or for the planet. Were we living in a world where each and every prospective parent was gripped by what Blackstone[2] has called 'the ecological attitude', and all made use of fool-proof contraceptive methods (or resorted to abortion or infanticide), individual preferences might well yield the desired, or near enough to the desired, socially optimum level of population. But the ecological attitude is uncommon and the desire for a small family, or for none at all, varies widely as between cultures. Such motivation is affected by general educational levels, the degree of urbanization, the status of women, land tenure and inheritance arrangements, the costs of raising and educating each child (especially after taking account of contributions which children make to the family income and work load), the availability of welfare schemes which provide some degree of financial security and independence in old age and so on.[3]

Family planning does not appear to be a measure sufficient to bring together individual preferences and aspirations and socially required goals. But in addition to its evident insufficiency as a precise regulator of population levels, there is a second difficulty. The present population situation is too urgent to permit our reliance on an approach which cannot guarantee a reduction in the time lag between a worldwide perception by individuals of the need to halt or slow population growth and the actual accomplishment of a stable population. The shorter is the time remaining before we reach the optimum level for the world's population the less can we put our confidence in family planning as a means to stabilisation at that level.

Some of those who have taken note of the urgent need to bring individual motivations more into line with the ecological attitude have urged the use of socio-economic incentives and disincentives to encourage voluntary restraints on family sizes. According to advocates[4] of this approach a two-pronged policy is needed:

1 a large-scale educational programme (e.g. through the schools, media and so on) to persuade people of the personal and social advantages of small families; and
2 an 'incentive' scheme with an anti-natalist bias – it would be more accurate to term many of the suggestions disincentives. Thus tax measures which would favour single people, working wives and small families; high marriage fees; taxes on baby goods and toys; removal of family allowances; and limitation of free education to two children, would basically operate as disincentives. Some 'positive' incentives have occasionally been advocated – they consist chiefly in giving direct monetary benefits. The main difficulty in administering them would be that changes in people's circumstances might necessitate recovery of the gift and the greater the incentive the more potentially disruptive this possibility would be.

There are many obvious criticisms to which such a scheme is open. In many countries people are not sufficiently affluent to pay taxes, let alone to be buying baby goods and toys in the way that we do. While, to be fair, advocates of such schemes have intended them to operate in affluent societies like ours, the relevance of such schemes is dubious once it is realized that in societies like ours population growth has largely been curbed voluntarily. Where the incentives are 'negative', a drawback would be that the proposals would operate regressively because their greatest impact would be on the poor who are least able to cope with such 'incentives'. Those most penalized would therefore be the children of poor families. . . . And there remains the serious objection that if we take the first two measures (family planning and a back-up scheme of financial incentives) together, unless the incentives were massive the scheme would almost certainly be insufficient to solve the problem for much the same reasons as given in criticism of family planning schemes taken on their own, the chief one being the sheer urgency of the present position. . . .

What, then, of the third proposal? Namely, that we should rely on the 'demographic transition' to stabilize population levels at or near the optimum. 'The demographic transition' is the term applied to a process in which improved well-being, effective programmes for reduction of infant mortality rates and making provision for social security have been found in a number of places to motivate people voluntarily to reduce fertility and, subject to short-term fluctuations, achieve long-term population stability. The process has been observed in many societies where *industrialization* has taken hold.[5]

Unfortunately, though, the general model just sketched has been shown by demographers to have equally many exceptions and variants. Furthermore, it has been shown that fertility is alterable *without* industrialization provided that other features like a high degree of literacy, a shift from subsistence to wage labour, protection of children from being caught up in the labour market and so on, are present in a society. And, even more significantly, as Borrie points out, it is changes in *mortality* patterns which provide the most reliable index of the change from a pattern of large families to one of small families. The upshot is that it is far too much of an oversimplification of the demographic evidence to suggest that if only we could promote demographic transitions in the developing world we would be well on the way to achieving a suitable world population.

But any appeal to the demographic transition as a way of avoiding spiralling of population numbers faces a further objection. Suppose a series of demographic transitions were to start as of today covering all of the developing countries and following patterns similar to those experienced in the developed countries in the past. It would still be several generations before one could expect population growth rates in the developing countries to be reduced to the range now found in developed ones. . . . But, more important still, there are not the opportunities available to the developing countries to duplicate the economic growth processes which made possible the demographic transition in the rich countries. Much of the prosperity of the rich countries was achieved at the expense of exploitation of the resources of the third world. . . .

It would seem, therefore, that we cannot afford to rely on any (or any combination) of the three preceding measures as a means to stabilizing world population at or near a level that would make possible a worthwhile life for every citizen. Nevertheless, many will find such a conclusion unacceptable not least because it suggests the need for some form of compulsion in order to control population levels.

[. . .]

III

. . . I want to elaborate on my remarks about the justifiability of coercion and defend the view that compulsory population control would be justifiable if implemented now, or, even if not right now, certainly up to any feasible time by which excessive population growth will voluntarily be curtailed. I will then go on, nevertheless, to propose an alternative strategy.

The clear and open use of such compulsory measures is generally dismissed out of hand as morally unpalatable. Now this is ambiguous as between whether the *particular* measures proposed to date are unpalatable or whether *any* conceivable measure of such a kind would be. Normally, I believe, it is both claims that are made, but the second, being the stronger one, is the one I shall concentrate on here. I am not unaware that there are 'political' barriers to the introduction of compulsory fertility control. . . . My concern here, however, will initially be with the morality of the measures, and, in particular, with whether our existing moral framework yields principles which will help us decide the issue. Subsequently, I shall consider the rather more political question of who, if anyone, should bear the burden of their imposition, and of how such a coercive arrangement might come about.

[. . .]

. . . Suppose a large number of people dump a little waste, in itself quite harmless, but which will at a future time react with everyone else's to create a huge threat. Surely here coercion to prevent this huge threat would morally be justified.[6] To the extent we would be justified in using coercion if need be to prevent the invasion by this toxic reaction of the interests of temporally distant people, so also would we be justified in using coercion if need be to prevent the invasion of the interests of future people by our uncontrolled population growth. Since I have argued that the commonly considered non-coercive measures are inadequate, there is a strong presumption in favour of the need for coercive measures to be tried.

The question that now has to be faced is: who, *if anybody*, is to be coerced into controlling population? For many, the answer will seem obvious – those with the highest birth rates, namely, people in the developing countries. This suggestion is, however, too simple by far. To begin with, resource utilization by an individual in a developing country is markedly lower than for those of us in the over-developed world. The U.S., for example, with about 5 percent of the world's population accounts for over 30 percent of the world's present resource utilization. Secondly, since the ecological crisis is largely of our making, injustice would be compounded by our holding the developing world mainly responsible for drastically cutting back population growth for, while it is a critical ingredient in the total crisis, it is not the only ingredient. Moreover, it is, in fact, the poorest who would most be affected by compulsory population control. One reason is that children provide about the only form of security for such people in their old age. Another reason is that children contribute significantly more to the household economy in, for instance, rural India than they cost. They provide agricultural help from an early age in peak periods; they mind younger siblings so that mothers may work outside the home; they are used in a marginal economy of odd jobs and petty trading, and may be apprenticed out to craftsmen. Rather than making family income inelastic, they make it more elastic.[7] Thirdly, despite the belief of many liberal-minded people that the efforts made in developing countries thus far to limit their soaring numbers have been of a sort to respect human freedom, I think it can plausibly be argued that the poor in the developing world have already been subjected to both subtle and unsubtle[8] compulsion to which we in the affluent consumer societies have not. A few brief remarks are in order concerning the more subtle forms such compulsion has assumed.

In some countries (such as India) offers in the form of 'economic incentives' have been employed in connection with sterilization programmes. I want briefly to argue that *for at least some people* these offers depend for success in their purpose, on being coercive.[9] It is important carefully to distinguish 'gifts' from 'offers'. Gifts don't have conditions attached, offers always do and sometimes these attached conditions render the offers coercive. In such cases coercion of a subject seems to involve taking an unfair advantage of his or her vulnerability, unless a certain offer is accepted. In particular, he or she will be treated in a way that is unfair and in a way which he or she lacks the power (and this need not merely be physical power) to prevent. I have already spoken of the element of unfairness. As regards the power to resist the offers made by governments, Schenk has argued convincingly that the evidence is that in India, at least, it is the poorest of the poor who have borne the brunt of sterilization programmes. A considerable degree of pressure must have been generated because, in agreeing to sterilization, parents risk the loss of benefits (like security) [see note 7] . . . .

My argument to date poses something of a dilemma. Compulsory control of the world's population would seem morally justified. But, it would, I have contended, be unfair to coerce the developing peoples into bearing the whole burden of cutting back population growth even though in terms of sheer numbers theirs is the major contribution to the problem. On the other hand our quite excessive use of resources is a major contribution to the ecological crisis and hence we must accept a large part of the burden of resolving the problem. Yet our population growth seems nowadays not to be excessive. How then are we to resolve the dilemma? . . .

First, population growth in the affluent world is not to be shrugged off as a non-problem. Even a doubling of the populations of those industrialized nations which have not achieved replacement levels of fertility, say over the next hundred years or so, will pose a serious problem. So a resort to compulsory fertility control may be needed even in the developed world.[10] But, secondly, it is obvious that what needs to be done in the over-developed world is to curb our mania for growth to satisfy artificial consumerist demands and, hence, our excessive resource utilization. If compulsion be justified to protect the interests of future peoples then compulsory cutting back on our wasteful consumerism would be justified. I cannot see this happening within existing political structures. But nothing short of the realization of an order in which production and consumption are in tune with and directed towards the satisfaction of genuine needs will do. . . .

Thirdly, the birth rates of the developing countries must indeed be cut back. But here we must change our whole relationship with the third world and promote a *just* solution. Instead of spending most of our resources (and theirs) on our pursuit of ever-greater affluence, we must channel some of these resources into providing the social benefits which children provide for people in the developing countries but without their needing to actually bear the children.

[. . .]

Notes

1 For an especially revealing case study see the discussion of automotive lead emissions in P. Ehrlich, A. Ehrlich and J. Holdren, *Human Ecology: Problems and Solutions* (San Francisco, 1973), p. 214. Cf. also B. Commoner's discussion of the eutrophication of Lake Erie in *The Closing Circle* (New York, 1971) ch. 6.

2 W.T. Blackstone, 'Ethics and Ecology' in Blackstone (ed.), *Philosophy and Environmental Crisis* (Athens, Georgia, 1974), p. 21.

3 J. Passmore, *Man's Responsibility for Nature* (London, 1974), pp. 140–57 discusses a number of other considerations of relevance, though it seems to me he exaggerates the importance of religion and ideology (his pet stalking horses) and omits even to refer to most of the demographically proven ones cited here.

4 Cf. K. Davis, 'Population Policy: Will Current Programs Succeed?', *Nature*, 158 (1967), pp. 730–39; E.M. Adams, 'Population Control: A Scientific or a Humanistic Approach?', *Journal of Value Inquiry*, 6 (1972), pp. 50–56. Adams' paper actually is a critique of an earlier paper by Davis, but Davis' later views are in common with those of Adams.

5 There is a useful discussion of the phenomenon in W.D. Borrie, *The Growth and Control of World Population*, (London, 1970), pp. 227ff. The burden of his argument is that transition theory is of little relevance to our present predicament.

6 Cf. Robert Nozick, *Anarchy, State and Utopia* (New York, 1974), ch. 4, p. 628.
7 Cf. Hans Schenk, 'India: Poverty and Sterilisation', *Development and Change*, 5 (1973–74), pp. 36–53, and the quite brilliant documentation of these points in Mahmood Mamdani's *The Myth of Population Control* (New York and London, 1972) esp. ch. 4. Mamdani, however, draws what I consider is the wrong conclusion from the data his study unearthed. He assumes a rather fatalistic attitude about the chances of preventing further spiralling of population levels, instead of seeking to supply these benefits from having children in ways that would place less strain on our environment and resources.
8 See e.g. Kai Bird, 'Indira Gandhi Uses Force', *The Nation*, 222 (June 19, 1976), pp. 747–49.
9 There have been several interesting discussions of this topic recently. See e.g. Virginia Held, 'Coercion and Coercive Offers' in J.R. Pennock and J.W. Chapman (eds.), *Coercion: Nomos XIV* (Chicago, 1972); Michael Bayles, 'Coercive Offers and Public Benefits', *The Personalist*, 55 (1974), pp. 139–44; Daniel Lyons, 'Welcome Threats and Coercive Offers', *Philosophy*, 50 (1975), pp. 425–36 and references thereto; and Jeffrie Murphy, 'Total Institutions and the Possibility of Consent to Organic Therapies', *Human Rights*, 1 (1975), pp. 25–45.
10 *Pace* M.P. Golding and N.H. Golding, 'Ethical and Value Issues in Population Limitation and Distribution in the United States', *Vanderbilt Law Review*, 24 (1971), pp. 395–523.

Source: Robert Young (1980) 'Population policies, coercion and morality', in D.S. Mannison, M. A. McRobbie and R. Routley (eds), *Environmental Philosophy*, Canberra, Australian National University, Monograph no. 2, pp. 356–63, 367–72. The version published here has been amended by the author.

5.4

Gary Malinas

On justifying and excusing coercion

[. . .]

I do not believe Young [Reading 5.3] has considered all the policies to reduce birth rates which warrant consideration. Nor do I find the reasons he gives for rejecting the three he considers convincing. I agree that the three he rejects "cannot guarantee a reduction in the time lag between a world wide perception" of the need to reduce birth rates and their actual reduction. . . . But coercive measures give no such guarantee either. The most we can ask for are policies which are feasible and likely to succeed.

Most governments have not addressed the problem of reducing birth rates as a high national priority or as posing a national emergency. If they did convince their populations that the need to reduce birth rates posed a national emergency, the problem of actually reducing them is much less foreboding. National commitments to programs combining education with family planning and social and economic incentives comparable in magnitude to the mobilization of nations for warfare have simply not been tried. There is no empirical evidence which shows programs of such magnitude are apt to fail while compulsory measures are apt to succeed. Secondly, highly industrialized consumer societies are not the only possible targets for demographic transition. Some developing nations aspire to become societies which maintain birth rates at or near replacement fertility, use available resources with discretion, and do not pollute their environments. Lastly, it is unclear that a policy of enforced sterilization would succeed in reducing birth rates. People's reactions to attempts to enforce such a policy surely could defeat its intentions and make them suspicious of other attempts to reduce birth rates.[1] Young's claim that coercive sterilization is morally justifiable, then, is not based on a cool and informed calculation of the consequences if such a policy is or is not adopted. Rather, it is based on the view that a possible world is *conceivable* in which coercive sterilization is morally justified on teleological grounds. This possible world, then, is said to be sufficiently like the actual world to cause us alarm. Young takes this to imply that such a policy is justifiable in the actual world. On the basis of this type of calculation, I cannot think of *any* policy recommendation which would not be morally justifiable. I can think of possible worlds alarmingly like our own in which nuclear scientists are a threat to the human race, and in which a policy of killing them was morally justified. Such exercises in fantasy are not guides to moral evaluation.

I agree with Young that a policy of coercive sterilization would most affect people in developing countries. I do not see it as unfair for *that* reason. Any program which is intended to reduce birth rates is going to most affect those people with the highest birth rates. Young sees such a policy to be unfair because people in affluent societies with relatively acceptable birth rates would be largely unaffected by it, whereas people in societies with high birth rates are often socially and economically dependent on producing children in excess of replacement fertility. Compensation of some sort is in order when children provide labour families need to sustain themselves, when progeny are relied upon to support parents in their old age, where women of child-bearing age are excluded from non-maternal social roles, etc. Young believes considerations of justice require affluent people to provide the needed compensation. In support of this, Young appeals to the following moral principle: "Where we are able without sacrificing anything of greater moral worth to ensure that all on board [i.e. the earth's inhabitants] have a worthwhile life we are obligated to ensure they do." . . . Young's reading of this principle takes the earth's inhabitants to include future people as well as current people. So affluent people now are obligated to make sacrifices to ensure that less affluent people and future people have worthwhile lives. The sacrifices which Young envisages are not just economic sacrifices. He also endorses the violation of deontological principles which taken collectively are denoted by our concept of respect for persons. He calls appeal to such principles in opposition to coercion glib, and says that they are reduced to the status of "a nasty joke" where many people lack the resources needed to sustain themselves. There is no doubt that the lives of many people are structured by social and economic conditions which violate respect for them as persons. Young's claim is that the political conditions which would sustain a policy of coerced sterilization and the application of that policy would no more violate respect for persons than current social and economic conditions do and will do if projected into the future. He believes that on balance such violations will be fewer.

Quite independently of whether Young's view is accurate in its empirical projections, it raises a question of general moral interest. Do our concepts of respect for persons and the roles they play in moral evaluation permit their violation when many people do or will lead lives in which such principles are not honoured? He believes appeal to such principles often serves the interests of people who enjoy their protection while perpetrating states of affairs wherein those very principles are violated in the lives of other people. I do not see the relevance of this latter claim to the general question his view raises. Christians believe the devil quotes scripture for his own ends. The fact that immoral purposes can be served by invoking moral principles does not vitiate the latter. So the general question remains: are actions morally justified which violate our concepts of respect for persons in circumstances wherein any other actions (including omissions) will result in a greater number of such violations?

From most consequentialist viewpoints, the answer is: Yes. The moral justification of actions is based upon projecting their consequences and then giving an evaluation of their consequences. An act is justifiable if, on balance, its consequences are preferable to the consequences of alternative courses of action. On act-utilitarian views, the deontological principles denoted by our concept of respect for persons are rules of thumb which have evolved to help us evaluate the consequences of actions. While they may be good rules of thumb, they have no binding moral force and may be overridden in particular cases. To subscribe to this view requires one to adopt criteria for evaluating the consequences of actions which do not themselves make appeal to the concept of respect for persons. Criteria which appeal to the maximization of happiness or pleasure or preferences or the realization of God's will meet this requirement. Young does not suggest any of these criteria, or similar ones. In fact, criteria for evaluating the consequences of actions he does suggest are based upon principles denoted by our concept of respect for persons. His view is that more people will enjoy the benefits of such respect if some people are deprived of some of its benefits.

. . . Nevertheless, I believe unsound Young's verdict that an action which violates some deontological principles is morally justifiable if, on balance, fewer violations of them occur by performing that action.

Young's condition for justified coercion is intended to apply to situations where people are confronted by moral dilemmas. These are situations where anything one chooses to do has some morally untoward aspect, or consequences, and not to do anything is not satisfactory either. No matter what one does in such situations, there is a *prima facie* case for saying that it is wrong to do that. This *prima facie* assessment can be overridden on Young's view. Actions proscribed on *prima facie* grounds are morally justifiable if their performance leads to fewer violations of people's rights and dignity than would otherwise occur. Young does not tell us what he means when he says an action is morally justifiable. This much is clear though: such actions are morally permissible, and any *prima facie* disapprobation of them can be cancelled if they have good enough consequences. If their consequences are good enough, we should view them with approbation.

I believe Young is confusing two categories of assessment. We sometimes are prepared to *excuse* people when they perform culpable acts. When people are confronted by moral dilemmas of the sort described above, we typically do excuse them when they choose the least immoral action of a set of immoral alternatives. The fact that it was the least culpable of the actions available to them does not thereby remove the action from the category of disapprobation. We excuse the agent of the action without approving of the action. Young has not shown actions of the kind he recommends to be morally justifiable. At most (if he were correct in his empirical details), the considerations he adduces show the agents of actions which violate the human rights and dignity of some people are excusable if on balance fewer violations occur.

There is more than a verbal difference between us on this point. On my view, all tokens of certain types of actions are culpable, though some agents who perform tokens of those types of action in certain circumstances may be excused for doing so. The onus falls on the agent to establish his or her claim to be excused.

[. . .]

I will conclude my comments with a general point about the way Young and some other writers on population formulate "the population problem." The problem, put simply, is how to reduce the birth rate of humans and maintain the reduced rate. We saw that Young proposes an incentive-based program buttressed by a policy of coercive sterilization. I believe responses like Young's [. . .] have been misled by the way the population problem has been posed. Both propose a program for reducing the human birth rate. But the human birth rate is an average taken from different birth rates for different populations which sustain their respective rates for different reasons and as a consequence of different social causes. While the population problem as a long-term problem has one solution only – reduced birth rates – there is no one policy or progam suitable and practicable for all populations. A patchwork approach to the problems of reducing birth rates is apt to stand a better chance of success than more simple, one-sided measures.

Note

1 E.g. India's "incentives" based program of sterilization produced suspicion of and resentment against other birth control programs.

Source: Gary Malinas (1980) 'Coercion, justification and excuses', in D.S. Mannism, M.A. Mc Robbie and R. Routley (eds) *Environmental Philosophy*, Canberra, Australian National University, Monograph no. 2, pp. 376–83.

5.5

Peter Saunders

Capitalism and the environment

Capitalism, we have seen, is a growth machine. Competition between capitalist producers stimulates perpetual innovation. The result is that basic resources such as food and clothing come to be produced in ever greater quantities at ever reduced costs while new luxury items become commonplace within the space of a generation. Average living standards are thus perpetually transformed. Across the world, per capita incomes rise even as total populations expand. It is as if we have come into possession of the alchemist's secret of how to turn base metal into gold.

During the last third of the twentieth century, however, there has arisen a critique of capitalism which sees this discovery of the secret of seemingly perpetual growth as a curse rather than a benefit for humankind. Finding its clearest expression in sections of the 'deep green' movement,[1] this critique accepts that modern capitalism is indeed a growth machine, but argues that this is precisely what is wrong with it. Where, it asks, is the advantage in a system which can sustain huge rates of population growth if the planet is already dangerously overcrowded? How can it make sense for Third World countries to follow a path of capitalist industrialization when existing levels of industrial production in the developed countries are already exhausting world supplies of energy? How can capitalism be permitted to go on expanding and growing when the ecosystem on which we all depend is already near the point of collapse? How can we tolerate capitalism's perpetual search for technological innovation when new technologies have already brought us to the precipice of global annihilation?

The environmental limits to growth

Economic development has always involved transformation of the natural environment, but in the last two hundred years the pace of development has quickened, and the relationship between human beings and the natural environment has changed fundamentally as a result.

One aspect of this change is that we have extended our control over the natural environment. Two hundred years of industrial capitalism have harnessed and tamed nature. This has brought obvious benefits, for people living in the developed capitalist nations no longer inhabit a world where their existence is daily threatened by famines, plagues, floods, droughts and other contingencies of nature, while even in the Third World, mortality rates have dropped dramatically. The environment in which we live has been made more controllable than ever before.

In another sense, however, nature now seems more threatening than ever before. The more we have extended our control over nature, the more we have brought about changes in the environment whose effects may be unpredictable. We have altered nature in ways which we do not fully comprehend, and as the pace of change speeds up, so the impact of our activities becomes ever more marked, for ecosystems cannot adjust quickly enough to accommodate all the changes wrought by industrialization over the last two hundred years. In many instances we simply do not know how much damage we are causing, how serious it is, and how long it might take to reverse it. We live in an age of high rewards but high risk.

The deep green critique of capitalism holds that this gamble with the future of the world is no longer worth the risks involved and that industrial capitalism must therefore now be replaced by an entirely different economic and social system underpinned by a very different set of values:

> Our current industrial way of life is too far gone. It is not a question of nearing the abyss; we daily look down into it . . . It is not so much decapitation that we should be aiming at as the decommissioning of the entire monster.[2]

In this view, the continued pursuit of growth in the West, coupled with the desire of the LDCs [less developed countries] to emulate Western living standards, can only lead to global disaster. The 1987 report of the World Commission on Environment and Development (the Brundtland Report) suggested that bringing LDCs up to the existing living standards in the developed countries would involve a fivefold increase in current levels of energy consumption, yet even a doubling of energy consumption would create a major crisis of global warming and acidification of rainfall.[3] The implication drawn by many greens is that the global ecosystem can only tolerate even limited Third World development if there is substantial and simultaneous deindustrialization in the West.[4]

Capitalism thus stands condemned, not for its failure to raise living standards, but for its success in so doing. The triumph of capitalism at the end of the twentieth century turns out to have been a Pyrrhic victory, for what capitalism does best – stimulation of rapid growth – is what the world can no longer afford. The future of our planet is apparently threatened by overpopulation, melting ice-caps, rising skin cancers, declining resources and a collapsing ecosystem, and most or all of this is the by-product of capitalism's success in generating economic growth.

Because nobody of goodwill could possibly desire such outcomes, greens tend to find a receptive audience for their ideas, particularly among the young. Green ideas, dimly understood, are nevertheless widely endorsed, for it seems only common sense to draw back from activities which threaten the destruction of the planet. The green agenda seems almost 'beyond politics', something which no sensible person could possibly oppose. Environmental education is even taught now in primary schools, and many teachers are apparently sympathetic to the green movement.[5] Yet this agenda is deeply political. Greens invite us to endorse policies which would lead to dramatic changes in our economic, social and political systems. So-called 'deep greens' in particular seek the power to control, regulate, limit, ban or even reverse the technological development which has made economic growth possible in the past. They seek to bury the Victorians' faith in progress and to foster a temerity about future development. Their enthusiasm is for turning back, not going on. The capitalist growth machine is represented by them as a monster created by Frankenstein, something powerful, out of control, destructive and seemingly uncontainable. Rather than celebrating it, as the Victorians did at the Great Exhibition, they exhort us to kill it off. For while it is true that not all greens are opposed in principle to capitalism, many aspects of the green programme represent a fundamental challenge to some of the most basic features of a modern, industrial, capitalist economy.

[. . .]

A technological fix?

In a capitalist market society, if demand for something rises and the supply begins to dwindle, this shows up in increased prices. These increased prices then induce businesses to raise production (e.g. in the case of energy by bringing coal, oil or gas fields on stream which were previously 'uneconomic'), to improve efficiency (e.g. by building lean-burn machines and improving insulation) and to search for substitutes. The market system thus contains its own feedback mechanisms. It is, indeed, much like an ecological system, for both involve perpetual adaptation to changing circumstances which is accomplished unconsciously without overall direction or control from above.

As a result of 200 years of industrialization, the world is richer in resources now than it has ever been. Furthermore, the development of new technologies (e.g. biotechnology, lasers and computer-based applications) means that modern industries often require fewer raw materials than in the past (fibre-optic cables and satellite communications, for example, have dramatically reduced our reliance on copper), they use materials which are in more plentiful supply (the raw material for computer chips, silicon, is found in sand which is effectively limitless in supply), and they create less pollution (the more prosperous capitalist countries become, the more 'environmentally friendly' their industries).

One major reason why so many gloomy green predictions in the past have not been realized is that growth and technological innovation have managed to solve the problems. Despite this, however, many greens remain deeply sceptical about any search for a 'technological fix' for the problems which confront us today, and they prefer instead to go down the route of political imposition or voluntary adoption of austerity programmes. Again, their principal objection concerns the risk factor, for they argue that the discovery of technical solutions in the past does not *guarantee* that appropriate solutions will also be found in the future.[6]

All human activity necessarily entails risk, for we can never be certain of outcomes. The green response to this uncertainty is to argue that we must insure against the worst outcome. Perhaps global warming is not occurring, but if it is, it is going to lead to serious consequences. Perhaps CFCs are not destroying the ozone layer, but if they are then we are all going to suffer in the future. Perhaps new technology can find solutions to many of the problems of continued global population expansion, but if it cannot then the implications of current growth rates look catastrophic. Better, then, to act on these problems now, while there is still time, than to wait until it is too late. As one green commentator puts it:

> We must be given good reason to think that solutions to each and all of the serious problems ahead will be found. Would it not be much wiser and safer to undertake social change to values and structures that do not generate any of these problems?[7]

There are three major objections to this kind of argument. The first is that it sets an impossible hurdle for industrial capitalism to jump over, for we cannot know in advance of its development what technology may become available to us in the future. All that can be said is that the catastrophes predicted by greens in the past have hitherto been avoided by technological developments, that capitalism contains within it an inherent spur to innovation, and that promising research is in progress today on all the problems which threaten danger in the future.[8]

The second objection is that if we allow the minimization of risk to become our overriding concern, then innovation, exploration and experimentation which could benefit future generations will all suffer. It is a unique feature of the so-called 'postmodern' period in which we are living that ours is the first generation seriously to consider whether we should try to stop economic and technological development in its tracks. It is by no means obvious that future generations will look back and thank us for such failings of nerve, any more than we would be grateful had our forebears decided not to open up coal mines or to travel in search of foreign lands. It is arguably no less irresponsible to abandon the pursuit of economic development than to allow it a free rein. So what makes us so sure that now is the time to relinquish the baton and give up the race?

The third objection is perhaps the most serious of all. Much deep green thinking seems simply to assume that a fundamental change of 'values and structures' can be engineered with few problems. All we have to do to avoid our current environmental problems is change society from top to bottom! This not only begs the question of how such a change is to occur and what kind of society will emerge as a result of it, but also betrays a frightening *naïveté* about the ease with which this new Utopia will overcome our problems.[9] Like radical socialists before them, radical greens today wish to drag us kicking and screaming into the Garden of Eden, but we are rarely given a glimpse of what lies on the other side of the fence.

Dismantling capitalism

Left to itself, capitalism clearly cannot resolve all the environmental problems which it creates. Problems of resource depletion can often be resolved through the market mechanism of rising prices, and provided there is money to be made, new technologies can often be relied upon to clean up the mess left by old ones. But what of the other problems identified by green critics? How can capitalist systems deal with environmental problems when there is no price stimulus inducing innovation and no obvious possibility for entrepreneurs to make profit?

Many environmental problems seem to demand governmental or even intergovernmental action if they are to be tackled, for capitalist enterprises seem unwilling or unable to sort out the problems themselves. As many economists have come to recognize, not everything can be left to the market.

In 1833, William Forster Lloyd wrote a pamphlet on the problem of over-exploitation of the common land in English villages prior to the enclosures. His argument was that when pasture land is made available to all, it will be in the interests of each herder to graze as many cattle as possible upon it. This is because the benefit to the individual herder of grazing one extra animal on the free pasture massively outweighs the cost in terms of the slight decline in quality of pasture which is then available for all the animals. But this same logic applies also to every other cattle owner, with the result that the pasture is rapidly exhausted and they all end up with nothing. Even if they recognize the dangers, the herders will not refrain from their damaging behaviour, for if one reduces the number of animals he or she grazes on the land, the benefit will simply be appropriated by others who do not follow suit (game theorists refer to this as the 'free-rider problem'). As Hardin explains in his discussion of Lloyd's original insight: 'Each man is locked into a system that compels him to increase his herd without limit – in a world that is limited. Ruin is the destination toward which all men rush'.[10]

Many of today's environmental problems exhibit the same features as this original 'tragedy of the commons'. Because the oceans are open to all, fishing fleets have every incentive to expand their catches to the point where fish stocks become exhausted and all fleets have to spend longer and longer catching fewer and fewer fish. Similarly, because access to the atmosphere is unlimited, it pays industries to expel waste untreated into the air even though we might all prefer a less polluted atmosphere. Where roads are provided free of charge, they too become part of the 'commons' and individual motorists pursue their own rational self-interest by seeking the convenience of driving everywhere even though they end up in nose-to-tail traffic jams because all the other motorists have followed the same option. And so on.

The tragedy of the commons poses a major problem for market-based capitalism, for it seems that competitive profit-seeking individuals will inevitably exploit the planet to the point of destruction unless the state steps in to stop them. This leads many greens to argue that the tragedy can only be resolved or averted by means of dramatic political solutions.[11] Even the relatively modest Brundtland Report called for 'decisive political action now . . . to ensure . . . human survival',[12] and for many greens, this is too mild in tone and too accommodating in content. Dismissive of such 'reformist thinking', they call for 'breathtakingly radical action' and do not shrink from mounting 'a radical, visionary and fundamentalist challenge to the prevailing economic and political world order'[13] Capitalism, in short, must be put to the sword.

This radical green agenda is implacably opposed to capitalism, and it is this which brings it close to more traditional socialist positions. Responding to socialist criticisms that they should identify capitalism rather than industrialism as the root cause of environmental problems, one radical green retorts: 'Greens will accept that the destruction of capitalism is indeed a necessary condition for restoring environmental integrity . . . The deeper green programme constitutes a serious threat to both the social relations and productive practices typical of capitalism.'[14] Like revolutionary socialism, deep green environmentalism offers an apocalyptic vision of the future which can only be avoided by wholesale destruction of the present social order and its replacement with a new one. Also, like revolutionary socialism, its adherents seem often to teeter on the brink of totalitarian political solutions and, as in the example of their dramatic prescriptions for population control, sometimes to topple over into them. More often, however, they avoid confronting the political implications of their programme, but it is difficult to see how radical changes designed to withdraw the comforts of modern consumerism and to plunge us all into bleak austerity could be achieved without resort to considerable force.

Although it shares much in common with old-style socialism, however, the radical green movement is more than simply socialism in a new guise. Environmentalists themselves generally claim that they are neither on the left nor the right, that they are 'neither red nor blue but green'. They are anti-capitalist, but they are also in one important sense anti-socialist, for they see both systems as contaminated by a faith in and reliance upon technological progress and economic growth. Most greens are well aware of the ecological disaster which was unleashed in Russia and eastern Europe under socialism and which only finally came to light after the 1989 revolutions,[15] and they know that nowhere in the Western capitalist nations is there evidence of environmental degradation of the scale or intensity which occurred throughout the eastern European socialist regimes.

Green capitalism: commodifying the environment

The deep green answer to the tragedy of the commons seems inevitably to end up in political coercion – we must be stopped from overpopulating the planet, from using up resources, from pumping carbon dioxide into the atmosphere, and so on. But there is an alternative green vision, for the need to find an answer to the tragedy of the commons need not always entail increased direction by public authorities.

The starting point for this alternative approach lies in the recognition that, if markets sometimes fail, so too do governments. Motivated by the best of intentions, government agencies often exacerbate the problems which they intend to resolve, for they can never have at their disposal all the information needed to work out all the implications of pursuing a particular course of action. This suggests that government controls should be tried only when market-based solutions have proved inadequate.

Consider as an example the problems of protecting endangered species of wildlife. In Kenya, the government responded to environmentalists' concerns about the declining number of elephants by agreeing to impose a ban on ivory trading. This, however, simply raised the black market price of ivory and therefore increased the incentives for poachers to kill even more elephants. In Zimbabwe, by contrast, the government introduced a licensing system for ivory trading under which revenue from tusks and from hunting permits went back to local communities who therefore had an interest in policing poaching themselves. Elephant numbers have been rising by 5 per cent per year as a result.[16]

The lesson to be drawn from this is that it is not capitalist trade, but the absence of clearly defined property rights, which results in overexploitation of natural resources. When resources have a market value and can be bought and sold as private property, they tend not to disappear, for owners then have an interest in maintaining and reproducing them. It is this that explains why the free-roaming American buffalo was wiped out while cows graze on the same land today in their thousands, or why crocodiles (which governments allow to be farmed for their skins) are in plentiful supply while rhinoceroses (which roam free in reserves and are poached for their horns) are on the endangered species list.[17] Seen in this light, the tragedy of the commons has less to do with the absence of government regulation on how the commons may be used than with the absence of any identifiable proprietorial interest in them.

There are three ways in which the capitalist profit motive may be harnessed to produce environmentally beneficial results. One is by means of 'green consumerism', for if there is an effective demand for products which do not damage the environment, capitalist producers will begin to compete to supply it. The second is by introducing charges for access to common resources which have hitherto been available free or at subsidized prices, for this builds the cost of resource depletion or despoliation into the economic calculations of those who use them. The third and most radical is by privatizing or commodifying public goods.

The growth of public awareness of green issues has led in recent years to a new market in environmentally friendly products in Europe and North America. Big industrial firms such as ICI and Shell have spent millions of pounds on environmental programmes and advertise their concern on national television. Car companies compete with each other to produce clean, energy-efficient models which can be recycled at the end of their useful lives. Financial institutions profitably specialize in 'ethical investments' which avoid companies which pollute the planet. Supermarkets now sell organic foodstuffs free of contaminating chemicals, eggs produced by free-range chickens, and toilet tissue manufactured from recycled paper. Manufacturers of washing powders market low-phosphorus products which will not choke the rivers with foaming suds, and even before the ban on CFCs, aerosol manufacturers were rushing to market products guaranteed not to damage the ozone layer. And retailers such as Body Shop have achieved remarkable success throughout the Western world by concentrating on product lines which they claim are not tested on animals, which are purchased at a 'fair price' from Third World producers, and which do not damage the natural environment.

All of this activity is proving very profitable. Having founded her first Body Shop in 1976, for example, Anita Roddick is now worth over £100 million and was listed by *The Sunday Times* in 1994 as the 106th richest individual in Britain.

Green consumerism by itself is, however, unlikely to be sufficient to meet all the problems and challenges, and capitalist market solutions to environmental problems must go beyond simply leaving green-minded consumers to demand environmentally sound products. Even if millions of us insist on buying liquids in glass bottles rather than plastic containers, this will not solve the problem of disposing of waste, and it is difficult to see how changes in purchasing decisions by individual consumers could help sort out problems such as the possible danger of global warming. Green capitalism will often require some degree of government action in order to enable market-based solutions to develop.

One form of government action which uses the market to bring about desired environmental objectives is the introduction of new pricing policies. An obvious example is the lower duty levied by many Western governments on unleaded than on leaded petrol. This has served as an effective inducement to motorists to switch their purchasing patterns (and for car manufacturers to change their engine requirements) without the need for political controls or bans.

The point about a policy like this is that, unlike blanket government regulation, it leaves it up to individual users to decide how best to adapt their behaviour. Drivers who can switch relatively easily to unleaded fuel have done so; those who cannot must now pay a premium which expresses the fact that their behaviour is generating a social cost (i.e. that they are polluting the commons).

It has been suggested that in practice there may turn out to be little difference between government pricing strategies and the use of political directives and bans to achieve the same result.[18] The crucial difference, however, is that political regulation imposes uniform standards on everybody in order to meet a specified target, whereas pricing strategies tend to bring about the biggest shift in behaviour among those for whom change is least costly. Not only is this latter strategy more sensitive to individual liberties, but it will also prove a more efficient way of achieving the desired aggregate result.

The introduction of charges has often proved effective as a means of pollution control in situations where polluting activities were previously a cost-free option. In England, for example, the discharge of farm and industrial waste into the river system had for years been controlled by a simple licensing system in which farmers and industrialists applied for 'discharge consents' which allowed them to pump a certain volume of waste into a specified river or stream. The problem with this system was that there was no incentive to reduce the volume or toxicity of discharge. Recognizing this, the National Rivers Authority has now introduced a new system in which those seeking to discharge waste into the rivers must pay for the right to do so. The more waste they pump out, the higher the price they must pay, and this helps meet the cost of cleaning up the rivers.[19]

This 'polluter pays' system does not seek to eradicate pollution altogether (for the rivers can handle a certain level of discharge and there is no other feasible way of disposing of large volumes of liquid waste), but it does ensure that the 'marginal external costs' of individuals' actions are properly taken into account when they decide between different courses of action. As with the example of lead-free petrol, those who can easily cut back on their polluting activities will find that it pays them to do so, while those who cannot will have to accept the higher charges which their activities incur.

The polluter pays principle is successfully being applied in many areas of environmental concern. In some German towns, for example, garbage trucks are now fitted with meters which weigh domestic refuse and automatically bill households according to the amount of waste they throw out. This system both encourages households to reduce waste and raises the revenue required to dispose of it. A similar strategy is being applied on a larger scale with the introduction by governments throughout Europe of carbon taxes as the means for tackling the (perceived) problem of global warming. Here again, consumers are encouraged to reduce emission of waste (in this case CO_2 emissions from coal-burning power stations) by raising the price they must pay for their electricity.

In California, this principle of charging for pollution has been taken one step further with the establishment of a market in tradable pollution permits.[20] In 1994, a Regional

Clean Air Incentives Market was established across four counties in southern California to allocate pollution credits among 390 companies which between them produce most of the industrial pollution in the area. Each credit allows a company to discharge 2.2 kilograms of nitrogen oxides or sulphur oxides, and each company was given a limit on its total emissions. Those which manage to stay below their limit will be able to sell their surplus credits at auction to those who continue to exceed theirs. As overall emission limits are progressively reduced by the government, so the price of traded credits will rise. It is expected that overall emissions of nitrogen oxides and sulphur oxides will fall by 75 per cent and 60 per cent, respectively, over a ten-year period.

Policies like tradable pollution permits take us into the third way of pursuing environmental benefits through capitalist market strategies, for what the Californian experiment has effectively done is to commodify the right of factory owners to use the air which was previously 'provided' to them free. As we saw in Chapter 1, for capitalism to function it is necessary for government to establish and enforce a clear system of private property rights. In many instances of environmental pollution and neglect, this is precisely what has been missing. Nobody owns the rainforest, or the ozone layer, or the North Sea, which is why everybody feels free to abuse them. Tradable pollution permits represent a particularly elegant market solution to the tragedy of the commons in the air, but this same market logic can also be applied in water.

In Australia, for example, overfishing of tuna was a problem because the ocean was a free resource for anybody with a boat and a net. Rather than trying to conserve fish stocks, as many European governments have done, by keeping boats in port for a specified number of days each month or by regulating the length of nets, the Australian government introduced tradable fishing quotas. Fishing boats are allocated a maximum quota, and these quotas can then be traded. The result is that the most efficient crews tend to buy up extra quotas from the least efficient. Environmental standards (in this case, fish stocks) are maintained, but not at the cost of economic efficiency.[21]

Of course, not all environmental problems are susceptible to market-based solutions like this, for it is not always possible clearly to specify the boundaries of a property holding, in which case common holdings cannot easily be privatized. Territory on land can usually be marked out and fenced in, and we have seen that rights over water and local air space may also be established and defended, but such solutions can become extremely difficult when applied to property rights in the atmosphere. Some important environmental resources, including the ozone layer and the stability of the global climate, cannot easily be privatized and therefore may not readily lend themselves to a capitalist market solution.

However, given a little political wisdom and the appropriate technology, property rights in the environment could be extended much further than they are at present. Just as the invention of barbed wire in the 1840s helped solve the overexploitation of the American prairies by enabling claimants to fence their land, so too modern technological developments could similarly help resolve some of today's environmental problems by enabling individuals to establish and defend clear property rights in the atmosphere against the claims or abuses of others.[22]

For example, chemical tracers could be added to emissions from chimneys which would allow those who suffer from atmospheric pollution from factories to charge those who are responsible for it. This would mean that owners of Scandinavian lakes could seek compensation from owners of power stations in England for the acidification of their water rights and the death of their fish. The owners of the power stations would then either pass on these external costs in the prices they charged for their electricity, or would invest in chimney-scrubbers in order to reduce the amount of compensation they had to pay. Either way, the value of the environment, which is currently ignored in the accounts, would properly be recognized and would enter into people's behaviour.

Road pricing is another way of cutting down pollution. Several Western governments seem likely to move soon to a computerized system for monitoring car use in cities and billing drivers on a monthly basis, but the principle could be extended by privatizing the highways and charging the firms which run them for the pollution which their customers are generating. Such charges would presumably be passed on to road users, with higher tolls at peak periods when exhaust emissions are at their worst, and this would in turn lead drivers to amend their behaviour (e.g. by travelling off-peak, by car pooling, or by switching to alternative modes of transport) by taking account of pollution costs which are currently socialized.

Taken to the extremes of technical and practical feasibility, we could even imagine a property rights solution to problems such as the decline in the whale population. It has been reported that Americans would on average be prepared to pay $8 each to prevent the extinction of the blue whale.[23] These creatures therefore have a value to people but there is no way at present that this can be translated into effective purchasing power. Rather than leaving them at the mercy of hunters for whom they represent a free good, it might make sense to sell particular whales to interested parties (such as individual Americans or members of Greenpeace), 'brand' them by genetic prints, and monitor them by satellite. In such a futuristic scenario, the plight of the world's whales could be relieved by the establishment of effective private ownership so that whaling fleets would have to pay market rates in compensation if they continued to catch creatures which would no longer be freely available to any who wished to go out with a harpoon gun.

Market capitalism and environmental quality are not inherently incompatible. It is all too easy to be seduced by the claim that big environmental problems demand draconian political solutions. Not all of the world's environmental problems can be resolved by pricing and privatizing resources, nor might we find technological solutions in every case, but lasting solutions are more likely to be found by governments working with the grain of individual self-interest as revealed in market behaviour, rather than against it. Some of the problems identified by the green movement are real, and they demand a response, but this does not mean that the time has come to turn off the capitalist growth machine. The growth of capitalist economies will continue to cause environmental problems, but continuing economic growth and technical innovation within a context of market relations and private property rights arguably offers our best hope for overcoming them. The deep green alternative of bleak austerity necessarily enforced through political coercion is not one which the world need, or is willing to, accept.

Notes

1 Not all greens are critical of capitalism; . . . some believe that the market mechanism and private property rights can be used to solve environmental problems. [Here] . . . I draw an admittedly crude distinction between 'deep greens', who seek solutions through deindustrialization and a

transcendence of capitalism, and 'green capitalists', who seek solutions within existing social, economic and political frameworks.

2 J. Porritt and D. Winner, *The Coming of the Greens* (London: Fontana, 1988), p. 233.
3 World Commission on Environment and Development (the Brundtland Report), *Our Common Future* (Oxford: Oxford University Press, 1987).
4 See L. Martell, *Ecology and Society* (Cambridge: Polity Press, 1994), p. 44.
5 Porritt and Winner, *The Coming of the Greens.*
6 See Martell, *Ecology*, pp. 72–75 for an outline of this argument.
7 Cited in ibid., p. 74.
8 Even seemingly huge and intractable problems such as global warming may have technical solutions. Not only is research continuing into non-carbon-based sources of energy such as wind farms, but it may also be possible to reduce current levels of carbon dioxide in the atmosphere through, for example, seeding the oceans with iron in order to stimulate the growth of algae which absorb CO_2 and take it to the seabed when they die.
9 In an otherwise sympathetic discussion of radical green ideas, Martell (*Ecology*) develops an unanswerable critique of those who believe that green objectives can be met through a move to decentralized, voluntary and largely self-sufficient communities. The basic problem is that, without a centralized, coordinating power, there will be nothing to oblige each community to follow green practice. This means that the use of centralized power is an unavoidable condition of effective green politics – but this then raises the spectre of state coercion used as an instrument against all those who do not come into line with the new austerity programme. This dilemma is, of course, already familiar from more than a century of socialist writing where futile attempts have been made to reconcile the desire for revolutionary social change with the desire to avoid totalitarian forms of politics – see P. Saunders 'When Prophecy Fails', *Economy and Society*, vol. 22, 1993, pp. 89–99.
10 G. Hardin, 'The Tragedy of the Commons' in G. Hardin and J. Baden (eds), *Managing the Commons* (San Francisco: W.H. Freeman, 1977), p. 20.
11 Though as we shall see, not all green thinkers draw this conclusion.
12 Brundtland Report, p. 1.
13 Porritt and Winner, *The Coming of the Greens*, pp. 263 and 11.
14 A. Dobson, *Green Political Thought* (London: HarperCollins, 1990), pp. 175 and 178. Similarly, Martell (*Ecology*) is in no doubt that greens share with socialists a commitment to the overthrow of capitalism, and he urges the green movement to form alliances with socialist parties (p. 13).
15 The Russian Environment Minister suggested at the end of 1993 that about 15 per cent of the country should be considered an environmental disaster zone. Half of all arable land was unsuitable for farming, four-fifths of industrial waste was inadequately treated, hundreds of rivers were polluted with nuclear waste, and 100,000 people were living on land which was dangerously irradiated (*The Independent*, 28 December 1993).
16 T. Anderson and D. Leal, *Free Market Environmentalism* (Boulder, CO: Westview Press, 1991). I. Sugg and U. Kreuter, *Elephants and Ivory* (London: Institute of Economic Affairs, 1994).
17 M. Sas-Rolfes, 'Trade in Endangered Species', *Economic Affairs*, vol. 14, 1994, pp. 10–12.
18 Martell, *Ecology.*
19 Saunders and Harris, *Privatization and Popular Capitalism*, (Milton Keynes, Open University Press, 1994), chapter 5.
20 *Independent on Sunday*, 2 January 1994.
21 Anderson and Leal, *Free Market Environmentalism.*
22 Ibid.
23 D. Pearce, E. Barber, A. Markandya, S. Barrett, R. Turner and T. Swanson, *Blueprint 2: Greening the World Economy* (London: Earth Scan Publications, 1991)

Source: Peter Saunders (1995) *Capitalism: A Social Audit*, Milton Keynes: Open University Press, pp. 52–5, 65–76, 127–30.

5.6

John Gray

An agenda for Green conservatism

> Man's conquest of nature, if the dreams of some scientific planners are realized, means the rule of a few hundreds of men over billions upon billions of men. There neither is nor can be any simple increase of power on man's side. Each power won *by* man is a power *over* man as well. Each advance leaves him weaker as well as stronger. In every victory, besides being the general who triumphs, he is also the prisoner who follows in the triumphal car.
>
> (C.S. Lewis)[1]

It is fair to say that, on the whole, conservative thought has been hostile to environmental concerns over the past decade or so in Britain, Europe and the United States. Especially in America, environmental concerns have been represented as anti-capitalist propaganda under another flag. In most Western countries, conservatives have accused environmentalists of misuse of science, of propagating an apocalyptic mentality, and of being enemies of the central institutions of modern civil society. Nor are these accusations always wide of the mark. Indeed, in considerable measure they show conservatives endorsing the self-image of the Greens as inheritors of the radical protest movements of earlier times, and as making common cause with contemporary radical movements, such as feminism and anti-colonialism. In other words, both the Greens themselves and their conservative critics have been happy to share the assumption that socialism and environmental concern go together.

The aim of this present argument is to contest that consensus. Far from having a natural home on the Left, concern for the integrity of the common environment, human as well as ecological, is most in harmony with the outlook of traditional conservatism of the British and European varieties. Many of the central conceptions of traditional conservatism have a natural congruence with Green concerns: the Burkean idea of the social contract, not as an agreement among anonymous, ephemeral individuals, but as a compact between the generations of the living, the dead and those yet unborn; Tory scepticism about progress, and awareness of its ironies and illusions; conservative resistance to untried novelty and large-scale social experiments; and, perhaps most especially, the traditional conservative tenet that individual flourishing can occur only in the context of forms of common life. All of these and other conservative ideas have clear affinities with Green thought, when it is not merely another scourge of the inherited institutions of civil society. The inherent tendency of Green thought is thus not radical but the opposite: it is conservative. At the same time, the absorption of Green concerns into conservative thinking will necessitate some radical changes within conservative philosophy and policy; particularly within those strands of conservative thought that have, during the past decade or so, come to be animated by neo-liberal doctrines whose origins are, in fact, not conservative at all, but rather in the classical liberal rationalist and libertarian ideologies which were spawned in the wake of the Enlightenment.

. . . On the view developed here, though human beings need a sphere of independent action, and so of liberty, if they are to flourish, their deepest need is a home, a network of common practices and inherited traditions that confers on them the blessing of a settled identity. Indeed, without the undergirding support of a framework of common culture, the freedom of the individual so cherished by liberalism is of little value, and will not long survive. Human beings are above all fragile creatures, for whom the meaning of life is a local matter that is easily dissipated: their freedom is worthwhile and meaningful to them only against a background of common cultural forms. Such forms cannot be created anew for each generation. We are not like the butterfly, whose generations are unknown to each other; we are a familial and historical species, for whom the past must have authority (that of memory) if we are to have identity, and whose lives are in part self-created narratives, woven from the received text of the common life. . . . [This essay is] an attempt to restore a balance within conservative philosophy which has in recent years shown signs of being lost.

It is also an attempt to correct some of the radical excesses of Green theory. On the whole, Green theory is inspired by an anti-capitalist mentality that neglects the environmental benefits of market institutions and suppresses the ecological costs of central planning. In the real world, environmental degradation has been at its most catastrophic where, as under the institutions of the former Soviet system, planners are unconstrained in their activities by clearly defined property rights or by the scarcities embodied in a properly functioning price mechanism. (The situation appears to be little different, or worse, in the People's Republic of China.) For reasons that are perfectly general, and which will be explored in greater detail later in the argument, environmental despoliation on a vast scale is an inexorable result of industrial development in the absence of the core institutions of a market economy, private property and the price mechanism. This is a vital truth as yet little understood by Green theorists, even though it is all too plain in the post-communist countries.

[. . .]

As against the neo-liberal strand within recent conservatism, on the other hand, my argument is that market institutions, although they are indispensably necessary, are insufficient as guarantors of the integrity of the environment, human as well as natural. They must be supplemented by governmental activity when, as with the restoration or preservation of the historic European city, private investment cannot by itself sustain the public environment of the common life. The environmental case against doctrinal neo-liberalism is yet stronger than this. The unfettered workings of market institutions may damage the natural and human environments, even if it is true . . . that in most cases they act to protect them. . . .

Neo-liberal ideas have been attractive to conservatives in many Western countries, and in parts of the post-communist world, partly in virtue of the real excesses of twentieth-century statism, to which they provide a healthy corrective. They are nevertheless a distraction from the central concerns of traditional conservatism, and they inhibit conservatives in addressing the problems that arise for them in an age in which economic growth on conventional lines has begun to come up against genuine environmental constraints. Conservative policy in the post-war world has been governed by the strategy of securing legitimacy for market institutions by so aligning the electoral and the economic cycles as to yield uninterrupted economic growth. This is a strategy which, in neglecting the deeper sources of allegiance, is risky when the economy turns sour. It offers nothing in an age – not so far off, and perhaps imminent in some Western countries – when economic growth on the old model is not sustainable, and has in any case come to a shuddering halt.

The prospect of open-ended growth in the quantity of goods, services and people is in any case hardly a conservative vision. Though the eradication of involuntary poverty remains a noble cause, the project of promoting maximal economic growth is, perhaps, the most vulgar ideal ever put before suffering humankind. The myth of open-ended progress is not an ennobling myth, and it should form no part of conservative philosophy. The task of conservative policy is not to spread the malady of infinite aspiration, to which our species is in any case all too prone, but to keep in good repair those institutions and practices whereby human beings come to be reconciled with their circumstances, and so can live and die in dignified and meaningful fashion, despite the imperfections of their condition. Chief among all of the objects of conservative policy, for this reason, is the replenishment of the common life; the shared environment in which, as members of communities and practitioners of a common culture, people can find enjoyment and consolation.

As against the values and policies of neo-liberalism, which tend to deplete further and even to destroy the resources of this common life, a Green conservatism animated by the concerns of traditional Toryism would seek, wherever this is feasible, to repair and renew the common life. It would acknowledge the vital role of the core institutions of civil society, private property and contractual liberty under the rule of law, in any civilised modern state, and, more particularly, in any polity which seeks to protect the common environment, human and natural. It would recognise that unlimited government has been the greatest destroyer of the common environment in our age, and would accordingly support measures for the limitation of government, often by the devolution or hiving-off of its activities. It would affirm that governmental monopoly, or near-monopoly, in a variety of vital services – perhaps even in the supply of money – has proved an evil against which the extension of market institutions may be the best remedy. It would nevertheless resist making a fetish of market institutions: it would be ready to extend them where their absence is a cause of environmental degradation (as when environmental depletion occurs through the occurrence of tragedies of the commons); but it would also be ready to curb them, when their workings are demonstrably harmful to the common life, and when the costs of such curbs are not prohibitive.

[. . .]

. . . Green thought – especially that variety of Green thought that is associated with the theory of deep ecology and with the Gaia hypothesis – embodies a challenge to the ruling world-view of the age, which is a sort of scientific fundamentalism allied with liberal humanism. This is a world-view which is thoroughly alien to conservative philosophy in virtue of the nihilism that it breeds in relations with nature and with fellow humans but which conservatives are nevertheless hesitant to resist, since it is associated with the prestige of science as the animating force of modernity. This scientistic world-view must be brought into question among conservatives, as it has already been among Greens, and it is part of the agenda of Green conservatism that it be so questioned. Green conservatism will, first and last, repudiate the hubristic rationalist ideology that suffuses neo-liberal thought and policy, which has captured and subjugated recent conservatism, but which is merely the most recent excrescence of modernist humanism – a creed that both genuine conservatives and Greens have every reason to reject.

Ecology, order and individualism

Ecological functions of market institutions

[. . .]

The central ecological function of market institutions is in the avoidance of the tragedy of the commons.[2] This occurs when, in the absence of private or several property rights in a valuable natural resource, separate economic actors – individuals, families, corporations or even sovereign states – are constrained to deplete the resource by over-use, given their realisation that, if they do not do so, others will. Tragedies of the commons occur because, in the absence of the institutions of private property, no one has an incentive to adopt a long-term view of the utilisation of resources. Tragedies of the commons also have features akin to those of Prisoner's Dilemmas,[3] in that each has an overriding incentive to do what is not in his or her own interest – in this case, to run down the resource to extinction. Examples of such tragedies are manifold in environmental literature, unfortunately, but two may suffice to illuminate the central point about them. If a forest, say, belongs to no one, then no one will have an interest in planting the next generation of trees, or, for that matter, in developing logging techniques that leave saplings standing. Since no one stands to benefit from such foresight, no one will exercise it. Hence the reckless deforestation, for agricultural and other purposes, including timber harvesting, that has occurred in parts of Latin America and South-East Asia. Or consider natural resources in fish. In so far as shoals of fish are in the commons – unowned assets in a state of nature – they will be harvested to extinction, since the only operative incentives on fisherfolk will be to catch the fish and reap a fast profit on them, before their competitors do. True, where fisherfolk live in isolated communities without competitors for their resources, they may develop traditions which limit the overexploitation of fish; but these will always go by the board when competitive fishing communities, or enterprises, come on to the scene. The moral of this example is a perfectly general one. Competition for natural resources, living or otherwise, in the absence of private property rights in them, spells inexorable ruin for such resources. The commons will always be doomed, and its resources fated to disappear, when there is a diversity of competing demands upon them. The lesson – rightly drawn by free market economists – is that the extension of private property institutions to cover resources, such as shoals of fish, hitherto in the commons, is a potent corrective to the over-exploitation of the natural environment.

Market institutions have another crucial ecological virtue: that of reflecting through the price mechanism shifting patterns of resource-scarcities. In the broadest terms, market pricing overcomes, at least partially, the otherwise insuperable epistemic dilemma facing all economic agents: that of utilising information which is dispersed throughout society and which, in virtue of its often fleeting and circumstantial nature, and the fact that it is sometimes embodied in dispositions and traditions whose content is not fully articulable, cannot be collected or gathered together by any planning agency.[4] By allowing this, often tacit and embodied, knowledge to be expressed in price information available to all, market institutions mitigate the ignorance in which we must all act in our capacity as economic agents. They in this way allow for a measure of rationality in the allocation of resources, and of efficiency in their uses, that is unavoidably denied to central planners and their agents.

[. . .]

Ecological limitations of market institutions

The ecological case for market institutions is undoubtedly a very strong one. If private property provides incentives for conservation of scarce resources, the price mechanism supplies a measure of the relative scarcity of different resources. Further, the price mechanism will encourage the search for alternative resources when existing resources of a certain sort grow too expensive, just as it will spur technological innovation in respect of the extraction and use of known resources. In these ways, market institutions embody the least irrational of available mechanisms for the allocation of resources, and, by comparison with socialist planning institutions, they are highly environmentally friendly.

Market institutions have, nevertheless, very serious ecological limitations. As they are described and defended by their most ideological advocates, market institutions are a sort of perpetual motion machine, an engine of unlimited growth, which only the ill-conceived interventions of government stalls. This conception is defective for several reasons. There are, in the first glance, forms of environmental market failure that it fails to capture. Consider the phenomenon of global warming. (From the point of view of my argument here, it does not matter whether this phenomenon exists, whether the evidence shows it to be a real danger, or whether there is no conclusive evidence in support of such propositions. It could be merely a hypothetical danger and still do the work I want it to do in my argument.) Global warming reveals the limits of market institutions, from an environmental perspective, in that it is a threshold phenomenon, coming about via billions of separate acts, each of which individually is innocuous or even imperceptible. Market pricing of each of these acts will not prevent the totality of them generating the phenomenon. In this, and in similar cases of a pure public bad, only prohibitive governmental intervention can prevent or alleviate the problem. The class of environmental market failures may be a larger one than the example of global warming suggests, if (as is surely plausible) there are areas where the extension of property rights is inviable or merely too costly to be reasonably envisaged. This may be true of endangered species that are migratory over vast distances and across several legal jurisdictions: only an intergovernmentally agreed and enforced ban, or a quota system similarly set up, can protect them from extinction. In other words, even where a pure public bad is not at issue, the public good of protection of endangered species will be underprovided whenever market institutions cannot sensibly be extended to create property rights in the species in question, and where market institutions are not supplemented by governmental institutions and policies.

It will often not be enough to supplement market institutions for the sake of environmental protection. Their workings will have to be constrained. Global markets, left to themselves, will often decimate local trades and modes of production and will destroy the ways of life that they support. (Conventional programmes of 'aid' to 'developing countries' often have the same effect.) Global markets in food, along with dumping from developed countries with artificial and unnecessary agricultural surpluses, have destroyed agrarian ways of life in many poor countries, promoting migration into unsustainably gigantist cities, with all of their familiar costs and hazards. Ending economic aid that is self-defeating or counterproductive, and curtailing agricultural protectionism in the developed countries, as advocated by free-marketeers, is *not* an adequate response to the dilemma of protecting otherwise sustainable agrarian communities in poor countries, even if such measures are part of the solution. Such countries may need protection (in the form of tariffs and subsidies) for their peasant farmers – whatever GATT, the IMF, or the World Bank may dictate. In this, and other contexts, market institutions must be restricted in their workings, and not merely supplemented.

We find another limitation of market institutions in their insensitivity to inherently public goods.[5] These are goods which do not necessarily satisfy the technical requirements of an economic public good, such as indivisibility and non-excludability, but which are ingredients in a worthwhile form of common life. Consider public parks in the context of a modern city . . . There are, of course, no insuperable technical obstacles to turning urban parks into private consumption goods. Fences can be set up, electronic ID cards printed for subscribing members, private security patrols hired, litter collected by profit-making agencies, and so on. Public parks are not, in the conventional economic sense, public goods. However, they are inherently public goods in the sense that I intend, inasmuch as public parks that are safe, well-tended, pleasing to the senses and easily accessible to urban dwellers are elements in the common form of life of the historic European city. The point is generalisable. Public spaces for recreation and for lingering, whether streets, squares or parks, are necessary ingredients in the common life of cities, as conceived in the European tradition, and elsewhere. Where such public places atrophy or disappear, become too dangerous or too unsightly to be occupied, and so vanish into a state of nature, the common life of the city has been compromised or lost. This is a nemesis, long reached in many American urban settlements and not far off in some British and European cities, which market institutions can do little to prevent. It is only one example, though perhaps a peculiarly compelling one, of the indifference of market institutions to inherently public goods.

If their workings are not to compromise the integrity of the environment, human as well as natural, market institutions must be both supplemented and constrained. They need such constraint and supplementation, in any case, if they and their various environmental benefits are to survive . . . Here, as elsewhere, market institutions must be curbed, or at least restrained in their workings, if a civilised and peaceful form of common life is to be preserved and transmitted across the generations.

. . . The office of government, in this connection, is the superintendence of market institutions, with the aim of ensuring that their workings are not self-defeating or such as to endanger themselves. This is a rudimentary tenet of conservative philosophy of which many contemporary conservatives, whose vision has been occluded by the empty vistas of neo-liberal dogma, appear ignorant.

Ecological theory and conservative philosophy

> Change is a threat to identity, and every change is an emblem of extinction. But a man's identity (or that of a community) is nothing more than an unbroken rehearsal of contingencies, each at the mercy of circumstances and each significant in proportion to its familiarity. It is not a fortress into which we may retire, and the only means we have of defending it (that is, ourselves) against the hostile forces of change is in the open field of experience; by throwing our weight upon the foot which for the time being is more firmly placed, by cleaving to whatever familiarities are not immediately threatened and thus assimilating what is new without becoming unrecognizable to ourselves. The Masai, when

> they were moved from their old country to the present Masai reserve in Kenya, took with them the names of their hills and plains and rivers and gave them to the hills and plains and rivers of the new country. And it is by some such subterfuge of conservatism that every man or people compelled to suffer a notable change avoids the shame of extinction.
>
> (Michael Oakeshott)[6]

One of my central theses is that Green thought and conservative philosophy converge at several crucial points, the very points at which they most diverge from fundamentalist liberalism. There are at least three deep affinities between Green thought and conservative philosophy that are important to my argument. There is first the fact that both conservatism and Green theory see the life of humans in a multi-generational perspective that distinguishes them from liberalism and socialism alike. Liberal individualism, with its disabling fiction of society as a contract among strangers, is a one-generational philosophy, which has forgotten, or never known, the truth invoked by David Hume against Thomas Hobbes: that, in our species, wherein sexual and parental love are intertwined, the generations overlap, so that we are *au fond* social and historical creatures, whose identities are always in part constituted by memories (such as those which are deposited in the languages that we speak) which cross the generations.[7] The forms of common life in which we find our identities are the environments in which we live and have our being: they are our human ecology.[8] . . . For conservative philosophy, therefore, as for ecological theory, the life of our species is never to be understood from the standpoint of a single generation of its members; each generation is what it is in virtue of its inheritance from earlier generations and what it contributes to its successors. In so far as one-generation philosophies prosper, the links between the past and the future are weakened, the natural and human patrimony is squandered, and the present is laid waste. The modernist idea that each of us is here only once, so we had better make the most of it, is a popular embodiment of the one-generational world-view, which finds expression in much liberal and socialist thought.

A second, connected idea, shared by conservative philosophy and Green theory, is the primacy of the common life. Both conservative and Green thinkers repudiate the shibboleth of liberal individualism, the sovereign subject, the autonomous agent whose choices are the origin of all that has value. They reject this conception, to begin with, because it is a fiction. Human individuals are not natural data, such as pebbles or apples, but are artefacts of social life, cultural and historical achievements: they are, in short, exfoliations of the common life itself.[9] Without common forms of life, there are no individuals: to think otherwise is to be misled by the vulgarised Kantian idea of the person which, shorn of the metaphysics that is its matrix in Kant and that gives it all the (slight) meaning it has, dominates recent liberal thought.[10] But liberal individualism also embodies a mistaken conception of the human good. For conservatives, as for Green thinkers, it is clear that choice-making has in itself little or no value: what has value are the choices that are made, and the options that are available – in short, what is chosen, provided it is good. . . . The ultimate locus of value in the human world is not, therefore, in individual choices but in forms of life. This should lead us to qualify, even to abandon, the ideal of the autonomous chooser . . . in favour of the recognition that the good life for human beings – as for many kindred animal species – necessarily presupposes embeddedness in communities. It is an implication of this point that Green theorists who extend to other animal species the legalist categories of individual rights are moving in precisely the wrong direction: what is required is the recognition that, among human beings, it is not individual rights but often forms of life that need most protection, if only because it is upon them that individual well-being ultimately depends.

A third idea shared by conservative and Green thinkers is the danger of novelty; in particular, the sorts of innovation that go with large-scale social (and technological) experimentation. It is not that conservatives (or sensible Green thinkers) seek to arrest change: that would be to confuse stability, which is achieved through changes that are responsive to the cycle of life and to the shifting environment, with fixity. It is rather that both Greens and conservatives consider risk-aversion the path of prudence when new technologies, or new social practices, have consequences that are large and unpredictable, and, most especially, when they are unquantifiable but potentially catastrophic risks associated with innovation. It is an irony that conservatives, whose official philosophy emphasises reliance on the tried and tested, should often embrace technological innovation, as if it were a good in itself. To be sure, there is little likelihood that the flood of technological innovation can in our time be stemmed; but that is no reason to welcome it or to refrain from curbing it, where this is feasible and there are clear dangers attached to it. It is at least questionable, for example, whether experimental advances in genetic engineering will on balance add to the sum of human well-being; whether their prohibition in any one country, or group of countries, can successfully halt their development, however, is another matter. It is more than questionable whether current high-tech policies in farming are defensible from a conservative perspective that is prudently risk-averse, since current farming technology, like other branches of industrial food technology, encompasses a myriad of interventions in natural processes, each of which has consequences that are unknown and whose effects, when taken together, are incalculable and unknowable.

A sound conservative maxim in all areas of policy, but especially of policy having large environmental implications, is that we should be very cautious of innovations, technological or otherwise, that have serious downside risks – even if the evidence suggestive of these risks is inconclusive, if the risks are small, or if their magnitudes cannot be known. A tiny chance of catastrophe may be a risk that can prudently be assumed, if all that is at stake is a human life or a few human lives. It is hard to see how any genuinely conservative philosophy can warrant risk-taking of this sort when the catastrophe that is being hazarded is environmental and millennial in its consequences. This is a truth which is acknowledged – and acknowledged as an element in a sane conservatism – by at least some Green thinkers.[11]

The conservative and Green aversion to risky change does not, of course, entail any policy of immobilism. It may indeed entail radical alternatives in current policy, if such policy encompasses substantial and unwarranted risks. Such alternatives will not, however, be animated by any conception of open-ended progress. It is a cardinal element in my argument for the consilience of conservative philosophy with Green thought that both reject the modernist myth of progress, and for very similar reasons. It is rejected by Green thought because it incorporates the idea of infinite growth – an idea alien to every tenet of ecology.[12] The characteristic that best distinguishes flourishing ecosystems is never growth, but rather stability (a conservative virtue in its own right). This is a truth which is acknowledged in the discipline of ecology in all of its varieties . . .

Conservatives have, or should have, their own good reasons for rejecting the idea of progress. Several come at once to mind. The idea of progress is particularly pernicious when it acts to suppress awareness of mystery and tragedy in human life. The broken lives of those who have been ruined by injustice or by sheer misfortune are not mended by the fact – if it were a fact – that future generations will live ever less under these evils. Meliorism, as embodied in the idea of progress, corrupts our perception of human life, in which the fate of each individual is – for him or her – an ultimate fact, which no improvement in the life of the species can alter or redeem. Again, the idea of progress presupposes a measure of improvement in human affairs which, except in limiting cases, we lack.[13] This is not to deny that we can meaningfully judge there to have been improvements in specific spheres of human life: no one who has read Thomas de Quincey's *Confessions of an English Opium-eater* can doubt that anaesthetic dentistry has made a not inconsiderable contribution to human well-being. But improvements in one sphere are accompanied by new evils in others: who is bold enough to affirm that the technological advances of modern medicine have, on balance, promoted human well-being? The facts of iatrogenic illness, of meaningless longevity and of the medicalisation of the human environment, well documented by Illich,[14] are telling evidences to the contrary. The deeper truth, however, is that, when assessing goods and evils across very different spheres of human life, we are trying to weigh incommensurables – longevity against the absence of pain, security against adventure, and so on. Although there are generically human evils – of torture, of constant danger of violent death, of human lives cut off in their prime – which are obstacles to any sort of human flourishing, even these universal evils cannot be weighed in the scales against each other. Like the goods of a flourishing human life, they are incommensurables. The conception of human history as a project of universal improvement, in so far as it is at all meaningful, is questionable, given that the eradication of one evil typically spawns others, and many goods are dependent for their existence on evils. At root, however, the idea of history as progressive amelioration is not so much debatable as incoherent . . .

. . . In this way the idea of progress reinforces the restless discontent that is one of the diseases of modernity, a disease symptomatically expressed in Hayek's nihilistic and characteristically candid statement that 'Progress is movement for movement's sake'.[15] No view of human life could be further from either Green thought or genuine conservative philosophy.

The modern conception of progress is only one symptom of the hubristic humanism that is the real religion of our age. As against that debased faith, both conservative and Green thought have as their ideal peace and stability. They seek a form of society that is sufficiently at ease with itself that its legitimacy does not depend on the illusory promise of unending growth. Neither Greens nor conservatives, if they are wise, are in any doubt as to the magnitude of the obstacles in the way of such a society. There can be no doubt that the project of a social order that does not rest on the prospect of indefinite future betterment creates problems for policy that have as yet been barely addressed by conventional thought, including the mainstream of conservative philosophy. Securing the legitimacy of political and economic institutions in a stationary-state society, which is without open-ended growth in population or production, is a hard and central problem for policy, which ought to concern Green thinkers as deeply as it should conservatives.

[. . .]

Notes

1 C.S. Lewis, *The Abolition of Man*, London: Macmillan, 1947. The citation occurs in Herman E. Daly (ed.) *Toward a Steady-State Economy*, San Francisco: W.H. Freeman, 1973, p. 323.

2 This is acknowledged in Hardin's book, *Nature and Man's Fate*, New York: Mentor Books, 1959, ch. 11.
3 For a discussion of the Prisoner's Dilemma, see Russell Hardin, *Collective Choice*, Chicago: University of Chicago Press, 1987.
4 The best version of this argument for market institutions is to be found in Michael Polanyi, *The Logic of Liberty*, Chicago: University of Chicago Press, 1951, ch. 8.
5 The term 'inherently public goods' I owe to Joseph Raz, who explains it in his *The Morality of Freedom*, Oxford: Clarendon Press, 1986, pp. 198–9.
6 Michael Oakeshott, *Rationalism in Politics*, London and New York: Methuen, 1977, p. 171.
7 There are some penetrating observations of the consequences of this point for moral and political thought in Stuart Hampshire, *Innocence and Experience*, London: Penguin, 1989.
8 That human individuals are tokens of which forms of life are the types is argued in the last chapter of my book, *Post-Liberalism: Studies in Political Thought*, London: Routledge, 1993.
9 See my *Post-Liberalism*, op. cit.
10 I have criticised the use of the Kantian conception of the person in recent Anglo-American political philosophy in my 'Against the New Liberalism', *Times Literary Supplement*, 3 July 1992.
11 See Arne Naess, 'Green Conservatism', in Andrew Dobson (ed.) *The Green Reader*, London: André Deutsch, 1991, pp. 253–4.
12 For a good statement of this truth, see Edward Goldsmith, *The Way: An Ecological World-View*, London: Rider, 1992, chs 22, 64 especially.
13 I discuss the incoherence of the idea of progress, in the context of incommensurabilities among human goods and evils, in chapter 20 of my *Post-Liberalism*, op. cit.
14 See Ivan Illich, *Limits to Medicine: Medical Nemesis: The Expropriation of Health*, London: Penguin, 1976, chs 1–3.
15 F.A. Hayek, *The Constitution of Liberty*, London: Routledge & Kegan Paul, 1960.

Source: John Gray (1993) *Beyond the New Right: Markets, Government and the Common Environment*, London: Routledge, pp. 124–30, 133–40, 185–6.

SECTION 6

Ecology and emancipatory strategies

Introduction

Throughout this section we shall look at approaches that seek to explore how environmental problems can be addressed as part of social and political transformation. In particular, these readings focus upon the relationship between ecological thinking and four closely connected sets of transformative social and political theories – socialism, Marxism, anarchism and feminism. In Reading 6.1, Ann Taylor draws an explicit connection between the concern for the health and safety of the British working classes and the emergence of environmental awareness. Whilst Taylor writes in the context of the UK Labour Party, her approach is representative of a broader orientation in Western centre-left politics and social movements in the twentieth century. In particular she highlights how the project of environmental improvement is closely tied to delivering social change to improve the quality of life of ordinary people, a concern which predates the contemporary ecological movements. In addition, this approach can point to a significant record of achievement for, as Taylor argues, there are a significant number of environmental problems where they made a difference. Nevertheless, the concern for improving material conditions is not easily reconciled with the recent environmental criticisms of industrialism. Taylor responds by trying to identify focused strategies for resolving environmental problems through a partnership of public (the state) and private interests (civil society) and without hitting those social groups that would be least able to absorb the costs of readjustment.

In Reading 6.2, Carolyn Merchant provides an outline of the central features of the Marxist approach to capitalism and the environment. Merchant identifies the dilemmas which characterize the relationship between Marx's materialist foundation of 'creative man' and the ecological insights that recent socialist ecologists have drawn out from his writings. This reading also introduces the social ecology approach developed by Murray Bookchin. This account, informed by anarchist assumptions, draws from but is also critical of both Marxism and deep ecology. Like Marxism it has a concern with the dialectical relationship between society and nature but it also raises questions about the authoritarian politics associated with communist regimes. Like deep ecology, social ecology starts from the premise that humankind is part of a complex web of life and its associated support systems but it also adopts a human-centred rather than ecocentric approach to the relationship between society and nature. Robyn Eckersley, in Reading 6.3, focuses on the anarchist legacy in Bookchin's writings and how this grounds the principle of evolutionary stewardship which combines the human-centred goals of selfhood and self-awareness with having a minimal impact on the environment. Eckersley's exploration of this approach generates a useful contrast with the impact of more localized strategies for ecologically informed social and political transformation. Here, she provides an account of the utopian socialist and anarchist theory and practice which underpin recent ecological movements, such as bio-regionalism and the eco-communalist approach of Rudolf Bahro.

The last three readings draw from an approach which seems to have fewer problems in integrating ecological consciousness – ecofeminism. Vandana Shiva, in Reading 6.4, provides a classic statement of the ecofeminist view of close connections between modern science as the domination of nature (also see David Pepper in Reading 1.3) and the emergence of patriarchy as the institutionalized domination of women. Shiva establishes these connections through a discussion of the impact of the reductionist science and the ecological perspectives generated in sustenance economies. Ynestra King (Reading 6.5) draws from Bookchin to explore the implications of rethinking the relationship between biology and culture for reassessing the connections between humankind and the environment. In addition, King outlines how the integration of ecological and feminist thinking has implications for knowledge construction, the social order and grassroots politics. In Reading 6.6, Joni Seager provides a powerful challenge to environmentalists and deep ecologists alike for their androcentric (male-centred) leanings. Seager's account of theory and practice demonstrates that the relationship between ecology and feminism is much more problematic than many ecofeminists assume and argues that broader issues of cultural difference should also be acknowledged.

6.1

Ann Taylor

The environment and the socialist ethic

Modern wisdom has it that the pressure of environmental collapse has forced upon us a new sense of responsibility towards future generations. This, like many important insights, is a half truth only. It is true that the burden of environmental damage has become so heavy that even the most blithe among us feel uncomfortable about the bill our children will have to pick up. It is not true that concern for future generations is new. On the contrary: such concern has always been the stuff of which socialism is made.

Nuclear waste provides an interesting case. Nuclear waste is rubbish: an unwanted residue. Throughout history we have dealt with unwanted residues by throwing them out. The nastier and more troublesome they are, the further we have thrown them: into middens, rivers and oceans. Even now, we tend to respond to rubbish with a 'chuck it as far as we can' attitude. Toxic chemicals are shipped off to the Third World. Nuclear waste, some have suggested, should be shipped even further: off into outer space.

The problem with nuclear waste is that it is not simply offensive, nor even just dangerous, but that it remains so for many years. Disposal, therefore, affects future generations as much as it affects us, or even more so. We can't simply throw the stuff outside the city gates, or dump it in the oceans and be done with it, because the safety of our children is at stake. This is new; and it also sums up many of the problems of modern pollution. For greenhouse gases are also wastes, rubbish of a different sort that we have thrown away into the atmosphere to the point where waste from our own consumption of energy and goods directly threatens the well-being, and even the prosperity, of future generations.

[. . .]

On an emotional and intuitive level, . . . the labour movement has always taken the future seriously; and in this sense it is well attuned to respond to the new threat posed by environmental degradation, which is in great measure a threat to the future. But there is also a further and more direct way in which 'environmental awareness' has long been embedded in socialist and labour movement campaigns to improve living and working conditions, through concern for the health of the working population.

It is important here to think about the related concepts of public and environmental health. In the international sphere, it is relatively easy to attract (Western) attention to problems such as deforestation and soil erosion, and their connection with issues of food security and climatic change. Environmental questions of this order are writ with an unmistakable capital E. However, in less developed countries the environmental issues which most immediately affect, and frequently cost, people's lives are very often to do with water supply and sewage disposal. These impinge upon the environment of many millions of poor people more directly and, to them, unignorably than the risk of future global warming.

It is easy to forget this from the perspective of a society with highly developed (albeit far from perfect) water supply and sanitation systems. Yet little more than a hundred years ago, when London was in the grip of cholera epidemics, the absence of such facilities was the most pressing issue of public, and therefore environmental, health. 'We ain't got no privez, no dustbins, no drains nor water splies and no drain or suer in the whole place' wrote 54 'poor and desperate residents of Soho' (with the alleged help of Charles Kingsley) in a letter to *The Times* in 1849. The environment for such people consisted wholly in crowded and filthy streets where, in Dickens' words, 'through the heart of the town a deadly sewer flowed, in place of a fine fresh river'; and a sewer, moreover, which provided untreated drinking water for the majority of the capital's population.

In aspiring and organising to overcome squalor, working class people in the early labour movement were making a perfectly clear statement about the kind of environment they wanted to live in; and if this doesn't seem much like a modern 'green' concern it should be set against the equally legitimate aspirations of the two-thirds of the world's present population who still don't have access to adequate water and sanitation facilities. This is not to say that modern green worries are the exclusive luxury of those who are no longer burdened by poverty. But it must be recognised that public health (and, indeed, poverty itself) is, and always has been, an environmental issue of foremost importance, precisely because it concerns the environment that people experience most intimately and immediately. The campaigns of the labour movement for medical provision, decent housing and services were, in this sense, all campaigns for a better environment.

On a different note it can be added that John Snow's discovery during the London epidemic, of the waterborne transmission of cholera was a breakthrough in our understanding of how the spread and nature of illness is profoundly affected by the way that society is organised. This concept is now familiar to us. It is readily appreciated that international trading routes permitted the spread of not just cholera but, in an earlier epoch, bubonic plague; that non-communicable illnesses such as lead, mercury or asbestos poisoning result directly from environmental contamination; and that even diseases whose exact causes are not yet understood, such as cancer, might be correlated with particular kinds of individual and social lifestyle. Many contemporary virologists also believe that the emergence of viruses such as that responsible for AIDS may be the result of changes in human population structure and mobility, creating optimal conditions for viruses that have long existed at a low level.

No-one could claim that these developments are not relevant to an environmental perspective. But applying the new, 'environmental' label to health issues of this kind should not mislead us into thinking that the issues themselves are new. The nineteenth-century cholera epidemics would now be considered a matter of environmental health, intimately connected with social conditions; and so too would the experience of the first generations of industrial workers, whose exposure to new industrial processes brought new, occupational health hazards. The early and enduring concern of the trade unions, and of organisations such as the Socialist Medical Association (established in 1930), with public health and with health and safety at work can therefore be seen as a forerunner of contemporary environmental concerns. Equally, the 1948 Factories Act and the 1974 Health and Safety at Work Act, both passed by Labour governments, have direct bearing on environmental health.

Health and safety in the workplace is a vital element of and in some ways a key to environmental protection for the simple reason that where industry pollutes, its workers will be the first to be exposed to risk. Clean, safe and healthy workplaces will, on the other hand, make for a cleaner, safer and healthier environment. Specific illnesses have always been associated with particular industries – pneumoconiosis in the mines, asbestosis among asbestos workers and, among textile workers, byssinosis, which despite hundreds of deaths and maimings was not officially recognised as an occupational illness until the 1970s. These occupational hazards usually have their counterparts in public health risks, as is the case for example in pulmonary disorders resulting from the burning of coal, and even more pointedly in public exposure to asbestos used in construction.

Similarly, it is estimated that every year in the Third World 10,000 people die and 400,000 suffer acutely from pesticide poisoning (Repetto, 1985, quoted in Winpenny, 1991; Hay *et al.*, 1991, quote even higher figures). Most of the casualties are farm-workers handling chemicals with inadequate training and precautions. The wider environmental damage caused by pesticides is well documented, as is their misuse; but the point to be made here is that environmentally hazardous substances often represent the most direct and serious threat to people who have to work with them.

Yet, of course, the risks from dangerous and polluting industries do not stop at the factory gates, as was demonstrated by the chemical explosions at Flixborough in 1974, and Seveso in 1977, where there are still no-go areas that were contaminated by highly toxic dioxins. A more recent leak of toxic gases from the Bhopal chemical works, in India, killed 2,000 people and injured 200,000 in 1983. Could such a calamity have happened if adequate health and safety standards were enforced? Or was the underlying truth that, as in so many cases of unclean and unsafe technologies, the absence of such standards itself acted as an incentive for Union Carbide to locate the factory in the Third World?

Even in highly developed countries, health and safety at work issues are still closely related to broader, public issues of environmental welfare. While the public at large is rightly concerned about emissions and spillages of toxic substances, the workers in those industries are the most likely to suffer the consequences of bad practice. (As can be seen from the record: the General, Municipal, Boilermakers and Allied Trades Union calculated in 1987 that 20,000 British people were dying each year from occupational diseases.) At the same time, it must be recognised that people are multidimensional: workers are also consumers, taxpayers and residents; they are also members of 'the public', and so ensuring a decent working environment is a crucial part of improving the total environment.

[. . .]

In 1970, three years before the Ecology Party was formed, the Labour Women's Conference endorsed a document on environmental policy which contained a wide-ranging programme to control pollution of air, rivers, sea and land. It called for mandatory lead-free petrol; various pollution taxes embracing the polluter pays principle; strict controls on dumping at sea; and encouragement of a shift towards organic farming. Two years later, the Party issued a further document which among other things called for a more environmentally sensitive approach to the calculation of GNP. These earlier papers were taken up in a 1986 Conference statement which was widely praised in the environmental movement for its comprehensive, radical analysis and the far-reaching solutions it proposed, and which has served as the foundation for subsequent policy reviews.

The establishment in 1973 of the Socialist Environment and Resources Association was another milestone in the emergence of this 'red green' caucus. Writing in the Spring 1984 issue of the Association's journal *New Ground*, Robin Cook gave a detailed critique, from an environmental perspective, of the Labour opposition's then Alternative Economic Strategy, which was based on a target of full employment and 3 per cent annual growth of GNP. At a time when Labour, and much of the country, was reeling from the effects of mass unemployment, he bravely attacked the notion of economic growth as a universal remedy, anticipating Labour's 1990s concern with the quality of life by insisting that 'We ought at least to ask what quality of growth is desirable at least as often as we ask what quantity is necessary.'

This was a direct attack on the primacy of growth as the governing economic concept, made from an environmental perspective but also harking back to the socialist critique of capitalism as a system of endlessly expanding capital and endless need to exploit new resources, both human and social. Such a view can no longer be said to exist only on the distant fringes of socialist politics, as SERA's 1990 affiliation to the Labour Party and the Party's commitment to appoint a Cabinet level Minister of Environmental Protection equally attest.

During the last two decades green undercurrents have also been felt in the trade unions, not only in the new understanding of health and safety as intimately related to environmental concerns, but also in specific issues, such as the National Union of Seamen's opposition to dumping radioactive waste at sea, the joint action of various unions on the control and use of pesticides, and the joint campaign against the privatisation of the water industry.

[. . .]

It should be stressed that this new environmental awareness did not just fall out of the sky or, for that matter, out of Ecology or Green Party pamphlets, but derives from the application of long-held values to new, environmental problems which have increasingly forced themselves upon our attention. There is no room for complacency: we must recognise that the scale and severity of global environmental problems do present a fundamental challenge to our past priorities and practice, because of the evidence that present levels of environmental consumption and degradation will result, if not in catastrophe, at least in severely curtailed opportunities for future generations. But in facing up to this challenge it is absolutely essential to have a framework of values with which to look for solutions to new problems. This, indeed, is the Labour Party's greatest environmental asset. It is reasonable to expect that, insofar as the problems really are new, meeting them will involve some changes in the framework itself, just as the framework was altered to accommodate racial and sexual equality. But, in the body of central, socialist values I have identified, we have an excellent starting point.

That does not mean there is no more thinking to be done. It is important, for example, that we reassess collectivism and be absolutely sure about what 'social welfare' is. Too often in the past we, along with very many non-socialists, have tended to identify social welfare with everyone in society having a high standard of individual – and primarily economic – welfare. This is wrong. We have always intuitively understood that wealth, however distributed, does not alone make for happiness; and environmental constraints now add an extra dimension to the argument, because of the newly perceived, real costs of creating wealth. Yet, properly understood, the welfare of society as a whole means much more than aggregated income, goods or services. We must, at the very least, enrich our idea of social welfare to embrace environmental benefits that are not only enjoyed collectively but can only be maintained through collective action.

A simple example illustrates the point: giving every adult a private car would, on one view, raise the level of social mobility, and thus well-being. But the consequences of accommodating so many millions of vehicles, in terms of congestion, noise, ugliness, loss of countryside, exhaustion of resources and atmospheric pollution, would certainly outweigh the benefits, so the total impact on social welfare would be negative. Yet this is exactly the direction in which we have been moving, and have been encouraged by government policy to move, for several decades. Obviously there is an alternative: the provision of an extensive and high quality public transport service. And, perhaps paradoxically, this also enhances choice.

This is not a new argument, and indeed it has recently come to be accepted in quite surprising circles. But the important point here is that it is an essentially socialist argument, containing a more far-sighted vision of social welfare than just a belief in raising the living standards of individuals, on an individual basis. Insofar as we seriously reconsider collective welfare, I believe we will move further away from reliance on increased consumption as the only means of human fulfilment.

At the same time we need to extend our idea of community. [. . .] We [. . .] need to look further, to people beyond our frontiers, and address the global inequities in the distribution of power and resources that we have allowed to exist throughout modern history. Environmental issues are pushing us in this direction, both because, in our new, small world, the activities of one nation impinge directly on the quality of the environment in other nations, and because the pressure for development in the poor world will create new strains upon resources and upon the global environment's absorptive capacities. The historic concern of socialists with distributional equality and justice – and our long solidarity with anti-imperialist struggles – uniquely qualify us to play a new and responsible role in the international community.

[. . .]

[Economics for our future]

In practical environmental decision making, one of the frequently perceived limiting factors of pollution abatement and environmental enhancement is the cost of taking the necessary measures. This preoccupation is indeed implicit in the way that pollution has long been considered an 'external' cost of production – one that needn't affect the producer's profitability, and that can be passed on to consumers either in financial terms (where government has to intervene to render wastes safe) or in terms of suffering a degraded environment.

Much less often mentioned is the cost of not preventing or cleaning up pollution, but of failing to do so: the cost of neglect – not just accidents but wilful damage. In terms of our experience of a degraded environment, the cost is often obvious and very high. But the financial costs can also be high. Expenditure on operations to clean up Prince William Sound after the *Exxon Valdez* disaster, to minimise the damage in the wake of Three Mile Island and Chernobyl, and the lost income from ruined environments like the American Great Lakes and the Black Sea are all examples of the way that allowing environmental degradation to happen in the first place can prove very costly in the long run. And this will be felt even more directly and painfully if global warming does result in a rise of sea levels, with whole Pacific islands disappearing and low-lying coastlands being rearranged. In this sense, we have clearly reached a stage where environmental degradation can actually be a limiting factor on economic activity. There is therefore an overwhelming need to reverse the mentality which sees pollution as a transferable cost – to be paid somewhere else by someone else – and to realise instead that environmental protection will not just result in a better quality of life but can represent genuine savings too.

Environmental efficiency is therefore a key to the approach that we must adopt. Energy efficiency is an obvious example where the case is clearly proven. The great advances made in methods of energy efficiency in recent years – although they have not all been put into practice – show how much can be achieved in this area. It has been estimated that our total consumption of energy could be cut by more than half if the best currently available energy efficiency technology were adopted in our homes, workplaces, generating industries and transport systems; and there is every likelihood that research and development of new technologies, stimulated by serious commitment to energy efficiency, would provide the possibility of even greater reductions in resource use and greenhouse gas emissions, without reducing consumer-end amenity. Even the most radical targets yet proposed for reductions in carbon dioxide emissions could be met without too much difficulty through energy efficiency measures on such a scale.

Needless to say, market forces won't stimulate this kind of change; and, unfortunately, privatisation of the generating industry was a decisive step in exactly the wrong direction. The private companies have a clear, vested interest in producing and selling as much energy as they can. Shareholders great and small are hardly likely to enthuse about efficiency programmes, or to encourage the companies to adopt policies aimed at reducing sales, given the ethos at the time of privatisation.

If the energy utilities are to remain in private hands, they must be subjected to strict efficiency standards, and required wherever possible to meet new demand not by increasing the amount of fuel burnt but by increasing efficiency so that the same amount of fuel goes further. Legislation of this kind already exists in the United States. At the same time we must begin to invest in 'combined heat and power' plants (usually, small plants which, instead of allowing heat to disappear up chimneys, use it for heating local houses and public buildings); and we must sponsor research and pilot schemes for renewable energy resources, against the day that we can begin to reduce our total dependence on fossil fuels. If a small amount of the research expenditure devoted to the nuclear power programme had been spent instead on research into renewables, we would now be much closer to meeting our energy requirements safely and cleanly.

It has been widely suggested that an energy tax is the best method of cutting energy demand, and some politicians have now accepted this view. Such a mechanism has an ideological appeal to non-socialists because it is 'market based'; that is, it attempts to influence behaviour by intervention in the market place rather than by direct regulation.

But the trouble with energy or carbon taxes is that they almost certainly hit the poorest hardest. Energy demand is relatively 'inelastic': that is, it is seldom a luxury item, and so increasing its price tends not to reduce demand significantly. We all need to keep warm in winter, and to light our homes; and many of us are dependent to a significant extent on private cars for personal mobility, getting to work, etc. So if we have to pay more for electricity or petrol, many of us will, reluctantly, do so; but this would clearly create major problems for those who cannot afford the difference. The wealthy will be relatively unaffected, with the fairly marginal reduction in demand being squeezed out of the poor, who almost invariably have the worst insulated houses and least efficient forms of heating, and who have less flexibility in the way that they use energy. People who are unemployed or dependent on state benefits cannot reduce their fuel bills by installing solar panels or buying a new, more energy-efficient car (if they have one at all).

A more promising approach would be to set up a programme of home insulation and energy efficiency, to introduce efficiency standards for construction and manufacturing industries and for energy-consuming goods (such as domestic appliances and, most importantly but most problematically, vehicles). Reducing consumer waste – unnecessary car journeys, and leaving house lights on all night – should be undertaken on the basis of public information campaigns rather than coercive indirect taxation. . . . General appeals to promote more environmentally conscious consumer behaviour must be backed by government action – to introduce product labelling schemes, for example, which must include information about energy efficiency, so that the choices of individuals will really make a difference.

This is not to say that financial incentives or market-based mechanisms have no role at all . . . [but] it may well be better to use financial incentives to encourage energy-efficient consumption rather than to punish inefficient consumption. . . .

It is generally accepted that much could be done to improve energy efficiency in a relatively short space of time, using existing technologies and well-defined policy instruments. Our use of energy is itself a fundamental environmental issue, but the example is of importance not just in signposting necessary changes in energy policy but as a model for our whole approach to environmental management (Jacobs, 1991 . . .).

For it may well be possible, following the course suggested, to bequeath to the next generation the same level of net energy capital that we now have available. In the literal sense, this will appear implausible: a barrel of oil burned today is gone for ever. But supposing we at present have known reserves of n billion barrels of oil equivalent, which under current technologies could generate x billion units of energy. More efficient technologies, already within our reach, would slow the rate of resource depletion and increase the amount of work done by each barrel of oil (or equivalent) consumed. So we might, even if no new reserves are discovered, be able to ensure that while gross reserves decline, the amount of work that can be done by those diminished reserves remains equivalent to what we can now achieve with present stocks under present conditions. In other words, we can not actually leave behind us the same stocks as now exist; but we could make sure that what is left is able to do as much work as x billion units of generated energy could do for us today. Future generations would therefore have an equal opportunity to generate and use energy (and an equal obligation to pass on the same potential).

A simpler, imaginary example shows how this could work in practice. In a world of 1 million families, where each family burns 2 tonnes of coal every year in open fires, there will be an annual energy demand of 2 million tonnes. Total coal stocks of 100 million tonnes would be enough for just 50 years on the open fire system. But, on this system, most of the heat from the fire goes up the chimney. If each family were able to acquire a simple stove, which produced the same amount of heat using only 1 tonne of coal per year, annual demand would drop to 1 million tonnes. So, after 50 years, there would be a remaining stock of 50 million tonnes: enough for another 50 years on the simple stove system. It would then be necessary, if it had not been done already, to develop a new stove that produced as much heat out of only half a tonne of coal a year, at the same time as investigating alternative ways of generating heat.

Here we have sustainability. It is not stocks that are sustained, but the ability of people to keep warm. Opportunity remains constant. Technology is not reviled as some kind of science-fiction nightmare, but it is put fairly and squarely in the service of people, rather than vice versa.

The pit, of course, is not bottomless. The day will come, in a world of a million people or in a world of more than 5,000 million, when declining reserves of fossil fuels, however carefully used, will no longer meet total needs. . . . Hence the need for investment, research, development and, hopefully, gradual transference to wind, wave and solar power. For sustainability also requires an eye to the future: an appreciation of the present outlook together with a sensitivity to long-term trends and possibilities. In other words, it requires intelligent planning of resources.

From this model of energy efficiency, resource substitution and investment in alternatives, we can derive principles that should govern our use not merely of energy but of all environmental goods and services. I am not just talking about resources. Our use of air, earth and water as dustbins for the byproducts of production and consumption relies upon their absorptive capacities which in many cases have already been pushed beyond reasonable limits. This has created a situation of environmental stress, which works in two directions: the environment itself is stressed (and none of us can actually say with any certainty how much more abuse it can withstand); and people too are experiencing stress as a result of environmental degradation. But if we make our industries and our consumption less polluting, reducing current levels of environmental stress, then we may be able to bequeath a constant surplus of environmental capacity, so that future generations with less polluting technologies will have the same opportunity as ourselves to produce and consume. Once again, opportunity will remain constant.

[. . .]

In the main, however, maintaining environmental opportunity would imply reducing pollution levels rather than completely banning certain activities. We can't simply ban the private car: but we can make ourselves less dependent on it (through investment in public transport), and we can make the machine itself less wasteful, less polluting. Meanwhile we should also be supporting more research into alternatives to the internal combustion engine.

Pollution abatement can best be achieved by setting targets, based on the principle of maintaining opportunity and informed by sound, scientific understanding of what the environment can tolerate. This must be the basis for agreeing targets, rather than considerations of how cheap and convenient it would be to meet them; and targets should have a strong, precautionary element, erring, if at all, on the side of underestimating the earth's capacities.

Once again, this sounds dramatic, and painful. But, as in the case of energy, we actually have much of the necessary means substantially to improve environmental efficiency. What is required is the political foresight and commitment to implement solutions that are already known about. So although greater control of industrial production is necessary, this is not inconsistent with maintaining, and even expanding, current levels of production itself. Provided each unit of production is made more environmentally efficient, it is technically feasible to sustain growth at the same time as reducing total use of environmental capacity and reserves: indeed, in the short term, opportunities for developing pollution abatement technologies, plus opportunities for employment in waste minimisation and insulation industries, and in the construction, operation and maintenance of treatment plants, etc., could stimulate a small green boom.

Reconciling growth with environmental protection has long been held by many environmentalists to be the equivalent of squaring the circle: but it is certainly not a logical impossibility and, if we are imaginative and inventive enough, it need not be a practical impossibility either. The extraordinary advances made in industrial and materials technology during the last 50 years are a testament to human creativity, and therefore a great source of hope; with the important caveat that such creativity must be put at the service of well-defined environmental criteria.

To keep the economy in line with pollution abatement targets a number of active government interventions will be necessary. In selecting policy instruments, we should not become preoccupied by a sterile debate about the rival merits of 'market-based mechanisms' as against direct regulation. These are often presented as the favoured instruments of right and left respectively; but while it is true that the right is generally fearful of anything that seems overtly regulatory, the ideological distinction between the two kinds of approach is in fact hollow. Regulating what industry can and cannot do has a direct effect on the marketplace; and intervening in the marketplace directly affects what industry does, and refrains from doing. Both are valid forms of intervention, and the choice between them must in each planning decision be made on the basis of what is most likely to achieve the desired ends, with due regard to cost effectiveness, ease of administration and social impact.

There is no doubt that there will have to be a strong, basic framework of regulations. Direct regulation has a substantial policy record, as a clear-cut and straightforward way of achieving objectives, and it is foolish to claim that further intervention of this kind is in principle undesirable or irreconcilable with profitable production. Health and safety legislation and regulations governing the food and water industries are solid examples of measures which are indispensable to public health, and which could not be achieved in any other way. They are not perfect, but they have got us as far as we are. There is an overwhelming case for direct regulations governing energy standards, pollution and waste control, and that those regulations must be more stringent than at present.

[. . .]

Market interventions through financial incentives or disincentives are often considered appealing because they do not require creating a watchdog body and because they have the appearance of being less coercive than direct regulation.

[. . .]

More far-reaching market mechanisms – such as an energy tax, . . . – would not escape the appearance of coerciveness at all. A tax of this kind would almost certainly be unpopular and would be very hard to justify, especially as it is doubtful whether it would actually succeed in significantly reducing fossil fuel demand. Indeed, the best argument for it is that it would provide a source of public revenue that could be reinvested in efficiency schemes, and used to compensate the lower income groups who would be hardest hit by the tax. . . .

But we must not reject financial incentives or market mechanisms out of hand. Clearly, some such incentives can work. Charging returnable deposits on bottles, for example, is a long-established mechanism for ensuring maximum reuse, and one that drinks manufacturers would do well to revive and extend. Perhaps there is a lesson here, that the simplest mechanisms are those which are most likely to get results.

[. . .]

References

Boys, P. (1983). 'Cholera, Class and Empire in the 19th Century,' in *Science for People*, No. 54.
Brockway, F. (1980). *Britain's First Socialists*. Quartet, London.
Hay, A., Hurst, P. and Dudley, N. (1991). *The Pesticide Handbook*. Pluto, London.
Jacobs, M. (1991). *The Green Economy*. Pluto, London.
Poulsen, C. (1984). *The English Rebels*. Journeyman, London.
Repetto, R. (1985). *Paying the Price: Pesticide Subsidies in Developing Countries*. WRI, Washington.
Thompson, E.P. (1977). *William Morris: Romantic to Revolutionary*. Merlin, London.
Williams, R. (1983). *Towards 2000*. Chatto & Windus, London.
Winpenny, J. T. (1991). *Values for the Environment*. HMSO, London.

Source: Ann Taylor (1992) *Choosing Our Future: A Practical Politics of the Environment*, London: Routledge, pp. 21–6, 39–40, 200–9, 219.

6.2

Carolyn Merchant

Emancipation and ecology

. . . [M]any people are searching for ways to resolve the contradiction between production and ecology. Calling themselves variously social ecologists, socialist ecologists, green Marxists, and red greens, they ground their approach in an ecologically sensitive form of Marxism. Social ecologists focus on the relations of production and the hegemony of the state in reproducing those relations. Their ethic is basically homocentric, inasmuch as social justice is a primary goal, but it is an ethic informed and modified by ecological and dialectical science. . . . [This] informs and draws on the actions of left greens, social and socialist ecofeminists, and many activists in the Third World sustainable development movement . . .

Marx and Engels on ecology

For most people, Marxism is synonymous with the rigidity and oppression of the bureaucratic states of the Soviet Union, Eastern Europe, and China. Moreover, Marx's prediction that capitalism would generate economic and social crises that would lead to socialist revolutions in capitalist countries, led by the working classes, has not been borne out. Marx's emphasis on the lawlike characteristics of a society's economy placed less stress on the role of social movements, politics, culture, and consciousness in transforming society than on the overthrow of the mode of production. Since the 1960s, however, Marxist theorists have emphasized the processes by which people are socialized through gender, race, and class and the ways in which social movements can identify and alter those patterns. Many groups, including the New Left, democratic socialists, socialist feminists, and racial and religious minorities, have found insights in the writings of Marx and Engels that promote goals of liberation, freedom, and economic equality. The same is true of ecological Marxists, who emphasize, not the control and domination of nature, but rather the ways in which ecological theories and green social movements can help to transform people's consciousness and practices toward nonhuman nature.[1]

Although Marx and Engels certainly argued that the domination of nature through science and technology would relieve humankind of the "tyranny" imposed by nature in procuring the necessities of life (food, clothing, shelter, and fuel), they were also acutely conscious of the "ecological" connections between humans and nonhuman nature. Like many critics today, they reacted against the mechanistic worldview of the seventeenth century. This mechanical materialism assumed that matter was made up of inert atoms and that all change was externally caused. Perception is explained as the result of corpuscles of light hitting an object such as a table or pencil, entering the eye, and being recorded as an impression on the brain. The individual is the passive receptor of information, just as the worker is the passive receptor of the capitalist's decision to offer minimal wages. Any worldview that casts the laborer as a powerless recipient of the ideas of a controlling elite is not healthy for her or him.

Similarly, the alternative view, prevalent in Marx's time, that the world was fundamentally spirit or idea, working itself out through history – the view of German philosopher Georg Hegel – was equally problematical. This worldview likewise rendered laborers powerless to change their destinies. What both the mechanists and the Hegelians had left out of their philosophies were social relations. People are born into a given type of society at a given time in history. Their place in that society is the perspective from which they view the world. Those in control of the society – the elite – will use the worldview to justify and maintain their hegemony. But laborers, artisans, minorities, and the poor have a choice of ways in which to view the world. They do not have to accept the mechanistic philosophy that renders them passive receptors of knowledge. More compatible with their social needs is a worldview that makes change, rather than *stasis*, central.

[. . .]

Seeing the world as fundamentally process and change, however, has implications not only for society, but also for nature. Marx, in his *Economic and Philosophical Manuscripts of 1844*, recognized the interdependence of humans and nature, an idea now central to the ecological vision. People, he asserted, were active natural beings who were corporeal and sensuous and who, like animals and plants, were limited and conditioned by things outside themselves. They were both different from these objects and yet dependent on them. "The sun is the object of the plant – an indispensable object to it, confirming its life – just as the plant is the object of the sun, being an expression of the life-awakening power of the sun." Like today's ecologists, Marx recognized the essential linkages between the materials that make up the human body and nonhuman nature. "Nature is man's inorganic body," he wrote. "Man lives on nature – means that nature is his body, with which he must remain in continuous interchange if he is not to die. That man's physical and spiritual life is linked to nature means simply that nature is linked to itself, for man is a part of nature."[2]

Humans, however, differed from other animals in the way in which they obtained the essential food and energy to continue living. What distinguished humans, thought Marx and Engels, was their capacity to produce, using tools and words. The tools of animals were, in most cases, parts of their bodies, with inconsequential effects on nature. Humans, by contrast, transformed external nature with instruments that were socially organized. In different periods in history, humans organized their instruments and labor into different modes of production. Gathering-hunting, horticulture, feudalism, capitalism, and socialism are different modes of production that transform nature in different ways.

Essential to the "ecological" vision of Marx and Engels is their study of the history of human interactions with nature. Early societies, they argued, had a different relationship to nature than do capitalist societies. While pastoral societies wander, taking from nature that which is necessary for life, horticultural societies settle down and appropriate the earth's resources for their own sustenance. Humans thus modify external nature, using the local climate, topography, and flora and fauna for their own purposes. The settled community uses the earth as "a great workshop," for its labor. Human labor, on the one hand, and the earth, with its soils, waters, and organic life as instrument of labor, on the other hand, are both necessary for the reproduction of human life. Humans, isolated from society, would live off the earth as do other animals.

For the earth to be appropriated as property, humans must settle on the land and occupy it. Under capitalism, the earth is bought and sold as private property. Here, according to Engels, the earth is peddled for profit. "To make the earth an object of huckstering," he wrote, "– the earth which is our one and all, the first condition of our existence – was the last step toward making oneself an object of huckstering." It is the ultimate in alienation. In the capitalist appropriation of the earth for profit, raw materials, to be taken from the earth, such as coal, oil, stone, and minerals, are the result of natural forces. They are the "free gift of Nature to capital." Nature produces them and the capitalist pays the laborer to transform them. Similarly, physical forces, such as water, steam, and electricity cost nothing. Science, likewise, costs capital nothing, but is exploited by it in the same manner as is labor.[3]

But these modes of transforming nature have unforeseen side effects. Like modern ecology, which is premised on the concept that everything affects everything else, Engels noted in his *Dialectics of Nature* that "in nature nothing takes place in isolation. Everything affects every other thing and vice versa, and it is mostly because this all-sided motion and interaction is forgotten that our natural scientists are prevented from clearly seeing the simplest things."

Engels warned that people should not boast about their ability to master nature because there were always harmful consequences of such conquests. . . .

"Thus at every step," Engels admonished, "we are reminded that we by no means rule over nature like a conqueror over a foreign people, like someone standing outside nature – but that we, with flesh, blood, and brain, belong to nature, and exist in its midst, and that all our mastery of it consists in the fact that we have the advantage over all other creatures of being able to know and correctly apply its laws." The more one understands the laws of nature and the consequences of human actions, he went on, the more humans will come to "know themselves to be one with nature," and that there is no inherent "contradiction between mind and matter, man and nature, soul and body." These dualisms originated in the philosophy of ancient Greece, were reinforced by Christianity in the Middle Ages, and codified by the philosophers and scientists of the seventeenth century. Their dissolution is one of the goals of the radical ecological and ecofeminist movements today.[4]

In *Capital*, Marx analyzed some of the "ecological" side-effects of the capitalist mode of production. He argued that capitalist agriculture, much more than communal farming, wastes and exploits the soil. In agriculture geared toward production for profit, the soil's vitality deteriorates because the competitiveness of the market fails to allow the large-scale owner or tenant farmer to introduce the additional labor or expense needed to maintain its fertility.

[. . .]

Industrialization, according to Marx, resulted in similar "ecological" problems. Wastes from industry and human consumption accumulated in the environment and were not reused by the capitalist unless the price of raw materials soared. Marx gave numerous examples of capitalist pollution: chemical by-products from industrial production; iron filings from the machine tool industry; flax, silk, wool, and cotton wastes in the clothing industry; rags and discarded clothing from consumers; and the contamination of London's River Thames with human waste. Yet this waste that clogged and polluted waterways was very valuable and had the potential to be recycled by industry. The chemical industry could reuse its own waste as well as that of other industries, converting it into useful products such as dyes and rugs. The clothing industry could improve its use of the waste through more efficient machinery. Human waste could be treated and used to build soil fertility. An "economy of the prevention of waste" that reused all waste to the maximum was required.[5]

Marx assumed a two-levelled structure of society: the economic base or mode of production (which consisted of the forces and relations of production) and the legal-political superstructure (Figure 11). Together these constituted the social formation. Different modes of production, such as primitive communism, ancient, asiatic, feudal, capitalist, and socialist, had different legitimating superstructures. Marx's theory of social change was based on a conflict between the material forces of production and the social relations of production. This dialectic initiates an era of social revolution in which the economic foundation breaks down, leading to a change in the superstructure. Today social ecologists envision a transformation of the global capitalist economy and its legitimating mechanistic worldview to a sustainable economy and a process-oriented ecologically-based science. It would be brought about by social movements, especially those concerned with environmental health and quality of life.

Anarchist social ecology

Current theories of social ecology draw on Marx and Engels' approach to "ecology" and society. Additionally, social ecologists draw their ideas from premodern tribal societies, eastern cultures, and from analyzing the ecological problems of capitalist, socialist, and Third World countries. For anarchist philosopher Murray Bookchin, social ecology is rooted in the balance of nature, process, diversity, spontaneity, freedom, and wholeness. His ideal society would eliminate all hierarchies in ecology and in society. The ecological

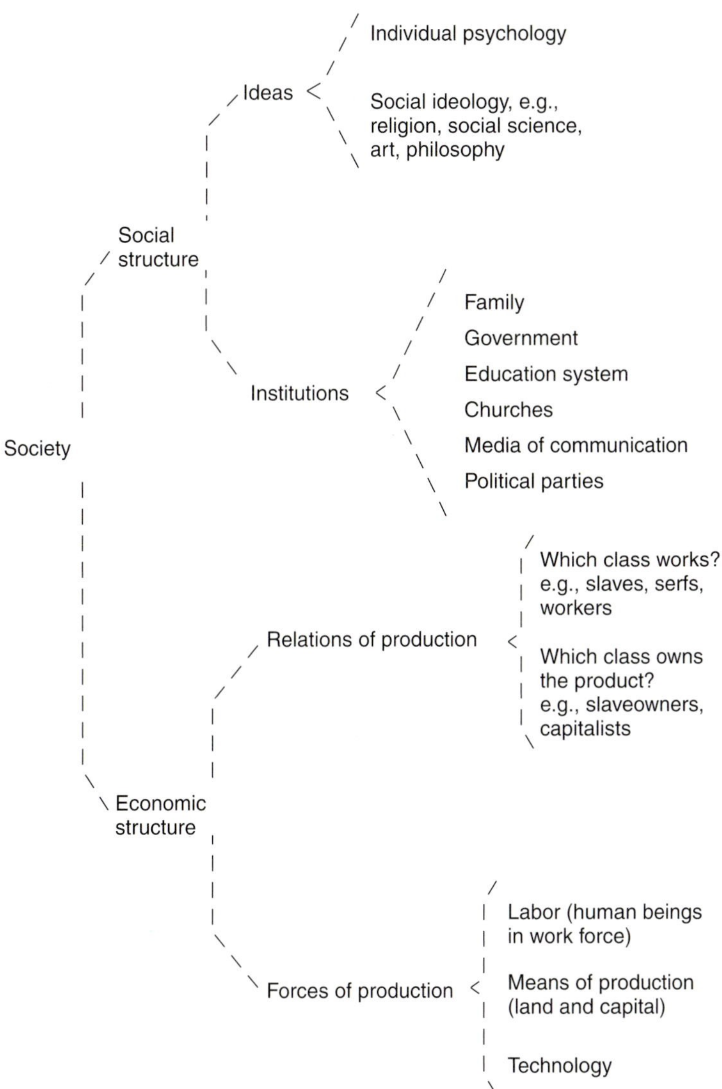

Figure 11 Marxist framework of social analysis
Source: Howard Sherman (1987) *Foundations of Radical Political Economy*, Armonk, N.Y.: M.E. Sharpe, p. 44, reprinted by permission.

society of the future would reclaim the fundamental organic non-hierarchical relationships of preliterate peoples.

[. . .]

. . . Scarcity and warfare escalated the problems created by the twin pillars of dominance and hierarchy, and non-egalitarian culture continued in all subsequent societies. Today dominance and hierarchy permeate all aspects of life, especially in the dominance of the intellectual over the physical, work over pleasure, and mental control over sensuous body. A major goal of social ecology is to abolish these dualisms.

In an ecological society, Bookchin argues, dominance and hierarchy would be replaced by equality and freedom. An "ecology of freedom" would reunite humans with nature and humans with humans. This would be achieved through an organic, process-oriented dialectic that would reclaim the outlook of preliterate peoples. The merging of their ecological sensibility with the analytical approach of Western culture would produce a new consciousness. Thus the advances of science and technology could be retained and infused with an ecological way of living in the world. This recognizes the mutual dependence of humans and nonhuman nature. The ecology of freedom is rooted in a concept of ecological wholeness that is more than the sum of its parts. "Unity in diversity" means the unfolding of the processes of life. Bud is replaced by flower and flower by fruit, as moments in an emerging unity. Spontaneity is the continual striving of nature toward change and of humans toward greater self-awareness and freedom.

Bookchin distinguishes between ecology and environmentalism. Environmentalism adopts the mechanistic, instrumental outlook of the modern world that sees nature as resource for humans and humans as resources for the economy. Nature consists of passive resource objects in habitats constructed for human benefit. Environmentalism does not question the *status quo*, but facilitates the domination of humans over nature and humans over other humans. Ecology, premised on interactions among the living and nonliving, contains the potential for an alternative. Social ecology incorporates humans and their interdependences with nonhuman nature. Bookchin uses the term ecosystem to mean "a fairly demarcatable animal-plant community and the abiotic or nonliving factors needed to sustain it." Extended to society, it becomes "a distinct human and natural community, [including] the social as well as organic factors that interrelate with each other to provide the basis for an ecologically rounded and balanced community."[6]

[. . .]

Social ecology has a deep commitment not only to reversing the domination of nature, but also to removing social domination. Hierarchical and class inequalities have resulted in homelessness, poverty, racial oppression, and sexism. Of particular concern are forced and insensitive methods of controlling populations, rather than restructuring and redistributing food, clothing, and shelter.

Bookchin argues that certain deep ecologists . . . are insufficiently sensitive to social issues, especially regarding population, race, class, and sex. This includes some, although by no means all, supporters of Earth First!, the spiritual Greens, some bioregionalists, and some spiritual ecofeminists. To speak of a global population problem as threatening wilderness and the entire biosphere is incorrectly to analyze the roots of ecological problems by disregarding the differential impact of economic growth, especially capitalist growth, on indigenous people, marginalized rural and urban people, people of color, and women.

Social ecologists decry the idea of involuntary methods of population control, the Malthusian idea that famine, disease, and war are positive checks on population expansion, and the policy that immigration of southern and eastern hemisphere people into northern countries should be tightly restricted. Instead they support an ecologically-based development policy that uses resources in a sustainable way while raising the quality of life and redistributing the means of fulfilling basic needs.

The debate between deep and social ecologists highlights differences of opinion on where to place the core of the analysis as well as approaches to solutions. Social ecologists tend to see the problem as rooted in the dialectic between society (especially economies) and ecology, whereas deep ecologists focus on the conflict between the ecological and mechanistic worldviews. Similarly, for social ecologists, action must be focused on ecodevelopment and social justice as opposed to the deep ecologists' goal of transforming the worldview and reclaiming spiritual connections to the earth.

Socialist ecology

Another alternative rooted in the Marxist tradition is socialist ecology. Socialist ecology offers an eco-economic analysis of the interaction between capital and nature and the transition to a post-capitalist society. Instead of Bookchin's emphasis on hierarchy and domination, a utopian anarchist society modelled on "nature," and a Hegelian dialectic, it envisions an economic transformation to ecological socialism, initiated by new green social movements.

Socialist ecology is spearheaded by economist James O'Connor, author of *The Fiscal Crisis of the State* and other books on economic crises. Rooted in Marx's conceptual framework, it nevertheless goes beyond Marxism to incorporate concepts of ecological science, the social construction of "nature," and the autonomy of nature. It argues that the environment and ecology are the key issues for the late-twentieth and twenty-first centuries, as evidenced by the global ecological crisis and the rapid growth of green social movements, ecofeminism, working-class anti-toxics crusades, and farm-worker anti-pesticide coalitions. It encourages an analysis of the dialectics between economy and ecology and between nature and history. Additionally, it offers a critique of existing socialist societies which have failed to address the ecological crisis and fosters thought about a reconstructive ecological socialism. In addressing the general problem of capitalism, nature, and socialism, it encourages dialogue among Marxists, Marxist-feminists, ecological Marxists, post-Marxists, left-greens, red-greens, and others.

O'Connor's theory of capital and nature is grounded in the traditional Marxian dialectic between the forces of production (technologies) and the relations of production (exploitation of labor by capital). This dialectic is the first contradiction of capitalism and leads to economic crisis and the breakdown of capitalism. But O'Connor equally emphasizes a second contradiction within capitalism, that between production and the environmental conditions of production (Figure 12). Marx and Engels used the term conditions of production to encompass human resources (labor), natural resources, and space. In ecological Marxist theory, these conditions of production come into conflict with the forces/relations of production. This second contradiction of capitalism leads to eco-economic crisis, initiating the transition to ecological socialism.[7]

Ecology is the basis of three conditions of production. First are the external physical conditions, what Marx called the natural elements entering into capital. Examples are the health and viability of ecosystems, such as the adequacy and stability of wetlands and the quality of soils, waters, and air. Second are the personal conditions of the laborers. Examples are the health of workers, as affected by the environment. Toxics and pesticides in the workplace, smoggy air and polluted water, unpleasant surroundings in the work environment, all affect the well-being of workers. Third are the social conditions of production, such as the means of communication among workers and managers.

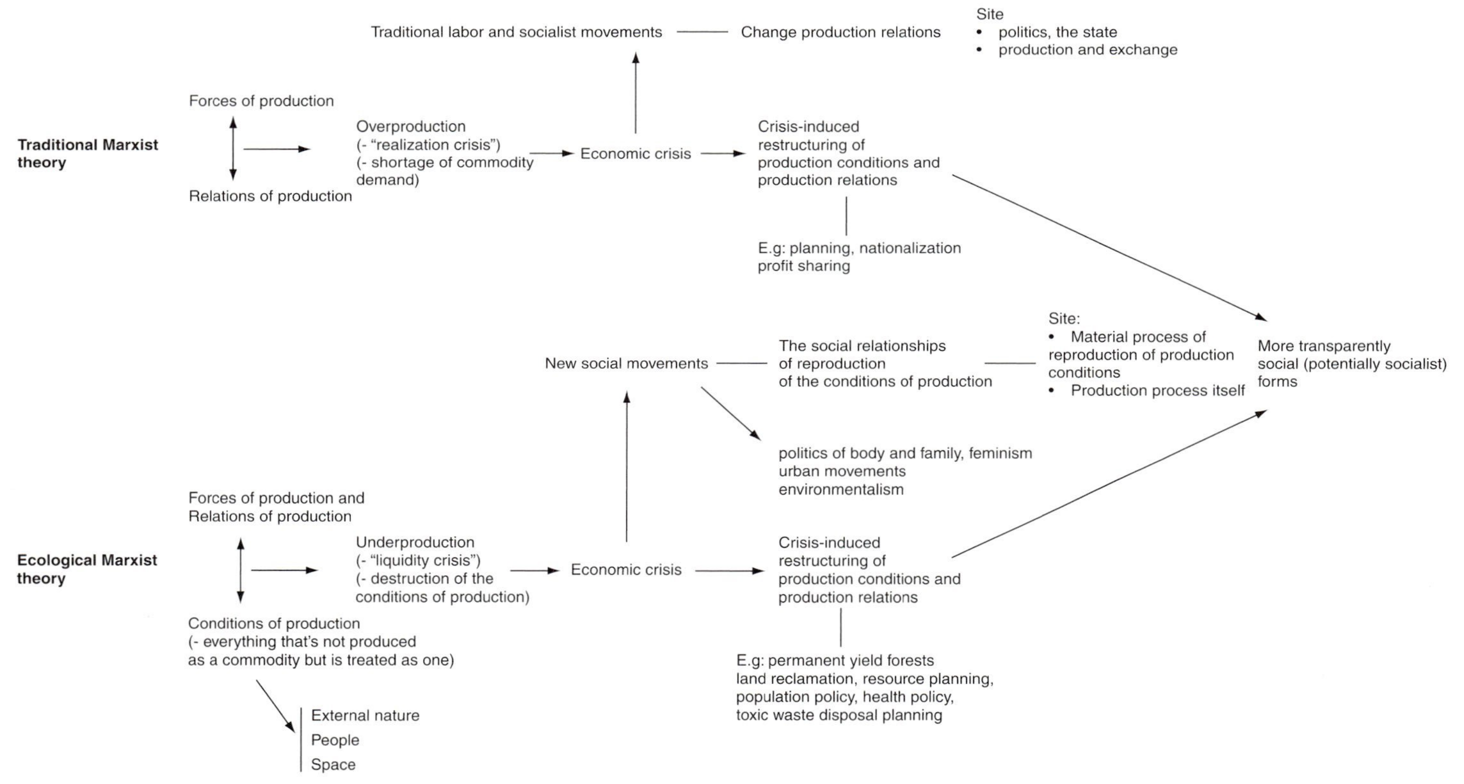

Figure 12 Socialist ecology
Source: Diagram by Yaakov Garb, based on James O'Connor, 'Capitalism, Nature, Socialism: A Theoretical Introduction', *Capitalism, Nature, Socialism*, Fall 1988, pp. 11–38.

In traditional Marxist theory, the first contradiction of capitalism leads to overproduction of goods. There is a decreased demand among consumers for the product. In ecological Marxist theory, however, the second contradiction of capitalism leads to underproduction. Capitalism creates its own barriers to growth by destroying its own environmental conditions of production. Ecologically destructive methods of agriculture, forestry, and fishing raise the costs of raw materials that lead to the underproduction of goods and the underproduction of capital. Soils are depleted, waters are polluted, workers' health fails, yields of produce, meat, wood, and textiles decline. In its hunger for profits, capital thus destroys its own ecological conditions of production. Rather than leaving Nature free and autonomous, capitalism recreates it as capitalized nature – a second nature treated as commodity and subjected to ecological abuse.

In traditional Marxism, the agencies of social transformation are the traditional labor and socialist movements that change the relations of production, through collective bargaining for example. Here economic crisis make it possible to imagine the transition to socialism. In ecological Marxism, instead, the agencies of social transformation are the new ecological social movements: environmental health and safety, farmworkers' antipesticide coalitions, ecofeminist protests over groundwater toxins, leftwing green parties, and so on. Here it is ecological crises that make it possible to imagine the transition to socialism. Such crises and social movements push capitalism to respond in more transparently social and potentially socialist ways. In turn, capitalism responds by introducing more environmental and natural resources planning – sustained yield forests, environmental health policies, toxic waste disposal practices, and so on.

But in imagining the transition to an ecological socialism, socialist ecology criticizes state socialism, arguing that this is not what the new vision entails. State socialist societies have created ecological crises and fostered ecologically destructive policies, as have capitalist societies. Their planning processes nationalize production rather than democratizing and socializing it. They stifle individual creativity and are bureaucratically inflexible. They abuse and deplete nature as do capitalist societies, but do so not because of the profit-motive, but because their commitment to full employment stifles appropriate technologies and permits polllution.[8]

In an ecological socialist society, Nature will be recognized as autonomous, rather than humanized and capitalized. Ecological diversity, an ecological sensibility, and a science of survival based on the interrelatedness of living organisms and the environment will all be needed and valued.

What is an example of such an ecology of survival? One such case history is the use of biological insect controls in Nicaragua. Before the Nicaraguan revolution of 1979, agricultural production was dependent on heavy applications of pesticides to produce high cotton exports.

Broad-spectrum chemicals destroyed natural insect enemies, created new chemically-resistant pests, and caused high numbers of pesticide poisonings among workers. A pesticide treadmill set in, in which a cotton export economy became dependent on increasing amounts of pesticides to maintain yields, fueling the profits of foreign chemical companies. After the overthrow of the Somoza regime, the new socialist government stepped up the use of Integrated Pest Management (IPM) techniques and revolutionized the forces of production.

Integrated Pest Management uses biological methods of controlling insect pests as its core. It depends on the careful monitoring of pest levels by trained field aides who assess when the economic threshold of pest damage has been reached, as opposed to spraying broad-spectrum chemicals on predetermined calendar dates. Pesticides are applied only in limited amounts and in narrow ranges. Plants are cut and plowed under the soil between seasons to avoid carryover of pests. Before the Sandinista revolution, regional IPM programs had been difficult to implement because not all capitalist growers in an area cooperated. The restructuring of farms under the new government created new relations of production which allowed new forces of production such as IPM to take root. These new productive forces fostered better conditions of production by improving both the health of the soil and the health of the workers. The government was able better to plan production, train IPM field hands, save on the enormous costs of pesticides, and achieve higher yields.

IPM as a force of production creates independence as opposed to chemical-company dependence and creates jobs for field workers. Yet the problem of outside markets for the sale of agricultural produce was obfuscated by trade embargoes imposed by the United States and by the continuing devastation of the country by the war with the Contras. Thus, despite increasing independence in production, dependence on world markets and politics hindered economic stability.[9]

[. . .]

Conclusion

Social ecology emphasizes the human implications of systems of economic production on the environment. Both capitalism and state socialism produce externalities that disrupt nature. Social ecology envisions a world in which basic human needs are fulfilled through an economic restructuring that is environmentally sustainable. While social ecologists would like to see world population stabilize at a level that is compatible with environmental sustainability, they deplore any programs that result in genocide, racism, or callous disregard for human rights in bringing about a demographic slowdown. Instead, economic programs that provide for basic needs, healthcare, security in old age, and employment are the pathways that will bring about a demographic transition in developing countries and equalize the quality of life in both developed and developing countries. Finally, social ecology advocates a science oriented toward social values and the recognition of change, rather than stability, as the basic premise on which to understand the natural world. It is similar to deep ecology in calling for a major transformation in worldviews and a process-oriented science, but differs from it in its emphasis on the human condition, the economic basis of transformation, and a homocentric as opposed to an ecocentric ethic.

Notes

1 Roger Gottlieb, *History and Subjectivity: The Transformation of Marxist Theory* (Philadelphia, Pa.: Temple University Press, 1987); Gottlieb, *An Anthology of Western Marxism: From Lukacs and Gramsci to Socialist-Feminism* (New York: Oxford University Press, 1989), pp. 1–25.

2 Howard Parsons, ed., *Marx and Engels on Ecology* (Westport, Ct.: Greenwood Press, 1977), p. 133.

3 Parsons, ed., *Marx and Engels on Ecology*, quotations on pp. 176, 172, 171.

4 Parsons, ed., *Marx and Engels on Ecology*, quotation on p. 179.

5 Parsons, ed., *Marx and Engels on Ecology*, quotation on p. 177.
6 Murray Bookchin, "The Concept of Social Ecology," *Coevolution Quarterly* (Winter 1981): 14–22, quotation on p. 17, Bookchin, *Ecology of Freedom* (Palo Alto, 1982).
7 O'Connor's second contradiction is similar to my first contradiction, that between ecology and production. In my framework, the second contradition is that between production and reproduction. James O'Connor *The Fiscal Crisis of the State*, New York: St. Martins Press, 1973.
8 James O'Connor, "Capitalism, Nature, Socialism: A Theoretical Introduction," *Capitalism, Nature, Socialism* 1 (Fall 1988): 11–38.
9 Sean Swezey and Daniel Faber, "Disarticulated Accumulation, Agroexport, and Ecological Crisis in Nicaragua: The Case of Cotton," *Capitalism, Nature, Socialism* 1 (Fall 1988): 47–68.

Source: Carolyn Merchant (1992) *Radical Ecology: The Search for a Liveable World*, London: Routledge, pp. 134–50, 153–4, 253–4.

6.3

Robyn Eckersley

Ecoanarchism: the non-Marxist visionaries

. . . [E]coanarchists divide in terms of the various theoretical explanations they offer to account for the ecological crisis (e.g., social ecology attaches greater theoretical importance to social hierarchy than does ecocommunalism); the types of ecocommunities they advocate (e.g., social ecology is more libertarian, whereas the ecomonastic strand of ecocommunalism tends to be relatively more ascetic); and the *degree* to which they are critical of the Western anthropocentric heritage (e.g., the ecocommunal tradition is generally more ecocentric than social ecology). A further difference is that social ecology is largely the work of one particular theorist – Murray Bookchin – who has developed a distinctive organismic ecophilosophical perspective whereas ecocommunalism is a more general category that I employ to encompass a variety of other kinds of ecoanarchist approaches of a relatively more ecocentric persuasion.

The social ecology of Murray Bookchin

Murray Bookchin stands as one of the early pioneers of Green political theory. Over the past three decades, Bookchin's numerous publications on "social ecology" have sought to restore a sense of continuity between human society and the creative process of natural evolution as the basis for the reconstruction of an ecoanarchist politics.[1] Bookchin describes his thought as carrying forward the "Western organismic tradition" represented by thinkers such as Aristotle, Hegel, and, more recently, Hans Jonas – a tradition that is process oriented and concerned to elicit the "logic" of evolution.[2] According to Bookchin, the role of an ecological ethics is "to help us distinguish which of our actions serve the thrust of natural evolution and which of them impede it."[3] For Bookchin, evolution is developmental and dialectical, moving from the simple to the complex, from the abstract and homogeneous to the particular and differentiated, ultimately toward greater individuation and freedom or selfhood. Social ecology – a communitarian anarchism rooted in an organismic philosophy of nature – is presented as the "natural" political philosophy for the Green movement because it has grasped this "true" grain of nature and can promise greater freedom or "selfhood" for both nonhuman nature and society. For Bookchin, an anarchist society, free of hierarchy, is "a precondition for the practice of ecological principles."[4]

Although ecosocialism and social ecology represent quite distinct schools of emancipatory thought, there is nonetheless a superficial resemblance between the two. For example, it is noteworthy that Bookchin has become a major voice of the "Green Left" in North America, sharing many of the European ecosocialists' criticisms of deep ecology (this is not surprising, given Bookchin's familiarity with Continental political theory and his former Marxist leanings). This shared critique arises from the fact that both social ecology and ecosocialism emphasize, as their names might suggest, the *social* origins of environmental degradation – an emphasis that has led both to criticize the deep ecology focus on anthropocentrism for deflecting attention away from inequities such as those based on class, gender, and race.[5] It is also interesting to note that Bookchin has claimed that the West German Green Party "has supplanted the traditional socialisms with a social ecology movement."[6] . . . [M]any ecosocialists also claim that the West German Green Party is essentially an *ecosocialist* party. The many factions within the West German Green Party are such that both claims have some degree of truth.

Beyond these superficial similarities, however, social ecology departs fundamentally from ecosocialism in terms of both analysis and ecophilosophical perspective. Whereas ecosocialists have singled out capitalism as the main driving force behind environmental degradation, Bookchin has conducted a more wide-ranging critique that regards capitalism as but a subset of a more deep-seated problem, namely, social hierarchy. Bookchin's social ecology, then, is not Marxist but rather *libertarian*. As Bookchin explains:

> To create a society in which every individual is seen as capable of participating directly in the formulation of social policy is to instantly invalidate social hierarchy and domination. To accept this single concept means that we are committed to dissolving State power, authority, and sovereignty into an inviolate form of personal empowerment.[7]

Bookchin is also critical of what he regards as the productivist and authoritarian legacy of socialist thought and its lack of a thoroughgoing ecological perspective.[8] . . .

Bookchin argues that social ecology, in contrast, *is* grounded in an ecological sensibility that rejects the instrumental posture toward nature that is characteristic of socialist thought. This follows from Bookchin's organismic philosophy, which recognizes "subjectivity" as present, however germinally, in all phenomena, not just humans. (By "subjectivity" Bookchin means any kind of purposive activity or striving, whether latent or advanced.[9]) Indeed, prior to Bookchin's much publicized attack on deep ecology at the National Greens Conference in Amherst, Massachusetts, in July 1987, many ecopolitical theorists regarded social ecology and deep ecology as complementary ecophilosophies. Both Naess's deep/shallow ecology distinction and Bookchin's social ecology/environmentalism distinction are critical of scientism and a purely instrumental orientation toward the nonhuman world;

both seek to re-embed humans in the natural world; and, at a more practical level, both support bioregionalism, small-scale decentralized communities, cultural and biological diversity, and "appropriate" technology.

Despite these important commonalities, however, there are also important differences in the ecophilosophical orientations of deep ecology and social ecology that have given rise to different perspectives concerning humanity's proper role in the evolutionary drama. . . . Not only are important ecophilosophical issues at stake here; there is also a struggle to influence the political priorities of the growing Green movement. What, then, are the distinctive claims of social ecology?

Bookchin's social hierarchy thesis

From as early as the 1960s Bookchin has maintained the thesis that the domination of non-human nature by humans arose from the domination of humans by humans. This argument finds its most developed expression in Bookchin's magnum opus *The Ecology of Freedom: The Emergence and Dissolution of Hierarchy*.[10] In this work, Bookchin has sought to develop this thesis by tracing the emergence of hierarchy and domination in human societies from the Paleolithic Age to modern times. According to Bookchin, the breakdown of "organic" or preliterate communities based on kinship ties into hierarchical and finally class societies, culminating in the State, was to gradually undermine the unity of human society with the nonhuman world. Bookchin argues that incipient domination arose originally in preliterate societies in the form of social hierarchies rooted in age, sex, and quasi-religious and quasi-political needs. These social hierarchies are presented as providing the conceptual apparatus of domination or what Bookchin calls the "epistemologies of rule" – the repressive sensibility of command and obedience that has enabled some humans to see others as objects of manipulation.

[. . .]

Social ecology . . . does not see the conquest of nature as the necessary "price" of human freedom. Rather, it looks to nature as the *ground* of freedom and seeks to re-embed humans in the natural world.[11] This, argues Bookchin, demands the creation of a libertarian, stateless society that is guided by Bookchin's description of the ecosystem: "the image of unity in diversity, spontaneity, and complementary relationships, free of all hierarchy and domination."[12] This translates politically into a society that is free of "gerontocracies, patriarchies, class relationships, elites of all kinds, and finally the State, particularly in its most socially parasitic form of state capitalism."[13]

The radical thrust of social ecology is seen to arise from its ecological sensibility, which recognizes what is seen to be the nonhierarchical interdependence of living and nonliving things. According to Bookchin, such an ecological sensibility implicitly undermines the conventional notion that hierarchy is part of the "natural" order of things in both nature and society.[14] In the case of the nonhuman world, he argues that "the seemingly hierarchical traits of many animals are more like variations in the links of a chain than organized stratifications of the kind we find in human societies and institutions."[15]

[. . .]

Bookchin's evolutionary stewardship thesis

Bookchin's anarchist ideal of freedom is one that sees any kind of other- or external-directedness, as distinct from self- or internal-directedness, as thwarting impulses that are deemed to be natural and good. Like many anarchists, Bookchin enlists evolutionary theory to support the notion of the inherent sociality of humanity and to claim that any form of higher or external authority is *against nature*. In this respect, George Woodcock's general observations on anarchism are pertinent to social ecology (although Bookchin speaks of evolutionary "directionality" rather than "natural laws"):

> The whole world-view within which anarchism is embraced depends on an acceptance of natural laws manifested through evolution, and this means that the anarchist sees himself as the representative of the true evolution of human society, and regards authoritarian political organizations as a perversion of that evolution.[16]

[. . .]

This creative role assigned to humans in fostering nature's evolution is the essential basis upon which Bookchin rejects asceticism, stoicism, biocentrism/ecocentrism, or any world-view that he interprets as involving the "quietistic surrender" or resignation by humans to the natural order. Bookchin interprets such approaches (quite wrongly, in the case of ecocentrism) as idolizing and reifying nature and setting it apart from a "fallen humanity" – an approach that Bookchin claims is an insult to humanity by denying us our creative role in evolution.[17] There must be an infusion of human values into nature, he argues, because humans are the fulfillment of a major tendency in *natural* evolution. Indeed, Bookchin claims that our uniqueness cannot be emphasized too strongly "for it is in this very *human* rationality that nature ultimately actualizes its own evolution of subjectivity over long aeons of neural and sensory development."[18] The clear message of Bookchin's ethics, then, is that humanity, as a self-conscious "moment" in nature's dialectic, has a responsibility to rationally direct the evolutionary process, which in Bookchin's terms means fostering a more diverse, complex, and fecund biosphere. Indeed, he suggests that we may "create more fecund gardens than Eden itself."[19]

Bookchin's view of humans as evolutionary stewards is at odds with an ecocentric orientation toward the world in two important respects. First, an ecocentric orientation is more concerned with "letting things be," that is, allowing all beings (human and nonhuman) to unfold in their own way (the means of achieving this goal, of course, require *active* political engagement in defence of the Earth and oppressed peoples). From a long-term ecological and evolutionary perspective, adaptation, change, innovation, destruction, and extinction are recognized as features of natural systems, but rather than being fostered or accelerated, they are allowed to unfold in accordance with natural successional and evolutionary time. This is because ecocentric theorists do not see later, more developed, or more complex life-forms as necessarily "higher" or "better" than earlier, less developed, or more simple ones. Second, an ecocentric perspective adopts a more *humble* position than social ecology insofar as it does not claim to know what the thrust or direction of evolution is. Rather, an ecocentric perspective remains open-minded toward what is seen as an essentially open-ended process. As Lorna Salzman points out:

> Extinction of species has been a fact – a second species of *homo* coexisted with *h. sapiens* until relatively recently. The fact that we are (or believe we are) the only self-aware species on Earth (which we cannot prove) does not mean that this was evolution's impulse or our "striving." We need not have survived at all; there was and is no "necessity" that we do so.[20]

From an ecocentric perspective, it is both arrogant and self-serving to make, as Bookchin does, the unverifiable claim that first nature is striving to achieve something (namely, greater subjectivity, awareness, or "selfhood") that "just happens" to have reached its most developed form in *us* – second nature.

[. . .]

The different ecophilosophical orientations of social ecology and ecocentrism give rise to different emphases when it comes to the "litmus issues" of wilderness preservation and human population growth (although these different emphases have become less marked in the wake of recent debates between deep ecologists and social ecologists). For example, although Bookchin has often emphasized the value of spontaneity, it is noteworthy that, *by comparison to ecocentric theorists* (and prior to recent debates and reconciliations), Bookchin has written very little on the subject of wilderness preservation as compared to, say, issues concerning social organization and the urban and agricultural or "domestic" environment . . .

Consistent with their strong stand on the need for wilderness preservation, ecocentric theorists advocate a population policy that seeks a long-term, gradual *reduction* in human numbers. This position flows directly from the ecocentric concern for biological and cultural diversity; from the ecocentric concern to provide space for all beings to unfold in their own ways. To the extent that Bookchin's approach to the population question can be discerned from his criticisms of deep ecology, his emphasis has been on the need to overcome social hierarchy, decentralize society, redistribute resources, and cultivate an "ecological technics" rather than on the need to address the problem of absolute numbers per se by means of birth control programs.[21]

On the population question, then, Bookchin has much more in common with ecosocialism than ecocentrism insofar as he points to social relations rather than absolute human numbers per se as the "real causes" of famine and environmental degradation. Indeed, Bookchin has gone much further than ecosocialism in criticizing ecocentrism as racist and misanthropic for ignoring inter-human inequities. While it is not my concern to present a detailed overview of Bookchin's accusations in this controversy, it is important to lay to rest one serious misconception of the ecocentric position that arises from Bookchin's critique. As should be clear from the discussion of the population question in the previous chapter, ecocentric theorists do not *only* advocate a long-term, gradual reduction in absolute human numbers in response to the population issue, as some of their critics suggest. They also advocate a more equitable inter-human distribution of resources, a lower overall level of resource consumption per capita, and the introduction of ecologically benign technologies.

[. . .]

Ecocommunalism

Ecocommunalism is a generic term that encompasses a diverse range of utopian, visionary, and essentially anarchist Green theories that seek the development of human-scale, cooperative communities that enable the rounded and mutualistic development of humans while at the same time respecting the integrity of the nonhuman world. Progress, according to ecocommunal theorists, is generally measured by the degree to which we are able to adapt human communities to ecosystems (rather than the other way around) and by the degree to which the full range of human needs are fulfilled. Like Bookchin, these theorists are generally critical of purely instrumental valuations of the nonhuman world and instead appeal to nature as a source of both inspiration and guidance.

[. . .]

Monasticism revisited

A common theme among ecocommunal theorists is the idea of disengagement or withdrawal from corrupt social and political institutions and the establishment of exemplary institutions and/or the pursuit of exemplary personal action. In view of the sweeping ecocommunal critique of most aspects of modern industrial society, such a strategy is, in many respects, the only reliable and authentic strategy that will maintain consistency between ends and means. In arguing for the establishment of ecological communities as the solution to the multifaceted crises facing modern society, a number of ecocommunal theorists have employed the analogy of the emergence of the medieval communalism of Saint Benedict of Nursia out of the ruins of the Roman Empire.

[. . .]

In contemporary ecocommunal literature, references to the monastic paradigm abound. For example, the ideal of "withdrawal and renewal" is central to Rudolf Bahro's ecofundamentalism, and he has often claimed that the establishment and growth of small-scale cooperatives or "Liberated Zones" (his ecocommunal solution to the ecological crisis) would lead society toward a better future in the same way that the communes founded by Saint Benedict were intended as a return to community and order after the chaos and social decay that had set in following the collapse of Rome.[22] Liberated Zones would ensure consistency between ends and means by providing both a supportive refuge from the destructiveness and alienation of industrialism and the nucleus of a new "biophile [i.e., life-loving] culture." Bahro is at pains to point out that the challenge of ecological degradation is primarily a cultural and spiritual one and only secondarily an economic one. Accordingly, we must direct our attention to cultural and spiritual renewal rather than structural or economic reform. Liberated Zones thus provide, in Bahro's view, a total solution to the multifaceted crises of modern times.

[. . .]

Bioregionalism

Although the exact source of the neologism "bioregionalism" (etymologically, *bioregion* means "life-place") is a matter of some uncertainty, its popularization as a unifying principle celebrating cultural and biological diversity and providing an ecological politics of living-in-place is generally traced to Peter Berg and Raymond Dasmann of the San Francisco Planet Drum Foundation.[23] The term *bioregion*, according to Berg and Dasmann, "refers both to a geographical terrain and a terrain of consciousness – to a place and the ideas that have developed about how to live in that place."[24] Geographically, bioregions are areas having common characteristics of soil, watershed, climate, native plants and nonhuman animals, and common human cultures. Culturally and psychologically, bioregionalism seeks the integration of human communities with the nonhuman world "at the level of the *particular ecosystem* and employs for its cognition a body of metaphors drawn from and structured in relation to that ecosystem."[25] This goal of adapting human communities to the local bioregion is facilitated through the practice of "reinhabitation":

> *Reinhabitation* means learning to live-in-place in an area that has been disrupted and injured through past exploitation. It involves becoming native to a place through becoming aware of the particular ecological relationships that operate within and around it. It means understanding activities and evolving social behavior that will enrich the life of that place, restore its life-supporting systems, and establish an ecologically and socially sustainable pattern of existence within it. Simply stated it involves becoming fully alive in and with a place. It involves applying for membership in a biotic community and ceasing to be its exploiter.[26]

Bioregionalism is principally a North American phenomenon that has grown into a significant tributary of the North American Green movement. . . . Consistent with the tradition of the ecological community, bioregionalism emphasizes decentralization, human scale communities, cultural and biological diversity, self-reliance, cooperation, and community responsibility (both social and biotic). In differs from ecomonasticism mainly in its greater emphasis on protecting, and rehabilitating if necessary, the characteristic diversity of *native* ecosystems. This is manifested in its concern to develop a sense of rootedness that is, as Morris Berman put it, "biotic, not merely ethnic."[27]

[. . .]

Kirkpatrick Sale articulates the view of most bioregionalists in arguing that self-government by the various human communities within a bioregion – possibly linked by a bioregional confederation – offers the best guarantee of social and ecological harmony.[28] (A *confederation* is a mutual association of many autonomous communities or states, each of which retains sovereignty. It should not be confused with a *federation*, which is a mutual association of semiautonomous states or provinces under one, central sovereign state.[29]

On the positive side, the idea that political communities should be based on bioregional contours has much to commend it, particularly in relation to land and water management (indeed, many existing management regimes for internal waters are already modelled along watershed lines). From an educational perspective, bioregionalism plays an invaluable role in underscoring the importance of thinking in terms of ecological relationships, asking where everything comes from and where everything goes, learning to identify, and become "respectful neighbors" with, the local species of flora and fauna. Such an orientation and understanding is crucial to the critical evaluation of existing development decisions. Indeed, some bioregionalists have suggested the formation of "ad hoc watershed shadow governments. Their function would be to serve as moral stewards for specific watersheds and bioregions and to help inhabitants learn the true ecological cost of any proposed development."[30]

At the more practical level, however, bioregionalism is confronted with the problem that linguistic, religious, and cultural boundaries do not necessarily follow bioregional lines. As Don Alexander argues, it is too simplistic to locate human communities on the basis of geography alone.[31] While the patterns of human settlement and movement, and the cultures of many traditional societies, have tended to be influenced quite strongly by geographical criteria, modern transport and communications have meant that regional consciousness in Western society is determined as much by *functional* (and, as already noted, linguistic, religious, and social) criteria rather than formal geographical ones.[32] This seriously challenges the basic assumption on which Sale's entire discussion of political forms is premised, namely, that there is a "clear identity of interest" among the various communities within a bioregion, that they are relatively homogeneous and organically bound together by an ecological identity.[33] What bioregional theorists should ask when considering the issue of institutional design is: what political forms will best promote bioregional goals, given that the many and varied human communities within the many and varied bioregions of the world (however determined – another vexing question) do *not* all possess a bioregional consciousness? Ceding complete political autonomy to the existing local communities that inhabit bioregions will provide no guarantee that development will be ecologically benign or cooperative. Nor will it provide any guarantee that they will form a confederation with neighboring local communities in their bioregion so as to enable proper bioregional management.

[. . .]

[Ecocentrism and ecoanarchism]

Unlike the individualistic liberal model of autonomy, ecoanarchists defend a social model of autonomy that is secured by voluntary cooperation with, and responsibility to, the human scale community of which the individual is part.[34] The dismantling of the State would not lead to social fragmentation, they argue, but rather to spontaneous cooperation and the strengthening of social bonds between people. Antisocial behavior would be dealt with via community censure (as in traditional, small-scale hunting and gathering and horticultural groups) rather than via the abstract and inflexible legal rules laid down by the remote nation State.[35] Moreover, the anarchist assumption that humans are naturally *cooperative*, but are presently corrupted by hierarchical institutions, also stands in contrast to the classical liberal view, which saw humans as naturally self-seeking and in need of restraint through, say, a limited government based on a social contract. It was on the basis of this latter model of human behavior that the survivalist ecopolitical theorists reached a conclusion that is diametrically opposed to that of ecoanarchists: that only a centralized, authoritarian government can rescue us from the ecological crisis and save us, as it were, from ourselves.

Ecoanarchists, in contrast, share a deeply felt desire for humans to cooperate more than they do and a conviction that they can do so in the appropriate social environment. The problem with the ecoanarchist model of human nature, however, is that it conflates people's *potential* nature with their *essential* nature. That is, ecoanarchists present our potential (i.e., better) nature *as* our essential nature and appeal to the reciprocity and mutual aid that they see in nature as evidence that their model of human nature is in alignment with "the natural order of things" and perforce "objectively" right. I have already criticized this ecoanarchist appeal to the natural as adding nothing to the *normative* force of ecopolitical argument and do not intend to repeat these points here. Instead, I want to address the more obvious problems associated with such a model when it comes to rethinking political forms. Specifically, the presumption that humans are "essentially" of a certain nature (i.e., cooperative) and that this nature can be "reawakened" under the right social and institutional circumstances (i.e., anarchism) leads to institutional designs that cannot adequately accommodate human behavior that defies this model of human nature.

[. . .]

. . . The general ecoanarchist approach of "leave it all to the locals who are affected" makes sense only when the locals possess an appropriate social and ecological consciousness. It also presumes that local bioregion A is not a matter of concern to people living in bioregion B and that these latter "outsiders" can have no effective input to development decisions made by the inhabitants of bioregion A. Moreover, the rejection of a vertical model of representative democracy in favor of a horizontal model of direct democracy underrates the innovative potential of what might be called the "cosmopolitan urban center" vis-à-vis the "local rural periphery." For example, historically most progressive social and environmental legislative changes – ranging from affirmative action, humans rights protection, and homosexual law reform to the preservation of wilderness areas – have tended to emanate from more cosmopolitan central governments rather than provincial or local decision-making bodies.[36] In many instances, such reforms have been carried through by central governments in the face of opposition from the local community or region affected – a situation that has been the hallmark of many environmental battles in the Australian federal system of government. At an even "higher" level, bodies such as the International Court of Justice and the World Heritage Committee are salutary reminders of the ways in which institutions created by international treaties can serve to protect both human rights and threatened species and ecosystems from the "excesses" of *local* political elites. Indeed, there is a large number of Green social and environmental reforms, ranging from the redistribution of resources from rich to poor countries to the abatement of the greenhouse effect, that can be effectively implemented *only* via international agreement between existing nation States. Successful ecodiplomacy of this kind is more likely to be achieved by the retention and reform of a democratically accountable State that can legitimately claim to represent in the international arena at least a majority of people in a nation. While unilateral action by "right minded" citizens in local bioregions is to be encouraged, it will have minimal effect for as long as recalcitrant neighboring local communities and regions continue to "externalize" their environmental costs. In view of the urgency and ubiquity of the ecological crisis, ultimately only a supraregional perspective and multilateral action by nation States can bring about the kind of dramatic changes necessary to save the "global commons" in the short and medium term. It must be emphasized that none of these arguments is intended to deny the innovative potential of local and municipal action and the importance of enhancing local autonomy, nor the many obstacles facing international agreement. I am merely concerned to point out the *two-edged* nature of the argument for the complete devolution of political and legal power to local assemblies. We need not only to act locally and think globally, but also to *act* globally.

Notes

1 Bookchin's publications are too numerous to list exhaustively here. His major books include *Our Synthetic Environment* (New York: Alfred A. Knopf, 1962), published under the pseudonym Lewis Herber; *Post-Scarcity Anarchism* (Berkeley: Ramparts, 1971); *Toward an Ecological Society* (Montreal: Black Rose, 1980); *The Ecology of Freedom* (Palo Alto, Calif.: Cheshire, 1982); *The Modern Crisis* (Philadelphia: New Society, 1986); *The Rise of Urbanization and the Decline of Citizenship* (San Francisco: Sierra Club Books, 1987); and *Remaking Society: Pathways to a Green Future* (Boston: South End, 1990). The major articles by Bookchin relevant to the present discussion are "Toward a Philosophy of Nature – The Bases for an Ecological Ethic," in *Deep Ecology*, ed. Michael Tobias (San Francisco: Avant, 1984), 213–35; "Freedom and Necessity in Nature: A Problem in Ecological Ethics," *Alternatives* 13 (1986): 29–38; and "Thinking Ecologically: A Dialectical Approach," *Our Generation* 18 (1987): 3–40. These three articles (some of which have been slightly revised) have been reprinted in *The Philosophy of Social Ecology: Essays on Dialectical Naturalism* (Montreal: Black Rose, 1990). See also John Clark, ed., *Renewing the Earth: The Promise of Social Ecology* (London: Green Print, 1990).

2 Bookchin, "Thinking Ecologically," 4.

3 Bookchin, *The Ecology of Freedom*, 342.

4 Bookchin, *Post-Scarcity Anarchism*, 70.

5 For a thoroughgoing response to these criticisms by a leading deep/transpersonal ecology theorist, see Warwick Fox, "The Deep Ecology-Ecofeminism Debate and its Parallels," *Environmental Ethics* 11 (1989): 5–25.

6 Murray Bookchin, "On the Last Intellectuals," *Telos* 73 (1987): 182.

7 Bookchin, *The Ecology of Freedom*, 340.

8 See, for example, Murray Bookchin, "Beyond Neo-Marxism," *Telos* 36 (1978): 5–28 and the chapter "On Neo-Marxism, Bureaucracy, and the Body Politic," in *Toward an Ecological Society*, 211–48. See also John Clark, *The Anarchist Moment: Reflections on Culture, Nature and Power* (Montreal: Black Rose, 1984).

9 See, for example, *Ecology of Freedom*, 275.

10 For an earlier statement of this thesis, see Bookchin, *Post-Scarcity Anarchism*, 63.

11 Bookchin, "Thinking Ecologically", 7, n.1.

12 Bookchin, *The Ecology of Freedom*, 352.

13 Ibid., 353.

14 Ibid., 36–37.

15 Ibid., 29.

16 George Woodcock, "Anarchism: A Historical Introduction," in *The Anarchist Reader*, ed. George Woodcock (London: Fontana, 1983), 27.

17 This is a misinterpretation . . . For example, we have seen that the central concern of deep/transpersonal ecology theorists such as Arne Naess, Bill Devall, George Sessions, Warwick Fox, Alan Drengson, and Freya Mathews is to cultivate a sense of identification or empathy with all of nature (*of which humans are part*). This identification or empathy stems from the realization of our *interdependence* with other life-forms. This can hardly be interpreted as an approach that "reifies" nature and sets it *apart* from humanity.

18 Bookchin, "Thinking Ecologically," 20.

19 Bookchin, *The Ecology of Freedom*, 343.

20 Salzman, "Politics as if Evolution Mattered: Some Thoughts on Deep and Social Ecology". Paper presented at the Ecopolitics IV Conference, University of Adelaide, South Australia, 21–24 Sept. 1989, 15.

21 On the vexed subject of human population, Bookchin's contribution has mainly been one of warning of the dangers of fascist solutions than of advocating specific measures to control human numbers. In particular, Bookchin's position has emerged largely as a response to, and critique of, certain blunt statements made by prominent members (such as Dave Foreman) of the U.S. environmental movement Earth First!, which Bookchin has wrongly taken as representative of the views of deep ecology theorists. For example, Foreman has remarked that it is better to leave

Ethiopian children to starve than "save these half dead children who will never live a whole life. Their development will be stunted." See Dave Foreman, "A Spanner in the Woods," interviewed by Bill Devall, *Simply Living* 2, no. 12 (n.d.): 43. Yet, as Fox points out, "it is as unreasonable for Bookchin to condemn the body of ideas known as deep ecology on the basis that he does as it would be for a critic of Bookchin to condemn the body of ideas known as social ecology on the basis of whatever personal views happen to be put forward by individual activists who support any environmental organization that claims to draw on social ecology principles" (Fox, "The Deep Ecology-Ecofeminism Debate," 20, n. 38). Needless to say, deep/transpersonal ecology theorists have made it clear that "faced with the problem of hungry children, humanitarian action is a priority." In addition to Fox, see Arne Naess, "Sustainable Development and the Deep Long-Range Ecology Movement," *The Trumpeter* 5 (1988): 141. . . . For a detailed account, see Murray Bookchin and Dave Foreman, *Defending the Earth: A Dialogue between Murray Bookchin and Dave Foreman*, (Boston: South End, 1991).

22 See Rudolf Bahro, *Building the Green Movement* (London: Heretic/GMP, 1986) especially 86–98. Ecocommunalism is something Bahro has arrived at in his more recent work. In his earlier publications (for example, *Socialism and Survival* [London: Heretic/GMP, 1982] and *From Red to Green* [London: Verso/NLB, 1984]) Bahro's position was closer to ecosocialism than ecocommunalism. For a general discussion of the trajectory of Bahro's thought since he left East Germany, see Robyn Eckersley, "The Prophet of Green Fundamentalism," review of *Building the Green Movement*, by Rudolf Bahro, *The Ecologist* 17 (1987): 120–22.

23 Peter Berg and Raymond F. Dasmann, "Reinhabiting California," in *Reinhabiting a Separate Country: A Bioregional Anthology of Northern California*, ed. Peter Berg (San Francisco: Planet Drum Foundation, 1978), 217–20. For a discussion of the possible origins of the term, see James J. Parsons, "On 'Bioregionalism' and 'Watershed' Consciousness," *The Professional Geographer* 37 (1985): 4.

24 Berg and Dasmann, "Reinhabiting California," 218.

25 Morris Berman, *The Reenchantment of the World* (Ithaca: Cornell University Press, 1981), 294.

26 Berg and Dasmann, "Reinhabiting California," 217–18.

27 Berman, *The Reenchantment of the World*, 294.

28 Kirkpatrick Sale, *Dwellers in the Land: The Bioregional Vision* (San Francisco: Sierra Club Books, 1985) 96. Sale also argues (108) that the bioregional emphasis on diversity is such that it does not ultimately matter what political forms are chosen within a particular bioregion – indeed, it is to be expected that not every bioregion will follow the American liberal tradition – provided they serve bioregional principles, namely, human scale, conservation and stability, self-sufficiency and cooperation, decentralization, and diversity. Nonetheless, he suggests that these bioregional principles would generally (though not always) impel the "polity in the direction of libertarian, noncoercive, open, and more or less democratic governance."

29 In a federation, political power is divided between the component states and a federal government under a federal constitution; moreover, the federal government can enact laws within its purview that apply directly to the citizens in the component states. See Roger Scruton, *A Dictionary of Political Thought* (London: Pan, 1982), 86 and 170.

30 Michael Helm, "Bioregional Planning," *RAIN*, October–November 1983, 23.

31 Alexander, "Bioregionalism," 167–69.

32 Ibid., 171.

33 Sale, *Dwellers in the Land*, 96.

34 Koula Mellos's reading of Bookchin's ecoanarchism as a petit bourgeois form of radicalism based on solitary or "asocial individual self-sufficiency" seems to completely miss Bookchin's emphasis on symbiosis and community (see Koula Mellos, *Perspectives on Ecology: A Critical Essay* [London: Macmillan, 1988], chapter 4).

35 For an example of how community censure operates in traditional societies, see Harold Barclay, *People without Government* (London: Kahn & Averill with Cienfuegos, 1982).

36 Stephen Rainbow has also criticized what he calls the "soft" Green, ultra-democratic approach for naively assuming that local people will always choose to attract ecologically sensitive industry. See Stephen Rainbow, "Eco-politics in Practice. Green Parties in New Zealand, Finland, and

Sweden." Paper presented to the Ecopolitics IV Conference, University of Adelaide, South Australia, 21–24 September 1989, 21.

References

Alexander, Don. (1990). "Bioregionalism: Science or Sensibility?" *Environmental Ethics* 12:161–73.

Bahro, Rudof. (1982). *Socialism and Survival*. London: Heretic Books.

—— (1984). *From Red to Green*. London: Verso/NLB.

Barclay, Harold. (1982). *People without Government*. London: Kahn & Averill with Cienfuegos.

Berg, Peter, and Dasmann, Raymond F. (1978). "Reinhabiting California." In *Reinhabiting a Separate Country: A Bioregional Anthology of Northern California*, 217–20. Edited by Peter Berg. San Francisco: Planet Drum Foundation.

Berman, Morris. (1981). *The Reenchantment of the World*. Ithaca: Cornell University Press.

Bookchin, Murray [Lewis Herber, pseud.]. (1962). *Our Synthetic Environment*. New York: Knopf.

—— (1971). *Post-Scarcity Anarchism*. Berkeley: Ramparts.

—— (1978). "Beyond Neo-Marxism." *Telos* 36:5 28.

—— (1980). *Toward an Ecological Society*. Montreal: Black Rose Books.

—— (1982). *The Ecology of Freedom: The Emergence and Dissolution of Hierarchy*. Palo Alto, Calif.: Cheshire.

—— (1984). "Toward a Philosophy of Nature – The Bases for an Ecological Ethic." In *Deep Ecology*, 213–35. Edited by Michael Tobias. San Francisco: Avant.

—— (1986). "Freedom and Necessity in Nature: A Problem in Ecological Ethics." *Alternatives* 13: 29–38.

—— (1986). *The Modern Crisis*. Philadelphia: New Society.

—— (1987). "On the Last Intellectuals." *Telos* 73: 182–85.

—— (1987). *The Rise of Urbanization and the Decline of Citizenship*. San Francisco: Sierra Club Books.

—— (1987). "Thinking Ecologically: A Dialectical Approach." *Our Generation* 18: 3–40.

—— (1988). "As If People Mattered." *The Nation*, 10 October, 294.

—— (1990). *The Philosophy of Social Ecology: Essays on Dialectical Naturalism*. Montreal: Black Rose.

—— (1990). *Remaking Society: Pathways to a Green Future*. Boston: South End Press.

Bookchin, Murray, and Foreman, Dave. (1991. *Defending the Earth: A Dialogue between Murray Bookchin and Dave Foreman*. Boston: South End.

Capra, Fritjof. (1983). *The Turning Point: Science, Society, and the Rising Culture*. London: Fontana. Reprint. 1985.

Clark, John. (1984). *The Anarchist Moment: Reflections on Culture, Nature and Power*. Montreal: Black Rose.

Clark —— ed. (1990). *Renewing the Earth: The Promise of Social Ecology*. London: Green Print.

Eckersley, Robyn —— (1987. "The Prophet of Green Fundamentalism." Review of *Building the Green Movement*, by Rudolf Bahro. *The Ecologist* 17: 120–22.

Foreman, Dave. n.d. "A Spanner in the Woods," Interviewed by Bill Devall. *Simply Living* 2(12): 40–43.

Fox, Warwick. (1989). "The Deep Ecology–Ecofeminism Debate and its Parallels." *Environmental Ethics* 11: 5–25.

Helm, Michael. (1983). "Bioregional Planning." *RAIN*, October–November, 22–23.

Mellos, Koula. (1988). *Perspectives on Ecology: A Critical Essay*. London: Macmillan.

Naess, Arne. (1988). "Sustainable Development and the Deep Long-Range Ecology Movement." *The Trumpeter* 5: 138–42.

Parsons, James J. (1985). "On 'Bioregionalism' and 'Watershed' Consciousness." *The Professional Geographer* 37: 1–6.

Rainbow, Stephen. (1989). "Eco-politics in Practice: Green Parties in New Zealand, Finland and Sweden." Paper presented at the Ecopolitics IV Conference, University of Adelaide, South Australia, 21–24 September.

Sale, Kirkpatrick. (1985). *Dwellers in the Land: The Bioregional Vision*. San Francisco: Sierra Club Books.

Salzman, Lorna. (1989). "Politics as if Evolution Mattered: Some Thoughts on Deep and Social Ecology." Paper presented at the Ecopolitics IV Conference, University of Adelaide, South Australia, 21–24 September.
Scruton, Roger. (1983). *A Dictionary of Political Thought*. London: Pan.
Woodcock, George. (1983). "Anarchism: A Historical Introduction." In *The Anarchist Reader*, 11–56. Edited by George Woodcock. London: Fontana.

Source: Robyn Eckersley (1992) *Environmentalism and Political Theory: Toward an Ecocentric Approach*, London: UCL Press, pp. 145–78, 226–34, 237–61.

6.4

Vandana Shiva

Science, nature and gender

The recovery of the feminine principle is an intellectual and political challenge to maldevelopment as a patriarchal project of domination and destruction, of violence and subjugation, of dispossession and the dispensability of both women and nature. The politics of life centred on the feminine principle challenges fundamental assumptions not just in political economy, but also in the science of life-threatening processes.

Maldevelopment is intellectually based on, and justified through, reductionist categories of scientific thought and action. Politically and economically each project which has fragmented nature and displaced women from productive work has been legitimised as 'scientific' by operationalising reductionist concepts to realise uniformity, centralisation and control. Development is thus the introduction of 'scientific agriculture', 'scientific animal husbandry', 'scientific water management' and so on. The reductionist and universalising tendencies of such 'science' become inherently violent and destructive in a world which is inherently interrelated and diverse. The feminine principle becomes an oppositional category of non-violent ways of conceiving the world, and of acting in it to sustain all life by maintaining the interconnectedness and diversity of nature. It allows an ecological transition from violence to non-violence, from destruction to creativity, from anti-life to life-giving processes, from uniformity to diversity and from fragmentation and reductionism to holism and complexity.

It is thus not just 'development' which is a source of violence to women and nature. At a deeper level, scientific knowledge, on which the development process is based, is itself a source of violence. Modern reductionist science, like development, turns out to be a patriarchal project, which has excluded women as experts, and has simultaneously excluded ecological and holistic ways of knowing which understand and respect nature's processes and interconnectedness *as science*.

Modern science as patriarchy's project

Modern science is projected as a universal, value-free system of knowledge, which has displaced all other belief and knowledge systems by its universality and value neutrality, and by the logic of its method to arrive at objective claims about nature. Yet the dominant stream of modern science, the reductionist or mechanical paradigm, is a particular response of a particular group of people. It is a specific project of western man which came into being during the fifteenth and seventeenth centuries as the much-acclaimed Scientific Revolution. During the last few years feminist scholarship has begun to recognise that the dominant science system emerged as a liberating force not for humanity as a whole (though it legitimised itself in terms of universal betterment of the species), but as a masculine and patriarchal project which necessarily entailed the subjugation of both nature and women. Harding has called it a 'western, bourgeois, masculine project',[1] and according to Keller

> Science has been produced by a particular sub-set of the human race, that is, almost entirely by white, middle class males. For the founding fathers of modern science, the reliance on the language of gender was explicit; they sought a philosophy that deserved to be called 'masculine', that could be distinguished from its ineffective predecessors by its 'virile' powers, its capacity to bind Nature to man's service and make her his slave.[2]

Bacon (1561–1626) was the father of modern science, the originator of the concept of the modern research institute and industrial science, and the inspiration behind the Royal Society. His contribution to modern science and its organisation is critical. From the point of view of nature, women and marginal groups, however, Bacon's programme was not humanly inclusive. It was a special programme benefiting the middle class, European, male entrepreneur through the conjunction of human knowledge and power in science.

In Bacon's experimental method, which was central to this masculine project, there was a dichotomising between male and female, mind and matter, objective and subjective, rational and emotional, and a conjunction of masculine and scientific dominating over nature, women and the non-west. His was not a 'neutral', 'objective', 'scientific' method – it was a masculine mode of aggression against nature and domination over women. The severe testing of hypotheses through controlled manipulations of nature, and the necessity of such manipulations if experiments are to be repeatable, are here formulated in clearly sexist metaphors. Both nature and inquiry appear conceptualised in ways modelled on rape and torture – on man's most violent and misogynous relationships with women.

[. . .]

. . . For Bacon, nature was no longer Mother Nature, but a female nature, conquered by an aggressive masculine mind. As Carolyn Merchant points out, this transformation of nature from a living, nurturing mother to inert, dead and manipulable matter was eminently suited to the exploitation imperative of growing capitalism. The nurturing earth image acted as a cultural constraint on exploitation of nature. 'One does not readily slay a mother, dig her entrails or mutilate her body.' But the mastery and domination images created by the Baconian programme and the scientific revolution removed all restraint and functioned as cultural sanctions for the denudation of nature.

> The removal of animistic, organic assumptions about the cosmos constituted the death of nature – the most far-reaching effect of the scientific revolution. Because

> nature was not viewed as a system of dead, inert particles moved by external, rather than inherent forces, the mechanical framework itself could legitimate the manipulation of nature. Moreover, as a conceptual framework, the mechanical order had associated with it a framework of values based on power, fully compatible with the directions taken by commercial capitalism.[3]

Modern science was a consciously gendered, patriarchal activity. As nature came to be seen more like a woman to be raped, gender too was recreated. Science as a male venture, based on the subjugation of female nature and female sex provided support for the polarisation of gender. Patriarchy as the new scientific and technological power was a political need of emerging industrial capitalism. While on the one hand the ideology of science sanctioned the denudation of nature, on the other it legitimised the dependency of women and the authority of men. Science and masculinity were associated in domination over nature and feminity, and the ideologies of science and gender reinforced each other. The witch-hunting hysteria which was aimed at annihilating women in Europe as knowers and experts was cotemporous with two centuries of scientific revolution. It reached its peak with Galileo's *Dialogue* concerning the Two Chief World Systems and died with the emergence of the Royal Society of London and the Paris Academy of Sciences.[4]

> The interrogation of witches as a symbol for the interrogation of nature, the courtroom as model for its inquisition, and torture through mechanical devices as a tool for the subjugation of disorder were fundamental to the scientific method as power. For Bacon, as for Harvey, sexual politics helped to structure the nature of the empirical method that would produce a new form of knowledge and a new ideology of objectivity seemingly devoid of cultural and political assumptions.[5]

The Royal Society, inspired by Bacon's philosophy, was clearly seen by its organisers as a masculine project. In 1664, Henry Oldenberg, Secretary of the Royal Society, announced that the intention of the society was to 'raise a *masculine philosophy*. . . whereby the Mind of Man may be ennobled with the knowledge of solid Truths'.[6]

[. . .]

For more than three centuries, reductionism has ruled as the only valid scientific method and system, distorting the history of the west as well as the non-west. It has hidden its ideology behind projected objectivism, neutrality and progress. The ideology that hides ideology has transformed complex pluralistic traditions of knowledge into a monolith of gender-based, class-based thought and transformed this particular tradition into a superior and universal tradition to be superimposed on all classes, genders and cultures which it helps in controlling and subjugating. This ideological projection has kept modern reductionist science inaccessible to criticism. The parochial roots of science in patriarchy and in a particular class and culture have been concealed behind a claim to universality, and can be seen only through other traditions – of women and non-western peoples. It is these subjugated traditions that are revealing how modern science is gendered, how it is specific to the needs and impulses of the dominant western culture and how ecological destruction and nature's exploitation are inherent to its logic. It is becoming increasingly clear that scientific neutrality has been a reflection of ideology, not history, and science is similar to all other socially constructed categories. This view of science as a social and political project of modern western man is emerging from the responses of those who were defined into nature and made passive and powerless: Mother Earth, women and colonised cultures. It is from these fringes that we are beginning to discern the economic, political and cultural mechanisms that have allowed a parochial science to dominate and how mechanisms of power and violence can be eliminated for a degendered, humanly inclusive knowledge.

The violence of reductionism

The myth that the 'scientific revolution' was a universal process of intellectual progress is being steadily undermined by feminist scholarship and the histories of science of non-western cultures. These are relating the rise of the reductionist paradigm with the subjugation and destruction of women's knowledge in the west, and the knowledge of non-western cultures. The witch-hunts of Europe were largely a process of delegitimising and destroying the expertise of European women. . . . By the sixteenth century women in Europe were totally excluded from the practice of medicine and healing because 'wise women' ran the risk of being declared witches. A deeper, more violent form of exclusion of women's knowledge and expertise, and of the knowledge of tribal and peasant cultures is now under way with the spread of the masculinist paradigm of science through 'development'.

I characterise modern western patriarchy's special epistemological tradition of the 'scientific revolution' as 'reductionist' because it reduced the capacity of humans to know nature both by excluding other knowers and other ways of knowing, and it reduced the capacity of nature to creatively regenerate and renew itself by manipulating it as inert and fragmented matter. . . .

The close nexus between reductionist science, patriarchy, violence and profits is explicit in 80 per cent of scientific research that is devoted to the war industry, and is frankly aimed directly at lethal violence – violence, in modern times, not only against the enemy fighting force but also against the much larger civilian population. In this book I argue that modern science is related to violence and profits even in peaceful domains such as, for example, forestry and agriculture, where the professed objective of scientific research is human welfare. The relationship between reductionism, violence and profits is built into the genesis of masculinist science, for its reductionist nature is an epistemic response to an economic organisation based on uncontrolled exploitation of nature for maximization of profits and capital accumulation.

Reductionism, far from being an epistemological accident, is a response to the needs of a particular form of economic and political organisation.[7] The reductionist world-view, the industrial revolution and the capitalist economy were the philosophical, technological and economic components of the same process. . . .

Commercial capitalism is based on specialised commodity production. Uniformity in production, and the uni-functional use of natural resources is therefore required. Reductionism thus reduces complex ecosystems to a single component, and a single component to a single function. It further allows the manipulation of the ecosystem in a manner that maximizes the single-function, single-component exploitation. In the reductionist paradigm, a forest is reduced to commercial wood, and wood is reduced to cellulose fibre for the pulp and paper industry. Forests, land and genetic resources are then manipulated to increase the production of pulpwood, and this distortion is legitimised scientifically as overall productivity increase, even though it might decrease the output of water from the forest, or reduce the diversity of life forms that constitute a forest community. The living and diverse ecosystem is thus violated and destroyed by 'scientific' forestry and forestry 'development'. In this way, reductionist science is at the root of the growing ecological crisis, because it entails a transformation of nature such that its organic processes and regularities and regenerative capacities are destroyed.

Women in sustenance economies, producing and reproducing wealth in partnership with nature, have been experts in their own right of a holistic and ecological knowledge of nature's processes. But these alternative modes of knowing, which are oriented to social benefits and sustenance needs, are not recognised by the reductionist paradigm, because it fails to perceive the interconnectedness of nature, or the connection of women's lives, work and knowledge with the creation of wealth.

[. . .]

The ultimate reductionism is achieved when nature is linked with a view of economic activity in which money is the only gauge of value and wealth. Life disappears as an organising principle of economic affairs. But the problem with money is that it has an asymmetric relationship to life and living processes. Exploitation, manipulation and destruction of the life in nature can be a source of money and profits but neither can ever become a source of nature's life and its life-supporting capacity. It is this asymmetry that accounts for a deepening of the ecological crises as a decrease in nature's life-producing potential, along with an increase of capital accumulation and the expansion of 'development' as a process of replacing the currency of life and sustenance with the currency of cash and profits. The 'development' of Africa by western experts is the primary cause for the destruction of Africa; the 'development' of Brazil by transnational banks and corporations is the primary cause for the destruction of the richness of Amazonian rainforests, the highest expression of life. Natives of Africa and Amazonia had survived over centuries with their ecologically evolved, indigenous knowledge systems. What local people had conserved through history, western experts and knowledge destroyed in a few decades, a few years even.

It is this destruction of ecologies and knowledge systems that I characterise as the violence of reductionism which results in:

(a) *Violence against women*: women, tribals, peasants as the knowing subject are violated socially through the expert/non-expert divide which converts them into non-knowers even in those areas of living in which through daily participation, they are the real experts – and in which responsibility of practice and action rests with them, such as in forestry, food and water systems.
(b) *Violence against nature*: nature as the object of knowledge is violated when modern science destroys its integrity of nature, both in the process of perception as well as manipulation.
(c) *Violence against the beneficiaries of knowledge*: contrary to the claim of modern science that people in general are ultimately the beneficiaries of scientific knowledge, they – particularly the poor and women – are its worst victims, deprived of their productive potential, livelihoods and life-support systems. Violence against nature recoils on man, the supposed beneficiary.
(d) *Violence against knowledge*: in order to assume the status of being the only legitimate mode of knowledge, rationally superior to alternative modes of knowing, reductionist science

resorts *to the suppression and falsification of facts* and thus commits violence against science itself. It declares organic systems of knowledge irrational, and rejects the belief systems of others without full rational evaluation. At the same time it protects itself from the exposure and investigation of the myths it has created by assigning itself a new sacredness that forbids any questioning of the claims of science.

Two kinds of facts

The conventional model of science, technology and society locates sources of violence in politics and ethics, in the *application* of science and technology, not in scientific knowledge itself. The assumed dichotomy between values and facts underlying this model implies a dichotomy between the world of values and the world of facts. In this view, sources of violence are located in the world of values while scientific knowledge inhabits the world of facts.

The fact–value dichotomy is a creation of modern reductionist science which, while being an epistemic response to a particular set of values, posits itself as independent of values. By splitting the world into facts vs. values, it conceals the real difference between two kinds of value-laden facts. Modern reductionist science is characterised in the received view as the discovery of the properties and laws of nature in accordance with a 'scientific' method which generates claims of being 'objective', 'neutral' and 'universal'. This view of reductionist science as being a description of reality *as it is*, unprejudiced by value, is being rejected increasingly on historical and philosophical grounds. It has been historically established that all knowledge, including modern scientific knowledge, is built on the use of a plurality of methodologies, and reductionism itself is only one of the scientific options available.

> There is no 'scientific method'; there is no single procedure, or set of rules that underlies every piece of research and guarantees that it is scientific and, therefore, trustworthy. The idea of a universal and stable method that is an unchanging measure of adequacy and even the idea of a universal and stable rationality is as unrealistic as the idea of a universal and stable measuring instrument that measures any magnitude, no matter what the circumstances. Scientists revise their standards, their procedures, their criteria of rationality as they move along and enter new domains of research just as they revise and perhaps entirely replace their theories and their instruments as they move along and enter new domains of research.[8]

The assumption that science deals purely with facts has no support from the practice of science itself. The 'facts' of reductionist science are socially constructed categories which have the cultural markings of the western bourgeois, patriarchal system which is their context of discovery and justification.

[. . .]

In the Third World, the conflict between reductionist and ecological perceptions of the world are a contemporary and everyday reality, in which western trained male scientists and experts epitomise reductionist knowledge. The political struggle for the feminist and ecology movements involves an epistemological shift in the criteria of assessment of the rationality of knowledge. The worth and validity of reductionist claims and beliefs need to be measured against ecological criteria when the crisis of sustainability and survival is the primary intellectual challenge. The view of reductionist scientific knowledge as a purely factual description of nature, superior to competing alternatives, is found to be ecologically unfounded. Ecology perceives relationships between different elements of an ecosystem: what properties will be selected for a particular resource element will depend on what relationships are taken as the context defining the properties. The context is fixed by priorities and values guiding the perception of nature. Selection of the context is a value-determined process and the selection in turn determines what properties are seen. There is nothing like a neutral fact about nature *independent of the value determined by human cognitive and economic activity*. Properties perceived in nature will depend on how one looks and how one looks depends on the economic interest one has in the resources of nature. The value of profit maximization is thus linked to reductionist systems, while the value of life and the maintenance of life is linked to holistic and ecological systems.

[. . .]

[Ecology, science and feminism]

Recent history has shown that in certain areas of human activity a return to ecological thought and action is possible and desirable. The primitive practice of breast-feeding had been discredited by the advertising and reductionist claims of the baby-food industry. The ecology of breast-feeding has, however, become appreciated once again, and the 'primitive' practice is enlightened practice today. Chemicalisation of health care seemed to be the only way to develop in the reductionist paradigm. Work in ethno-medicine is again bringing back wholesome drugs and treatment. Sustainable organic farming which created 'farmers of forty centuries' is on its way back, in all the diversity and plurality of its traditional base. Each of these steps towards ecological thought and action has been possible because contact was made with an ethno-scientific tradition. If the world is to be conserved for survival, the human potential for conservation must be conserved first. It is the only resource we have to foresee and forestall the destruction of our ecosystems.

Contemporary women's ecological struggles are new attempts to establish that steadiness and stability are not stagnation, and balance with nature's essential ecological processes is not technological backwardness but technological sophistication. At a time when a quarter of the world's population is threatened by starvation due to erosion of soil, water and genetic diversity of living resources, chasing the mirage of unending growth, by spreading resource-destructive technologies, becomes a major source of genocide. The killing of people by the murder of nature is an invisible form of violence which is today the biggest threat to justice and peace.

The emerging feminist and ecological critiques of reductionist science extend the domain of the testing of scientific beliefs into the wider physical world. Socially, the world of scientific experiments and beliefs has to be extended beyond the so-called experts and specialists into the world of all those who have systematically been excluded from it – women, peasants, tribals. The verification and validation of a scientific system would then be validation in practice, where practice and experimentation is real-life activity in society and nature. Harding says:

> Neither God nor tradition is privileged with the same credibility as scientific rationality in modern cultures . . . The project that science's sacredness makes taboo is the examination of science in just the ways any other institution or set of social practices can be examined. If we are not willing to try and see the favoured intellectual structures and practices of science as cultural artifacts rather than as sacred commandments handed down to humanity at the birth of modern science, then it will be hard to understand how gender symbolism, the gendered social structure of science, and the masculine identities and behaviours of individual scientists have left their marks on the problematics, concepts, theories, methods, interpretation, ethics, meanings and goals of science.[9]

The intellectual recovery of the feminine principle creates new conditions for women and non-western cultures to become principal actors in establishing a democracy of all life, as countervailing forces to the intellectual culture of death and dispensability that reductionism creates.

Ecology movements are political movements for a non-violent world order in which nature is conserved for conserving the options for survival. These movements are small, but they are growing. They are local, but their success lies in non-local impact. They demand only the right to survival yet with that minimal demand is associated the right to live in a peaceful and just world. With the success of these grassroots movements is linked the global issue of survival. Unless the world is restructured ecologically at the level of world-views and life-styles, peace and justice will continue to be violated and ultimately the very survival of humanity will be threatened.

Notes

1 Susan Harding, *The Science Question in Feminism*, Ithaca: Cornell University Press, 1986, p. 8.
2 Evelyn F. Keller, *Reflections on Gender and Science*, New Haven: Yale University Press, 1985, p. 7.
3 Carolyn Merchant, *The Death of Nature: Women, Ecology and the Scientific Revolution*, New York: Harper & Row, 1980, p.193.
4 Brian Easlea, *Science and Sexual Oppression: Patriarchy's Confrontation with Woman and Nature*, London: Weidenfeld and Nicolson, 1981, p. 64.
5 Merchant, op. cit., p. 172.
6 Easlea, op. cit., p. 70.
7 J. Bandyopadhyay and V. Shiva, 'Ecological Sciences: A Response to Ecological Crises' in J. Bandyopadhyay, *et al.*, *India's Environment*, Dehradun: Natraj, 1985, p. 196; and J. Bandyopadhyay and V. Shiva, 'Environmental Conflicts and Public Interest Science,' in *Economic and Political Weekly*, Vol. XXI, No. 2, Jan. 11, 1986, pp. 84.90.
8 Paul Feyerband, *Science in a Free Society*, London: New Left Books, 1978, p. 10.
9 Harding, op. cit., p. 30.

Source: Vandana Shiva (1989) *Staying Alive: Women, Ecology and Development*, London: Zed Books, pp. 14–28, 36–7.

6.5

Ynestra King

Ecology and feminism

All human beings are natural beings. That may seem like an obvious fact, yet we live in a culture that is founded on the repudiation and domination of nature. This has a special significance for women because, in patriarchal thought, women are believed to be closer to nature than men. This gives women a particular stake in ending the domination of nature – in healing the alienation between human and nonhuman nature. This is also the ultimate goal of the ecology movement, but the ecology movement is not necessarily feminist.

For the most part, ecologists, with their concern for nonhuman nature, have yet to understand that they have a particular stake in ending the domination of women. They do not understand that a central reason for woman's oppression is her association with the despised nature they are so concerned about. The hatred of women and the hatred of nature are intimately connected and mutually reinforcing. Starting with this premise, this article explores why feminism and ecology need each other, and suggests the beginnings of a theory of ecological feminism: ecofeminism.

What is ecology?

Ecological science concerns itself with the interrelationships among all forms of life. It aims to harmonize nature, human and nonhuman. It is an integrative science in an age of fragmentation and specialization. It is also a critical science which grounds and necessitates a critique of our existing society. It is a reconstructive science in that it suggests directions for reconstructing human society in harmony with the natural environment.

Social ecologists are asking how we might survive on the planet and develop systems of food and energy production, architecture, and ways of life that will allow human beings to fulfill our material needs and live in harmony with nonhuman nature. This work has led to a social critique by biologists and to an exploration of biology and ecology by social thinkers. The perspective that self-consciously attempts to integrate both biological and social aspects of the relationship between human beings and their environment is known as *social ecology*. This perspective, developed primarily by Murray Bookchin,[1] to whom I am indebted for my understanding of social ecology, has embodied the anarchist critique that links domination and hierarchy in human society to the despoliation of nonhuman nature. While this analysis is useful, social ecology without feminism is incomplete.

Feminism grounds this critique of domination by identifying the prototype of other forms of domination: that of man over woman. Potentially, feminism creates a concrete global community of interests among particularly life-oriented people of the world: women. Feminist analysis supplies the theory, program, and process without which the radical potential of social ecology remains blunted. Ecofeminism develops the connections between ecology and feminism that social ecology needs in order to reach its own avowed goal of creating a free and ecological way of life.

What are these connections? Social ecology challenges the dualistic belief that nature and culture are separate and opposed. Ecofeminism finds misogyny at the root of that opposition. Ecofeminist principles are based on the following beliefs:

1 The building of Western industrial civilization in opposition to nature interacts dialectically with and reinforces the subjugation of women, because women are believed to be closer to nature. Therefore, ecofeminists take on the life-struggles of all of nature as our own.
2 Life on earth is an interconnected web, not a hierarchy. There is no natural hierarchy; human hierarchy is projected on to nature and then used to justify social domination. Therefore, ecofeminist theory seeks to show the connections between all forms of domination, including the domination of nonhuman nature, and ecofeminist practice is necessarily antihierarchical.
3 A healthy, balanced ecosystem, including human and nonhuman inhabitants, must maintain diversity. Ecologically, environmental simplification is as significant a problem as environmental pollution. Biological simplification, i.e., the wiping out of whole species, corresponds to reducing human diversity into faceless workers, or to the homogenization of taste and culture through mass consumer markets. Social life and natural life are literally simplified to the inorganic for the convenience of market society. Therefore we need a decentralized global movement that is founded on common interests yet celebrates diversity and opposes all forms of domination and violence. Potentially, ecofeminism is such a movement.
4 The survival of the species necessitates a renewed understanding of our relationship to nature, of our own bodily nature, and of nonhuman nature around us; it necessitates a challenging of the nature–culture dualism and a corresponding radical restructuring of human society according to feminist and ecological principles. . . .

The ecology movement, in theory and practice, attempts to speak for nature – the "other" that has no voice and is not conceived of subjectively in our civilization. Feminism represents the refusal of the original "other" in patriarchal human society to remain silent or to be the "other" any longer. Its challenge of social domination extends beyond sex to social domination of all kinds, because the domination of sex, race, and class and the domination of nature are mutually reinforcing. Women are the "others" in human society, who have been silent in public and who now speak through the feminist movement.

[. . .]

The recognition of the connections between women and nature and of woman's bridge-like position between nature and culture poses three possible directions for feminism. One direction is the integration of women into the world of culture and production by severing the woman–nature connection. . . .

Other feminists have reinforced the woman–nature connection: woman and nature, the spiritual and intuitive, versus man and the culture of patriarchal rationality.[2] This position also does not necessarily question nature–culture dualism or recognize . . . women's ecological sensitivity . . .

Ecofeminism suggests a third direction: a recognition that although the nature–culture dualism is a product of culture, we can nonetheless *consciously choose* not to sever the woman–nature connection by joining male culture. Rather, we can use it as a vantage point for creating a different kind of culture and politics that would integrate intuitive, spiritual, and rational forms of knowledge, embracing both science and magic insofar as they enable us to transform the nature–culture distinction and to envision and create a free, ecological society.

Ecofeminism and the intersection of feminism and ecology

The implications of a culture based on the devaluation of life-giving and the celebration of life-taking are profound for ecology and for women. This fact about our culture links the theories and politics of the ecology movement with those of the feminist movement. Adrienne Rich has written:

> We have been perceived for too many centuries as pure Nature, exploited and raped like the earth and the solar system; small wonder if we now long to become Culture: pure spirit, mind. Yet it is precisely this culture and its political institutions which have split us off from itself. In so doing it has also split itself off from life, becoming the death culture of quantification, abstraction, and the will to power which has reached its most refined destructiveness in this century. It is this culture and politics of abstraction which women are talking of changing, of bringing into accountability in human terms.[3]

The way to ground a feminist critique of "this culture and politics of abstraction" is with a self-conscious ecological perspective that we apply to all theories and strategies, in the way that we are learning to apply race and class factors to every phase of feminist analysis.

Similarly, ecology requires a feminist perspective. Without a thorough feminist analysis of social domination that reveals the interconnected roots of misogyny and hatred of nature, ecology remains an abstraction: it is incomplete. If male ecological scientists and social ecologists fail to deal with misogyny – the deepest manifestation of nature-hating in their own lives – they are not living the ecological lives or creating the ecological society they claim.

The goals of harmonizing humanity and nonhuman nature, at both the experiential and theoretical levels, cannot be attained without the radical vision and understanding available from feminism. The twin concerns of ecofeminism – human liberation and our relationship to nonhuman nature – open the way to developing a set of ethics required for decision-making about technology. Technology signifies the tools that human beings use to interact with nature, including everything from the digging stick to nuclear bombs.

Ecofeminism also contributes an understanding of the connections between the domination of persons and the domination of nonhuman nature. Ecological science tells us that there is no hierarchy in nature itself, but rather a hierarchy in human society that is projected onto nature. Ecofeminism draws on feminist theory which asserts that the domination of woman was the original domination in human society from which all other hierarchies – of rank, class, and political power – flow. Building on this unmasking of the ideology of a natural hierarchy of persons, ecofeminism uses its ecological perspective to develop the position that there is no hierarchy in nature: among persons, between persons and the rest of the natural world, or among the many forms of nonhuman nature. We live on the earth with millions of species, only one of which is the human species. Yet the human species in its patriarchal form is the only species which holds a conscious belief that it is entitled to dominion over the other species, and over the planet. Paradoxically, the human species is utterly dependent on nonhuman nature. We could not live without the rest of nature; it *could* live without us.

Ecofeminism draws on another basic principle of ecological science – unity in diversity – and develops it politically. Diversity in nature is necessary, and enriching. One of the major effects of industrial technology, capitalist or socialist, is environmental simplification. Many species are simply being wiped out, never to be seen on the earth again. In human society, commodity capitalism is intentionally simplifying human community and culture so that the same products can be marketed anywhere to anyone. The prospect is for all of us to be alike, with identical needs and desires, around the globe: Coca Cola in China, blue jeans in Russia, and American rock music virtually everywhere.

Few peoples of the earth have not had their lives touched and changed to some degree by the technology of industrialization. Ecofeminism as a social movement resists this social simplification through supporting the rich diversity of women the world over, and seeking a oneness in that diversity. Politically, ecofeminism opposes the ways that differences can separate women from each other, through the oppressions of class, privilege, sexuality, and race.

The special message of ecofeminism is that when women suffer through both social domination and the domination of nature, most of life on this planet suffers and is threatened as well. It is significant that feminism and ecology as social movements have emerged now, as nature's revolt against domination plays itself out in human history and in nonhuman nature at the same time. As we face slow environmental poisoning and the resulting environmental simplification, or the possible unleashing of our nuclear arsenals, we can hope that the prospect of the extinction of life on the planet will provide a universal impetus to social change. Ecofeminism supports utopian visions of harmonious, diverse, decentralized communities, using only those technologies based on ecological principles, as the only practical solution for the continuation of life on earth.

Visions and politics are joined as an ecofeminist culture and politics begin to emerge. Ecofeminists are taking direct action to effect changes that are immediate and personal as well as long-term and structural. Direct actions include learning holistic health and alternate ecological technologies, living in communities that explore old and new forms of spirituality which celebrate all life as diverse expressions of nature, considering the ecological consequences of our lifestyles and personal habits, and participating in creative public forms of resistance, including nonviolent civil disobedience.

[. . .]

Notes

1 Murray Bookchin, *The Ecology of Freedom: The Emergence and Dissolution of Hierarchy*, Cheshire Books, Palo Alto, 1982.

2 Many such feminists call themselves ecofeminists. Some of them cite Susan Griffin's *Woman and Nature* (Harper & Row, San Francisco, 1978) as the source of their understanding of the deep connections between women and nature, and their politics. *Woman and Nature* is an inspirational poetic work with political implications. It explores the terrain of our deepest naturalness, but I do not read it as a delineation of a set of politics. To use Griffin's work in this way is to make it into

something it was not intended to be. In personal conversation and in her more politically explicit works such as *Pornography and Silence* (1981), Griffin is antidualistic, struggling to bridge the false oppositions of nature and culture, passion and reason. Both science and poetry are deeply intuitive processes. Another work often cited by ecofeminists is Mary Daly's *Gyn/ecology* (1978). Daly, a theologian/philosopher, is also an inspirational thinker, but she is a genuinely dualistic thinker, reversing the "truths" of patriarchal theology. While I have learned a great deal from Daly, my perspective differs from hers in that I believe that any truly ecological politics, including ecological feminism, must be ultimately antidualistic.

3 Adrienne Rich, *Of Woman Born*, W. W. Norton, New York, 1976, p. 285.

Source: Ynestra King (1989) 'The ecology of feminism and the feminism of ecology', in J. Plant (ed.) *Healing the Wounds: The Promise of Ecofeminism*, London: Green Print, pp. 18–25, 27–8.

6.6

Joni Seager

Deep ecology and ecofeminism

Deep ecology

For a hopeful moment in the mid-1980s, an environmental wave sweeping Europe and North America seemed to offer a new vision and a *counterculture*, in the fullest sense of that word: "deep ecology" was an appealing, puzzling, and exotic environmental movement. Deep ecology was the pin set to burst the bubble of environmental hubris on which we build our human privilege. Its philosophers demanded that we ask probing questions of ourselves, of the nature of "being." Deep ecology represents an environmental *philosophy*, but at the same time it was a philosophy that actually spawned an activist wing with a distinct identity – deep ecology was not just an ideology, it was also a practice. The principles of deep ecology seemed to offer a challenge to patriarchal attitudes toward nature; its practice suggested a potential challenge to patriarchal methods of environmental organizing. Deep ecology offered hope and a refreshing vision to people who were concerned about the environment but who were disillusioned with the bureaucratic, reformist, and presumably coopted mainstream environmental groups.

The term "deep ecology" was coined in the early 1970s by a Norwegian philosopher, Arne Naess. Naess articulated an ecological approach that posed "deeper" questions about life on earth than mainstream environmentalists allowed themselves to ask.[1] The deep ecology he articulated was rooted in recasting the religious and philosophical interpretation of human relations with the natural world, starting with the necessity of shifting from human-centrism into biocentrism, a commitment to revaluing humanity's oneness with nature, and an appreciation of the intrinsic worth of all life forms.

Working from Naess's starting point, American deep ecologists in the 1980s elaborated an eight-point manifesto of "basic principles," among the most important of which are these: a reification of "biocentrism," which is a philosophy that nonhuman Life on Earth (capitals in original) has intrinsic value *in itself* independent of its usefulness to humans; that humans are too numerous and that a "substantial decrease" in human populations is required for the well-being of the earth; that humans must change their basic economic, technological, and ideological structures; and that everyone who subscribes to deep ecology has an obligation to try to implement these necessary changes.[2] The imperative to take direct action in defense of the earth is central to the philosophy of deep ecology. In the US, a loosely structured national group (with international affiliates) called Earth First! emerged as the organizational vehicle for translating the broad philosophical principles of deep ecology into an operational environmentalism. Although Earth First! disbanded in 1990, several EF! splinter groups remain in place as the organizational foci for the deep ecology philosophy.

Deep ecology is a big tent, under which many environmentalists gather – many of whom may disagree with one another on specific tactics or campaigns, but all of whom would broadly ascribe to the basic principles outlined above. As a Western environmental movement deep ecology is also characterized by a distinctive tone and a particular set of associations:

- Deep ecology environmentalism is suffused with a ritualized vision of the Earth as Mother, and of the Earth as a independent, self-regulating female organism. . . .
- Many Earth Firsters consider themselves a "tribe," and many of the American deep ecologists posit an affinity with indigenous, Native American ecological sensibilities.
- At heart, deep ecology is concerned with the preservation and protection of wilderness. Deep ecologists revere undisturbed wilderness as the pinnacle ecological state. Their militant defense of "Mother Earth" is rooted in an unflinching opposition to human attacks on wilderness. To the committed Earth Firster, the preservation of wilderness takes precedence over all human need.

Earth Firsters quickly established a high profile among environmentalists as guerrilla-theater, lay-your-body-on-the-line aficionados. In Australia, Earth First! protesters buried themselves up to their necks in the sand in the middle of logging roads to stop lumbering operations; in the American Southwest, Earth Firsters handcuffed themselves to trees and bulldozers to prevent logging; in California, they dressed in dolphin and mermaid costumes to picket the stockholders' meeting of a tuna-fishing company. Earth First! actions were often choreographed with a beguiling sense of humor, and carried out with daring and panache – their most endearing and environmentally useful characteristics.

Deep ecology is not confined by a single script. It is meant to be defined through its actions as well as its philosophy.[3] If, then, the measure of deep ecology is to be taken in the actions of Earth First!-style environmentalists, the conclusions are troubling. Deep ecology, in practice, has been transformed into a paramilitary, direct action ecology force. Some of the tactics employed by Canadian, Australian, and American Earth First! contingents are questionable: for example, pouring graphite (or sand or sugar) into the fuel tanks of bulldozers and road-clearing equipment involved in logging and mining operations to seize the engines; or, more controversially, tree-spiking, a practice of hammering long nails into trees to "booby-trap" them (spiked trees are dangerous to cut, and most loggers won't work a forest that has been spiked.) While many Earth First! groups renounced the use of destructive tactics, others embraced a "no compromise" environmental fundamentalism.

[. . .]

. . .[D]eep ecology, at first blush, appears to offer a philosophy that speaks to feminist values. The call by deep ecologists for a major overhaul of the political, economic, and ideological system is a necessity that feminists have been arguing for years. Deep ecologists speak of the interconnectedness of life, a reverence for nature, a nonexploitative relationship with wilderness, a valuing of intuition over rational, "anthropocentric" linear thought – all essentially women-identified ideas.

Deep ecology is virtually the only wing of the environmental movement to make specific overtures to women in general and to feminists in particular. In the abstract, if not in the practice, deep ecologists are astute enough to recognize that it is not possible in the 1990s to make credible claims of a "radical" agenda without some nod to feminist analysis. A number of male deep ecologists argue that the philosophy and theory of deep ecology are in sympathetic harmony with feminism. Others, too, assume an affinity: Kirkpatrick Sale, a noted bioregionalist, for example, referred to deep ecology as "that form of environmentalism that comes closest to embodying a feminist sensibility," continuing by saying "I don't see anything in the formulation of deep ecology that contravenes the values of feminism or puts forward the values of patriarchy."[4]

Deep machismo

One commentator, musing on deep ecology, noted, "Freely mixing pseudo-scholarly tomes and spit-in-the-can barroom philosophy, there is something in Earth First! to offend just about anyone."[5] To this assessment, I would only add that women have been especially offended. Despite its surface overtures to feminists, the transformation of deep ecology into an environmental force has been characterized by deeply misogynistic proclivities. Macho rhetoric of the most conventional and offensive sort riddles the written record of deep ecologists. Their common practice of using women-identified terms as taunts, such as calling their critics "wimps," "sops," or "effetes," panders to a blatant sexist and homophobic bias – as though the worst thing in the world is to be womanly! . . .

Deep ecology is saturated with male bravado and macho posturing. The American Earth First! movement is particularly symptomatic of the masculinist ethos that suffused representations of deep ecology's philosophy. With very few exceptions, the self-styled leaders and spokespeople of Earth First! were all men, as was a considerable proportion of its membership (in contrast with all other environmental groups).

[. . .]

Moreover, Earth Firsters are not just men; they are "men's men." Dave Foreman, one of the founders of the American Earth First! movement, represented the tone and tenor of the group when he said that, "I see Earth First! as a warrior society."[6] The leaders of Earth First reveled in an image of themselves as beer-swilling, ass-kicking, "dumb-cowboy rednecks" coming to the rescue of a helpless female – in this case, Mother Earth. Throughout most of the 1980s, the logo of the US organization was a clenched fist encircled by the motto "No Compromise in Defense of Mother Earth" – a bewildering mixed metaphor if ever there was one. (See Figure 13.)

[. . .]

While deep ecologists represent themselves as forging a radical new relationship with nature, they give no credit to the women who broke this path for them. Deep ecologists speak reverently of rediscovering intuition in relating to nature; of breaking through the

THE COMPLEET
RADICAL ENVIRONMENTALIST

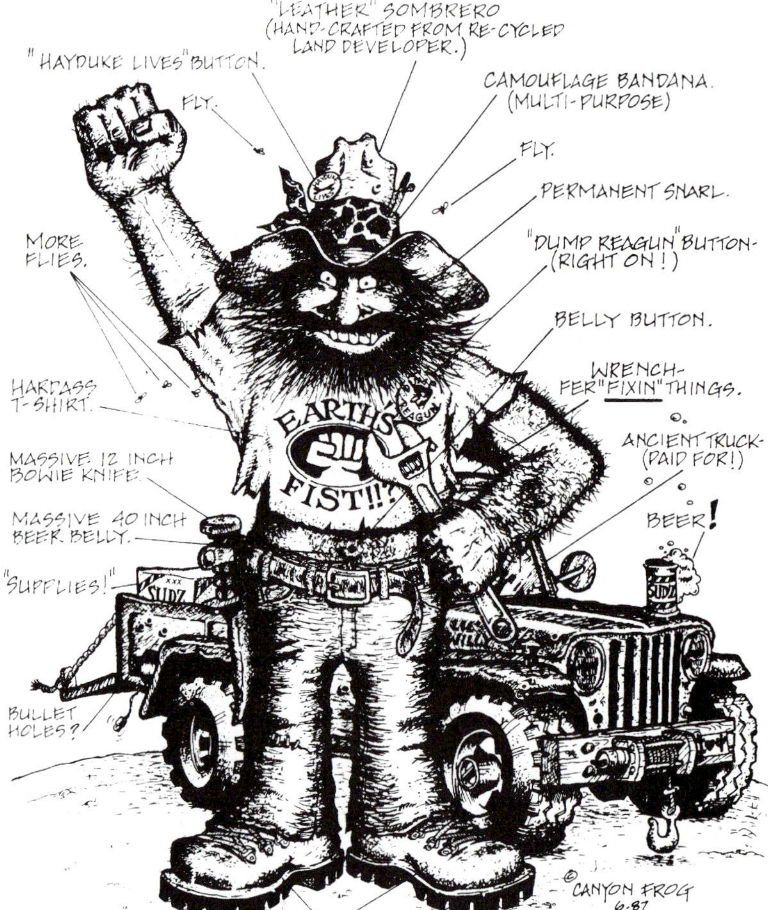

Figure 13 'The compleet radical environmentalist'

barriers of hierarchical thinking; of the necessity of viewing life on earth as an unbroken continuum; of celebrating the interwoven connectedness of us all – sensibilities that have been scorned by men for years as female-identified traits. But deep ecologists (mostly men) never say that. They exalt as though they are the first to discover this cooperative, noninvasive, and holistic life philosophy.

[. . .]

Disregarding difference

Despite their putative tilt toward feminism, deep ecologists are unwilling to include gender analysis in their analytical tool kit. Deconstructing and then *re*-constructing the human relationship to nature is absolutely central to deep ecology environmentalism. Yet, most deep ecologists are not interested in the social construction of attitudes toward nature, nor are they curious about the divergence in Western history (if not universally) of male and female attitudes toward wilderness and nature. Thus while there is an explicit criticism of destructive cultural attitudes toward nature, there is no apparent curiosity about the extent to which those "cultural" attitudes may be gender, race, or class specific.

And yet there is, now, a rich literature that explores the cultural and gender differences in human relationships to wilderness. Highly respected (and widely available) research by Annette Kolodny, for example, blazes a trail for gender-specific landscape/wilderness study.[7] Kolodny rewrites the history of the American frontier, establishing that images of "conquering the wilderness," "taming nature," "mastering the wild," and the like – images that North Americans take to be the standard fare of the European encounter with new lands, and images that continue to be the standard fare of popular culture – were, in fact, *male* fantasies and *male* imagery, and were not shared by their women counterparts in the wilderness. A recent study of women's attitudes toward nature and landscape in the American Southwest reinforces the argument made by Kolodny that, whereas European men saw the American West as a virgin land, ready to be raped and exploited, women typically regarded the landscape as "masterless." Rather than seeking conflict with the wilderness, women sought accommodation and reciprocity.[8] The conventional wisdom about the American wilderness experience is that "for most of their history, Americans [of European extraction] regarded the wilderness as a moral and physical wasteland, suitable only for conquest";[9] Kolodny's persuasive reply is that massive exploitation and alteration of the continent do not seem to have been part of women's fantasies.

[. . .]

The generalizations of deep ecologists blur distinctions not only of gender, but of race, class, and nationality too. Many deep ecologists portray the human race, as one species, as a sort of "cancer" on the earth that has devoured its resources, destroyed its wildlife, and endangered the biosphere. This sweeping misanthropy lacks social perspective – it is analytically unsound to make no distinction among peoples, nations, or cultures in assigning accountability for ecological destruction. Humanity is not an undifferentiated whole, and it is not credible to lay equal "blame" for environmental degradation on elites and minorities, women and men, the Third World and the First, the poor and the rich, the colonized and the colonizers.[10] Nonetheless, many American deep ecologists are insistent on this point. Dave Foreman, the US Earth First! founder, explains the disinterest in social analysis: "We are not opposed to campaigns for social and economic justice. We are generally supportive of such causes. But Earth First! has from the beginning been a wilderness preservation group, not a class-struggle group."[11] Paul Watson, a well-known American ecoiconoclast, casts Foreman's sentiments in stronger language: "My heart does not bleed for the third world. My energies point toward saving one world, the planet Earth, which is being plundered by one species, the human primate. . . . All human political systems developed to date, be they right or left, are anthropocentric in philosophy and support the exploitation of the Earth."[12] In his germinal article, Arne Naess, the "father" of deep ecology, expressed concerns about inequalities within and between nations. But his concern with social cleavages and their impact on resource utilization patterns and ecological destruction appears to have gotten lost in the translation, because it is all but invisible in the later writings of deep ecologists.[13]

[. . .]

The "population problem"

It is their stance on population, "overpopulation," and overpopulation "solutions" that has most alienated women and short-circuited an intellectual alliance between feminists and deep ecologists. A deep belief of deep ecology is that there are "too many people" on the planet. "Substantial" and fast *reductions* in the human population (not just a stabilization of population growth rates), deep ecologists say, are essential for the survival of the earth.[14] Some deep ecologists have tried to put actual figures to these reductions – Arne Naess, for example, proposed that for the health of the planet, "we should have no more than 100 million people";[15] an Earth First! writer using the pseudonym "Miss Ann Thropy" suggested that the US population would have to decline to 50 million.[16] (Given that the current population of the earth is just over 5 billion, and of the US almost 300 million, it is clear that the reductions called for by deep ecologists are drastic and would require catastrophic action to implement.) Other deep ecologists have proposed a 90 percent reduction in human populations to allow a restoration of pristine environments, while still others have argued forcefully that a large portion of the globe must be immediately cordoned off from human beings.[17]

The logistics of achieving such population reductions don't daunt deep ecologists. Deep ecology spokespeople have proposed a number of solutions to the "population problem" – "solutions" that range along a short spectrum from tame, vague, and muddled at one end to racist, sexist, and brutish at the other end. At the tame end, some deep ecologists have issued vague calls for widespread birth control programs – which is neither new, nor radical, nor much different from the population policies of most mainstream environmental groups.[18] There is no evidence of the much-vaunted feminist sensibility in the discussion of deep ecology population programs: women are once again rendered invisible, there is no linkage made between "population policies" and the daily lives of women around the world, nor is there any discussion or even acknowledgment of the fact that birth control policies inevitably bear disproportionately on Third World women. There is also no consciousness that, in issuing this call for population control, deep ecologists are preaching the same gospel as other men before them: that controlling female reproduction by technical means will solve the problems of "nature." Despite these "oversights" in the ideology of deep ecology population control, the call for birth control is not what has roused so much controversy.

Many of the other population solutions proposed by deep ecologists are not so benign. David Foreman, the Earth First! founder, proposes reducing the human population by halting life-saving medical interventions and aid for famine and disaster victims around the world. Speaking of the famines in Ethiopia, Foreman said, "The worst thing we could do in Ethiopia is to give aid. The best thing would be to just let nature seek its own balance, to let people there just starve."[19] This is an unconscionable and arrogant argument for a well-fed American to make, and is also based on the false premise that the famines in Ethiopia are somehow "natural." Mass starvation in Ethiopia derives not from a natural proclivity to famine, but from years of internal warfare, military spending bloated at the expense of social and environmental programs, corrupt governance, and regional environmental degradation.[20]

Edward Abbey, the guiding light of the American Earth First! movement, set off another firestorm of controversy by linking environmental population pressures with immigration policy. In the mid 1980s, Abbey started off by advocating that the US close its borders to Central American and Latin American immigrants, and went on from there with an escalating racism that was only thinly wrapped in a concern for the environment. . . .

Abbey's racist remarks were never repudiated by Earth First!, and were celebrated and repeated by many.[21] Some Earth Firsters argued that allowing Central Americans to use the "overflow valve" of fleeing to the US has two consequences: a) it increases the US population, and b) it allows Central American governments to continue their irresponsible ways. Once again, there is no discussion of social context or historical reality: if people are fleeing Central America in increasing numbers, it is in some measure because large parts of the region have been rendered all but uninhabitable by US-backed large landowners mismanaging the land and by US-backed military destruction. Nor is there discussion of the century of US intervention in and manipulation of the governments and economies of virtually every Central American country.

Racism lurks just beneath the surface of most discussions of "the population problem" in the deep ecology literature, and sometimes doesn't even lurk. Unfortunately, most deep ecologists have not taken on the issue of racism seriously, and instead of considering the issue anew, the anti-immigrant, and in particular anti-Hispanic, sentiment has been given considerable visibility in Earth First! literature.

[. . .]

Building a politics from a movement

Over the course of the past century, most radical movements for social change have been shaped by preexisting political agendas. In contrast, grassroots environmentalism is a movement of locally specific responses to particular circumstances, with no "core" doctrine; it is activism shaped more by internal realities than external politics. However, if grassroots environmentalism *is* to become the leading edge of eco/social transformation, we need to draw larger lessons from the experiences of thousands of locally-specific, often idiosyncratic grassroots actions.

Because of the gender skew in grassroots movements, we can only distill these larger lessons if we listen to women's voices and take seriously women's experiences. There is a certain repetitiveness to women activists' stories – particular patterns and themes turn up over and over again in the narratives of these women, almost no matter where in the world they live.

First and foremost, women who take the lead in community-based environmental causes do not generally have prior experience as political activists; many describe themselves in self-deprecating terms, as "mere housewives." In the great majority of cases, women become involved in environmental issues because of their social roles: as sustainers of families, it is often women who first notice environmental degradation. Many women grassroots activists report that their role as mothers and family caretakers is the key catalyst in their concern for protecting the environment. Because of those same social roles (and their related reproductive roles), women suffer the effects of environmental degradation first and longest – whether in Bhopal, the South African homelands, the Himalayas, or Vietnam, women are hardest hit by a diminished resource base, by exposure to toxins, and by localized pollution. Women, therefore, often have a vested interest in environmental protection.

Most women activists face tremendous resistance to their activism from the men they oppose (who enter the fray as representatives of industry, the military, the government), and also from men who are part of their daily lives, their husbands, sons, brothers, and friends. This resistance is patently based in sexist assessments of appropriate roles for women; for many men, the notion of a woman activist is an oxymoron. Women activists are stepping outside the bounds of sanctioned feminine behavior, and the techniques which men invoke to put women back in their place are often entirely based on sexist "policing" – there can hardly be a woman environmental activist in the world, for example, who has not been called a "hysterical housewife." It is clear that when women walk out of their homes to protest a planned clear-cutting scheme, toxic-waste dump, or highway through their community, their gender and sex identity go with them – in a way that is not true for men. Many women take up grassroots environmentalism not just as activists, but self-consciously as *women* activists – often as mothers. But whether they encourage it or not, they are perceived by men not just as environmental activists, but as *women* activists.

These commonalities in women's grassroots experience underscore the extent to which priorities and practices in the environmental movement will have to shift if the grassroots model is going to become the new paradigm for environmentalism. And even though many grassroots women may not be individually feminist, their collective experience points the way to constructing a *feminist* set of principles for coming to terms with the environmental tasks ahead of us.

The experience of women on the environmental front-lines should help us change our notion of what environmental destruction looks like: It's usually not big, flashy, or of global proportions – or, if a global problem, it manifests itself locally. Environmental degradation is pretty mundane; it occurs in small measures, drop by drop, tree by tree. This fact is discomforting to big scientific and environmental organizations whose prestige depends on solving "big" problems in heroic ways.

Because environmental destruction shows up in small ways in ordinary lives, we need to change our perception of who are reliable environmental narrators. Women, worldwide, are often the first to notice environmental degradation. Women are the first to notice when the water they cook with and bathe the children in smells peculiar; they are the first to know when the supply of water starts to dry up. Women are the first to know when the children come home with stories of mysterious barrels dumped in the local creek; they are the first to know when children develop mysterious ailments.

When the environment starts to suffer, signs of its degradation show up first in the water, food, and fuel women have to work with on a daily basis. Very early on, environmental decay starts to subtly impinge on the daily lives of women. In whatever context women are living, environmental degradation makes women's daily and ordinary lives more difficult. For Mrs. Woodman in Jacksonville Florida, the earliest sign of "something going wrong" was the fact that she couldn't get the clothes in her laundry to wash out white, no matter how hard she tried.[22] For Sithembiso Nyoni of Zimbabwe, environmental deterioration shows up first in her daily workload: "As a woman, it means that I have to walk long miles to fetch firewood and water. I have very little time, therefore, to grow vegetables and other food."[23]

Because women in all countries populate the most vulnerable segments of society, a disproportionate share of the impact of environmental problems falls on women. Women are confronted daily with having to care for children and go about their household chores without access to safe water, food, and sanitary facilities. It is often the case that community women are first spurred into environmental action when they perceive a threat to their children's health, and/or their own reproductive health. Reproductive disorders are an early-warning indicator of environmental deterioration. The presence of poisoned water, air and food often becomes evident first in unusually high numbers of miscarriages, still-births, low birth weights, birth defects, and occurrences of unusual diseases among infants and young children. Low birth rates, for example, are commonly found in the vicinity of hazardous waste sites; birth defects frequently occur when pregnant women are exposed to high levels of dioxins and other chemicals; childhood leukemia crops up in unusual clusters around nuclear power plants. These patterns of reproductive disorders and diseases are usually first detected not by a trained epidemiologist or local doctor, but by neighborhood women as they gather over communal coffee or child care and compare notes about sickness and health in their daily lives.

When women start to act on their environmental concerns, their actions often revolve around issues of the quality of daily life and the environmental safety of home and neighborhood. Knowledge of the wider links between "the local and the global" should not undercut a certainty, forged by women from their daily experience, that the appropriate forum for environmental action is often the local community itself.

But, the other side of the grassroots coin is that a home-based, local-defense environmentalism can also be mired in parochialism. The environmental report card on the American NIMBY movement, for example, gives a mixed review. While it is true that much environmental reform has come from people who face pollution in their own backyards and *then* go on to care about everyone else's, this ripple effect is not always present. Instead, the NIMBY movement often sets off an elaborate geographical shell game, wherein environmentally hazardous activities are shifted from one region to another (or one country to another). NIMBY movements do not necessarily present a radical challenge to the underlying structures of corporate and industrial enterprise – they challenge the right of businesses to locate dangerous activities in a *particular* neighborhood without necessarily challenging the presumed right of business to engage in environmentally unsound practices in the first place. Because of this, NIMBY movements thus often have the effect (perhaps unintended, but predictable) of shifting the burden of environmental hazards from the community that organizes the fastest – often predominantly white communities with an educated, middle-income population – to poorer, already-disadvantaged communities.[24] A number of recent studies of hazardous waste dumps in the US establish clearly that, among other things, the majority of the nation's largest hazardous waste dumps are located near African American or Hispanic communities; communities that host one toxic waste landfill or incinerator have, on average, a population that is 24 percent minority, while towns that have two or more dumps have an average minority population of 38 percent.[25]

In response to a NIMBY challenge, rather than change their environmentally unsound practices, corporations typically try to second-guess where hot spots of local activism are likely to flare up. In order to facilitate this sort of predictive planning, a large American waste disposal firm in the mid-1980s contracted with a research team to compile a profile of communities that were most and least likely to be resistant to the local siting of hazardous activities.[26] It is a tribute to the success of the NIMBY movement that "housewives" turned up in the "most resistant" occupational category! Not surprisingly, the communities described as "least resistant" to environmentally hazardous projects were small, rural communities in the South or Midwest with conservative politics and a middle-aged or elderly, low-income population – the sort of place like Emelle, Alabama.

The class and race subtext of the environmental agenda is written clearly on the landscape of Emelle, home of the largest hazardous dump in the US, probably in the world. Waste from 46 of the 50 states is transported (not all of it legally) to Emelle for "disposal." Emelle also receives wastes from other countries through shipments from military bases,[27] and current plans call for expanding the dump to include a hazardous waste incinerator. Sixty-nine percent of the families in the county live below the official poverty line; seven out of ten residents are African American; illiteracy and unemployment rates are high. And despite state assurances that no acutely toxic chemicals would be dumped at the site, residents have recently become aware of the dumping of benzene, PCBs, and possibly dioxins – substances with effects ranging from skin disorders to birth defects – and recent testing established that deadly chemicals have contaminated the groundwater around Emelle.[28]

The fact that the toxic trail often starts when relatively well-off communities prevent hazardous wastes and activities from locating in their "backyards" fuels accusations that the environmental community, as it is now constituted, is not taking seriously issues of racism in the environmental struggle. This tension then in turn undermines the effectiveness of environmental activists in making alliances with people of color in poorer communities – so their voices are missing from the environmental chorus, and the cycle of exclusion and privilege perpetuates itself.

Because of the nature of their daily lives, women often provide early warning on pollution problems and environmental degradation. They are also thus well placed to serve as agents for change – and because of their subordinate status they have the most to gain from movements for social, political, and environmental change. Successful environmental projects start with a recognition that women and men may have different priorities, agendas, and needs, and that men's needs must not be privileged over women's.

[. . .]

The use of science and scientific expertise in environmental impact assessment is becoming particularly contentious, and it may be the single most important factor in widening the schism between women grassroots activists and men in the environmental mainstream. As grassroots environmental watchdogs, women see it as an advantage that they have been socialized to listen to "their gut feelings"; men are socialized to veer away from intuition, and to formulate judgments only on the basis of irrefutable "facts." On an academic level, there is a growing gap between feminist and masculinist presumptions about the appropriate applications of scientific inquiry to environmental issues. Feminists have formulated sophisticated critiques of science, specifically focusing on the extent to which science *is* a values-based activity, arguing that there is no "neutral," objective science. Many scientists, especially male scientists, resist this premise; many male environmentalists are reluctant to cast aside the prop of presumed scientific neutrality and expertise.

It is very difficult to pinpoint "scientifically" or "absolutely" the particular cause of an illness or death. To "neutral" observers, just because a family's water supply is inadvertently poisoned by leachate from a chemical dump doesn't mean that the child they lost to cancer was poisoned by that water; even if it is proven that toxins did enter the water supply, it can seldom be proven that *those* substances *directly* caused the cancer or illness of *that* child. Corporations rely on this scientific uncertainty to save them from culpability. Many families and many community groups have fought local environmental battles through the media and the courts for years, only to be denied justice in the end because "the experts" can't "prove" causality.

[. . .]

The sexual division of emotional and intellectual labor on the environment – women care about it, men think about it – may be somewhat descriptive of present realities; without guarding against it, this could also become *pre*scriptive. If women are consigned to "care" about the environment, this lets men off the hook (why do men have to care about the environment if women are doing so?). If we are looking to build a new environmental agenda based on the lessons from grassroots activism, we must be willing to critically examine the assumptions about gender roles *before* they get inscribed in social movements. If a grassroots-informed environmentalism tells us anything, it is that both men and women must *care* – and think – about the environment.

[. . .]

The participation of women in social-justice movements (of which the environmental movement is one) must be allowed to develop free from male attitudes about women's appropriate place. Specifically, male violence against women is often used to "keep women in their place." It is often the *form* of women's activism – women acting as independent agents outside the home – not necessarily the *content* of women's activism which sparks male resentment and violence in the first place. If there is a single lesson to take from the world of grassroots organizing, it is this: women are absolutely key players in the environmental arena, but for women to act as agents of environmental change, they must be freed from narrow male assumptions about appropriate gender behavior, and they must be free to act without the threat of male violence. Men need to change what is considered to be the "normal" pattern of male exercise of power. The struggle to forge an environmentally just and sound future is inextricable from the struggle for gender justice and equality. Environmental relations are embedded in the larger gender relations that shape modern life.

Notes

1 Arne Naess, "The shallow and the deep, long-range ecology movement: a summary," *Inquiry*, Vol. 16, 1973.

2 Bill Devall and George Sessions. *Deep Ecology: Living as if Nature Mattered*. Salt Lake City, Utah: Peregrine Smith Books, 1985. Readers interested in exploring the literature of deep ecology might also read: Rik Scarce. *Eco-Warriors: Understanding the Radical Environmental Movement*. Chicago: Noble Press, 1990; George Bradford. *How Deep Is Deep Ecology?* Ojai, CA: Times Change Press,

1989; Michael Tobias, ed. *Deep Ecology*. San Marcos, CA: Avant Books, 1988 (2nd. revised printing); John Davis and Dave Foreman eds. *The Earth First Reader*. Salt Lake City, Utah: Peregrine Smith Books, 1991.

3 John Davis and Dave Foreman eds. *The Earth First Reader*. Salt Lake City, Utah: Peregrine Smith Books, 1991.

4 Quoted in Janet Biehl, "Ecofeminsm and Deep Ecology: Unresolvable Conflict?" *Green Perspectives*, 3.

5 Brian Tokar, "Exploring the new ecologies: Social ecology, deep ecology and the future of green political thought," *Alternatives*, Vol. 15, No. 4, 1988.

6 Quoted in Jim Robbins, "The environmental guerillas," *Boston Globe Magazine*, March 27, 1988.

7 Annette Kolodny. *The Lay of the Land* (Chapel Hill: University of North Carolina Press, 1975) and *The Land Before Her* (Chapel Hill: University of North Carolina Press, 1984).

8 Vera Norwood and Janice Monk, eds. *The Desert Is No Lady: Southwestern Landscapes in Women's Writing and Art*. New Haven: Yale University Press, 1987.

9 A quote from noted American historian, Roderick Nash, *Wilderness and the American Mind* (New Haven: Yale University Press, 1967). Nash's book continues to be used as the standard source for understanding the American encounter with wilderness. Nash himself now appears to acknowledge the andro-centric bias in his earlier work (see, for example, his article, "Rounding out the American revolution," in the Michael Tobias anthology cited above), but he does not revise his basic presumptions about the nature of the American wilderness experience.

10 A point made by Ynestra King, quoted in "Social ecology vs. deep ecology," *Utne Reader*, Nov. Dec. 1988, p. 135.

11 Dave Foreman and Nancy Morton, "Good luck, darlin', it's been great," in John Davis and Dave Foreman eds. *The Earth First Reader*.

12 Paul Watson in *Earth First!*, December 22, 1987, p. 20.

13 Ramachandra Guha, "Radical American environmentalism and wilderness preservation: a Third World critique," *Environmental Ethics*, Vol. 11, Spring 1989, p. 72.

14 This call for severe population reductions has, surprisingly, recently been taken up by some representatives of the UK Green Party. A conference paper at the 1989 annual conference proposed a reduction of the population of the UK to "between 30 and 40 million." These population reductions, it was proposed, could be achieved through public education (to reverse attitudes that portray childless people as unfulfilled, for example), and through family planning. Victor Smart, "Greens aim to halve population," *The Observer* (London) Sept. 17, 1989.

15 Quoted in Bill Devall and George Sessions, *Deep Ecology*, p. 76.

16 Miss Ann Thropy, "Population and AIDS," *Earth First!* May 1, 1987.

17 Ramachandra Guha, "Radical American environmentalism," p. 72.

18 See the population discussion in the chapter 4 of Joni Seager, *Earth Follies: Feminism, Politics and the Environment*, Earthscan Publications, London, 1993.

19 Quoted in Chris Reed, "Wild men of the woods," *The Independent* (UK), July 14, 1988. Similar sentiments were expressed in an article by Tom Stoddard, "On Death," and an accompanying editorial in *Earth First!* February 6, 1986.

20 For example, the Brundtland Commission on the Environment estimated that the Ethiopian government could have reversed the advance of desertification threatening its food supplies in the mid-1970s by spending no more than about $50 million a year to plant trees and fight soil erosion. Instead, the government in Addis Ababa pumped $275 million a year into its military machine between 1975 and 1985 to fight secessionist movements in Eritrea and Tigre. Figures cited in Michael Renner, "What's Sacrificed when we arm," *World Watch Magazine*, Vol. 25, September/October 1989.

21 See again, Tom Stoddard, *Earth First!* February 1986; see also Miss Ann Thropy, "Overpopulation and industrialism," originally published in *Earth First*! 1987, reprinted in John Davis and Dave Foreman, eds. *The Earth First Reader*.

22 See the discussion of the Jacksonville environmental crisis in Chapter 1 of Joni Seager, *Earth Follies: Feminism, Politics and the Environment*, Earthscan Publications, London, 1993.

23 Sithembiso Nyoni, "Africa's food crisis: price of ignoring village women?" in *Women and the Environmental Crisis*, United Nations Environment Liaison Centre, Nairobi, 1985.

24 Zack Nauth, "How toxic pollution can break down racial barriers," *In These Times*, Dec. 14–20, 1988. There is a burgeoning literature on the race and class matrix of environmental issues. Readers interested in pursuing this further might start with Robert Bullard, *Dumping in Dixie*. Boulder, Co: Westview Press, 1990 and Charles Lee, ed. *Toxic Wastes and Race in the United States*. NY: United Church of Christ, 1987.
25 "Toxic racism," *In These Times*, August 5–18, 1987.
26 Described in William Glaberson, "Coping in the Age of 'Nimby'," *New York Times*, June 19, 1988 and also in "Sign of success?" Citizen's Clearinghouse for Hazardous Waste, *Action Bulletin*, October 1987.
27 "Environmental rights in Emelle," *Greenpeace*, V. 13 2, 1988.
28 Reported in "On the Brink," *New Internationalist*, No. 157, March 1986; also, the *Greenpeace* article, in note 27.

Source: Joni Seager (1993) *Earth Follies: Feminism, Politics and the Environment*, London: Earthscan, pp. 223–34, 270–9, 312–21.

SECTION 7

Prospects for ecological citizenship

Introduction

This concluding section presents a selection of contemporary readings which attempt to think through the arguments addressed throughout the whole book. The section opens with Tim Hayward's exploration of the relationship between ecologism and environmentalism. In Reading 7.1 he demonstrates how ecological thought poses substantive challenges for the desired ends of existing social and political arrangements (that social practices should be based upon 'ecologically sensitive techniques and values') but also asks difficult questions about the institutional means for achieving these ends. In short, ecologism seeks to change the institutional arrangements that are taken for granted in environmentalism. By drawing together arguments from ecoanarchism and bio-regionalism with ethical standpoints like that of Aldo Leopold and Arne Naess, Hayward examines and evaluates the variety of ways in which ecological politics can make a difference.

This section considers whether the relationship between society and nature has changed considerably in the last twenty years. If so, then we need to be careful about the theories we adopt to explain and understand the human relationship with the natural world. By the 1980s, some of the more apocalyptic claims of the 1960s had given way to a recognition of complexity and uncertainty involved in human impacts on the environment. In Reading 7.2 Maarten A. Hajer outlines the theories of ecological modernization which increasingly inform environmental policy-making. This approach draws lessons from the diverse accounts of the natural world within the environmental movement. It also draws our attention to the succession of environmental problems caused by industrial society (from deforestation in the nineteenth century and air pollution in the mid-twentieth century, to chemicals in the 1960s, nuclear power in the 1970s, acid rain in the 1980s and global warming and ozone depletion in the 1990s). Ecological modernization occupies the niche between ecologism, with its plans for radical change, and the state regulation of the unacceptable side-effects of proven hazards. The advocates of ecological modernization try to find a way of working within a modern industrial society using institutional innovation to achieve environmental objectives (developing clean technologies, integrated traffic systems, eliminating inefficiencies in the production of waste).

In Reading 7.3 David Goldblatt provides an outline of Ulrich Beck's 'Risk Society'. Environmental degradation has always accompanied industrial society; however, recent social transformations have changed the ground rules for understanding risks. In industrial society risks were visible and localized hazards with a collective institutionalized response, the 'safety state'. In a risk society, the perception and determination of risk, the attribution of a definite cause, and the identification of responsibility are no longer guaranteed. For Beck, the big problem is that we still respond to risks as if we still live in an industrial society – a climate of organized irresponsibility where the origins and consequences of danger are masked and concealed. Finally Goldblatt considers the alternative futures, including the idea of ecological democracy. Both of these approaches are taken to task by Andrew Blowers (Reading 7.4) who highlights the weaknesses of ecological modernization and risk society approaches, chiefly their neglect of political considerations and the role of flexible planning strategies.

In Reading 7.5 by Mike Mills we address the theme of the place of democracy in the light of ecological arguments about the expansion of the moral community (and its associated conceptions of rights, obligations, citizenship, etc.). Mills presents a case for green democracy in the light of the conflicting alternatives of ecoauthoritarianism (focusing on the writings of Hardin) and ecoradicalism (ecologism) as a way of challenging the present anthropocentric policy and its associated liberal conceptions of citizenship. He also recognizes that democracy may not deliver what ecologists want. Finally, Peter Christoff (Reading 7.6) draws upon the political theory of David Held to rethink citizenship in the light of the displacement of the nation-state and the demands for a 'double democratization' to revitalize civil society and restructure the state. Christoff attempts to defend environmental values by incorporating the rights of future generations and animals within the moral community, creating a new form of citizenship. Whether ecological citizenship can be achieved without challenging the distinction between public (the state) and private (civil society) is an unresolved issue (see Smith, 1998).

Reference

Smith, M.J. (1998) *Ecologism: Towards Ecological Citizenship* (Milton Keynes: Open University Press).

7.1

Tim Hayward

Ecologism as a political ideology

A claim that there is a specifically green political ideology has been advanced by a number of writers. One of the more systematic arguments to this effect comes from Dobson, who claims that ecologism constitutes a political ideology in its own right. Dobson's presentation of specifically green political thought is organized around a central distinction between ecologism and environmentalism. 'Ecologism', he says, 'seeks radically to call into question a whole series of political, economic and social practices in a way that environmentalism does not.'[1] Whereas environmentalism may be compatible with any number of political forms, and may also encourage technocratic solutions to ecological crises, ecologism is to be contrasted on both points: it advocates institutions which embody and guarantee egalitarian and participatory politics; and its practices must be based on ecologically sensitive techniques and values. Hence, Dobson argues, whereas environmentalism can be coopted with relative ease into more traditional political ideologies, ecologism represents a distinct and new radical political ideology. This is the case, he claims, because it contains a description of the political and social world, a programme for political change, and a picture of the kind of society that political ecologists think we ought to inhabit – the sustainable society. . . .

Ecological utopia and the politics of the good society

There is a view of the good society which, according to Dobson, is distinctively green and cannot readily be identified in any traditional political ideology. Although the literature on ecological utopias offers some contrasting visions of the kind of social and political life to be aimed at,[2] there are certain fairly common features; and if we concentrate specifically on those aspects bearing on questions of political organization, there is something approaching consensus. Given the key values of an ecological society it appears, in particular, that its political organization would have to be non-hierarchical and decentralized. In what follows I shall briefly note why decentralization is thought desirable and how it is claimed to be connected with ecology. . . .

Among the major reasons why decentralization is so important is that it means hierarchies are broken down and people are empowered by being members of small political communities. The root ideas here are that people must feel a part of their community in order to participate meaningfully; to do so they must be able to meet face to face, confident that their participation might make some material difference; moreover, they must be able to comprehend what is actually going on in their community well enough to estimate how different policies might affect it; finally they must be able to survey the community as a whole, not just their own corner of it, if they are to judge the general good rather than pursue narrow sectional interests.[3] In these respects, of course, the ideal of decentralization is not novel or peculiar to ecological politics. It also figures in certain branches of liberal, socialist and communist politics; more particularly though, it embodies concerns which are very much those of anarchist traditions. It is therefore not surprising that a strong current of opinion maintains that ecological politics must in fact be a form of anarchism. Certainly, from a historical perspective, it seems to be the case that the most searching ecological questions were raised – long before questions of environment and ecology were of widespread public concern – by anarchists more than by thinkers of other political colours.[4] In more contemporary debates too, Eckersley, for instance, sees concerns with ecology and decentralization as combined in 'ecoanarchism', and she takes seriously the claim 'not only that anarchism is the political philosophy that is most compatible with an ecological perspective but also that anarchism is grounded in, or otherwise draws its inspiration from, ecology'.[5]

This claim was already forcefully advanced in Murray Bookchin's seminal writings of the 1960s, which argue that ecological and anarchist principles are mutually reinforcing. Bookchin believes there is no hierarchy in nature and that a society free of hierarchy is a precondition for putting ecological principles into practice; decentralization is therefore not only desirable but even necessary for human survival. Specifically ecological reasons why decentralization is necessary have to do with how centralization reverses the direction of organic evolution and causes ecological imbalance. Bookchin believes the 'validity of the decentralist case can be demonstrated for nearly all the "logistical" problems of our time';[6] for instance, he argues, sustainable agriculture and the use of energy resources as well as the fight against pollution require a far-reaching decentralization of society and a truly regional concept of social organization. This also means that metropolitan areas need to be broken down so that a 'new type of community, carefully tailored to the characteristics and resources of a region, must replace the sprawling urban belts that are emerging today'.[7] Bookchin thereby invokes what he calls a 'true regionalism' as an antidote to the ecologically destructive concentrations of economic and political power which characterize the modern world.

More recently this idea of 'true regionalism' has been worked up into the influential idea of 'bioregionalism', in which the aim of decentralization receives an even more distinctively green colouring. A bioregion, on the definition given by Kirkpatrick Sale, 'is a part of the earth's surface whose rough boundaries are determined by natural rather than human dictates, distinguishable from other areas by attributes of flora, fauna, water, climate, soils and land-forms, and the human settlements and cultures those attributes have given rise to.'[8] The boundaries of a bioregion should be drawn so as to encapsulate certain self-contained physical and biological processes of nature, such as airsheds and watersheds. The term 'bioregion', though, has been used to refer not only to the place but also to ways of living in that place. Living bioregionally means living with the resources which are found within one's own territory, or, more accurately, becoming aware of the ecological relationships that operate within and around it so that humans see themselves, in Leopoldian fashion, as members of a biotic community rather than its exploiters. Within bioregions people would live in communities, which Sale sees as the basic units of the ecological world: these 'more-or-less intimate groupings' of somewhere between 1,000 and 10,000 people would be the primary locus of decision-making.[9] They would seek to minimize resource use, emphasize conservation and recycling, avoid pollution and waste. The aim of all this is to achieve sustainability through self-sufficiency. This involves reducing both the spiritual and material distance between humans and the land they live on:

> We must somehow live as close to it [the land] as possible, be in touch with its particular soils, its waters, its winds; we must learn its ways, its capacities, its limits; we must make its rhythms our patterns, its laws our guide, its fruits our bounty.[10]

The guiding principle of bioregionalism, and of green politics in general, according to Dobson, 'is that the "natural" world should determine the political, economic and social life of communities'.[11] Thus Sale, for instance, believes that 'by a diligent study of nature we can guide ourselves in constructing human settlements and systems'.[12]

Certainly, one must always be duly cautious on the question of what is to be learnt from nature, and criticisms that can be made of bioregionalism will be considered at the end of this section, but I believe that the broad principles underlying an ecological utopia constitute an ideal worth striving for. To live in communities whose economies and politics are of human scale, whose principles embody the aim of living in a closer relation to nature, and where belonging is a relation of reciprocity and membership rather than ownership, are desirable ends. The question to consider next is that of strategies for attempting to achieve a society embodying these principles.

Strategies for getting there: green parties and alternatives

Whatever the desirability of the green utopia, whatever its internal coherence and its consistency with ecological realities, there remains the question of if and how it might be achieved. This is to raise the question of what forms and strategies of politics are appropriate in the here and now; it is also to raise the question of how greens can exercise power, what sort of power is possible for them, and what sort is consistent with their ends. New sorts of answer to these questions have to be sought because, viewed from a radical green perspective, political power in the contemporary world is fatally compromised. More often than not it is grounded in vested interests which are unecological or give low priority to ecological values. Certainly, to the extent that modern political systems support and are supported by capitalist economies, they reproduce the characteristic ecological shortcomings of compulsively pursuing economic growth, discounting the future and, in allowing market forces full rein, having no mechanisms for dealing with the problems they generate.[13] Hence it is arguable that the liberal democracies tend to share capitalism's own defects. On the other hand, more centralized alternatives have defects which are no less serious. The manifest failure of central planning to cope with ecological problems in actually and formerly existing socialist states, for example, may very well be due to inherent features of that form of organization: in particular, those at the centre will often not have the political will, and almost certainly not have the ability, to respond with sufficient sensitivity to the complex and novel problems that continually arise.[14]

Green politics aspires to new political forms which are more sensitive to both human and ecological needs than are those oriented either to the free market or to centralized organization. A distinctive feature of green politics being its basis in grassroots social movements which are committed to extra-parliamentary, non-conventional and decentralized forms of political action, it aims to organize these into a force for change. Pursuit of this aim does involve potentially conflicting demands, though. The imperative of decentralization means seeking to develop a genuinely participatory polity which provides for a grassroots democracy that will remain true to the aims of the various constituent new social movements. Another imperative, however, is that of actually transferring power, and if this is not to be sought through violent revolution (and no greens to my knowledge think it should be), then it will mean working within and against existing institutions, attempting to render them more democratic, that is, more participatory and genuinely transparent.

The need for strategies to mediate between these potentially conflicting demands has been the *raison d'etre* for the formation of green political parties. The idea of a 'new kind of party' which would seek to bring the influence of new social movements to bear on the existing machinery of power was summed up by Petra Kelly of *die Grünen*: 'We can no longer rely on the established parties . . . and we can no longer depend entirely on the extra-parliamentary road. The system is bankrupt, but a new force has to be created both inside and outside parliament.'[15] When *die Grünen* won seats in the Bundestag, they saw this as part of a dual strategy which they described with the metaphor of 'two legs': one leg moving freely through the parliamentary institutions but the other having a firm foot in the extra-parliamentary movement.

This dual strategy of green party politics needs to be looked at in two dimensions: the role of oppositional groupings within and against the state; and the internal organization of a green party itself.

A constructive role for green representatives in parliament, as outlined by Joachim Hirsch, includes offering careful criticism of the administration's bills and countering them with concrete alternatives. Such critical labours aim at uncovering distortions and the systematic exclusion of certain issues. The objective, though, is not just to force concessions, but also to stimulate political development within the opposition movements themselves. This, he argues, establishes political and social legitimacy and demonstrates that there is *no* area where the opposition is not present. Parliamentary activity is thus seen as creating legal, organizational and financial conditions for developing and strengthening practical measures for changing society.[16] It provides a *support* for the political praxis whose real basis and vital context remains outside, with the movements. On the other hand, however, there remains the question of whether the greens' achievements within parliament will not be of too piecemeal and pragmatic a nature really to satisfy demands, pointing to the need for a more radical transformation of the political system itself. This in turn raises questions about the role of the green *party* as a mediator of grassroots politics. As Hülsberg notes, there appears to be a contradiction between the anti-institutional character of the new social movements on the one hand, and their coagulability into a political party on the other. Moreover, it also seems to be the case that green parties virtually everywhere (perhaps like any political grouping with radical aims) tend to be wracked by internal strife, with a central line of tension invariably being that between accepting the demands of party discipline and the need for leadership, and allowing a genuine voice for the grassroots membership.

A central question, therefore, is whether green parties, through innovations in their internal organization, can overcome these contradictions and tensions. *Die Grünen* set out to develop a new type of party organization, the fundamental idea of which was continuous control over all officials and elected representatives in parliament and their recallability at any time. As stated in their programme of 1983, they sought to establish a party which was half citizens' initiative and half party. Rank and file democracy was supposed to be guaranteed by a number of means, including a system of rotation in the Praesidium. Nevertheless, in practice, the weight tended to shift in favour of the parliamentary party. So whilst the German Greens went into parliament with the aim of changing the system, the system also changed them, as one by one the measures designed to ensure grassroots control were dropped.[17] One cannot, of course, conclude from this example, or even from the lack of counter-examples, that these kinds of measures designed to ensure democratic party structures will necessarily fail. On the other hand, though, nor should this perhaps be seen simply as a story of a careerist or corrupt centre disregarding the views or wishes of the grassroots. It does in fact appear that some of the problems are intrinsic to the project of grassroots democracy itself.

Political analysts have identified basic problems which appear to be intrinsic to grassroots democracy. For one thing, the openness of all meetings to ordinary members means that rank and file democracy easily degenerates into endless discussion: the more people who take part in meetings, and the more that meetings strive for unanimity, the longer it takes to reach any decision, which makes for inefficiency. This inefficiency is only compounded by the principle of rotation in office which, argues Goodin, generally means more in costs (in terms of lost expertise and effectiveness) than in benefits (which he sees as amounting to little more than the self-satisfaction of maintaining the principle of rotation itself).[18] Worse still, though, is how the green vision of grassroots democracy threatens to undercut the democratic character of the process it was intended to ensure: because no decision will be taken until nearly everyone has been talked around, observes Goodin, 'green theorists guarantee that only a small and unrepresentative sample of party members will be left in the room by the time the final decision is taken.'[19] Moreover, this open structure means whoever is there can vote for whatever measures they wish, without having to bear any direct responsibility for the consequences. These problems are not peculiar to *green* theories of grassroots democracies, Goodin notes, but nor do the latter have any novel answers to offer.

In practice, parties like *die Grünen* have responded to these problems by developing a kind of professionalism with informal leadership structures. In doing so they have been taken to confirm a widely recognized tendency of parties in general to succumb to what has been called the 'iron law of oligarchy', whereby they end up being run by small groups of unrepresentative leaders.[20] From the perspective of someone like Goodin, this may be no bad thing; for radical greens, however, it points up the question of whether engagement in conventional politics is consistent with green objectives at all, and whether a single green party is even necessary or desirable. Certainly, in the light of these problems it does appear that the ideal of the green party being a radically different sort of party, a non-party party, might not be workable. Hence the choice may be the starker one of either accepting the role of a more or less conventional party *or* returning to strategies of more fundamental opposition – of strategies, in other words, of a movement *rather than* a party.

If it is the case, as Dobson suggests, that the institutions of representative democracy are effectively designed to preclude the possibility of massive regular participation, then the pursuit of participatory politics will demand the radical restructuring, or even abolition, of existing institutions.[21] Arguably, then, to be consistent, fundamentalists should abandon the party altogether.[22] And even though others may grant a residual case for considering participation in parliamentary politics as necessary or useful, there is agreement among greens that parliamentary politics is not sufficient to bring about the kind of change they seek.

So we turn to consider alternative green politics – that is, practices over and above those associated with parliamentary activity, with local government, or with political parties. Under the general heading of alternative politics may be included citizens' initiatives, networking, changes in individual and family lifestyle, peace camps, cooperatives, and communities experimenting with ecological lifestyles such as rural self-sufficiency farms or city farms. These are generally localized and particularistic types of activity, aimed not at seizing levers of existing power but at creating a new society in the interstices of the old. Something distinctive about these types of practice, therefore, is that they are to some extent pursued as ends or goods in themselves, and not only as means to separate goals. To some extent they constitute an attempt to create a radical alternative ecological culture in the here-and-now, an anticipatory practice in a quite different spirit from a reformism which would allow the use of any means necessary in the pursuit of green policy outcomes.

Accordingly a good deal of importance has been placed within green movements on the strategy of seeking social and political change by means of a change in one's lifestyle. Schumacher concluded his book, *Small Is Beautiful*, with words whose spirit is a notable influence in green politics: 'Everywhere people ask: "What can I actually *do*"? The answer is as simple as it is disconcerting: we can, each of us, work to put our own inner house in order.'[23] A guiding lifestyle value has been captured by Bill Devall in the slogan 'simple in means, rich in ends'.[24] And as Arne Naess argues, this aim is not to be confounded with appeals to be Spartan, austere, or self-denying, for in one's lifestyle one can appreciate opulence, richness, luxury, and affluence so long as these are defined in terms of quality of life rather than standard of living. On this view of it, the politics of lifestyle would follow quite directly from the ecologically enlightened conception of the good life which was described in chapter 3. It would also have a greater importance than allowed it by reformists like Goodin. Arne Naess offers reasons for a different order of priorities: 'Attempts at a change in lifestyle cannot wait for the implementation of policies which render such change more or less required. The demand for "a new system" *first* is misguided and can lead to passivity.'[25]

Of course, as Naess also notes, changes in lifestyle alone are unlikely to be sufficient, and must therefore proceed simultaneously with more directly political changes. The point is, though, that they are necessary.

Examples of the kinds of practice envisaged as lifestyle would be, according to Dobson, 'care with the things you buy, the things you say, where you invest your money, the way you treat people, the transport you use and so on'.[26] Lifestyles may incorporate ends which are communicative, educational, exemplary, or even involve direct action. Naess describes how they can operate as part of a cultural renewal: 'The movement encourages the reduction of individual total consumption, and will through information, increased awareness, and mutual influence attempt to free the individual and society from the consumer pressures which make it very difficult for politicians to support better policies and a healthier society.'[27]

It should be noted, though, that a good deal of the emphasis is placed on the power of *consumers*, and, as Dobson points out, there are pros and cons with this: 'The positive aspect of this strategy is that some individuals do indeed end up living sounder, more ecological lives . . . The disadvantage, though, is that the world around goes on much as before, unGreened and unsustainable.'[28] Dobson notes that the consumer strategy may even be counterproductive at a deeper level of green analysis: for one thing, 'there are masses of people who are disenfranchised from this exercise of power by virtue of not having the money to spend in the first place'; and, for another, 'the underlying aim of this green consumerism is to *reform* rather than fundamentally restructure our patterns of consumption.' But although these are valid criticisms of consumerist strategies, it is nevertheless arguable that lifestyle is not reducible without remainder to these, especially when it is oriented to new forms of production and exchange. A second line of criticism, though, is potentially more serious: such strategies do not take account of the problem of political power, argues Dobson, 'they mostly reject the idea that bringing about change is a properly "political" affair – they do not hold that Green change is principally a matter of occupying positions of political power and shifting the levers in the right direction.'[29] However, without denying that there are always likely to be individual lifestylers who are too naïve about political realities, I think this line of criticism risks missing the potentially radical force of alternative politics, for it is liable to beg the very question of principle which is central to any grassroots form of politics – namely, that of identifying, or even creating, new points of leverage and positions of political power. It is worth appreciating that lifestyle change can be of positive value not only to those individuals who end up living sounder, more ecological lives, nor just because it provides examples that others might be encouraged to follow, but also, and maybe more importantly, because it can highlight – as has already been the case for some time in feminist practice – that many aspects of daily life can indeed be considered 'properly political'. In this way, lifestyle changes can be seen as part of a broader strategy to transform and enrich the very concept of the political, and thereby to increase the sense, and reality, of empowerment of those hitherto excluded from politics in its mainstream definition.

Still, I speak of lifestyle as part of a broader strategy, for even if political change can start with a change in individual consciousness, it clearly does not end there. If a major question with lifestyle change is how the individualism on which it is based will convert into the communitarianism which is central to the ecological utopia, then it may in fact be, as Dobson says, more sensible to subscribe to forms of political action that are already communitarian. What he refers to as community strategies would be an improvement on lifestyle strategies 'because they are already a practice of the future in a more complete sense than that allowed by changes in individual behaviour patterns. They are more clearly an alternative to existing norms and practices, and, to the extent that they work, they show that it is possible to live differently – even sustainably.'[30] One notable proponent of this sort of strategy is Rudolf Bahro, who speaks of creating alternative 'basic communes' in which people opt for non-industrial lifestyles outside conventional patterns of employment and consumption. These would be 'liberated zones' oriented to genuine human needs and fulfilment. They would be self-sufficient communities of around 3,000 people 'which would agree on a mode of simple, non-expanded reproduction of their material basis'.[31] Robyn Eckersley locates Bahro's suggestion in the tradition of monasticism, and summarizes its appeal:

> this tradition has shown that it is possible to create relatively self-sufficient and stable domestic economies from very small and humble beginnings. Moreover, it is a tradition that fosters a community that is 'simple in means and rich in ends,' provides an economics of permanence, offers egalitarian fellowship, and is able to synthesize qualities that have become polarized in modern life such as the personal/social and the practical/spiritual.[32]

The pursuit of community strategy also harmonizes with the aim of bioregional politics, already discussed above, to achieve communal autarky, that is, the self-sufficiency and independence of constituent communities.

There are, however, questions as to how well community strategies will serve the ends of either the good life or justice. On the one hand, the quality of life in communities will not necessarily be an improvement in all respects on life in more anomic societies. For instance, as Naess has noted, communities have a potential for oppressiveness, a pressure towards conformity, which reduces both personal initiative and self-determination.[33] This is echoed by Gorz, who writes in quite a contrasting spirit to Bahro: 'communal autarky always has an impoverishing effect: the more self-sufficient and numerically limited a community is, the smaller the range of activities and choices it can offer to its members. If it has no opening to exogenous activity, knowledge and production, the community becomes a prison.'[34] On the other hand, connected to these concerns about the parochialism and personalism of community life, there are also concerns to do with justice, both retributtive and distributive. One of these is the question of how recalcitrant members of the community are to be dealt with: Edward Goldsmith has suggested that 'crime' might be controlled 'through the medium of public opinion' by subjecting the offender to 'ridicule', but, as Dobson rightly points out, there is good reason to worry about this sort of suggestion,[35] and to prefer the impersonal impartiality of more formal systems of justice. On the distributive side, another set of questions concerns intercommunal or interregional justice and the problems of disparities of resources between regions which are differently endowed. If the principle of independence and self-sufficiency is paramount, then it can always be invoked by the better endowed as a reason for their non-beholdenness to the less well endowed.

If community strategies in and of themselves do not necessarily meet the desiderata either of good life or of justice, it is also a moot point how ecological they are: for it would not seem to be quite consonant with ecological principles to suppose that any region, let alone community, is wholly independent of others and therefore able to be entirely self-sufficient. Indeed, if communal decentralization means a more uniform distribution of human settlement, there are ecological arguments against it: for instance, as Paehlke points out, 'urban settlements are a less ecologically stressful and more energy efficient way of accommodating large numbers of people on the land than dispersing the human population more thinly and widely throughout existing wilderness and rural areas.'[36]

These observations lead us to appreciate that there are both social and ecological limits to how far decentralist strategies in general, whether communal or individualistic, can successfully be pressed. The appropriateness of decentralization as either strategy or objective depends at least in part on assumptions which are made about the nature of people and their relations. It presupposes people can and will work together cooperatively without a need for coercion from above or even coordination from a centre. This in turn presupposes that a consensus can normally be arrived at, and that therefore there are no irreconcilably conflicting interests between or among individuals, communities and regions, or related problems such as that of chronic discrimination against permanent minorities.[37] In short, purely decentralist strategies are ultimately dependent on extremely sanguine assumptions about human cooperativeness, yet as has already been said, there are good reasons for caution about this sort of assumption. These reasons are compounded when such assumptions are claimed to derive support from ecological considerations. When the idea that nature knows best, and that we should therefore follow nature, is appealed to as a political principle, it should be remembered that nature teaches notoriously equivocal lessons: decentralists are impressed by the lack of hierarchy in nature, but social Darwinists, for instance, read quite other messages in nature. In the light of these considerations, finally, one should be extremely wary that if, as Sale suggests, it is not necessarily the case that each bioregional society 'will construct itself upon the values of democracy, equality, liberty, freedom, justice, and other suchlike desiderata',[38] then this may well be a good reason for wanting to maintain or develop some overarching institutions which *do* embody these values.

I should emphasize that these criticisms of decentralization apply to the carrying of it to an extreme; they do not require us to deny it as an important political principle. Kemp and Wall try to put the matter in perspective:

> Critics of decentralization argue that tight national control and strong central powers are needed in case local communities decide to dump PCBs in their drinking water or build nuclear power stations, but a far more real danger is that strong central government, supported by big business, distant from its electorate and protected by laws of official secrecy, will create pollution on a national scale . . . The ecological dangers of centralization are proven and dangerous; the positive contribution decentralization could make to a cleaner environment and a better democracy are clear.[39]

. . . Yet if the pursuit of green politics in its more radical forms is rife with paradoxes and contradictions, this in itself need not constitute an objection, since they have arisen as a response to a reality which itself is rife with contradictions; what it does mean though, is that contradictions need to be worked through rather than peremptorily foreclosed. To preserve the cutting edge of radical green ideas, it is necessary to find ways of bringing them to bear on existing power relations. Effective strategies depend not only on utopian goals, but also on a theory of the given: it is necessary to understand existing power relations and forces for change. . . .

Notes

1 Dobson, *Green Political Thought*, p. 205.
2 For an overview of the range, see Boris Frankel, *The Post-Industrial Utopians*.
3 This description is a summary of Goodin, *Green Political Theory*, pp. 149–50.
4 Kropotkin's *Mutual Aid*, in particular, is frequently referred to as a seminal tract in this connection.
5 Eckersley, *Environmentalism and Political Theory*, p. 145.
6 Murray Bookchin, *Ecology and Revolutionary Thought*, p. 14.
7 Ibid.
8 Quoted in Goodin, *Green Political Theory*, p. 149n.
9 Ibid., p. 150.
10 Sale quoted in Dobson, *Green Political Thought*, p. 119.
11 Ibid.
12 Ibid.
13 Cf. John S. Dryzek, 'Ecology and Discursive Democracy'.
14 Cf. Dryzek, *Rational Ecology*, chapter 8.
15 Quoted in Hülsberg, *The German Greens*, p. 78.
16 Joachim Hirsch, 'Die Grünen: Zwischen Fundamentalopposition und Realpolitik', p. 11.
17 Hülsberg, *The German Greens*, p. 123.
18 Goodin, *Green Political Theory*, pp. 144–6.
19 Ibid., p. 142.
20 Ibid., p. 144.
21 Dobson, *Green Political Thought*, p. 135.
22 This is the conclusion Rudolf Bahro came to: see Bahro's resignation speech, *Building the Green Movement*, pp. 210–11.
23 E.F. Schumacher, *Small Is Beautiful*, pp. 249–50.
24 See Bill Devall, *Simple in Means, Rich in Ends*.
25 Arne Naess, *Ecology, Community and Lifestyle*, p. 89.
26 Dobson, *Green Political Thought*, p. 140.
27 Naess, *Ecology, Community and Lifestyle*, p. 91.
28 Dobson, *Green Political Thought*, p. 141.
29 Ibid., pp. 142–3.
30 Ibid., p. 146.
31 Quoted in Frankel, *The Post-Industrial Utopians*, p. 140.
32 Eckersley, *Environmentalism and Political Theory*, p. 165.
33 Naess, *Ecology, Community and Lifestyle*, p. 159.
34 Gorz quoted in Frankel, *The Post-Industrial Utopians*, p. 59.
35 Dobson, *Green Political Thought*, p. 124.
36 Paehlke in Eckersley, *Environmentalism and Political Theory*, p. 174.
37 Cf. Anthony Arblaster, *Democracy*.
38 Quoted in Dobson, *Green Political Thought*, p. 122.
39 P. Kemp and D. Wall, *A Green Manifesto for the 1990s*, p. 177.

References

Arblaster, Anthony, *Democracy*, Open University Press, Milton Keynes, 1987.
Bahro, Rudolf, *Building the Green Movement*, trans. Mary Tyler, Heretic Books, London, 1986.
Bookchin, Murray, *Ecology and Revolutionary Thought*, with an introduction by Howard Hawkins, Green Program Project, Vermont, undated (first published in 1964).
Devall, Bill, *Simple in Means, Rich in Ends: Practising Deep Ecology*, Green Print, London, 1990.
Dobson, Andrew, *Green Political Thought*, Unwin Hyman, London, 1990.
Dryzek, John S., *Rational Ecology: Environment and Political Economy*, Blackwell, Oxford, 1987.

Dryzek, John S., 'Ecology and Discursive Democracy: Beyond Liberal Capitalism and the Administrative State', *Capitalism, Nature, Socialism*, 10 (1992), pp. 18–42.
Eckersley, Robyn, *Environmentalism and Political Theory*, UCL Press, London, 1992.
Frankel, Boris, *The Post-Industrial Utopians*, Polity Press, Cambridge, 1987.
Goodin, Robert E., *Green Political Theory*, Polity Press, Cambridge, 1992.
Hirsch, Joachim, 'Die Grünen: Zwischen Fundamentalopposition und Realpolitik', *Grüne Hessenzeitung*, no. 9/10 (December/January 1982/3), pp. 8–11.
Hülsberg, Werner, *The German Greens: A Social and Political Profile*, trans. Gus Fagan, Verso, London, 1988.
Kemp, Penny and Wall, Derek, *A Green Manifesto for the 1990s*, Penguin, Harmondsworth, 1990.
Kropotkin, P., *Mutual Aid: A Factor in Evolution*, Porter Sargent, Boston, 1914.
Naess, Arne, *Ecology, Community and Lifestyle*, trans. and ed. David Rothenberg, Cambridge University Press, Cambridge, 1989.
Sale, Kirkpatrick, *Human Scale*, Secker and Warburg, London, 1980.
Schumacher, E.F., *Small Is Beautiful*, Abacus, London, 1974.

Source: Tim Hayward (1994) *Ecological Thought: An Introduction*, Cambridge: Polity Press, pp. 187–99, 233–50.

7.2

Maarten A. Hajer

Ecological modernisation as cultural politics

In his celebrated study of the US conservation movement around the turn of the century, Samuel Hays (1979) describes how the popular moral crusade for conservation of American wilderness paved the way for a group of experts that, under the veil of working for conservation, advanced their own particular programme. These 'apostles of efficiency' did not share the somewhat sentimental attitude towards wilderness that was typical of the predominantly urban movement for conservation. Above all, they were interested in applying new techniques of efficient resource management in introducing new forestry practices or in constructing and experimenting with the latest hydroelectric dams. For the American urbanites wilderness had a deeply symbolic meaning. Trees and mighty rivers were the icons of the alleged moral superiority of nature that stood in sharp contrast to the bitter reality of a rapidly industrialising society. For experts like Gifford Pinchot and his colleagues, in contrast, wilderness was a nuisance and nature was a resource: trees were merely crops and rivers were to be tamed and tapped. For the urbanites nature had to be preserved; for the experts nature had to be developed.

The story is instructive in several respects. Firstly, it shows that 'our' ecological 'problematique' most certainly is not new. The negative effects of industrialisation for nature have been thematised time and again over the last 150 years. Yet characteristically the public outcry focuses on specific 'emblems': issues of great symbolic potential that dominate environmental discourse. Examples of emblems are deforestation in the mid-nineteenth century, wilderness conservation (USA) and countryside protection (UK) at the turn of the century, soil erosion in the 1930s, urban smog in the 1950s, proliferation of chemicals in the early 1960s, resource depletion in the early 1970s, nuclear power in the late 1970s, acid rain in the early 1980s, followed by a set of global ecological issues like ozone depletion or the 'greenhouse effect' that dominate our consciousness right now. Given this sequence of issues, it is better to refrain from speaking of today's predicament in terms of 'our ecological crisis' (which suggests it is time and space specific) and to speak of the *ecological dilemma* of industrial society instead.

Secondly, if we accept the thesis that environmental discourse is organised around changing emblems, we should investigate the repercussions of these subsequent orientations of the debate. After all, emblems mobilise bias in and out of environmental politics. They can be seen as specific discursive constructions or 'story lines' that dominate the perception of the nature of the ecological dilemma at a specific moment in time. Here the framing of the problem also governs the debate on necessary changes. In the case of the US conservation issue the prevailing story line framed the environmental threat as a case of 'big companies' that tried to destroy the American wilderness and rob 'the American people' of something that was constitutive of its national identity. This then paved the way for the state-controlled technocrats who established 'national parks', and seized control over rivers and pastures in the name of the common good. Hays's reinterpretation of the history of the conservation movement illuminates the often disregarded fact that technocrats subsequently used their brief to implement a comprehensive scheme of 'scientific resource management' in which wildlife and nature were largely made subordinate to their concern about achieving optimal yields, thus directly going against the original intentions of the popular movement. The word 'conservation' remained central, yet its institutional meaning changed radically. Ergo: ecological discourse is not about the environment alone. Indeed, the key question is about which social projects are furthered under the flag of environmental protection.

Thirdly, the story of the US conservation movement illustrates the complex nature of what is so often easily labelled the 'environmental movement'. Here the term 'movement' leads astray. Hays's narrative is in fact about not so much a movement as a bizarre coalition that comprised at least two rather distinct tendencies: a popular tendency that was morally motivated, and a technocratic tendency organised around a relatively confined group of experts, administrators and politicians. The important thing is that both had their own understanding of what the problem 'really' was and what sort of interventions could or should be considered as solutions. Nevertheless, together they constituted the social force behind the changes that were made. Hence, instead of speaking of a movement, we would be better to think in terms of 'coalitions'. And, as the above indicates, these coalitions are not necessarily based on shared interests, let alone shared goals, but much more on shared concepts and terms. We therefore call them 'discourse-coalitions' (see Hajer, 1995).

Fourthly, in environmental debates we can often identify implicit ideas about the appropriate role and relationship of nature, technology and society that structure implicit future scenarios. Hays sees a dialectical relation between the public outcry over the destruction of the American wilderness and the implicit critique of industrial society. Nature symbolised the unspoiled, the uncorrupted or the harmonious which was the mirror-image of the everyday reality of Chicago, Detroit or Baltimore at that time. The popular movement wanted to save nature from the effects of industrialisation but did not address the practices of industrial society head on, focusing instead on the effects on nature. In the end it thus paved the way for a programme that focused on the application of new technologies and scientific management techniques to 'conserve nature'. Here the concern about the immorality of society was matched by a renewed appeal to forms of techno-scientific management that were very similar to those industrialistic practices that had motivated the moral outcry in the first place.

[. . .]

What is Meant by Ecological Modernisation?

Ecological modernisation is a discourse that started to dominate environmental politics from about 1984 onwards.[1] Behind the text we can distinguish a complex social project. At its centre stands the politico-administrative response to the latest manifestation of the ecological dilemma. Global ecological threats such as ozone layer depletion and global warming are met by a regulatory approach that starts from the assumption that economic growth and the resolution of ecological problems can, in principle, be reconciled. In this sense, it constitutes a break with the past. In the 1970s environmental discourse comprised a wide spectrum of – often antagonistic – views. On one side there was a radical environmentalist tendency that thought that the 'ecological crisis' could be remedied only through radical social change. Its paradigmatic example was nuclear power. On the other side of the spectrum was a very pragmatic legal-administrative response. The 'Departments for the Environment', erected all over the Western world in the early 1970s, worked on the basis that pollution *as such* was not the problem; the real issue was to guarantee a certain environmental quality. Its paradigmatic example was the end-of-pipe solution. Where ecological damage was proven and shown to be socially unacceptable, 'pollution ceilings' were introduced and scrubbers and filters were installed as the appropriate solution. Moderate NGOs or liberal politicians would subsequently quarrel about the definition of the height of ceilings and whether 'enough' was being done, but they shared with the state the conviction that ecological needs set clear limits to economic growth.

Ecological modernisation stands for a political project that breaks with both tendencies. On the one hand it recognises the structural character of the environmental problematic, while on the other ecological modernisation differs essentially from a radical green perspective. Radical greens or deep ecologists will argue that the 'ecological crisis' cannot be overcome unless society breaks away from industrial modernity. They might maintain that what is needed is a new 'place-bound' society with a high degree of self-sufficiency. This stands in contrast to ecological modernisation, which starts from the conviction that the ecological crisis can be overcome by technical and procedural innovation. What is more, it makes the 'ecological deficiency' of industrial society into the driving force for a new round of industrial innovation. As before, society has to modernise itself out of the crisis. Remedying environmental damage is seen as a 'positive sum game': environmental damage is not an impediment for growth: quite the contrary, it is the new impetus for growth. In ecomodernist discourse environmental pollution is framed as a matter of inefficiency, and producing 'clean technologies' (clean cars, waste incinerators, new combustion processes) and 'environmentally sound' technical systems (traffic management, road pricing, cyclical product management, etc.), it is argued, will stimulate innovation in the methods of industrial production and distribution. In this sense ecological modernisation is orientated precisely towards those forces that Schumpeter once identified as producing the 'fundamental impulse that sets and keeps the capitalist engine in motion' (Schumpeter, 1961: 83).

The paradigmatic examples of ecological modernisation are Japan's response to its notorious air pollution problem in the 1970s, the 'pollution prevention pays' schemes introduced by the American company 3M, and the U-turn made by the German government after the discovery of acid rain or *Waldsterben* in the early 1980s. Ecological modernisation started to emerge in Western countries and international organisations around 1980. Around 1984 it was generally recognised as a promising policy alternative, and with the global endorsement of the Brundtland report *Our Common Future* and the general acceptance of Agenda 21 at the United Nations Conference of Environment and Development held at Rio de Janeiro in June 1992, this approach can now be said to be dominant in political debates on ecological affairs.

Making sense of ecological modernisation

How should we interpret ecological modernisation? Is it just rhetoric, 'greenspeak' devoid of any relationship with the 'material' reality of ongoing pollution and ecological destruction? Here we have to differentiate. The empirical evidence of the developments in environmental policymaking and product-innovation in Germany and Japan . . . shows that the least we can say is that ecological modernisation has produced a real change in *thinking* about nature and society and in the *conceptualisation* of environmental problems in the circles of government and industry. . . . One of the core ideas of ecological modernisation, 'integrating ecological concerns into the first conceptualisation of products and policies', was an abstract notion in the early 1980s but is by now a reality in many industrial practices. Especially in OECD countries, ecomodernist concepts and story lines can now be seen to act as powerful structuring principles of administration and industrial decision-making from the global down to the local levels.[2] It has produced a new ethics, since straightforward exploitation of nature (without giving thought to the ecological consequences) is, more than ever before, seen as illegitimate.

Yet one should also assess the extent to which the discourse has produced non-discursive social effects (the condition of discourse institutionalisation). Here one has to define a way to assess social change. There seems to be a consensus that in terms of classical indicators (such as energy consumption, pollution levels) one cannot come to a straightforward conclusion. There have been marked successes in some realms (say, curbing SO_2 emissions), but mostly they have been cancelled out by other developments (such as rising NO_2 levels). Likewise, where energy consumption has gone down, one may legitimately wonder whether these changes are the result of the new discourse or whether the 'achievements' should be attributed to some other processes (such as economic restructuring). Hence in terms of *ecological* indicators it is difficult to come to an assessment.

The question that we focus on here is the sort of *social* change that ecological modernisation has produced, a question that is neglected only too often in social-scientific research on environmental matters. Is ecological modernisation 'mercantilism with a green twist'? Has it led to a new form of 'state-managerialism'? Does ecological modernisation produce a break with previous discourses on technology and nature, or is it precisely the extension of the established technology-led social project? Or should the 'ecological question' be understood as the successor of the 'social question', and ecological modernisation as the new manifestation of progressive politics in the era of the 'risk society'?

My approach here is to first sketch three different interpretations of ecological modernisation. They are ideal-typical interpretations in the sense that one will not find them in real life in this pure form. Almost inevitably all three of them draw on certain social-scientific notions. In this respect it is important to see them merely as heuristic devices that should help us define the challenges that the social dominance of the discourse of ecological modernisation produces. Each ideal-type has its own structuring principles, its own historical narrative, its own definition of what the problem 'really' is and its own preferred socio-political arrangements.

Ecological Modernisation as Institutional Learning

The most widespread reading of the developments in environmental discourse interprets the course of events as a process of institutional learning and societal convergence. The structuring principle of the institutional learning interpretation of ecological modernisation is that nature is 'out of control'. The historical account is framed around the sudden recognition of nature's fragility and the subsequent quasi-religious wish to 'return' to a balanced relationship with nature. . . . Global environmental problems like global warming or the diminishing ozone layer call for decisive political interventions.

Typically, the political conflict is also seen as a learning process. 'We owe the greens something', it is argued. The dyed-in-the-wool radicals of the 1970s had a point but failed to get it through. This was partly due to the rather unqualified nature of their *Totalkritik*. The new consensus on ecological modernisation is here attributed to a process of maturation of the environmental movement: after a radical phase the issue was taken off the streets and the movement became institutionalised as so many social movements before it. With the adoption of the discourse of ecological modernisation its protagonists now speak the proper language and have been integrated in the advisory boards where they fulfil a 'tremendously important' role showing how we can design new institutional forms to come to terms with environmental problems. Likewise, the new consensus around ecological modernisation has made it possible that the arguments of individual scientists that found themselves shouting in the dark during the 1970s are now channelled into the policymaking process.

The central assumption of this paradigm is that the dominant institutions indeed *can* learn and that their learning can produce meaningful change. Following that postulate, the ecological crisis comes to be seen as a primarily *conceptual* problem. Essentially, environmental degradation is seen as an 'externality' problem, and 'integration' is the conceptual solution: as economists we have too long regarded nature as a 'sink' or as a free good; as (national) politicians we have not paid enough attention to the repercussions of collective action and have failed to devise the political arrangements that could deal with 'our' global crisis. Likewise, scientists have for too long sought to understand nature in a reductionist way; what we need now is an integrated perspective. Time and again nature was defined 'outside' society, but further degradation can be prevented if we integrate nature into our conceptual apparatus. . . .

. . . The institutional learning perspective would insist that we have to consider which alterations in scale and organisation we have to make to the existing institutional arrangements to improve 'communication' and make ecological concerns an 'integral part' of their thinking. On the one hand, that implies changes on the level of the firm and the nation state (that is, the stimulation of so-called 'autopoietic' or self-organising effects – for instance mineral or energy accounting in the firm, or the 'greening' of GNP and taxes on the national level). On the other hand, the need for integration finds its political translation in an increased demand for coordination which results in a preference for 'centralisation' of decision-making. Global ecological problems have to be brought under political jurisdiction so what we need, above all, are new forms of global management. On the local level, ecological modernisation implies that the scenarios that have been devised to further the ecologisation of society have to be protected against the – inevitable – attacks from particular interest groups. Hence the possibilities for essentially selfish NIMBY (Not In My Back Yard) protests might have to be restricted.

The sciences should in this perspective search for the conceptual apparatus that can facilitate instrumental control over nature and minimise social disturbances. They should, first and foremost, devise a language that makes ecological decision-making possible. What is required is a specific set of social, economic and scientific concepts that make environmental issues calculable and facilitate rational social choice. Hence the natural sciences are called upon to determine 'critical loads' of how much (pollution) nature can take, and should devise 'optimal exploitation rates', as well as come up with ratings of ecological value to assist drawing up of development plans. Engineering sciences are called upon to devise the technological equipment necessary to achieve the necessary ecological quality standards respecting existing social patterns. In a similar vein, the social sciences' role in solving the puzzle of ecologisation is to come up with ideas of how behavioural patterns might be changed and to help understand how 'anti-ecological' cultural patterns might be modified.

In all, in this interpretation ecological modernisation appears as a moderate social project. It assumes that the existing political institutions can internalise ecological concerns or can at least give birth to new supranational forms of management that can deal with the relevant issues. . . .

Ecological modernisation as a technocratic project

The interpretation of ecological modernisation as a technocratic project holds that the ecological crisis requires more than social learning by existing social organisations. Its structuring principle is that not nature but technology[3] is out of control. In this context it draws upon the dichotomies dominant–peripheral and material–symbolic. It holds that ecological modernisation is propelled by an elite of policymakers, experts and scientists that imposes its definition of problems and solutions on the debate. An empirical example is the UN Brundtland Report. It is a 'nice try' but, as the Rio Conference and its aftermath show so dramatically, it falters because it is only able to generate global support by going along with the main institutional interests of national and international elites as expressed by nation-states, global managerial organisations like the World Bank or the IMF, and the various industrial interests that hide behind these actors. Hence ecological modernisation is a case of 'real problems' and 'false solutions'. The material–symbolic dichotomy surfaces in the conviction that there is a deeper reality behind all the window dressing. Behind the official 'rhetoric' of ecological modernisation one can discern the silhouette of technocracy in a new disguise that stands in the way of implementing 'real solutions' for what are very 'real problems'.

Its historical narrative starts with the emergence of the 'counterculture' in the 1960s. The environmental movement is essentially seen as an offspring of that broader wave of social criticism. Environmentalism was driven by a critique of the social institutions that produced environmental degradation. Important icons are the culture of consumerism that forces people to live according to the dictum 'I shop, therefore I am', or nuclear power that would not only create a demand for more energy consumption but would also enhance the tendency towards further centralisation of power in society.

[. . .]

. . . The technocracy critique fiercely challenges the assumption that the dominant institutions can learn. How can it be that we try to resolve the ecological crisis drawing on precisely those institutional principles that brought the mess about in the first place: efficiency, technological innovation, techno-scientific management, procedural integration and coordinated management? Who believes that growth can solve the problems caused by growth? . . .

This interpretation would also point at the 'structural' aspects of the problem that are left unaddressed in the discourse of ecological modernisation. What ecological modernisation fails to address are those immanent features of capitalism that make waste, instability and insecurity inherent aspects of capitalist development. . . .

This interpretation opens the black box of society and argues that the emergence of ecological modernisation was to be seen in the context of the increasing domination of humanity by technology, where technology refers not merely to technical 'artefacts' or machines but to social techniques as well. Consequently, the *real* problem at issue is how to stop the 'growth machine'. Only then can one set about trying to remedy the very real environmental problems.

The technocracy critique argues that the sciences have in fact to a large degree been incorporated in this technocratic project. The institutional history of the discipline of systems ecology is used as a case in point (see Kwa, 1987). . . . Likewise, the consequence of the prevailing institutional framework is that engineers develop only those technologies that enhance control over nature and society rather than achieve ecological effects while making society more humane. The social sciences are similarly implicated and are called upon as 'social engineers' who only work to help achieve preconceived policy goals. Alternatively, new institutional arrangements in academia and 'science for policy' should be developed. 'Counterexperts' should be able to illuminate the 'technocratic bias' in the official scientific reports. Likewise, more attention, credit and space should be given to those engineers who have been working on 'soft energy paths' that would show the viability of decentralised alternatives. Finally, the social sciences should not work on puzzle-solving activities like changing individual consumer patterns but on the analysis of the immanent forces that keep the juggernaut running towards the apocalypse, so that it might be possible to steer it, or preferably to stop and dismantle it.

The preferred socio-political arrangements of this technocracy critique are those that can correct the prevailing bias towards hierarchisation and centralisation. Its initiatives to further a more democratic social choice centre on 'civil society' rather than on the state. Social movements and local initiatives need protection and attention. New political institutions that would facilitate this correction are the introduction of 'right-to-know' schemes (in Europe), the widespread use of referendums, and, above all, the decentralisation of decision-making and the right to self-determination. Here the differences with the institutional learning perspective come out clearly. The fight to circumvent local NIMBY protests through centralisation and 'increased procedural efficiency', indeed the mere construction of complaints as 'NIMBY' protests, are now seen as illustrations of the tendency to take away democratic rights under the veil of environmental care. Here NIMBY protests are recognised as a building stone for an anti-technocratic coalition. After all, protests that are initially motivated by self-interest often lead to a increased awareness of the ecological problematique. Hence NIMBYs may become NIABYs (Not In Anybody's Back Yard) (see Schwarz and Thompson, 1990). . . .

Ecological modernisation as cultural politics

The interpretation of ecological modernisation as cultural politics takes the contextualisation of the practices of ecological modernisation one step further. Here one is reminded of Mary Douglas's classic definition of pollution as 'matter out of place'. Her point was that debates on pollution are essentially to be understood as debates on the preferred social order. In the definition of certain aspects of reality as pollution, in defining 'nature', or in defining certain installations as solutions, one seeks to either maintain or change the social order. So the cultural politics perspective asks why certain aspects of reality are now singled out as 'our common problems' and wonders what sort of society is being created in the name of protecting 'nature'.

Ecological modernisation here appears as a set of claims about what the problem 'really' is. The cultural politics approach argues that some of the main political issues are hidden in these discursive constructs and it seeks to illuminate the feeble basis on which the choice for one particular scenario of development is presently made. The structuring principle in this third interpretation is that there is no coherent ecological crisis, but only story lines problematising various aspects of a changing physical and social reality. Ecological modernisation is understood as the routinisation of a new set of story lines (images, causal understandings, priorities, etc.) that provides the cognitive maps and incentives for social action. In so doing ecological modernisation 'freezes' or excludes some aspects of reality while manipulating others. Of course, reductions are inevitable for any effort to create meaningful political action in a complex society. The point is that one should be aware that this coherence is necessarily an artificial one, and that the creation of discursive realities are in fact moments at which cultural politics is being made. Whether or not the actors *themselves* are aware of this is not the point. Implicitly, metaphors, categorisations, or definition of solutions always structure reality, making certain framings of reality seem plausible and closing off certain possible future scenarios while making other scenarios 'thinkable'.

To be sure, in this third interpretation there is no implicit assumption of a grand cultural design. Quite the contrary, environmental discourse is made up of 'historically constituted sets of claims' (John Forester) uttered by a variety of actors. Yet in interaction these claims 'somehow' produce new social orders. Foucault speaks in this respect of the 'polymorphous interweaving of correlations'. The analytical aim of this approach is, firstly, to reconstruct the social construction of the reductions, exclusions and choices. Secondly, it tries to come to a historical and cultural understanding of these dispositions. . . .

The historical narrative of this third perspective takes up the themes touched upon at the beginning of this chapter. It emphasises that the ecological problem is not new. It observes that the ecological dilemma of industrial society is almost constantly under discussion, be it through different emblems. What these discussions are about, it argues, are in fact the social relationships between nature, society and technology. For that reason this perspective calls attention to the 'secondary discursive reality' of environmental politics: there is a layer of mediating principles that determines our understanding of ecological problems and implicitly directs our discussion on social change. Hence it would investigate what image of nature, technology and society can be recognised in the 'story lines' that dominated environmental discourse . . . during the confrontation between the state and radical social movements in the 1970s, or in the consensual story lines that dominate ecological modernisation in the 1990s. What is the cultural meaning of the biospheric orientation that is central to present-day environmental discourse? In this respect it argues that ecological modernisation is based on objectivist, physicalist and realist assumptions, all of which are highly arbitrary. Story lines of global warming, biodiversity or the ozone layer suggest the presence of the threat of biological extinction and assert that these problems should be taken as the absolute basis for an ecological modernisation of society. But do these story lines really have the same meaning and implications for all regions? Are they as relevant for the farmers of the Himalaya as for the sunbathers of the coasts of Australia? Should we not understand the global environmental story lines as the product of 'globalised local definitions', as intellectuals from the South have suggested, since the problems have mainly been caused by the North while the solutions apparently have to come mainly from the South (Shiva, 1993)?

Rather than suggesting that there is an unequivocal (set of) ecological problem(s), the third interpretation would argue that there are only implicit future scenarios. . . .

Whereas the previous two interpretations in fact shared a clear idea of the ecological problem, and both had their own idea about a possible remedial strategy (respectively conceptual or institutional change, and more coordination or more decentralisation), the third interpretation holds that there can be no recourse to an 'objective' truth. It suggests that the ecological crisis is first and foremost a discursive reality which is the outcome of intricate social processes. It is aware of the ambivalences of environmental discourse and would, in the first instance, not try to get 'behind' the metaphors of ecological discourse. It would try to encircle them to be able to challenge them scientifically, and to enhance consciousness of the contingency of knowledge about ecological matters. What is more, it would investigate the cultural consequences of prevailing story lines and would seek to find out which social forces propel this ecomodernist discourse-coalition. Once the implicit future scenarios have been exposed, they might lead to a more reflective attitude towards certain environmental constructs and perhaps even to the formulation of alternative scenarios, the socio-political consequences of which would present a more attractive, more fair, or more responsible package. . . .

The role of academia follows from this commitment to choice and open debate. They have to help to open the black boxes of society, technology and nature. The cultural politics perspective would resist the suggestion that nature can be understood and managed by framing it in a new 'ecological' language, as for instance by giving priority to economics and systems ecology, on the basis that a pure language does not exist. Its aim would rather be to pit different languages and knowledges (for example expert knowledge versus lay knowledge) against one another to get to a higher understanding of what ecological problems could be about. Here it would assume that this interplay would lead to the recognition of the wide diversity of perspectives.

Notes

1 For a more elaborate analysis of ecological modernisation, see von Prittwitz, 1993; Hajer, 1995.
2 This interpretation is now more widely supported. See for example Weale, 1992; Spaargaren and Mol, 1992; von Prittwitz, 1993; Harvey, 1993; Liefferink *et al.*, 1993; Healey and Shaw, 1993; Teubner, 1994.
3 Whereby technology is conceptualised in the Schelskyan sense of the term – that is, including both technology as artefacts and 'social technologies'. See Schelsky, 1965.

References

Hajer, Maarten A. (1995) *The Politics of Environmental Discourse: Ecological Modernisation and the Policy Process*. Oxford: Clarendon Press.
Harvey, David (1993) 'The nature of environment: Dialectics of social and environmental change', in Ralph Miliband and Leo Panitch (eds) *Socialist Register 1993*. London: Merlin Press, pp. 1–51.
Hays, Samuel P. (1979 (1959)) *Conservation and the Gospel of Efficiency: The Progressive Conservation Movement, 1890–1920*. New York: Atheneum.
Healey, P. and Shaw, T. (1993) *The Treatment of 'Environment' by Planners: Evolving Concepts and Policies in Development Plans*. Working Paper no. 31, Department of Town and Country Planning. University of Newcastle upon Tyne.
Kwa, C. (1987) 'Representations of nature mediating between ecology and science policy: The case of the international biological programme', *Social Studies of Science*, 17: 413–42.
Liefferink, J.D., Lowe, P. and Mol, A.P.J. (eds) (1993) *European Integration and Environmental Policy*. London: Belhaven.
Schelsky, H. (1965) 'Der Mensch in der wissenschaftliche Zivilisation', in *Auf der Suche nach Wirklichkeit*. Dusseldor/Koln. pp. 439–80.
Schumpeter, J.A. (1961 (1943)) *Capitalism, Socialism and Democracy*, London: George Allen and Unwin.
Schwarz, M. and Thompson, M. (1990) *Divided We Stand: Redefining Politics, Technology and Social Choice*. London: Harvester Wheatsheaf.
Shiva, Vandana (1993) 'The greening of the global reach', in W. Sachs (ed.), *Global Ecology*, pp. 149–56.
Spaargaren, G. and Mol, A.P.J. (1992) 'Sociology, environment, and modernity: ecological modernisation as a theory of social change', in *Society and Natural Resources*, 5: 323–44.
Teubner, Gunther (ed.) (1994) *Ecological Responsibility*, London: Wiley & Sons.
von Prittwitz, V. (ed.) (1993) *Umweltpolitik als Modernisierungsprozess*, Opladen: Leske & Budrich.
Weale, Albert (1992) *The New Politics of Pollution*. Manchester: Manchester University Press.

Source: Maarten A. Hajer (1996) 'Ecological modernisation as cultural politics', in S. Lash, B. Szerszynynski, and B. Wynne (eds) *Risk, Environment and Modernity. Towards a New Ecology*, London: Sage, pp. 246–59.

7.3

David Goldblatt

Risk society and the environment

Beck's work[1] has a particular importance for anyone concerned with the response of social theory to environmental degradation and environmental politics. The distinguishing feature of his work is to place the origins and consequences of environmental degradation right at the heart of a theory of modern society, rather than seeing it as a peripheral element or theoretical afterthought. Beck's sociology and the societies it describes are dominated by the existence of environmental threats and the ways we understand and respond to them. Indeed, one could go so far as to argue that the *risk society* is predicated on and defined by the emergence of these distinctively new and distinctively problematic environmental hazards.

[. . .]

Risk society: the modernization of modernity

In *Risk Society* . . . Beck outlines the characteristics and consequences of the threats and dangers generated by the processes of modernization and industrialization, focusing on the ways in which they alter the dynamic and constitution of the classical industrial society that has generated them. In short, the process of *reflexive modernization* – exemplified by the emergence and interpretation of new risks and hazards – is ushering in a *risk society* from the corpse of a decaying industrial society. . . . Beck connects this widening penumbra of risk and insecurity with complementary processes of reflexive modernization, detraditionalization and individualization in the spheres of work, family life and self-identity. . . .

The characteristics of modern risks

While Beck uses the idea of risks and hazards to refer to many areas of social life, it is in the equation of risks and hazards with environmental degradation that these ideas are most closely examined. Of course, dangers and hazards have always threatened human societies. What makes them risks, as Giddens has argued, is that they are dangers and hazards which are known, whose occurrence can be predicted and whose likelihood can be calculated.[2] To be in danger is one thing. To know that one is in danger is quite another. To know that one is in danger and to feel essentially powerless to alter the course of events which generate that danger is another again. Alongside the shift from danger to risk, contemporary environmental problems possess further distinctive characteristics which elicit and demand very particular patterns of political and psychological response. These demands and responses are of such magnitude that their emergence can be said to herald the emergence of a distinctive form of modernity. There are at least three sets of arguments in Beck's work to suggest that the production and implications of contemporary risks differ – qualitatively and quantitatively – from earlier forms of risk and hazard.

First, while the risks that threatened industrial societies were locally significant and often personally devastating, they were ultimately spatially limited in their effects. They did not threaten entire societies. Neither spatial nor social limitations apply to contemporary hazards. Consider, for example, the pollution produced by a nineteenth- or mid-twentieth-century steel mill or foundry. The emissions and wastes produced would have had significant consequences for people who worked within the mill, for the local community who breathed waste from its stacks, and for those who drank from the local water supplies it polluted with scrap and solvents. However, those threats, indeed the sum of the threats generated by all the steel industries of the industrialized world, did not threaten entire populations, or in fact the planet as a whole. The environmental problems of the steel industry have not disappeared, although in the West some of its older manifestations have been mitigated. They have, however, been eclipsed by new types of environmental degradation from new industrial processes. The contemporary forms of environmental degradation that Beck focuses on are not spatially limited in the range of their impact or socially confined to particular communities. They are potentially global in their reach. The reasons for this are complex and Beck is less than specific about them, though he does suggest a number of features of contemporary risk that account for these qualitative changes. First, the toxicity of contemporary forms of environmental degradation is quantitatively greater than that of industrial forms of degradation. Second, the impact of those toxins on human bodies and the wider ecosystem is irreversible and cumulative in its effect. Thus the consequences of modern risks outlast their generators. They are risks that accumulate in intensity and complexity across the generations. In addition to transcending the spatial and social limits of purely industrial risk, they also exceed their temporal boundaries.

Alongside these threats of increasing toxicity, Beck effectively denotes a second set of environmental risks which, without hyperbole, can genuinely be described as catastrophic in their potential. The environmental dangers posed by large-scale nuclear accidents, large-scale chemical release and by the alteration and manipulation of the genetic make-up of the planet's flora and fauna actually pose the possibility of self-annihilation. The character of these potential apocalypses is not spelt out in any great length. It is almost as if Beck assumes that we accept and agree with his estimations of the dangers we face – a somewhat surprising elision given Beck's acknowledgement of the relative and contested character of risk perception and definition.[3]

A third distinctive set of features of modern environmental risks is, according to Beck, that their point of impact is not obviously tied to their point of origin and that their transmission and movements are often invisible and untrackable to everyday perception. This social invisibility means that, unlike many other political issues, environmental risks must first clearly be brought to consciousness, and only then can it be said that they constitute an actual threat, and that involves a process of scientific argument and cultural contestation. Thus the politics of risk is intrinsically a politics of knowledge, expertise and counter-expertise.

The transition from industrial society to risk society

In both *Risk Society* and *Ecological Politics* Beck uses this model of risk and hazard to underpin his threefold model of social development, running from preindustrial society to industrial society to risk society. In preindustrial societies, risk takes the form of natural hazards – earthquakes, drought, etc. These are not contingent on decisions made by individuals, thus they cannot be considered voluntary or intentionally created and they are, therefore, effectively unavoidable. The spatial and social range of hazards can be both highly localized and quite extensive, the Black Death, for example, which affected the trajectory of an entire civilization. Preindustrial societies are openly insecure. In cultural terms, the origins of risks are invariably assigned to external, supernatural forces and the help of those same forces must be sought in mitigating or avoiding the worst effects of hazards and contingencies.

With the emergence of classical industrial societies, the origins, consequences and characteristics of risk change. The way in which risks are socially understood and responded to also changes. Risks and accidents become clearly contingent on the actions of both individuals and wider social forces, be they dangers at work from machinery and poisons, or the dangers of unemployment and penury induced by the uncertain dynamics of the business cycle and structural economic change. Given that risks are no longer solely attributable to external agency or individual fecklessness, industrial societies develop institutions and rules for coping with and mitigating the impact of localized risks and dangers. Beck goes so far as to argue that the welfare state can be seen as a collective and institutionalized response to the nature of industrialized risks, based on principles of rule-governed attribution of fault and blame, legally implemented compensation, actuarial insurance principles and collectively shared responsibility. In industrial society, blame or culpability for threats can be confidently asserted and their statistical likelihood calculated. This makes possible the construction of reliable actuarial schedules of the volume and impact of risks against which financially solvent institutions of collective burden-sharing can be maintained. The classic example of this would be the creation of compensation and insurance schemes for accident, injury and unemployment at work.

However, under the impact of modern risks and hazards, these modes of determining and perceiving risk, attributing causality and allocating compensation have irreversibly broken down. In so doing, they have also thrown the functioning and legitimacy of modern bureaucracies, states, economies, and science into question. Risks that were calculable under industrial society become incalculable and unpredictable in the risk society. Compared to the possibility of adjudging blame and cause in classical modernity, risk society possesses no such certainties or guarantees.

[. . .]

Ecological politics in an age of risk: organized irresponsibility and the social explosiveness of hazard

While *Risk Society* cast its sociological net very widely, the substantive focus of its successor, *Ecological Politics in an Age of Risk*, is much narrower. In this book Beck retained most of the conceptual apparatus of *Risk Society*, but chose to focus on the most systemic and encompassing of risks and hazards that modern societies have generated: environmental degradation. . . . At the heart of the book are two interconnected ideas; organized irresponsibility and the social explosiveness of hazard. Beck uses that former term to describe the ways in which the political and legal systems of risk societies, intentionally and unintentionally, render the social origins and consequences of large-scale environmental hazards invisible. While the risks and dangers of industrial societies could be adequately captured with the models of social causation and risk available to them, this is no longer possible in risk societies. The industrial fatalism that Beck excoriates describes the dominant response of Western publics to the paradoxical situation: that we must live with the obvious threats of uncontrolled industrial development, but are unable to account for their existence or accurately determine the culpability of either individuals or organizations for those threats. However, such is the magnitude of the risks we face and so lamentable are the ways in which we have politically and institutionally attempted to cope with them, that the thin veneer of tranquillity and normality is constantly broken by the harsh overwheening reality of unavoidable dangers and threat. It is this that Beck describes as the *social explosiveness of hazard*. The first casualties of this are conventional, industrial notions of risk perception, safety insurance, and compensation, and with them the legitimacy and utility of the state institutions that have notionally secured, or at the very least convincingly promised, safety and security – what Beck often refers to as the *safety state*. Beck then goes on to examine this social explosiveness in a number of main areas, including the impact of risk on the bureaucracy of the safety state, and the impact on conventional, industrial lines of political conflict and political organization. I examine these in turn below.

Organized irresponsibility and the relations of definition

[. . .]

Organized irresponsibility denotes a concatenation of cultural and institutional mechanisms by which political and economic elites effectively mask the origins and consequences of the catastrophic risks and dangers of late industrialization. In doing so they limit, deflect and control the protests that these risks engender. 'Thus what is at issue is an elaborate labyrinth designed according to principles, not of non-liability or irresponsibility, but of simultaneous liability and unaccountability: more precisely, liability as unaccountability, or organized irresponsibility.'[4] To put it another way, risk societies are plagued by the paradox of more and more environmental degradation, perceived and possible, and a greater weight of environmental law and regulation. Yet simultaneously, no one individual or institution seems to be held specifically accountable for anything. How can this be? The key to explaining this state of affairs is the mismatch that exists in the risk society between the character of hazards and dangers produced by late industrialism and the prevalent *relations of definition* which date in their construction and content from an earlier and qualitatively different epoch. The relations of definition, as I described them above, are the rules, institutions and capacities that structure the identification and assessment of environmental problems and risks; they are the legal, epistemological and cultural matrix in which environmental politics is conducted.

The relations that Beck focuses on are fourfold.

1. Who is to determine the harmfulness of products or the danger or risks? Is the responsibility on those who generate those risks, those benefiting from them, those affected or potentially affected by them or public agencies?
2. To whom does that proof have to be submitted? Who, after all, are the proper arbiters of risk assessment and who should have to defend or interrogate those claims?
3. What is to count as sufficient proof? In a world in which we necessarily deal with contested knowledge, and probabilities and possibilities, what are the grounds if any for accepting or rejecting different claims about risks and hazards?
4. If there are dangers and damages, who is to decide on compensation for the afflicted and appropriate forms of future control and regulation? If there is to be compensation, who is to be compensated, and by how much relative to others affected?

In relation to each of these questions, risk societies are currently trapped in a vocabulary that befits the risks and hazards interrogated by the relations of definition of industrial societies. They are singularly inappropriate for modern catastrophes. Consequently, we face the paradox that at the very time when threats and hazards seem to become more dangerous and more obvious, they simultaneously slip through the net of proofs, attributions and compensation that the legal and political systems attempt to capture them with.

The responsibility to demonstrate proof, Beck argues, currently lies overwhelmingly with the afflicted rather than potential polluters. The legacy of industrial society's faith in progress is that the legal system assumes that industrial production will be benign unless demonstrated otherwise. Given that companies are the only actors likely to have any sense of the likely risk implications of any given process or product in development, no one else is likely to consider the environmental implications before pollution has begun, and attempts to demonstrate harm will occur only after people have been exposed to the danger. Thus the prevailing relations of definition are weighted towards the polluter. First, because the legal system demands proofs of *post hoc* toxicity rather than *pre hoc* non-toxicity or safety. Second, because those who must prove toxicity are inevitably less endowed with the detailed skills and information necessary to make a convincing case.

The arbitration of such disputes invariably lies with the legal system and courts and judges. However, Beck does not argue that the social origins or political inclinations of the legal system's senior figures can explain the consistent capacity of polluters to be acquitted or treated lightly. Nor does he claim that the quantity of laws and the weight of regulatory agencies are too small. Indeed they have radically increased as levels of pollution and degradation have increased. Rather, it is the qualitative difference between the nineteenth-century relations of definition and twentieth-century risks and hazards that accounts for the problem. Most legal cases attempt to determine whether a single substance is responsible for a particular set of pathological effects. However, we live in world in which there is an enormous number of potentially damaging pollutants, and many of them must have complex interactions with and consequences for human bodies and the wider ecosystem. Further, most cases attempt to determine whether a particular source of a given pollutant is responsible for a particular form of degradation. However, pollutants almost invariably come from many sources. Thus a collective danger from collective origins must be legally pursued through the fiction of the individual legal person or corporation.

The social explosiveness of hazard: risk and the safety state

Through the Byzantine labyrinth of organized irresponsibility, protest still breaks out and the claims and legitimacy of the prevailing relations of definition come into question. It is this that Beck describes as the *social explosiveness of hazard*. In short, he sees the existence of large-scale ecological disasters, of which Chernobyl is his favourite example, as the central casual factor in delegitimizing and destabilizing state institutions with responsibilities for pollution control, in particular, and public safety in general. This occurs because the pledges and commitments of modern states to the security of the population are, despite frenzied attempts to patch up the old models of industrial safety and security, overwhelmed by the enormity of contemporary risks. Their models of risk perception and risk insurance are swept away. Governments and bureaucracies, of course, have well-worn routines of denial. Data can be hidden, denied and distorted. Counter-arguments and counter-experts can be mobilized in the face of ecological hazards. Maximum permissible pollution levels can be raised and transformed to accommodate new waves of unexpected pollution. Human error rather than systemic risk can be cast as the villain of the piece. However, states are, according to Beck, fighting a losing battle, for they offer nineteenth-century pledges of security to a world that is unequivocally engaged with risks and hazards of a qualitatively different order.

[. . .]

What hazards repeatedly demonstrate is that the cultural and institutional legacies of classical industrial society are inadequate to the problems that now face them. We have already seen how the models of causality and attributability of industrial societies have been undermined and replaced by the Kafkaesque construct of organized irresponsibility. Alongside this, industrial society's modes of risk perception and assessment are rendered irrelevant, its systems of risk control and mitigation appear powerless, its models of insurance and compensation are inadequate. The certainties of scientific models are shaken by the occurrence of the worst-case scenario. Simultaneously this demonstrates that societies are locked into a very high-consequence gamble rather than a position of security. Models of insurance and compensation under the industrial safety state were tied to an actuarial and calculable basis. The statistical likelihood of accidents and dangers could be predicted and a schedule of payments from individuals into a collective fund could be calculated. The accumulation of payments could also be predicted in advance and relatively clear rules established as to who would pay, who would be eligible for compensation, and at what rate. All these calculations are rendered meaningless by large-scale hazards. The premise of nuclear power's legitimacy was that it would not go wrong, and that therefore the whole schedule of compensation and insurance was not necessary. However, as Beck argues, Chernobyl rendered the worst-case scenario possible, even probable. The absence of insurance and compensation became more problematic. However, even if there was an admission of danger on the part of the safety state, how could a meaningful insurance scheme for the possibility of very large-scale ecological disruption be calculated, and who would pay for it? 'The shining achievement of nuclear power plants is not only to have made redundant the principle of assurance in the economic sense, but also in its medical, psychological, cultural, and religious senses. Residual risk society is a society without assurance, whose insurance cover paradoxically diminishes in proportion to the scale of the hazard.'[5]

[. . .]

Learning to live with risk

In *Risk Society* Beck sketches three broad scenarios for the future direction of politics: back to industrial society; the democratization of technological development; the emergence of differential politics. In a rather crude manner they all have their counterparts in contemporary German politics. The first roughly equates with the world-view of the Christian Democratic right in Germany. The second option, which seeks to address the problem of risk from within the institutional framework of industrial society, corresponds to mainstream social democratic thinking. The third option, which calls for new political institutions as well as new rules of political decision-making about risk, corresponds to the politics of the greens, and the broader environmental movement in German civil society.

The return to industrial society captures the basic political orientation of orthodox politics to the risk society. It remains tied to notions of progress and benign technological change. It believes that the risks we face can still be captured by nineteenth-century, scientific models of hazard assessment and industrial notions of hazard and safety. Simultaneously, the disintegrating institutions of industrial society – nuclear families, stable labour markets, segregated gender roles – can be shored up and buttressed against the waves of reflexive modernization sweeping the West. However, if Beck is right, then this attempt to apply nineteenth-century ideas to the late twentieth century is doomed to failure, for risks and dangers will constantly spill out of the control of bureaucracies, safety pledges will be compromised, and the malign face of technology will inevitably be displayed. Similarly, the personal lives of individuals will constantly contradict claims that traditional social roles and institutions retain their efficacy and legitimacy. The consequent irrelevance and delegitimization of orthodox politics could then open up calls for strong leadership and authoritarian political change.[6]

Beck calls the second response to the runaway train of risk society the democratization of techno-economic development. In the same sense that science brought its inherent scepticism to reflexively bear on its epistemological claims, modernity can bring its commitment to self-determination to bear on the consequences of socioeconomic progress and technological development. The institutional model for doing this is varied but rests on bringing the decision-making process in scientific debate and technological research under public scrutiny and parliamentary checks. This would be accompanied by an *ecological variant of the welfare state*. Beck is, once again, rather imprecise as to what this grand term means, but one can assume, I think, that he is describing a project in which the resources of the state and state agencies take responsibility for curtailing environmental degradation, and pick up the pieces – environmentally and socially – once degradation has occurred.

However, both elements of this strategy apply the arguments and deploy the institutional solutions of nineteenth- and early twentieth-century industrial society to late twentieth-century risk societies. In so doing they are inadequate and inappropriate responses – though ones with some mitigating effect – as Beck's third variant of subpolitical activity seeks to demonstrate. Moreover, they bring their own peculiar side-effects and pathologies. They entail an excessive centralization of power in welfare states and parliaments, when the needs of a risk society require the greatest devolution of power and resources to the sites of risk creation and exposure to threats. They entail an excessive bureaucratization. This locates further powers and decision-making capacities in institutions which have hitherto consistently aided and abetted the creation of risks, as well as supervising subsequent public detoxification. Finally, this project does not tackle the displacement of power into the hands of scientific expertise and institutions where actual decisions about safety and permissibility are made. This can only be tackled by a *differential politics*.

The idea of a differential politics turns on Beck's . . . distinction between the conventional politics of parliamentary states and the subpolitics of civil society and democratized para-state institutions.[7] The three key conditions of this politics of control are a strong and independent legal system, a free and critical mass media and what Beck calls the widespread opportunity for self-criticism. 'That means that the preserving, settling, discursive functions of politics – which quietly are already dominant, but remain overshadowed by fictitious power constructions – could become the core of its tasks.'[8] This is a differential politics in that it sees a role for the application of political power in different forms on different institutions. It recognizes a continuing role for state and para-state institutions in democratized forms, and their application to the recognition and mitigation of risk, but it sees them as only a component, and a subsidiary component at that, of a broader programme of political change and democratization. This involves opening up the private sphere of economic decision-making to political debate and control, not merely by the state but by the institutions of civil society.

In *Ecological Politics* Beck returns to the ways in which we might respond to the politics of hazard and how we might learn to live in the risk society. The choices have narrowed to two: *authoritarian technocracy* and *ecological democracy*. Beck express the path to authoritarian technocracy:

> Performing on the high-wire act near catastrophes of unimaginable extent, risk societies tend to undergo basic political mood swings, phases of forced normality alternate with dazed states of emergency. Calls for the 'strong hand' [of the state], that is supposed to avert ecological catastrophe, tie in with calls for the 'strong hand' that is supposed to counteract the collapse of state power and rationality.[9]

This line of argument clearly needs expanding on. Beck, it would seem, is arguing that the nature of contemporary threats and hazards leads to occasional crises of insecurity and threat, of which Chernobyl is perhaps the paradigmatic example. These situations appear to demand an intensification of state activity and an expansion of state power to ensure security and guard against threats. . . .

Ecological democracy is, by contrast, the utopia of a responsible modernity. Beck is much more specific about the preconditions of such an arrangement. He can be because the idea of an ecological democracy rests on overturning the relations of definition that Beck has outlined and explored in *Ecological Politics*. First, it envisions a society in which the consequences of technological development and economic change are debated before key decisions are taken. Second, the burden of proof regarding future risks, hazards and current environmental degradation would lie with the perpetrators rather than the injured party: from the polluter *pays* principle to the polluter *proves* principle. Third, there must be established a new body of standards of proof, correctness, truth and agreement in science and law. Beck suggests some lines of attack on the current complacency of the safety state and the system of organized irresponsibility in which the politics of risk is currently mired. He argues for strategies that pursue the *denormalization of acceptance*, in which not only are the limits of probabilistic safety claims and science's own understanding of its limitations ruthlessly probed and turned against them, but in which the very notion of safety is superseded by a baseline commitment to the preservation of life.

[. . .]

Notes

1 Beck, *Risikogesellschaft: Auf dem Weg in eine andere Moderne* (1986); *Risk Society: Towards a New Modernity* (1991), henceforth *RS*, *Ecological Politics in an Age of Risk* (1995), henceforth *EPAR*; Beck, Giddens and Lash, *Reflexive Modernisation: Politics, Tradition and Aesthetics in the Modern Social Order* (1994); Beck and Beck-Gernsheim, *The Normal Chaos of Love* (1995).
2 Extensive use of Beck's work has been made by Giddens, see esp. *The Consequences of Modernity and Beyond Left and Right*. See also Lash and Urry, *Economies of Signs and Space*.
3 See the discussions in *RS*, pp. 27–8, 31–2.
4 *EPAR*, p. 61.
5 *EPAR*, p. 85.
6 *RS*, pp. 223–35.
7 Beck pursues a similar set of arguments under the heading of subpolitics, see Beck, Giddens and Lash, *Reflexive Modernisation*, pp. 13–23.
8 *RS*, p. 235.
9 *EPAR*, p. 168.

References

Beck, U., *Risk Society: Towards a New Modernity*, trans. M. Ritter, Sage, London, 1991; originally *Risikogesellschaft: Auf dem Weg in eine andere Moderne*, Suhrkamp, Frankfurt, 1986.
—— *Ecological Politics in an Age of Risk*, trans. Amos Oz, Polity, Cambridge, 1995.
Beck, U. and Beck-Gernsheim, E. *The Normal Chaos of Love*, trans. Mark Ritter and Jane Wiebel, Polity, Cambridge, 1995.
Beck, U., Giddens, A. and Lash, S., *Reflexive Modernisation: Politics, Tradition and Aesthetics in the Modern Social Order*, Polity, Cambridge, 1994.
Ewald, F., *L'Etat de Providence*, Grasset, Paris, 1986.
Giddens, A. *The Consequences of Modernity*, Polity, Cambridge, 1990.
—— *Beyond Left and Right*, Polity, Cambridge, 1994.
Lash, S. and Urry, J., *Economies of Signs and Space*, Sage, London, 1993.

Source: David Goldblatt (1996) *Social Theory and the Environment*, Cambridge: Polity Press, pp. 154–60, 165–73.

7.4

Andrew Blowers

The way forward: ecological modernisation or risk society?

Opposing perspectives

The phenomenon of environmental change has begun to occupy a central role in social science discourse and analysis at the level of both theory and policy. In charting the possible way forward for society in its relationship with the environment, alternative theoretical perspectives offer different and, at first encounter, apparently irreconcilable insights. On the one hand, exponents of ecological modernisation offer what appears to be a celebration of modernity and its ability to adjust to the ecological problems created by industrial society. The environmental problem provides a challenge for further innovation and the transition to an industrial society in which ecological protection and economic development are complementary rather than in conflict (Mol, 1996). On the other hand, critics of the whole project of modernisation hold that the extension and continuation of industrial progress inevitably spell disaster for the environment and, with it, for humanity itself. The hope here lies in the process of 'reflexive modernisation' whereby society confronts the crisis it has created and a transformation ensues at the levels of production, institutions and, consequently, in values and life-styles.

These are idealised models, combining various strains of thought and posing a dichotomy which offers analytical clarity at the expense of empirical validity. They are partial models. Ecological modernisation focuses largely on the technical and economic aspects of change and, even here, ignores the implications of a continuation of high-consequence risky technologies. The alternative approach, exemplified by its most developed ideal type, risk society, exaggerates the importance of such technologies whilst underestimating the significant changes in the direction of environmental protection already occurring and potentially available. Each perspective projects an inevitability in the trends it identifies, although the directions of change they foresee are entirely divergent. Both theories are conceived in a Western context but their focus on the aggregate society and economy ignores the fundamental issue of social inequality which is likely to act as a barrier to progress in either direction. Furthermore, in a sense they are devoid of politics. Whereas ecological modernisation presumes a continuation of the politics of liberal democratic societies, the risk society regards such politics as irrelevant to deal with the coming crisis. Thus, neither theory has very much to say about the nature of the political institutions necessary to deliver a society that is able to develop sustainably. The theoretical debate is clearly at an early stage and further elaboration of the links between the economy, society and the environment and a detailed analysis of social institutional change may be expected as the debate proceeds.

Meanwhile, other social scientists, political scientists in particular, have detected two broad trends that potentially provide the political context for responding to environmental change. The first of these is at the level of the state, where the transboundary nature of environmental processes has been paralleled by the creation of transboundary political institutions. At the same time, the localised nature of environmental issues requires a response at the sub-state level. Quite aside from the environmental pressures, the state is undergoing change as an integral part of the globalisation of economic forces and the dominance of the market economy. All this, according to some scholars, has led to some surrender of sovereignty upwards and the withdrawal of the state's boundaries internally opening up political space which is occupied by various interests and actors.

Environmental movements are one set of interests in civil society which have begun to thrive in this emerging space. The rise of such movements, their influence on policy and their impact at the level of consciousness-raising and value-shift has been a feature of political change. In particular their contribution (with other actors) to a process of social and institutional learning (List, 1996; Breyman, 1993; Hajer, 1996) is arguably a major factor in the incorporation of the environmental dimension into all levels of policy-making.

The urban dimension

The theoretical debates about global environmental change and its political and social consequences have implications at the urban level. There are, perhaps, four areas where the urban dimension provides illustration for the debates and reveals the limitations of the analysis.

1 *The local/global* Neither ecological modernisation nor risk society theories address the urban dimension and it may be thought that political concern with global environmental issues diverts attention away from the immediate problems of development, deprivation, environmental degradation, transport and energy which are currently faced by cities worldwide (Christoff, 1996). On the other hand, there is ample evidence that concern for the urban dimension remains vigorous at the policy level. The post-Rio process has placed emphasis

on the development of local Agenda 21 and has stimulated widespread participation in developing ideas and plans at the urban level. Sustainable development has been at the heart of several significant national policy statements on transport and retail development. At a more local level, structure and local plans have taken sustainability as their theme and have included detailed policies for transport (reducing car travel and emphasising public transport, cycling and walking), urban development (concentrating development in urban areas or major transport corridors), energy (conservation, use of renewable energy and combined heat and power and reduction of CO_2 levels), resource depletion (recycling and re-use and conservation of land and heritage) and waste management (minimisation and reduction). Although the focus of attention for all these policies is local, the connections are readily made to the global implications of action taken (or not taken) at the local level.

2 *The social question* Ecological modernisation focuses entirely on the economic sphere and, while Beck draws attention to the problem of insecurity and social fragmentation engendered by modernised societies, he concentrates on the psychological rather than the political consequences of the phenomenon. Yet, the need to address issues of social inequality (and especially the problems of the South), in order to encourage social cohesion becomes especially evident when viewed in the urban context. Problems of poverty, alienation and environmental deprivation which are especially evident among the underclass of major cities (and especially rife in the burgeoning cities in the developing world) constitute a significant social and political issue and threat, if only because of their sheer spatial concentration. In terms of sustainability, it must be recognised that the disadvantaged are hardly likely to co-operate in programmes for environmental protection so long as they see the rich and powerful enjoying and consuming a disproportionate share of resources and producing a burden of pollution which degrades already-impoverished environments.

3 *Institutional implications* Neither ecological modernisation nor the risk society theory have much to say about the form of institutional adaptation or change required to inaugurate sustainable development. Ecological modernisation sublimates the 'enabling state' as the institutional response that will secure the efficient functioning of the market economy within a framework of state regulation. However, the role of the state as an enabling institution for individuals through the provision of welfare, housing, social services, good education and an accessible health service tends to be diminished in this model. By contrast, Beck turns his attention to the demise of integrating social institutions which the process of modernisation has undermined or swept away. At the urban level, it is the very decline of support structures, the absence of opportunity and the weakening of integrating institutions that contributes, among other impacts, to the deteriorating environmental conditions [Beck, 1992, 1995, 1996].

4. *Political participation and governance* Ecological modernisation is based on the presumption of corporatist relationships primarily between government and industry but also incorporating environmental movements where mutually acceptable. Questions of political participation are absent from the analysis. Beck's observations on questions of political participation go little beyond identifying the 'zone of sub-politics' as a fertile arena for political participation. Indeed, both theories, as was observed earlier, are, in a sense,

apolitical. Looked at in terms of urban governance, the idea that environmental or urban social movements are given space to operate in a vigorous civil society does not adequately deal with the problem of legitimating and implementing decisions. Environmental movements are not representative, nor are they accountable and, consequently, their influence must be secured ultimately through the formal political process.

One of the worrying features of modern political institutions is the tendency to depoliticise large areas, including the environment. In the UK, recent reforms have largely completed a process whereby environmental regulation and control is in the hands of regulatory bodies appointed by central government. Local government reform has had a mixed impact. While the creation of unitary authorities can be seen as a positive move towards providing greater integration and authority at the urban level, elsewhere many county councils have been severely weakened through the loss of financial resources and the surrender of their function in co-ordinating urban and rural development.

The role of planning

Appropriate institutional frameworks and representative political structures are both necessary but not sufficient conditions for securing sustainable development. The long timespans and detailed coordination across policy areas that are necessary for environmental policies to be made effective require a form of decision-making that transcends the prevailing short-term, sectoral structures. Under ecological modernisation, the state engages in 'partnership' with business, the voluntary sector and other interests while retaining a residual coordinating and regulatory role. The notion of risk society certainly implies the need for more imperative coordination, but does not spell it out. In a word, the missing dimension is the need for planning.

The case for planning goes well beyond the current limited role of urban and land-use planning. First, there is the need for a holistic approach which combines both land-use and environmental aspects. This necessary integration is emphasised in many planning statements but, at the institutional level, powers remain diffused and fragmented. Secondly, the local/global relationship between environmental problems must be recognised in the articulation of integrated plans at different levels (local, regional, national and international). These plans would establish indicators, set targets and identify measures of implementation. Thirdly, planning needs to revivify its *social* purpose interpreting the relationships between social inequality and environmental sustainability. In the past planning has engaged in a social vision which has been physically manifest in such ideas as new towns, neighbourhood units, green belts, national parks and urban regeneration and, in a contemporary context, there is a need for concepts which combine social purpose with sustainable living (Blowers, 1997).

All this would require greater intervention and public investment but also commitment to a participatory framework of decision-making which provides people with effective power. The dangers are obvious. More intervention and subsidy will revive the spectre of the overpowering state; more participatory democracy will introduce lengthy, untidy and contradictory decision-making leading to outcomes that endanger rather than protect the environment. There is the possibility that it will merely reinforce patterns of power and inequality. On the other hand, it is very difficult to see how change in the direction of sustainability can be achieved without changes at the institutional, political and social level as well, in which people are given both freedom and responsibility. Unless this is achieved at the local level, there is little chance that sustainability at the global level can ever be secured.

Prospects for change

Fundamental social changes and the introduction of long-term planning would appear hopelessly unrealistic and idealistic. It must be doubted whether the environmental challenge yet has the potential to bring about fundamental changes in the relationship between economy and society. The state and the market economy remain the twin pillars of power, mutually dependent and mutually reinforcing (Lindblom, 1977). Although states vary in size, competence and degree of legitimacy, all, to a greater or lesser degree, possess territorial monopoly, bureaucratic administration and coercive power (Conca, 1993; Giddens, 1985). The capitalist economy, in its turn, provides production, wealth and employment. Its leading actors, the TNCs, 'are nationally based and trade multinationally on the strength of a major national location of production and sales' (Hirst and Thompson, 1996, p. 2).

Given the remarkable adaptability and persistence of state and market power, a gradual transition in the direction of ecological modernisation appears to be the most likely direction of change, at least for the foreseeable future. With the apotheosis of the capitalist market economy on an almost global basis, environmental protection will emphasise the sustenance of the resource base necessary for economic development. It will be achieved by a combination of market-based economic incentives and state regulation and standard setting. The environment will be a more significant constraint on the market, the more so in advanced industrial societies where the environment is increasingly valued as a component of well-being than in developing or impoverished countries where environmental resources are ravaged for economic growth or to maintain a subsistence economy. The problem with ecological modernisation is that it may become the victim of its own propaganda, promising more than it can conceivably perform.

The risk society portends catastrophe if present trends continue, either sooner (through nuclear proliferation or accident), or later (through global warming or ozone depletion). But its dire pronouncements also have a rhetorical, almost biblical, ring, warning us of the tyranny of technology, the concentration of power among elites, the fragmentation of society and the inevitability of disaster. Risk society is a contemporary version of the radical critiques that are an enduring element of theoretical and political debate. The problem with the approach (as with its forbears) is that the appeal to the notion of reflexive modernisation, the social self-criticism and perception that can arrest the trend and pave the way to a new Enlightenment is vague and offers no practical prescription or notion of how a transformation will proceed.

There is, perhaps, the possibility that the response to environmental change can be managed without the high risks attached to strategies of ecological modernisation or the dire consequences of the risk society. The task for social scientists is to engage in the analysis of change and speculation about outcomes. This will involve the elaboration of perspectives which take account both of political feasibility and the avoidance of risk. Such perspectives will also focus on the claims of political participation, the problem of social inequality, and the development of forms of planning that ensure a sustainable environment. To do this, it will be necessary to imagine what a sustainable society will look like in terms of production, consumption, institutions and living patterns.

It is conceivable that, as the problems of environmental change become more evident, a common interest in the need to protect resources and avoid the overloading of ecosystems will prevail. In such an analysis, a countervailing power must develop, perhaps motivated by environmental movements, but spreading its influence beyond into government, business and the society at large; a power sufficient to bring about the abandonment of those technologies that are not, ultimately, capable of ecological modernisation.

This suggests a conception of sustainable development far different from the present dispensation, one that integrates environment and economy. In the political realm, it is likely to require a revival of concern about social equality and a commitment to long-term planning as a method of environmental management. Such an analysis has scarcely begun, but the debate about environmental change and its social implications has, in the past few years, entered the centre stage of social scientific concern. The outcome of the debate in terms of policy is as yet unclear; but thought is the prelude to action.

References

Beck, U. (1992) *Risk Society: Towards a New Modernity*. London: Sage.

Beck, U. (1995) *Ecological Politics in an Age of Risk*. Cambridge: Polity Press.

Beck, U. (1996) 'Risk society and the provident state', in: S. Lash *et al. Risk, Environment and Modernity: Towards a New Ecology*, pp. 27–43. London: Sage.

Blowers, A. (1997) 'Society and sustainability: the context of change for planning', in: A. Blowers and B. Evans (eds) *Town Planning into the 21st Century*. London: Routledge.

Breyman, S. (1993) 'Knowledge as power: ecology movements and global environmental problems', in: R. Lipschutz and K. Conca (eds) *The State and Social Power in Global Environmental Politics*, pp. 124–157. New York: Columbia University Press.

Christoff, P. (1996) 'Ecological modernisation, ecological modernities', *Environmental Politics*, 5(3), pp. 476–500.

Conca, K. (1993) 'Environmental change and the deep structure of world politics', in: R. Lipschutz and K. Conca (eds) *The State and Social Power in Global Environmental Politics*, pp. 306–326. New York: Columbia University Press.

Giddens, A. (1985) *The Nation-state and Violence*. Cambridge: Polity Press.

Hajer, M. (1996) 'Ecological modernisation as cultural politics', in: S. Lash *et al.* (eds) *Risk, Environment and Modernity: Towards a New Ecology*, pp. 246–268. London: Sage.

Hirst, P. and Thompson, G. (1996) *Globalization in Question*. Cambridge: Polity Press.

Lindblom, C. (1977) *Politics and Markets: The World's Political-Economic Systems*. New York: Basic Books.

List, M. (1996) 'Sovereign states and international regimes', in: A. Blowers and P. Glasbergen (eds) *Environmental Policy in an International Context, 3 Prospects*, pp. 7–24. London: Arnold.

Mol, A. (1996) 'Ecological modernisation and institutional reflexivity: environmental reform in the late modern age', *Environmental Politics*, 5(2), pp. 302–323.

Source: Andrew Blowers (1997) 'Environmental policy: ecological modernisation or risk society?', *Urban Studies*, vol. 34, nos 5–6, pp. 864–71.

7.5

Mike Mills

Green democracy: the search for an ethical solution

Green democracy has, perhaps surprisingly, now become controversial. When such controversies appear the tendency is to search for principles or values that, when consistently applied, may properly guide our behaviour and our thoughts. . . . By looking, very briefly, at two branches of green political theory which do, indeed, compete in their use of democracy (ecoauthoritarianism and what I will call ecoradicalism), I will suggest that both suffer by failing adequately to consider two things. First, and most importantly, both advance policy prescriptions without purposefully expanding the moral community to which that policy should be addressed. My argument will be that however we characterise what greens believe or what they want to do, the question of the moral community – its expansion and the implications of its expansion – is logically prior to all others. Second, I will argue that green political theory (in both its ecoauthoritarian and its ecoradical sense) has been perhaps too concerned with outcome, and could risk being more concerned with process. . . .

Ecoauthoritarianism and ecoradicalism

Ecoauthoritarianism

Whether it argues the perils of over-population (Hardin 1968) or of scarcity more generally . . . the thrust of ecoauthoritarianism appears to remain the same. There are ecological imperatives which have to be addressed and the political organisation necessary to resolve them may not be particularly democratic. Primarily, this disorientation towards democracy is a function of two things; first, a strongly Hobbesian conception of human nature which implies that individuals will not, of their own free will, make selfless, co-operative personal choices; and second, an apparently overwhelming orientation towards ends-based policies, structures and institutions, rather than means-based ones. In other words, ecoauthoritarians take a very consequentialist moral line in which stark and mutually exclusive choices exist based upon either avoiding the ecological crisis through authoritarian measures, or suffering it. In Hardin's case he argues for 'mutual coercion mutually agreed upon' (1968). . .

. . .[A]rguments given by ecoauthoritarians to justify the circumvention of democracy will do quite as well in arguing the complete opposite – the greater the problem, the more severe the risk, the more pressing the imperative, the more necessary it becomes that democracy becomes extended, entrenched and practised. As Sagoff has argued:

> Democracy, as everyone knows, is susceptible to abuse and has all kinds of problems, but I know of no other mechanism for making policy decisions that has this ethical underpinning (citizens having the opportunity to present their views to legislators).
>
> (Sagoff 1988: 115)

The counter-argument to this given by Saward (directed against the ecoradicals, 1993) is that even if democracy is desirable (whether for intrinsic or instrumental reasons), it is simply incompatible with other green goals. . . . [T]his point depends entirely on what these goals are. If green political theory is goal or ends oriented (as the ecoauthoritarians are), then this argument has some validity although it will depend on the extent to which we accept that democracy can exist within ecological constraints. However, it is not only possible, but also perfectly reasonable, to argue that green political theory should be more process oriented, in which case democracy becomes not only compatible with, but also essential to, green political theory.

Without labouring the critique, let me just make one more point, the full importance of which will also become more evident as we go on. For ecoauthoritarians the ecological crisis presents us with the need to manage the finite resources of the planet effectively to trust in political or technocratic decisions that will, presumably, secure and distribute goods and control the ecological effects of their consumption. I would argue, however, that our ability to manage on the scale envisaged by ecoauthoritarians must be limited. In this respect then, it may be better to ensure that whatever solution we find to the ecological crisis (and to the centrality of democracy to green political theory), it had better be a process-oriented one, that is, one which minimises the need for management on the basis of inadequate information.

Ecoradicalism

Ecoradicals, on the other hand, are a much messier proposition. My central point will simply be that ecoradicals (if we take ecoradicals to belong in part to green parties and radical green movements) could afford to be more concerned with political processes (as opposed to political ends) and this in turn would help to resolve problems of green democracy. I will not be able to resolve these arguments, however, until later on. Goodin (1992) suggests that while there is a great deal of policy prescription within green programmes (as you would expect), there is actually very little on institutional change and on political processes. He notes that 'it is ecological values that form the focus of the green programme' (1992: 183) and in this respect, green politics and political theory posit a largely holistic view of both problems and their solutions. In this, Goodin suggests that it is the non-discriminatory nature of green values and their push for diversity which lead them 'positively [to] embrace pluralism' (1992: 199) and 'cherish diversity in its social every bit as much as in its biological form' (1992: 199).

There is, apparently, some gap between principles (values) on the one hand and policies on the other because there does not seem to be much on the question of policy or political processes to join the two, even at a theoretical level. In the economic programme there is an onus on individuals to act, as Goodin notices (and as rallying calls like 'think globally, act locally' imply), and for policy and lifestyle to become one and the same thing in places. In economic relations, as in politics, the green message appears to be the same and appears, still, to imply the type of political (democratic) relations necessary for individuals to take control.

Dobson (1990, 1993) sees decentralisation as the central green prescription and sees the guiding principles of ecologism as subsumed under the broad headings of limits to growth (which implies interdependence, finite resources, the paucity of technological solutions) and ecologism's commitment to non-anthropocentric principles and policies. As far as the principle goes, Dobson, quite rightly, says that: 'much of ecologism's momentum is controversially engaged in widening the community of rights holders to animals, trees, plants and even inanimate nature' (1993: 223). This provides us with our first tentative link between democracy on the one hand and philosophy on the other.

[. . .]

One observation made by Saward (1993) concerns the relationship between direct democracy and other green goals such as intrinsic value and holism. Here, he makes the point, rightly, that it is difficult to work to imperatives on the one hand and have a fairly arbitrary decision-making process such as democracy on the other hand. The fact that the green movement does have a perception of a Good Life (Dobson 1990: 14) would seem to reinforce the point that the ecoradicals have a problem – how can you go for the Good Life when your democracy may not take you there? Given, in addition, that there clearly is a bottom line green commitment to political liberalism as seen in 'diversity' and 'direct democracy' for example, then Sagoff's point is relevant:

> Liberal political theory cannot commit a democracy beforehand to adopt any general rule or principle that answers the moral questions that confront it; if political theory could do this, it would become autocratic and inconsistent with democracy.
>
> (Sagoff 1988: 162)

Of course, we are not dealing with an exclusively liberal political theory, but green theory does have liberal political elements which make the principle the same in both theories. In fact liberal political theory does commit us to certain rules and principles which are thought to be, in some sense, prior to the democratic process itself. Mostly, these revolve around who or what might be considered as worthy of participation in such a process – who decides over these moral questions? Liberal political theory does, then, have something to say about the nature of the democratic process, it is simply reticent about policy outcomes. Indeed, although Paehlke (1988) has quite rightly argued that democracy can be enhanced by environmental policy, nevertheless, it is also true to say that the ecoradicals have few safeguards against the triumph of one laudable principle, say, sustainability, over another, decentralisation for example. Dobson observes that:

> It has been suggested that ecologism's commitment to principles such as liberty and democracy is compromised by apparently laying such great emphasis on the *ends* rather than the *means* of political association.
>
> (Dobson 1993: 234)

I will be arguing that both of these points are correct, and that they are related. In other words, philosophically, greens have been concerned to expand the community of rights holders (or something similar) and they have also been concerned with ends rather than means. A commitment to some form of liberalism may help in counteracting some forms of strongly prescriptive ideology but it is likely that democracy will suffer unless other safeguards are built in.

Of all the difficulties with the ecoradical's view of democracy, that is by far the most difficult to surmount. Yet in a broader philosophical sense, ecoradicals are far from as consequentialist as their programmes suggest; they lean quite strongly towards doing what is 'right', as much as they do towards what is ultimately 'good'. Certainly in terms of their views on the political system, we can see fairly clearly that their support for participatory democracy and decentralisation denotes a view of how people can develop, grow and take control which is independent of what outcomes that might entail.

Ecoauthoritarians and ecoradicals do display some of the same problems, but in different degrees. Primarily, we can see that both do not necessarily protect (or even advance) a position on democracy which is defensible against strongly asserted and pressing political, economic and social goals. While in the case of ecoauthoritarians I would argue that they are perhaps a little more democratic than we sometimes give them credit for, nevertheless, there is little room theoretically for a consistent commitment to democracy. Ecoradicals, on the other hand, do emphasise democracy but fail to reconcile it with broader imperatives. Such a reconciliation is possible if we change the emphasis of their thinking away from goals and direct it more towards political mechanisms.

Both approaches, to my mind at least, have not taken on board the consequences of their very starting point sufficiently. Before all else, for both, comes the assertion that the ecological crisis is primarily a crisis of our ethical system. Although we may perceive our current problems as those of over-population, resource depletion, and food scarcity, these are first and foremost symptoms of an ethical crisis. Environmental philosophers have been criticised for not providing the types of guidelines to political action which political theorists prefer (see, for example, Dobson 1989) but it has to be said that there is plenty to be getting on with if an ethically based green political theory is what theorists are after.

The expansion of the moral community

If we consider what greens argue is distinctive about their ideology, their political theory and their practical concerns, it is invariably the case that these can be reduced to a concern to expand the moral community.[1] Eckersley (1992) argues that the fundamental characteristic of green political theory is the fact that it is ecocentric and this ecocentricity is logically prior to all other political considerations:

> A non-anthropocentric perspective is one that ensures that the interests of non-human species and ecological communities . . . are not ignored in human decision making simply because they are not human or because they are not of instrumental value to humans.
>
> (Eckersley 1992: 57)

By accepting that humans are not the only ones with value or the only measure of value, we accept that our moral community expands because it is now necessary to accommodate others within our ethical choices. Eckersley, here, has made two interrelated points – one is that the non-human world should be given consideration and the second is that such consideration can be ensured only when a non-anthropocentric perspective is taken. This is a position I would support, indeed, my argument makes it essential that a non-anthropocentric approach is adopted – without this there is only a contingent expansion of the moral community, and if that happens, then democracy cannot be secured either. Goodin (1992) continues this line of argument when he says that it is 'naturalness' that greens value. If it is naturalness that has value, then presumably natural things should be given moral consideration. Dobson too (1990), as we saw earlier, took as a fundamental axiom of ecologism the idea that we should look toward a biocentric, or ecocentric (non-anthropocentric), basis for our political theory.

In terms of theory, we can see our ethical responsibilities shifting away from a largely human-centred approach to be ecocentric. These have been the primary terms of the debate and it is perhaps because of this that the political consequences of expanding the moral community – which I take to be a corollary of switching to an essentially ecocentric ethic – have not received the attention they might deserve. The 1992 General Election Manifesto of the British Green Party, for example, can illustrate this point quite well, not because there is no evidence of a shift, but rather because it surfaces in a rather ad hoc way and is not, to any large extent, directed at political variables.

From what we have said so far, we would expect to find the core axiology – that the Green Party was ecocentric and this entailed greater moral considerability of non-human species – represented somewhere within the manifesto itself, or at least an indication of such. And, indeed, it is perfectly possible to find such indications, but they tend to come in very specific forms. Largely, these are either in the forms of policy prescriptions (for example, rapidly phasing out factory farming or the greater protection of the soil), or in terms of changing the basis upon which future decisions will be made:

> [The Green Party] would revolutionize the system of national accounts by rigorously identifying real costs and real benefits in our industrial society. In so doing we would attribute equal value to the natural capital on which we depend (topsoil, water, clean air, fossil fuels etc.) as we do to the financial capital which greases the wheels of the world economy.
>
> (Green Party 1992: 11)

To the extent that decision-making would change (in terms of values and outputs at least), then clearly this change is in line with a broadening of the moral community. However, green political theory might suggest that concomitant changes might also occur in the political system more broadly – for example, in the attribution of rights to non-humans on the basis of their having value. This type of analysis is not in the manifesto, the overtly political provisions are aimed largely, although not exclusively, at resolving problems of the British political system. Without wishing to labour the point, the critique of Saward (1993) and the warning of Sagoff (1988) still remain. If there is to be an essentially goal-oriented policy-making process (and one based on ecological imperatives), then the Green Party may have problems with democracy. The commitment that the manifesto shows to *ends* might need to be matched by a similar commitment to the reform of the *process* of politics in a non-anthropocentric way which would ensure that green goals did not take precedence over due process.[2]

A more consistent application of the holistic principle, and the greater incorporation of non-human interests into the political features of the state would make green positions defensible against the accusation that highly deterministic, imperative-driven, consequentialistic policy runs the risk of forsaking democracy. If we accept that value exists non-anthropocentrically, then it clearly is the case that what political movements (and philosophies) do should reflect this change and should find more things morally considerable. We can see in these programmes the obvious wish to expand the moral community but this is invariably associated with ends-based policies, rather than means-based processes.

So, greens propose the expansion of the moral community for very particular reasons (non-anthropocentric) and although such a position (which will be expanded below) does exist theoretically, in practice there are two problems. The first is that programmes are not necessarily informed by ecocentrism and holism and the second is that only certain formulations of this ethic will actually serve the dual function of both securing the democratic basis of green politics and the ethical basis that greens want. Importantly, greens appear to prefer, as you might expect, to argue their case in terms of policy ends, rather than in terms of the more deontological (axiological) premises of the policy process.

Achieving the expansion of the moral community

The expansion of the moral community can be achieved in a number of different ways depending, to a great extent, upon how far we want to go and what arguments we are going to use along the way. I have already argued that the type of expansion associated with transpersonal/deep ecology is not one that I am going to pursue here but, rather, will follow Dobson and Eckersley in suggesting that it is difficult to formulate politically.[3] I am further restricted by the critique of green politics offered by Saward and the observations of Sagoff, both of which suggest that too great an emphasis on the ends of policy rather than the means will undermine the democratic basis of green politics. This inclines me to believe that it is best to avoid ethical arguments that are entirely consequentialist. Lastly, I have argued that I am concerned with an ecocentric ethic, one which finds value in the non-human world and leads to the expansion of the moral community.

In expanding the moral community the most obvious first step would be to include animals or sentient beings. For Regan (1984), it is the ability to be 'the experiencing subject of life' which denotes whether we get 'rights' or not. Once sentience is established, Regan argues that all sentient beings are equal in having value and that it is only just that claims which these beings make upon us morally should be seen as valid (i.e. they have rights). For Regan then our moral community expands to incorporate sentient beings, although these rights do not have to be the same rights as those held by humans (because in some cases, such as the vote, it would be silly) and to establish a moral right is not the same as establishing a legal right (the classic case is that of keeping a promise).

So, the boundaries are pretty clear here, and Regan is looking very much to replicate for animals the existing ethic for humans. In this sense, it is not an 'environmental ethic' (Rolston 1987) because it does not look to the environment for the source of its value. Neither, then, is it holistic because Regan is concerned with individual sentient beings, not with species, nor with the systems which sustain the animals and to which the animals contribute. Consequently, we could not use this as the only basis from which to expand the moral community, primarily because it could not hope to resolve all of the ecological problems we might be interested in.

Another possibility is the 'reverence for life' literature which draws the line of moral considerability differently. Here the fundamental axiom, according to Goodpaster, is not whether something is sentient but whether it has life or not: 'Nothing short of the condition of *being alive* seems to me to be a plausible and non-arbitrary criterion (for moral considerability)' (1983: 31). This distinguishes Goodpaster both from those who take a more holistic approach (and would value the systems of those who lived above the individuals themselves) and from animal rights/liberation authors who do construct distinctions (which Goodpaster appears to believe are 'arbitrary') between living things, usually on the basis of sentience, the ability to suffer (Singer 1975) or having inherent value which it would be unjust to ignore (Regan 1984). Reverence for life theorists do not, therefore, distinguish between plants and animals as far as moral considerability goes, although this does not mean, of course, that each is as morally *significant*.

Those who are holistically minded have criticised the reverence for life approach as an ethical system because it takes a conventional view of ethics (that it should be concerned with discrete individuals) and extends it into the non-human world (Callicott 1983: 301) – in this respect the argument is very similar to those heard against animal rights. . . . More important, though, is the argument that reverence for life is difficult if not impossible to live by because we would not be able to do anything any harm (Callicott 1983: 301). Interestingly, then, reverence for life (or life-principle) theorists provide an ethic which is ecocentric but which is not holistic. . . .

An environmental ethic which is going to be of any use is going to have to allow us to value systems as well as individuals – otherwise, as Sylvan (1984) pointed out, we may save the tree and spoil the forest. Equally though, we cannot risk having an ethic which does not protect individuals, otherwise we are all dispensable in the scheme of things. Similarly, this ethic (which will allow us to expand our moral community) must find the source of worth or value of that which will be morally considerable, in nature, rather than in any instrumental value for people.

[. . .]

. . . Green political theory, by virtue of it being 'green', emphasises holism, that is, it emphasises the value of the wholes of which individuals are a part, rather than the individuals themselves. . . .

Process and democracy

Let me begin by briefly returning to the argument that greens should be seen as more process oriented than is the case at the moment.[4] Greens want to secure, for example, sustainability and through this the circumstances of future generations. It is perfectly reasonable to see this as a goal towards which any green polity should work. Following the arguments of Saward and Sagoff, making such an 'end' integral to the form of a green democracy would undermine democracy itself. Now if, as I argued earlier, greens are concerned with expanding the moral community, this need not be a problem – it has been perceived as such only because the implications of such an expansion have not been taken to their logical conclusion. If we ensure that those future generations, non-human species, and ecosystems are afforded the political consideration that we might expect for a member of the moral community, then we do not have to prescribe as many of the policy outcomes in advance (i.e. we don't have to be as ends oriented). We simply have to construct our political institutions (which would include rules, structures, basic laws) in a way which guarantees that the political process will be 'considerate' of all those interests which are represented. If we take for example the four basic principles of the German greens (ecology, social responsibility, grassroots democracy and non-violence: Spretnak and Capra 1985) it is perfectly possible to see all of these as principles which are as much guides to the way the political system might operate, as they are ends which the system must achieve.

An anthropocentric polity, such as we have at the moment, displays certain characteristics which may give us clues as to how we may proceed to reconstruct democracy on the basis of green principles. Primary amongst these is that the polity is constructed for, by and of people – in other words there is some congruence between the nature of the political and the moral communities. This is not to say they are, or have to be, identical, but rather that there is a relationship between what, or who, is thought morally considerable, and what, or who, is politically represented. Presumably, then, a green polity would want to do the same thing – it would want to make the nature of its (expanded) moral community more congruent with the political community. To my mind, if we are following both the logic of the philosophical basis of green political theory and the idea that moral and political communities are similar, then we end up, broadly, with four central political areas which would need to be changed to accommodate the new moral community: standing, quality of democracy, decision-making and political representation.

Standing

Under such arguments, which are largely based upon liberal conceptions of citizenship and the rule of law, it would be possible to allow various non-human entities (perhaps, a river, marsh, brook, beach, national monument, commons, tree, species)[5] to have action taken on their behalf against those who injure them. Although the 'liberalness' of this type of approach has been criticised because it reinforces current perceptions of individualism in law and in nature I would argue, first, that such redress *should* be available no matter what type of political theory we construct (because it would be arrogant to believe our theory to be so good that redress would never be necessary)[6] and, second, that it would be wrong to see this as an approach which could only reinforce individualistic stereotypes of our place in nature. Indeed, it is perfectly possible to argue that the idea of a whole 'system' being damaged and complaining about it could do a great deal to change cultural perceptions of ecosystems.

There is no reason why it would be only individual members of sentient species that would have standing – it is perfectly possible to have many non-individualistic aspects of ecosystems given legal and moral standing. It may be difficult to do this in a strictly 'holistic' way but on some (limited) versions of holism such an approach could be helpful if standing was not restricted to individual examples of species or ecosystems. I think the obvious qualification to make is not that liberal-legal solutions are unnecessary but, rather, that they are insufficient. Their problem, from a green point of view, is that all other aspects of the political, economic and social system remain unchanged and hence moral considerability, significance and community membership become far too contingent upon purely legal processes.

We could make a similar case on the basis of attributing rights to the non-human world. These may be specific rights, say, to thrive, exist without threat of damage, and may exist independently of standing in other areas which may offer more general protection. If we are to take the idea of rights to their logical conclusion we could, for example, consider whether we might extend or redefine the notion of social rights (Marshall 1963), extending them from the civil and political to include the quality of life. Here, we might, for example, have to adjust either the legal or constitutional standing of non-humans or perhaps see their welfare in the same 'entitlement' framework as we view our own. The possibilities for using such a system to change or promote consciousness is quite formidable. . . .

Quality of democracy

In existing (anthropocentric) political systems the ability of subjects to participate between elections is, in part at least, a measure of democracy. Equally, analysis which considers the role of the institutional arrangements of the state has increasingly suggested that how the state is organised can, and does, affect political outcomes (see, for example, Evans 1984; Hall 1986). . . .

The same judgements will be applicable to a green democracy. In other words we will have to consider whether such political systems represent a real expansion of the moral community. Precisely what is represented will, of course, depend upon the extent to which we expand our moral community. Nevertheless, it is possible to imagine that consultation exercises might be required to provide opportunities for human representatives of non-human interests to give an opinion; that licensing authorities for industrial plants may have similar constraints placed upon them; that the state may have a statutory responsibility to promote, fund and consult representatives of non-humans in the same way that some political parties are state funded; that regulatory agencies may be required to promote the interests of non-humans' entities and that the representative basis of these agencies should reflect this, and so on.

Decision making

. . . Given that we do have to operationalise our philosophical base, then Goodin (1983) provides a very good starting place. Working from the assumption that we must now consider (if not apply moral significance to) non-humans, . . . Goodin argues that rather than pursue a utility-maximising strategy, we should opt for decision-making criteria which are biased against irreversible decisions; in favour of protecting the vulnerable; in favour of sustainability; and against causing harm (1983: 16). . . . In particular, the principle of 'avoiding harm' would now seem to be a central one given our commitment to some notion of ecocentrism and the moral dispersion of value. This would also tend to accord with an orientation toward ecosystems which is more humble and one which was as concerned with doing the 'right' thing, as achieving a 'good' result. . . .

Political representation

It has been my intention all along to use the issue of political representation as a means of circumventing arguments about green democracy. To be process oriented *and* to change the nature of political representation (in accordance with an expanded moral community) would mean we could expect green(ish) outcomes, without prescribing the ends of the policy process. We would simply be making a morally defensible case for changing the political process itself. Precisely how this may work out in practice is, I concede, a very difficult question. It is rather unlikely, for example, that individual species would, say, be represented in parliament – since this does seem to lead us into some peculiar possibilities.

Having said this, I think we could make a more plausible case for multi-member constituencies in which some of the representatives were expected to represent the interests of their non-human constituency members. Certainly there could be great benefits from a system like this at national or regional level where representatives are not always (or perhaps often) confronted with the ecological consequences of their actions. This would also fit with the idea that we should be concerned with systems (seen as areas or regions in this case) rather than individual species that might be threatened. It may eventually lead to political boundaries being drawn on ecological lines if, for example, ecological problems are seen as more pressing.

In fact, Kavka and Warren (1983) rehearse many of the arguments as far as the representation of future generations is concerned. They believe it is meaningful to argue that a being is representable if it has interests and if the representative takes instruction from the being, or, has a better than random judgement of their interests (1983: 25). Under these circumstances it should, on Kavka and Warren's formulation, be possible to represent a very diverse range of non-human interests within a green political system because their arguments appear to apply as much to non-humans as they do to future generations. Most arguments in favour of intrinsic value (or other holistic, ecocentric theories) do suppose some idea of the interests of those under consideration, so this is not in itself a problem. Once we have established that something has value or interests, then it is representable. I do not propose to go into the questions of how an equitable representation might be achieved, but would say that if we are to make the moral and the political community similar, then such an approach is valuable and we cannot dismiss this as an idea, particularly given that it would have a profound and positive effect on the nature of green democracy.

[. . .]

Notes

1 By 'the expansion of the moral community' I simply mean the increase in the number of individuals, species or systems which become morally considerable. I do not expect, as some do (Moline 1986) that 'extending' the community means that the same ethic applies. I would say that the expansion of the community presupposes a change in the ethic.

2 Personally, I would not have much objection to surrendering some, or perhaps most, of my democratic rights to a loving and trustworthy green council which would pursue ends with which I agreed and which would benefit me and my family. It is quite on the cards that such sacrifices may be necessary at a practical level and we will have to risk the abuse of the democratic process. This does not mean, though, that, theoretically, we have to be happy at the prospect. Nor does it mean that a little more attention to the anthropocentric nature of the political process would go amiss.

3 I believe that the problems reside principally in translating the spiritual into the political with all the associated problems of accommodating things such as 'faith' and 'intuition' within a political framework. I have declined to do this, but believe that it is a job well worth trying nevertheless.

4 By process oriented, I mean a concern with the way the political system operates, the nature of political representation, the values which are embodied in the system, the opportunities there are for political participation, and the constraints there are on political action (e.g. what the state may legitimately do).

5 These are all examples of real complaints that were filed in the USA.

6 Gandhi cautioned the West against thinking we could construct social systems which were so perfect that people no longer had to be good. Perhaps the same applies to green political theory?

References

Callicott, J. Baird (1983) 'Non-Anthropocentric Value Theory and Environmental Ethics', *American Philosophical Quarterly* 21, 4: 299–309.
Dobson, A. (1989) 'Deep Ecology', *Cogito* spring.
—— (1990) *Green Political Thought*, London: HarperCollins.
—— (1993) 'Ecologism', in R. Eatwell and H. Wright (eds) *Contemporary Political Ideologies*, London: Pinter.
Eckersley, R. (1992) *Environmentalism and Political Theory: Toward an Ecocentric Approach*, London: UCL Press.
Evans, P. (ed.) (1984) *Bringing the State Back In*, Cambridge: Cambridge University Press.
Goodin, R. (1983) 'Ethical Principles for Environmental Protection', in R. Elliot and A. Gare (eds) *Environmental Philosophy*, Milton Keynes: Open University Press.
—— (1992) *Green Political Theory*, Oxford: Polity.
Goodpaster, K. (1983) 'On Being Morally Considerable', in D. Scherer and T. Attig (eds) *Ethics and the Environment*, Englewood Cliffs, NJ: Prentice Hall.
Green Party (1992) *New Directions: The Path to a Green Britain Now*, General Election Campaign Manifesto, London: The Green Party.
Hall, P. (1986) *Governing the Economy*, Oxford: Polity.
Hardin, G. (1968) 'The Tragedy of the Commons', *Science* 162: 1,243–8.
Kavka, G. G. and Warren, V. (1983) 'Political Representation for Future Generations', in R. Elliot and A. Gare (eds) *Environmental Philosophy*, Milton Keynes: Open University Press.
Marshall, T. H. (1963) *Sociology at the Crossroads*, London: Heinemann.
Moline, J.N. (1986) 'Aldo Leopold and the Moral Community', *Environmental Ethics* 8, 2: 99–120.
Paehlke, R. (1988) 'Democracy, Bureaucracy, and Environmentalism', *Environmental Ethics* 10, 4: 291–309.
Regan, T. (1984) *The Case for Animal Rights*, London: Routledge.
Rolston III, Holmes (1987) *Environmental Ethics*, Philadelphia, Pa: Temple University Press.
Sagoff, M. (1988) *The Economy of the Earth*, Cambridge: Cambridge University Press.
Saward, M. (1993) 'Green Democracy', in A. Dobson and P. Lucardie (eds) *The Politics of Nature*, London: Routledge.
Singer, P. (1975) *Animal Liberation*, New York: Random House.
Spretnak, C. and Capra, F. (1985) *Green Politics*, London: Paladin.
Sylvan, R. (1984) 'A Critique of Deep Ecology, Parts I and II', *Radical Philosophy* 40 and 41.

Source: Mike Mills (1996) 'Green Democracy: the search for an ethical solution', in B. Doherty and M. de Geus (eds) *Democracy and Green Political Thought: Sustainability, Rights and Citizenship*, London: Routledge, pp. 97–114.

7.6

Peter Christoff

Ecological citizens and democracy

Issues of political participation and representation are especially challenging when one considers environmental concerns. Consider the case of a chemical factory to be built on the banks of a river flowing through five countries. A serious accident at this factory would affect not only people living in the country in which it is sited but also inhabitants of the other countries, future generations of humans and other species. Who, then, should participate in decision making about the factory's construction?

By virtue of their regional and global impacts, environmental issues have expanded both temporally and spatially beyond the conventional borders of political decision making. They point to the need for new approaches to the protection of the environment and environment-related rights. As David Held argues,

> the very idea of consent, and the particular notion that the relevant constituencies of voluntary agreement are the communities of a bounded territory or a State, become deeply problematic as soon as the issue of national, regional and global interconnectedness is considered and the nature of a so-called 'relevant community' is contested.
>
> (Held 1991: 203)

So how are we to respond to the problem for 'green democracy' that 'democratic theory can no longer be elaborated as a theory of the territorially delimited polity alone, nor can the nation-state be displaced as a central point of reference' (Held 1991: 223)? Ideally, ecologically sensitive decision making would encompass a well-developed public recognition of the implications and impacts of human activities over time and over large distances. Such decision making would therefore depend on active citizens and a state better organised to facilitate democratic participation. For both democracy and the environment to flourish, we now need to elaborate further upon what Held (1989: 167ff.) has called 'double democratisation' – the revitalisation of civil society and the related restructuring of the state.

[. . .]

Problems for citizenship and democracy increase as one moves outward to the international level. The role of the citizen has, to date, been institutionalised only at the level of the sovereign nation-state: the main actors in the international community are states, not citizens. Similarly, the notion of political community is assumed to work only up to the level of the nation-state: at present, the rest depends on increasingly remote and unpopular technocratic administration. To date, citizens seem to have little direct purchase or influence on transnational regulatory and administrative institutions which are governing or reshaping their lives. They have no effective formal means of debating international decisions or influencing decision-making processes at this level, other than through their national government.

[. . .]

Permanent quasi-federal regional parliaments – such as the European Parliament – could be established, in which elected representatives of regions or states voted on single issues of significance upon the advice of their local electorates or cantons. Separately, or additionally, states considering mutual environmental concerns could initially determine their positions on specific environmental issues through plebiscites of their populations. Alternatively, decision making could be based on plebiscites or referenda in which the total populace pooled across nation-states participated. Or, more radically, decisions could be made on the basis of direct democracy within a mobile or 'flexible' electorate, changing in composition according to the problem, including and enfranchising all its 'residents' on the basis of recognised vital interest and (usually) aggregated in terms of ecological units such as river catchments or airsheds.

[. . .]

One key issue is that of agenda-setting. How would transnational associations of voters determine which environmental issues would be discussed?

An expanded system of national and international environmental laws, treaties and conventions – such as those governing protection of the North Sea, reduction of greenhouse gas emissions, elimination of CFCs, trade in wastes, harvesting of endangered species, which are now regarded as the responsibilities of signatory nation-states – could trigger action by encoding criteria for issue identification. This approach would rely upon the successful international mobilisation and integration of interests, and consequent co-operation between states, to encourage, enable or require individual states to recognise that actions within their borders have significant consequences for other citizens who therefore become formally recognised and empowered participants in decisions.

As the negotiations around the Montreal Protocol on ozone-depleting substances showed, identification of issues and effective co-operation to achieve their resolution is notoriously time-consuming and awash with political compromise but not impossible. It is equally clear that – with regard to many relatively 'localised' issues of ecological concern . . . – political processes would need to be flexible yet precise, rapid and also inexpensive. The constant redefinition of 'political communities' in relation to democratic decision making in areas of ecological concern merely adds to these challenges.

To begin to answer what are essentially questions of power and political will, it is helpful to look at notions of citizenship from a completely different angle, and turn to conceptions of citizenship based on moral responsibility and participation in the public sphere rather than those defined formally by legal relationships to the state. It is also necessary to consider additional dimensions of the environmental problem relating to other species and future generations.

[. . .]

The extent, intensity and multi-dimensional nature of environmental destruction revealed since the 1960s have magnified existing problems for democracy and for citizenship, making them both more elaborate and more urgent. That such problems and threats require urgent resolution, and efficient international mechanisms for doing so, are well recognised. Environmental degradation – such as the pollution of groundwater aquifers, radioactive contamination, or induced climate change – may take decades to reveal itself and may also persist for hundreds or thousands of years, affecting many generations of humans and other species and altering the time-frame over which the consequences of decisions must be assessed. We decide not only for ourselves and our children, but often for our children's children. Yet the information required for ecologically sustainable decisions about production, consumption and environmental protection increase in complexity as the intensity or scale of human intervention increases. Decisions must be informed by evolving scientific understandings of the intricate behaviour of fragile ecosystems, and of the environmental implications of human activity. These informational demands exacerbate tensions relating to the limited capacity of representative democracy adequately to reflect informed environmental choice.

There is an emerging consensus, underlying the arguments presented in the Brundtland Report (WCED 1987) and Agenda 21, that several principles need to be observed in decisions with potential environmental impacts if ecological sustainability is to be achieved and maintained. These principles include:

- the precautionary principle
- the principle that biological diversity must be preserved, for ecological, economic and ethical reasons
- the principle of intergenerational equity
- a procedural principle relating to the need for reflexivity in decision making.

Realisation of these principles depends on expanding both legal and practical notions of citizenship to require a duty of care towards non-human species and unborn generations, and a corresponding reconfiguration of the state to provide widespread guarantees of environmental rights.

Extending citizenship: citizens as ecological trustees

. . . What to do about the rights and needs of non-humans? Fish cannot raise their fins to vote nor the unborn express their potential desires. So how then can their needs and rights be included in democratic discourse? Whether representative or deliberative, democracy remains dependent on decisions by humans who are capable of articulating and considering their individual and collective opinions, needs, rights and interests. Yet to shy away from the epistemological challenge of 'representing' non-humans and future generations is to shy away from taking responsibility for their fate. Indeed, we can, as Dryzek (1993) suggests, move some way beyond our species' limitations if we consider 'signals' emanating from the natural world – even if, ultimately, we do not escape our human bounds.

The principles of ecological sustainability require that we defend ecological values and the rights of future generations and other species just as we are morally obliged, and increasingly legally required, to consider and protect the rights of those humans who cannot be defined as 'morally competent' (children, intellectually disabled people, and so on). To become ecological rather than narrowly anthropocentric citizens, existing humans must assume responsibility for future humans and other species, and 'represent' their rights and potential choices according to the duties of environmental stewardship.

This concept of ecological citizenship – of *homo ecologicus* – adds challenges to those noted earlier in relation to the consequences of globalisation. The apparent paradox caused by the increasing de-linkage of the citizen from the modern nation-state is accentuated by environmental concerns. We have seen that the citizen's 'political community' – which, for other issues, may remain that of the nation-state – is profoundly reshaped by an ecological emphasis which generates additional and occasionally alternative transnational allegiances ranging from the bio-regional through to the global, as well as to other species and the survival of ecosystems. . . .

Ecological citizens, the green movement and civil society

The creation of ecological citizens depends on material preconditions – the impetus to social change caused by the deteriorating environment. It also depends upon the related emergence of a culture of environmental solidarity with its new forms of association (in particular, the green movement), as well as upon the changing opportunities afforded by the state.[1]

Since the 1970s, new social movements have sought to 'rescue' civil society from the administrative and regulative incursions of liberal-capitalist and 'actual-socialist' states.[2] At the same time, they have sought to revitalise the public sphere and to democratise both the state and (occasionally) the economic sphere, to make them more responsive, transparent and intelligible to emancipatory demands relating to issues of environment, race, gender, sexual desire and so on.[3]

In most Western industrial nations, the green movement has transformed the public sphere by enabling citizens to present the state with ecological-ethical demands which are increasingly seen as an extension of existing civil, political and social rights. It has forced ecological concerns on to the formal political agenda either through existing political parties or through the creation of new green parties. Its critique of the colonisation, exploitation and destruction of nature has sought recognition of the importance of the biological world in the calculus of political and economic decision making. This emancipatory impulse represents an attempt to define or redefine, for the first time, human aspirations in an ecological context. The movement has also encouraged a critique of liberal-democratic and socialist states as institutions articulated around modernist notions of industrialising progress, and has challenged the legitimacy of specific capitalist and socialist governments over their promotion of resource exploitation and their inability to resolve environmental problems.

Despite the transnational aspect of ecological citizenship, the state remains an exceptionally important focus of concern for ecological citizens and their organisations seeking to refashion its activities; to have it enshrine protection of generalisable environmental interests in legislation guaranteeing environmental standards, the protection of ecosystems and species; and to provide the legal and material support for further (ecological) democratisation.

However, partly because of the nation-state's territorial boundedness, ecological citizens also increasingly work 'beyond' and 'around' as well as 'in and against' the state. For instance, ecological citizens focusing on community action to halt or repair environmental degradation have deliberately mobilised within civil society, with the state as the secondary rather than primary focus of their attention. And through its use of the media, the green movement has sought to create a public space apart from the state, in which ecological issues – alongside and in relation to other concerns of green politics, such as social welfare – might be debated, and directly influence private life and the economic sphere (e.g. by changing consumer behaviour and industry's investment patterns).

Environmentalists have, with limited success, sought to use international forums, treaties and conventions to articulate environmental rights and to create tools to strengthen opportunities for action at the level of the nation-state. Increasingly, there is also a 'shadowing' of international government agencies by the organisations of the developing international public sphere. The United Nations Environment Program (UNEP) is increasingly ambiguously placed between government and non-government organisations at the international level. The Earth Summit was also the focus for the first major gathering of non-government environmental organisations – the Global Forum. The G7 Economic Summit is regularly accompanied by a parallel meeting of activists and representatives of interested non-government bodies, at what is now called The Other Economic Summit (TOES). Rosenau

(1993) sees a trend here toward the 'bifurcation of world politics', with the traditional state-centred structure of the international system now coexisting with a (weaker), more decentralised, poly-centric system comprised of non-governmental organisations and other transnational actors.

Indeed, recent technological innovations have enabled the creation of 'virtual communities' which are combinations of face-to-face and abstract networks, transnational and linked through their interests by computers, telephones, video, television, faxes, magazines and jet travel. The new global communication technologies have become increasingly important determinants of the efficacy of those engaged in political activity, whether in social movements or in political parties. This is particularly the case for the green movement, which is increasingly reliant upon rapid transfer of information between global networks of activists, scientists and environmental organisations and consequently increasingly capable of transnational political interventions.

Together, these developments reshape the definition of the 'relevant community' and the 'relevant actors' for democratic participation and representation of environmental issues. They emphasise the growing disjuncture or dislocation observed earlier between moral citizenship (as practised in individual and 'community' action and moral responsibility) and legal citizenship as defined by the nation-state. They force a redefinition of what constitutes global political action, operating as they do on several levels simultaneously. And they increase pressure for more universalistic, inclusive constitutional guarantees of citizens' rights and for definitions of ecological responsibility which tie local and international levels together through the conduit of the nation-state's apparatuses.

It is now possible to draw together some of the essential characteristics of ecological citizenship, including those which – depending on which form such citizenship takes – challenge, extend or alter existing notions of social and political citizenship.[4] Ecological citizenship is centrally defined by its attempt to extend social welfare discourse to recognise 'universal' principles relating to environmental rights and centrally incorporate these in law, culture and politics. In part, it seeks to do so by pressing for recognition of the need actively to include human 'non-citizens' (in a territorial and legal sense) in decision making. It also promotes fundamental incorporation of the interests of other species and future generations into processes of democratic consideration. This leads to challenges to extend the boundaries of existing political citizenship beyond the formerly relatively homogeneous notions of the 'nation-state' and 'national community' that to date have determined 'formal' citizenship.

The focus on a broadly ecological notion of welfare also increases demands for appropriate institutions for delivering such welfare. The state is under increasing pressure both to provide environmental education, regulation, expenditure and remediation, and to reduce its contradictory facilitation of resource exploitation for traditional economic growth. As an extension of social citizenship, ecological citizenship establishes demands for environmental welfare.[5] These demands at minimum reshape, and often work against, the requirements of capital reproduction and accumulation (capitalist or state socialist). It therefore has an impact on the capacity of the welfare state to pursue its social welfare functions where these conflict with longer-term social and ecological needs. Nevertheless, it remains unclear whether ecological citizenship in practice opposes capitalism, stands in tension with it by merely inhibiting the market, or supports it by believing that capitalism can be made truly 'green'. All three strands may presently be seen in competition with one another in the green movement. This ambiguity perhaps relates more to the different tactics of green and wider environmental movements and their different political and economic analyses rather than the normative construction of ecological citizenship as such.

For its success, the emancipatory project which is shaped by – and in turn constitutes – ecological citizens depends on the revitalisation and extension of civil society. It depends upon the active transformation of private life through creation of a 'green conscience', and increased democratic influence over the economic sphere through the actions of 'green workers', 'green producers' and 'green consumers'. This is reflected in the high value which green theorists and activists place on self-rule. This value includes the moral priority given to 'self-restraint' within civil society and also to active citizenship as defined by individual (self-) development beyond a merely instrumental relationship to the public sphere; a sense of active responsibility for representation and protection of environmental rights, and the individual and collective use of the public sphere and the state to provide the formal opportunities and protection for the institutions of ecologically guided democracy.

[. . .]

Notes

1 Talking about 'the environment movement' as if it were a homogeneous social actor is, of course, problematic. As Falk (1992: 129) writes, 'the new movements are exploratory and include quite a wide range of outlooks among their adherents. Perhaps it is questionable to group disparate initiatives within an issue area (say, environment or human rights) in an aggregate of the sort implied by the seeming coherence and solidity of the term "movement".'

2 The term is used in the Habermasian and Gramscian sense of public sphere, to mean – 'a social realm in which all cultural institutions within which opinion is formed are included' (Honneth 1993: 20).

3 See, for instance, Keane (1988) and Cohen and Arato (1992).

4 See, for instance, Turner (1993) and Hindess (1993).

5 It is also possible to argue that the lasting recognition and protection of environmental rights require concomitant action to address the issue of economic inequality (particularly between First and Third World nations), and therefore incorporate and depend on achieving demands for economic justice.

References

Cohen, J. and Arato, A. (1992) 'Politics and the Reconstruction of the Concept of Civil Society', in A. Honneth, C. Offe and A. Wellmer (eds) *Cultural-political Interventions in the Unfinished Project of Enlightenment*, Cambridge, Mass.: MIT Press.

Dryzek, J. S. (1993) 'Green Democracy', unpublished MS.

Falk, R. (1992) *Explorations at the End of Time: The Prospects for World Order*, Philadelphia, Pa: Temple University Press.

Held, D. (1989) *Political Theory and the Modern State*, Stanford, Calif.: Stanford University Press.

— (1991) 'Democracy, the Nation-State and the Global System', in D. Held (ed.) *Political Theory Today*, Cambridge: Polity.

Hindess, B. (1993) 'Citizenship in the Modern West', in B. S. Turner (ed.) *Citizenship and Social Theory*, London: Sage.

Honneth, A. (1993) 'Conceptions of "Civil Society"', *Radical Philosophy* 64, summer: 19–22.

Keane, J. (1988) *Democracy and Civil Society*, London: Verso.

Rosenau, J. N. (1993) 'Environmental Challenges', in R. D. Lipschutz and K. Conca (eds) *The State and Social Power in Global Environmental Politics*, New York: Columbia University Press.
Turner, B. S. (1993) 'Contemporary Problems in the Theory of Citizenship', in B. S. Turner (ed.) *Citizenship and Social Theory*, London: Sage.
United Nations Environment Programme (UNEP) (1992) *Saving Our Planet*, Nairobi: United Nations.
World Commission on Environment and Development (WCED) (1987) *Our Common Future* (Brundtland Report), London: Oxford University Press.

Source: Peter Christoff (1996) 'Ecological citizens and ecologically guided democracy', in B. Doherty and M. de Geus (eds) *Democracy and Green Political Thought: Sustainability, Rights and Citizenship*, London: Routledge, pp. 151, 155–62, 168–9.

Mark J. Smith

Epilogue: Thinking through ecological citizenship

The conceptions of justice and citizenship which have prevailed in Western social and political thought for centuries no longer provide us with an adequate set of tools for resolving the difficulties created by contemporary environmental problems. We face a situation where universal answers are no longer trusted and technological solutions are regarded with suspicion. The incidence and consequences of environmental hazards cannot be effectively predicted nor responses planned in advance. Ulrich Beck argues that the hazards of a technologically driven society now 'penetrate every region and level of society'. Yet human beings remain wedded to the responses to environmental degradation which were more appropriate in the nineteenth century, when risks were calculable and responsibility could be clearly identified. There are no longer any clear rules of conduct for attributing causal responsibility and awarding compensation. The scale and scope of current human impacts upon the environment generates complex and unanticipated consequences which cannot be contained effectively within the earlier guarantees and safety mechanisms (Beck, 1992, 1995). According to recent sociological argument, we are entering a new phase of social existence characterized by uncertainty and anxiety where science can offer few concrete reassurances (Giddens, 1990; Beck *et al*., 1994). Yet political agenda-blenders blind themselves and everyone else to the perils of environmental degradation, where the enormous risks now produced by human activities are translated into acceptable costs.

So what alternatives are open to us? One option is to continue to apply solutions appropriate to an industrial society within a 'risk society', with all the potential dangers involved in maintaining present human activities. Ecological thought clearly rules this out. Two further possibilities are open to us: according to Beck, these are *authoritarian technocracy* and *ecological democracy*. We could establish a 'strong state' approach, whereby the existing political institutions exert greater control over potential hazards. With each batch of regulations and legislative measures, we would witness an extension of state power and control as well as its centralization in the hands of an elite which controlled knowledge and expertise. The alternative, ecological democracy, would require a substantial reorientation of social existence and the rules of conduct which organize it. Like other transformative projects, it is defined through its negation of existing arrangements and the principles which underpin them: an insistence that technological change should emerge only from a thorough discussion of the possible consequences, rather than being imposed from above; a transformation of the basis of proof required in the legal rules applied to environmental risk assessment; and the presumption that potential polluters would have to prove the safety of productive processes, rather than the afflicted having to prove that they have been adversely affected. Both of these possibilities demand that we readdress the ethical codes which provide the rules of conduct in everyday life.

Just as we have witnessed endless disputes about anthropocentric moral communities, we are likely to encounter a variety of ways in which ecocentric moral communities can be defined (with membership defined by sentiency, the possession of life or one's place in the biotic community). This would, as argued by Mike Mills (see Reading 7.5), demand a reappraisal of the way we define the moral community and a fundamental reappraisal of the grounds for ethical consideration. By extending the moral community in the manner suggested by Aldo Leopold's plea to respect the biotic community (see Readings 4.3a and 4.3b), we are attributing intrinsic value to creatures and other natural things as ends in themselves rather than as the means to some set of human ends. This displaces the human species from its dominant position within the ecosystem, so that any set of moral rules should prompt consideration of obligations towards non-human animals, the land, forests and woodland, the oceans, mountains and the biosphere. If such an ethical system were to be adopted, then moral governance shifts from the application of principles to the application of moral frameworks: that is, the 'network of mutually supportive principles, theories, and attitudes toward consequences' (Stone, 1987). So, by focusing upon the ethical context of each situation, we can identify the appropriate moral guidance for particular historical and cultural locations, in a way which is capable of acknowledging the complex ecological conditions which prevail in a specific biome and the distinctive cultural practices in each situation.

How can the idea of citizenship help us to understand the relationship between the political possibilities raised by Beck and the ethical transformation proposed by ecological thought? Citizenship refers to the framework of complex interlocking relations which exist between obligations and entitlements in any ethical and legal system. Western societies have witnessed three phases of citizenship: civil citizenship, political citizenship and social citizenship. Civil citizenship established property ownership rights and a corresponding set of obligations to respect the possessions of other property owners. Political citizenship involved an extension of voting rights to all members of society on the grounds that as members of the political, legal and moral community, they should have some say in the decision-making processes which affected their lives. Hence the emergence of free speech, free association and freedom of mobility (with corresponding obligations not to abuse them). Social citizenship involved the establishment of welfare rights and corresponding obligations to contribute to the funds established for the provision of the needs of strangers (to ameliorate the effects of unemployment, sickness, destitution and squalor). Each form of citizenship presumed a different criterion for membership of the moral community and its associated relations between entitlements and obligations. Each was also founded upon the liberal distinction between private and public spheres: that a space should exist where political authority has a limited role.

So how does ecological citizenship differ from these earlier forms of citizenship? In each of these earlier forms, the boundary between what is private and what is public moves but the distinction itself remains unquestioned. Ecological citizenship, however, questions the efficacy and appropriateness of this distinction. Extending the membership of the moral community in the way suggested earlier means that human beings should acknowledge their obligations to animals, trees, mountains, oceans, and other members of the biotic community. As a result, we should exercise extreme caution before embarking upon any project which is likely to have adverse effects upon the ecosystems we inhabit. The limits that this places upon human actions are severe, and no existing political vocabulary has managed to capture this transformation in the relationship between society and nature. Many personal ('private') choices which were previously considered inviolable would be open to serious 'public' challenges. The increased realization of the complexity, uncertainty and interconnectedness of all living things and their life-support systems makes this reassessment of human obligations even more imperative.

In such a situation, the institutional embodiments of 'private' and 'public' – that is, civil society and the state – will be in doubt. As we have seen, we face two choices: authoritarian technocracy (the absorption of civil society by the state) or ecological democracy (the absorption of the state by civil society). In Beck's account of ecological democracy, the liberal distinction between public and private becomes questionable but the moral community remains human-centred. Human beings become more sensitive to the unanticipated consequences of their actions, but the basis of the social order remains fixed upon human needs and desires. In Peter Christoff's account of ecologically guided democracy (see Reading 7.6), the moral community is transformed but the distinction between state and civil society remains. In each case, we can see the preconditions for ecological citizenship; but each account is partial.

In summary, ecological citizenship not only challenges the distinction between public and private, it also presumes that this is closely connected to an ethical transformation, by displacing the human species from the central ethical position it has always held. In short, the adoption of an ethical standpoint which embraces ecocentrism involves a shift in social and political thought to a new 'politics of obligation'. This would involve enormous changes in human assumptions and behaviour as well as in institutional structures and social relations. We do not require a 'blueprint', an ideal ecotopia worked out to the last detail. Instead we need to work towards a 'greenprint' which acknowledges the complexity, uncertainty and interconnectedness of the relations between society and nature, to establish a set of working principles which can be used for developing flexible strategies for change.

References

Beck, U. (1992) *Risk Society: Towards a New Modernity*, London: Sage.

Beck, U. (1995) *Ecological Politics in an Age of Risk*, Cambridge: Polity.

Beck, U., Giddens, A. and Lash, S. (1994) *Reflexive Modernisation: Politics, Tradition and Aesthetics in the Modern Social Order*, Cambridge: Polity.

Giddens A. (1990) *The Consequences of Modernity*, Cambridge: Polity.

Stone, C. D. (1987) *Earth and Other Ethics: The Case for Moral Pluralism*, New York: Harper & Row.

Acknowledgements to copyright holders

MARKHAM, Adam *A Brief History of Pollution*, Earthscan, London (1994). Copyright © Adam Markham. Reproduced by kind permission of the publisher.

THOMAS, Keith *Man and the Natural World 1500–1800*, Penguin, London (1984). Copyright © Keith Thomas.

PEPPER, David *Modern Environmentalism*, Routledge, London (1996). Copyright © David Pepper. Reproduced by kind permission of the publisher.

O'RIORDAN, Tim *Environmentalism*, Pion Books, London (1981). Copyright © Pion Books. Reproduced by kind permission of the publisher.

CARSON, Rachel *Silent Spring*. Copyright © 1962 Rachel Carson. Reproduced by kind permission of Laurence Pollinger Limited and the Estate of Rachel Carson.

GOLLEY, Frank B. 'Environmental attitudes in North America', in R. J. Berry (ed.) *Environmental Dilemmas: Ethics and Decisions*, Chapman & Hall, London (1992). Reproduced by kind permission of Kluwer Academic Publishers.

MABEY, Richard *The Common Ground*, J. M. Dent, London (1980). Reproduced by kind permission of The Orion Publishing Group Ltd.

GOLDING, Michael P. 'Obligations to future generations', in *The Monist*. Copyright © 1972 The Monist, La Salle, Illinois 61301. Reprinted by kind permission.

CALLAGHAN, Daniel 'What obligations do we have to future generations?', in *American Ecclesiastical Review*, 164, pp. 265–80.

SMITH, Mark J. *Ecologism: Towards Ecological Citizenship*, Open University Press, Buckingham (1998). Copyright © Mark J. Smith. Reproduced by kind permission of the publisher.

BARRY, Brian, 'Justice between generations: power and knowledge', in P. Hacker and J. Raz (eds), *Law, Morality and Society*, Clarendon Press, Oxford (1978).

GOODIN, Robert E. *Green Political Theory*, Polity, Cambridge (1992). Reproduced by kind permission of Blackwell Publishers.

KAVKA, Gregory S. and WARREN, Virginia 'Can future generations be represented?', in R. Elliot and A. Gare (eds), *Environmental Philosophy*, Open University Press, Buckingham (1983). Reproduced by kind permission of the publisher.

FIDDES, Nick *Meat: A Natural Symbol*, Routledge, London (1991). Reproduced by kind permission of the publisher.

THOMAS, Keith *Man and the Natural World 1500–1800*, Penguin, London (1983).

SINGER, Peter, *Applied Ethics*, Oxford University Press, Oxford (1986).

REGAN, Tom 'Animal rights: the Kantian case', in P. Singer (ed.) *The Case for Animal Rights*, Blackwell, Oxford (1984). Reproduced by kind permission of the publisher.

SCRUTON, Roger *Animal Rights and Wrongs*, Demos, London (1996).

GARNER, Robert *Animals, Politics and Morality*, Manchester University Press, Manchester (1993).

MUIR, John *Our National Parks*, Houghton Mifflin, Boston (1901).

LEOPOLD, Aldo 'The chitchat of the woods' and 'The land ethic', in *A Sand County Almanac – and Sketches Here and There*, Oxford University Press, New York (1949). Reproduced by kind permission of the publisher.

NAESS, Arne 'The shallow and the deep, long-range ecology movement', in *Inquiry*, 16 (1973).

DEVALL, Bill and SESSIONS, George, *Deep Ecology*, Peregrine Smith Books, Utah (1985).

ROLSTON, Holmes III 'Valuing the environment'. Adapted from *Environmental Ethics* by Holmes Rolston, III. Reprinted by permission of Temple University Press. Copyright © 1988 by Temple University Press. All rights reserved.

STONE, Christopher D. 'Should trees have standing?: Toward legal rights for natural objects', in *Southern California Law Review*, 45 (1972), pp. 450–501. Reprinted with the permission of the Southern California Law Review.

STONE, Christopher D. *Earth and Other Ethics: The Case for Moral Pluralism*, Harper and Row, New York (1987).

MARTELL, Luke *Ecology and Society*, Polity, Cambridge (1995). Reproduced by kind permission of Blackwell Publishers and The University of Massachusetts Press.

SAGOFF, Mark *The Economy of the Earth: Philosophy, Law and the Environment*, Cambridge University Press, Cambridge (1988). Reproduced by kind permission of the publisher.

HARDIN, Garrett 'The tragedy of the Commons', in *Science*, 162 (1968).

YOUNG, Robert 'Population policies, coercion and morality', in D. Mannison, M. McRobbie and R. Routley (eds), *Environmental Philosophy*, Australian National University (1980). Reproduced by kind permission of the author.

MALINAS, Gary 'On justifying and excusing coercion', in D. Mannison, M. McRobbie and R. Routley (eds), *Environmental Philosophy*, Australian National University (1980). Reproduced by kind permission of the author.

SAUNDERS, Peter *Capitalism: A Social Audit*, Open University Press, Buckingham (1995). Reproduced by kind permission of the publisher.

GRAY, John *Beyond the New Right: Markets, Government and the Common Environment*, Routledge, London (1993). Reproduced by kind permission of the publisher.

TAYLOR, Ann *Choosing our Future: A Practical Politics of the Environment*, Routledge, London (1992). Reproduced by kind permission of the publisher.

MERCHANT, Carolyn *Radical Ecology: The Search for a Liveable World*, Routledge, London (1992). Reproduced by kind permission of the publisher.

ECKERSLEY, Robyn *Environmentalism and Political Theory: Toward an Ecocentric Approach*, UCL Press, London (1992). Reproduced by kind permission of the UCL Press Ltd and State University of New York Press.

SHIVA, Vandana *Staying Alive: Women, Ecology and Development*, Zed Books, London (1989). Reproduced by kind permission of the publisher.

KING, Ynestra 'Ecology of Feminism and the feminism of ecology', in J. Plant (ed.), *Healing the Wounds: The Promise of Ecofeminism*, Green Print, London (1989).

SEAGER, Joni *Earth Follies: Feminism, Politics and the Environment*, Earthscan, London (1993). Reproduced by kind permission of the publisher.

HAYWARD, Tim *Ecological Thought: An Introduction*, Polity, Cambridge (1994). Reproduced by kind permission of Blackwell Publishers.

HAJER, Maarten A. 'Ecological modernisation as cultural politics' in S. Lash, B. Szerszynynski and B. Wynne (eds), *Risk, Environment and Modernity: Towards a New Ecology*, Sage, London (1996). Reprinted by permission of Sage Publications Ltd.

GOLDBLATT, David *Social Theory and the Environment*, Polity: Cambridge (1996). Reproduced by kind permission of Blackwell Publishers.

BLOWERS, Andrew 'Environmental policy: ecological modernisation or risk society?', in *Urban Studies* 34 (1997), Carfax Publishing Ltd, Abingdon. Reproduced by kind permission of the publisher.

MILLS, Mike 'Green democracy: the search for an ethical solution', in B. Doherty and M. de Geus (eds), *Democracy and Green Political Thought: Sustainability, Rights and Citizenship*, Routledge, London (1996). Reproduced by kind permission of the publisher.

CHRISTOFF, Peter 'Ecological citizens and democracy', in B. Doherty and M. de Geus (eds), *Democracy and Green Political Thought: Sustainability, Rights and Citizenship*, Routledge, London (1996). Reproduced by kind permission of the publisher.

The publishers have made every effort to contact copyright holders of material reprinted in *Thinking through the Environment*. However, this has not been possible in every case and we would welcome correspondence from those individuals or companies we have been unable to trace.

Index